D1058636

**Kruger, Paul,** 1925–
    Principles of activation analysis.   New York, Wiley-
Interscience [1971]

   xi, 522 p.   illus.   24 cm.

   Includes bibliographies.

1. Radioactivation analysis.    I. Title.

QD606.K75                          545′.822                    72–137108
ISBN 0–471–50860–8                                             MARC

Library of Congress                        71 [4]

# Principles of
# Activation Analysis

# Principles of
# Activation Analysis

PAUL KRUGER
Department of Civil Engineering
Stanford University

WILEY—INTERSCIENCE
A Division of John Wiley & Sons, Inc.

New York     London     Sydney     Toronto

Library of Congress Catalog Card Number: 72-137108
ISBN 0-471-50860-8

Printed in the United States of America

10  9  8  7  6  5  4  3  2  1

This Book is Dedicated to
Sharon and Kenneth

# Preface

*Activation Analysis* is paradoxically a very special and a very general subject. It is a special subject in that it is a specific application of nuclear science which requires elaborate nuclear sources for activation and electronic instrumentation for measurement. It is a general subject in that the principles and applications are pertinent to problems in almost all fields of science and technology where the measurement of a chemical element is the means to a solution. Thus activation analysis is primarily a form of analytical chemistry, but it also is useful in other ways, such as in tracer technology in which tracer elements can be identified, followed, or measured, even in complex systems or processes.

*Activation Analysis* has proven a successful method of measurement in a great many diverse disciplines. It has been used extensively in the biological sciences, in the earth sciences, in the physical sciences, and in industrial plants to name but a few. It is also being developed in such fields as environmental sciences and forensic sciences. Its role in stable-isotope tracing is growing in hydrology, petroleum, and other engineering sciences.

*Acitivation Analysis* has grown from a specialized technique practiced by a few skilled nuclear chemists to a generalized technique practiced by scientists in all of the disciplines referred to earlier. It has grown in scope from an area of research where each new successful analysis of a different element or of a different matrix material resulted in publication to an established business in which several commercial firms offer competitive and routine services. It has grown from a technology requiring access to elaborate nuclear reactors and accelerators (available primarily only at large universities or the national laboratories) to a technology where "packaged" computer-controlled activation-radiation detection systems are commercially available at reasonable cost.

*Activation Analysis* has amassed a considerable literature. Several extensive bibliographies have been prepared. A great many review articles have appeared. Several international conferences devoted solely to activation analysis have been held. The proceedings of these conferences constitute a significant contribution to the literature. A number of professional books have been published to complement the initial handbook of R. C. Koch.

*Activation Analysis* has grown to the status of an academic subject. Short and long-term courses are being offered at national laboratories, industrial companies, and universities to train scientific personnel at all levels in the use of the method. Activation analysis laboratory centers exist at an increasing number of universities. Activation analysis is being included as a major topic in college courses in analytical chemistry, biochemistry, or radio-chemistry, and in many complimentary laboratory courses. Activation Analysis is a full course at several universities.

*Activation Analysis* was introduced as a one-quarter course at Stanford University in 1962. It has been offered annually since. The course is designed to prepare engineering and physical and life sciences students to study and use the method. The course covers the fundamentals of the radioactivation method, including basic nuclear science, nuclear reactions, radioisotope production, radiochemical separations, and radiation detection and measurement. The fundamentals are supplemented with discussions of experimental practices, laboratory demonstrations and experiments, and individual student study of method development or solution of specific problems.

*The Principles of Activation Analysis* has been written from the course material developed over the past seven years. It has been designed as a textbook for use by students of diverse scientific backgrounds, such as engineers, biologists, chemists, and metallurgists, and is suitable for a one-quarter or one-semester course at the senior or first-year graduate student level. Because of the diversity of student backgrounds, little prior education in nuclear sciences is assumed other than basic courses in chemistry, physics, and mathematics. The nuclear science required to understand the principles of activation analysis are developed in the text. The first six chapters of the textbook analyze the component parts of activation analysis. These consist of the characteristics of the atomic nucleus (its structure, energetics, stability, radiations, and reactions), the process of radioactivation, the description of irradiation sources, the properties of radionuclides and their radiations, the methods of radiochemical separations, and the detection and measurement of radiation. The synthesis of these components in the many available practices developed as activation analysis techniques are reviewed in Chapter 7. The limitations to the methods from the several components are discussed in Chapter 8. A survey of the many applications of activation analysis techniques for trace analysis, as special analytical methods, and as a tracer technology are categorized by discipline in Chapter 9. Examples of ac-complished investigations are given for the many diverse scientific fields enumerated earlier.

The author hopes that this textbook will, in a small way, contribute to the growth of activation analysis as an academic subject, and assist in the education of our future scientists who will use radioactivation as a means of studying biological, physical, engineering, and industrial problems. The book was completed with the assistance of a great number of colleagues. Special thanks are due to Irwin Gruverman and Robert C. Koch for their early advice and encouragement, and to James De Voe and Richard E. Wainerdi for their review of the manuscript.

PAUL KRUGER

*Stanford, California*
*October, 1970*

# Contents

Chapter 1

# Stable and Radioactive Nuclides

Activation analysis is most commonly considered a method of analysis of the chemical elements. It is accomplished by the production of radioactive isotopes by nuclear reactions with stable isotopes of the elements in the sample, followed by the measurement of the radiations emitted by the desired radioisotopes. The principles of radioactivation analysis are thus derived from the principles of atomic and nuclear structure, of stable and

1

radioactive isotopes, of nuclear transformations, of the radiation characteristics of radioactive isotopes, and of the interactions of these radiations with matter. It is these interactions that permit quantitative measurement of the radioactivation products.

The method of activation analysis, however, is not necessarily confined to analytical chemistry but may also be considered to encompass a broader scope which includes the production and measurement of radioisotopes in materials of known composition. Such conditions are met, for example, when radioactivation is used for nuclear reaction studies, for flux and beam-intensity measurements, for tracer experiments with stable isotope tracers, and for process quality control.

The principles of radioactivation and radiation measurement remain the same for the analytical, dosimetry, and tracer aspects of the technique. The formulation of successful applications of activation analysis in these areas requires an understanding of the pertinent principles of nuclear science. Thus we begin by reviewing those features of nuclear science pertinent to our subject: nuclear structure, nuclear energetics, nuclear stability, nuclear radiations, and nuclear reactions.

## 1.1  NUCLEAR STRUCTURE

### 1.1.1  The Atom

Matter is anything that occupies space and constitutes all substances. Substances, in turn, are homogeneous matter that possesses definite chemical composition and constitutes the chemical elements and their compounds. Each substance is characterized by specific properties, such as melting and boiling points and solubility. Since at least 500 B.C. man has wondered what would happen if an object were divided into two indefinitely. The Greeks postulated that eventually one would come to the smallest unit of matter and called it ατομοσ, from which the word *atom* evolved. Some 2000 years later, in 1803, Dalton gave arithmetic reality to the chemical molecule and the atom. We know the molecule as the smallest particle of matter with properties characteristic of a chemical compound and consisting of electron-bonded atoms. An atom is the smallest particle of a chemical element.

The cataloging of the chemical characteristics of the elements resulted in the periodic table, which includes 105 elements, the last of which was discovered in 1970. The periodicity of the chemical elements has been successfully explained on the basis of atomic structure, the knowledge of which has developed primarily in this century. The now classical model of the atom as a tightly bound positive-charged nucleus surrounded, in the neutral atom, by the appropriate number of orbiting negatively charged electrons in definitive (quantized) shells has evolved in the last 60 or so years. The periodic filling of electron shells about the nucleus is the foundation for the chemical properties of the elements and the concepts of valence electrons and ionic and covalent bonds. With this model we also understand spectroscopic data, x-ray emission, the photoelectric effect, and other phenomena of atomic physics.

### 1.1.2 The Nuclear Model

The nuclear model of the atom emerged in 1911, when, from his "historical" scattering experiments, Rutherford postulated an atomic model which required that the positive charge and most of the mass of the atom be contained in a nucleus of less than $10^{-12}$ cm diameter, with a sufficient number of electrons to balance the positive charge of the nucleus, distributed over atomic dimensions known to be of the order of $10^{-8}$ cm in diameter. It was not until 1932, the year that Chadwick discovered the neutron, that the now "practical" model of the nucleus evolved. The practical model, which accounts for all the nuclear physics required in the study of activation analysis (and neglects the refinements of meson and strange-particle physics), assumes only the proton and the neutron as the unit building blocks of nuclear matter. By considering protons and neutrons as equivalent building blocks (nucleons) and the volume of the nucleus proportional to the number of nucleons present ($A$) we can express the radius of nuclei by the empirical formula

$$R = R_0 A^{1/3} \qquad (1)$$

where $R_0$ is a constant. Values of $R_0$ obtained by several independent experimental methods range from 1.2 to $1.6 \times 10^{-13}$ cm. An average value of $1.4 \times 10^{-13}$ cm is adequate for our purposes.

### 1.1.3 Conservation of Nucleons

Many of the classical conservation laws of physics have been modified during this century; notable among them are the independent conservations of mass and energy. Others involve concepts originating in the refinements of nuclear physics alluded to earlier. They are covered in many textbooks of nuclear physics. However, one of the most successful has been the integer conservation of nucleons (protons and neutrons) in all atomic nuclei, in all radioactive decay processes, and in all nuclear reactions of interest. It is this conservation that allows us to build a table of isotopes from the "practical" nuclear model, somewhat analogous to the periodic table of the elements.

**Table 1.1**   Glossary of Nuclear Structure Terms

| | |
|---|---|
| Nucleon | A constituent of the atomic nucleus; for the practical nuclear model either a proton or a neutron. |
| Atomic number | The chemical identity of an atom, given by the number of protons $(Z)$ in the nucleus. |
| Mass number | The total number of nucleons $(A = Z + N)$ in the atom. The term is used interchangeably with atomic weight in integer value. |
| Atomic weight | The actual mass of an atom relative to $^{12}C$ which has a mass of 12.0000000 atomic mass units (amu) by definition; for example, $^4He$ has a mass of 4.0026036 amu. |
| Nuclide | A composition of nucleons that is stable or exists for a measurable length of time. A nuclide is characterized by its atomic number, mass number, and energy content. The term is now used interchangeably with isotope; for example, $^2_1H$, $^{12}_6C$, and $^{90}_{38}Sr$ are nuclides. |
| Isotopes | Nuclides of a given atomic number but different mass number, that is, nuclides with the same number of protons, but different number of neutrons; for example, $^{12}_6C$, $^{13}_6C$, and $^{14}_6C$ are isotopes of carbon. |
| Isobars | Nuclides of a given mass number but different atomic numbers; for example, $^{90}_{36}Kr$, $^{90}_{37}Rb$, and $^{90}_{38}Sr$ are isobars. |
| Isomers | Nuclides of the same nucleon composition, that is, the same number of protons and neutrons, which differ only by their energy content. The isomer with the greater energy content is referred to as the metastable isomer, with an $m$ placed after the mass number; for example, $^{60}_{27}Co$ and $^{60m}_{27}Co$ are isomers. |

### 1.1.4   Chemical Identity

The phenomenon of radioactive decay, in which an atom of one chemical element changes spontaneously into an atom of another element, and the phenomena of nuclear reactions, in which such changes can be induced, emphasize the separation of the concepts of chemical identity and chemical behavior. The chemical behavior of an atom is related to the number of valence electrons involved in its chemical combinations, whereas the chemical identity of an atom is related to the number of electrons present in the neutral atom. From the nuclear viewpoint the chemical identity of an atom may be given by the number of protons in the nucleus. In the neutral atom these numbers are both referred to as the atomic number of the element. In nuclear processes a change in the number of protons in a nucleus results in the corresponding change in chemical identity of the atom, even before the orbital electrons have had time to readjust to the change in nuclear charge. Thus it is

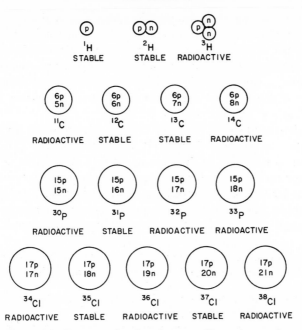

**Figure 1.1** Isotopes of a few elements, illustrating nuclear composition of some stable and radioactive nuclides. The nuclei are drawn with radius proportional to $R_0 A^{1/3}$. The nuclide $^{31}P$ is an example of a monoisotopic stable nuclide of an element.

convenient to define the chemical identity (atomic number) of an atom or element by the number of protons ($Z$) in the nucleus.

The charge of the nucleus ($Q$) is given by

$$Q = Ze \tag{2}$$

where $e$ is the electronic charge with a value of $4.803 \times 10^{-10}$ esu. The value of $e$ is determined experimentally, and because it has an associated experimental error the charge of a nucleus cannot be known precisely. Since the atomic number of a nucleus is given by the integer number of protons in the nucleus, it is one of the characteristic parameters of the nucleus that is known precisely. Another is the neutron number ($N$). The mass number ($A$) is given by the number of nucleons in the nucleus:

$$A = Z + N \tag{3}$$

For convenience some of the more common terms of nuclear structure are given as a glossary in Table 1.1.

The neutron may be considered the first "element" with $Z = 0$, $N = 1$, and $A = 1$. The proton is the nucleus of the hydrogen atom with $Z = 1$,

$N = 0$, and $A = 1$. Heavy hydrogen, the stable isotope deuterium, is composed of a proton and neutron bound together. Chemically it is hydrogen with $Z = 1$, but with $N = 1$ and $A = 2$. It is thus about twice as heavy as ordinary hydrogen. The heaviest isotope of hydrogen, the radioactive isotope tritium, is still chemically hydrogen with $Z = 1$, but with $N = 2$ and thus $A = 3$. These isotopes are shown in Figure 1.1 together with some isotopes of carbon, phosphorus, and chlorine.

Nuclear shorthand shows the neutron as $_0^1 n$, the proton as $_1^1 p$, the neutral hydrogen atom as $_1^1 H$, deuterium as $_1^2 H$, and tritium as $_1^3 H$. This formalism gives the chemical symbol of the element, the atomic number ($Z$) as a presubscript, and the mass number ($A$) identifying the particular isotope of the element as a presuperscript, according to international convention. In some countries and in much of the older literature the mass number is given as a post-superscript; for example, the isotope tritium is most often found in the form $H^3$.

Since the atomic number defines the chemical symbol and is therefore redundant, the subscript is usually omitted in nuclear publications. Table 1.2

**Table 1.2**  Nuclear Composition of the Monoisotopic Stable Nuclides of the Elements ($f = 1.00$)

| Chemical Element | Atomic Number $(Z)$ | Neutron Number $(N)$ | $N/Z$ Ratio | Mass Number $(A)$ | Nuclide Symbol |
|---|---|---|---|---|---|
| Beryllium | 4 | 5 | 1.250 | 9 | $^9$Be |
| Fluorine | 9 | 10 | 1.111 | 19 | $^{19}$F |
| Sodium | 11 | 12 | 1.091 | 23 | $^{23}$Na |
| Aluminum | 13 | 14 | 1.077 | 27 | $^{27}$Al |
| Phosphorus | 15 | 16 | 1.067 | 31 | $^{31}$P |
| Scandium | 21 | 24 | 1.143 | 45 | $^{45}$Sc |
| Manganese | 25 | 30 | 1.200 | 55 | $^{55}$Mn |
| Cobalt | 27 | 32 | 1.185 | 59 | $^{59}$Co |
| Arsenic | 33 | 42 | 1.273 | 75 | $^{75}$As |
| Yttrium | 39 | 50 | 1.282 | 89 | $^{89}$Y |
| Niobium | 41 | 52 | 1.268 | 93 | $^{93}$Nb |
| Rhodium | 45 | 58 | 1.289 | 103 | $^{103}$Rh |
| Iodine | 53 | 74 | 1.396 | 127 | $^{127}$I |
| Cesium | 55 | 78 | 1.418 | 133 | $^{133}$Cs |
| Praesodymium | 59 | 82 | 1.390 | 141 | $^{141}$Pr |
| Terbium | 65 | 94 | 1.446 | 159 | $^{159}$Tb |
| Holmium | 67 | 98 | 1.478 | 165 | $^{165}$Ho |
| Thulium | 69 | 100 | 1.450 | 169 | $^{169}$Tm |
| Gold | 79 | 118 | 1.495 | 197 | $^{197}$Au |
| Bismuth | 83 | 126 | 1.520 | 209 | $^{209}$Bi |

gives as additional examples of nuclear composition those nuclides that constitute the only stable isotope of the elements listed. For these isotopes, the isotopic abundance, $f$, = 100%. In general, monoisotopic stable nuclides present simpler problems in the activation analysis for such elements. A complete listing of the known stable and radioactive isotopes of the elements is given in many available nuclide charts (see the bibliography in Section 1.6).

## 1.2 NUCLEAR ENERGETICS

### 1.2.1 Atomic Masses

The atomic weights of the elements were determined initially from calculations of chemical combinations and later by physical methods. The natural isotopic mixture of oxygen was adopted as a standard with a value of exactly 16 mass units. Natural oxygen is about 16 times as heavy as natural hydrogen, the lightest element known. Thus hydrogen, by comparison, is considered to have an atomic weight of one unit. However, elements such as chlorine, which have several prominent stable isotopes, have nonintegral atomic weights. Chlorine, consisting in nature of 75.53% $^{35}$Cl and 24.47% $^{37}$Cl, would be expected to have an atomic weight of

$$(0.7553 \times 35) + (0.2447 \times 37) = 35.4894 \tag{4}$$

The actual atomic weight of chlorine is 35.453; that is, 0.036 units lighter than calculated above.

The apparent discrepancy is due in part to the fact that $^{35}$Cl does not weigh exactly 35 mass units compared with 16 for oxygen, nor does $^{37}$Cl weigh exactly 37 mass units, and in part to the fact that natural oxygen contains about 0.0374% $^{17}$O and about 0.2039% $^{18}$O isotopes. This composition varies slightly for different sources of oxygen. A physical atomic-weight scale was also introduced in which the mass of the $^{16}$O nuclide alone was assigned the value of exactly 16 mass units. Because of the $^{17}$O and $^{18}$O isotopes in natural oxygen the numerical value of the atomic weights on the chemical scale was smaller (by about 0.032%) than the values given on the physical scale. To eliminate the resulting confusion between the two scales a new standard for atomic weights was adopted internationally in 1961 in which an atom of $^{12}$C is defined as weighing exactly 12 atomic mass units (amu). Since $^{12}$C accounts for 98.89% of natural carbon, the chemical scale of atomic weights is only about 0.005% different from the new $^{12}$C scale and the two, for all practical purposes, are considered the same. Care should be exercised in using mass values from literature published before 1961.

The atomic mass unit is very small on the ordinary weight scale; 1 amu is numerically equivalent to the reciprocal of Avogadro's number (defined in Section 2.1.3):

$$1 \text{ amu} = \frac{1}{6.023 \times 10^{23}} = 1.660 \times 10^{-24} \text{ g} \tag{5}$$

Thus a single $^{12}$C atom weighs but $2 \times 10^{-23}$ g. Each nuclide has both a characteristic

mass number   $A$, the number of nucleons in the nucleus

and

isotopic mass   $M$, in amu, relative to $^{12}$C $= 12$ amu.

On the $^{12}$C scale

the mass of the hydrogen atom $^{1}_{1}$H $= 1.0078252$ amu,

the mass of the neutron $^{1}_{0}$n $= 1.0086654$ amu,

the mass of the electron e $= 0.0005486$ amu.

The mass number $A$ is the integer value of the isotopic mass $M$, and for many purposes the two are interchangeable. A noted exception is in the calculation of mass changes due to nuclear reactions or radioactive decay for which the exact mass values must be used. Values of atomic mass, based on the $^{12}$C scale, are given in many sources, some of which are listed in the bibliography. Table 1.3 gives the masses of some of the nuclides of interest in this book.

### 1.2.2  Equivalence of Mass and Energy

In 1905 Einstein in developing his special theory of relativity came to the conclusion that the properties of mass $M$ and energy $E$ were equivalent to one another. He showed that this equivalence could be expressed by the equation

$$E = Mc^2 \tag{6}$$

where $c$ is the speed of light ($3.0 \times 10^{10}$ cm/sec in vacuum). It follows from this equation that potential energy is stored as mass and that in a given system a change in mass is accompanied by an equivalent change in energy

$$\Delta E = \Delta Mc^2 \tag{7}$$

The energy contained in 1 amu is given by

$$E = 1.66 \times 10^{-24} (3 \times 10^{10})^2 = 1.49 \times 10^{-3} \text{ erg} \tag{8}$$

In nuclear science, energy values are more conveniently expressed in electron volts (eV). An electron volt is the energy acquired by an electron in passing through a potential of 1 V. Nuclear events usually involve energy changes of

**Table 1.3** Nuclear Masses for Selected Nuclides[a]

| Atomic Number (Z) | Chemical Element | Mass Number (A) | Nuclide | Nuclear Mass (amu) |
|---|---|---|---|---|
| 0 | Neutron | 1 | $^1$n | 1.008665 |
| 1 | Hydrogen | 1 | $^1$H | 1.007825 |
| 1 | Hydrogen | 2 | $^2$H | 2.014102 |
| 1 | Hydrogen | 3 | $^3$H | 3.016049 |
| 2 | Helium | 3 | $^3$He | 3.016030 |
| 2 | Helium | 4 | $^4$He | 4.002604 |
| 3 | Lithium | 6 | $^6$Li | 6.015126 |
| 6 | Carbon | 11 | $^{11}$C | 11.011433 |
| 6 | Carbon | 12 | $^{12}$C | 12.000000 |
| 6 | Carbon | 13 | $^{13}$C | 13.003354 |
| 6 | Carbon | 14 | $^{14}$C | 14.003242 |
| 7 | Nitrogen | 14 | $^{14}$N | 14.003074 |
| 8 | Oxygen | 16 | $^{16}$O | 15.994915 |
| 11 | Sodium | 23 | $^{23}$Na | 22.989773 |
| 11 | Sodium | 24 | $^{24}$Na | 23.990967 |
| 12 | Magnesium | 24 | $^{24}$Mg | 23.985045 |
| 12 | Magnesium | 27 | $^{27}$Mg | 26.984345 |
| 13 | Aluminum | 27 | $^{27}$Al | 26.981535 |
| 15 | Phosphorous | 32 | $^{32}$P | 31.973908 |
| 16 | Sulfur | 32 | $^{32}$S | 31.972074 |
| 17 | Chlorine | 35 | $^{35}$Cl | 34.968854 |

[a] Adapted from L. A. König, J. E. H. Mattauch, and A. H. Wapstra, 1961 Nuclidic Mass Table, *Nuclear Physics* **31**, 18 (1962).

the order of kilo electron volts (1 keV = $10^3$ eV) and million electron volts (1 MeV = $10^6$ eV).

$$1 \text{ eV} = 1.602 \times 10^{-12} \text{ erg} = 3.829 \times 10^{-20} \text{ cal} \tag{9}$$

$$1 \text{ MeV} = 1.602 \times 10^{-6} \text{ erg} = 3.829 \times 10^{-14} \text{ cal} \tag{10}$$

From these conversions the energies associated with changes in mass may be expressed in MeV. Thus the energy of 1 amu is

$$\Delta E = 1.493 \times 10^{-3} \text{ erg} \times 6.25 \times 10^5 \text{ MeV/erg} = 931.4 \text{ MeV} \tag{11}$$

for example, the rest mass of an electron (an electron with zero kinetic energy is $m_0 = 0.005486$ amu and the potential energy of an electron is therefore

$$E = 0.005486 \text{ amu} \times 931.4 \text{ MeV/amu} = 0.51 \text{ MeV} \tag{12}$$

Energy changes associated with ordinary chemical reactions result in equivalent mass changes too small to detect with even the most sensitive balances. The vigorous chemical reaction

$$H + F \rightarrow HF + \Delta E \tag{13}$$

results in the liberation of $\Delta E = 64{,}200 \text{ cal/mole}$ of HF. The loss in weight of the 20 g of HF would be

$$\Delta M = -\frac{\Delta E}{c^2} = -\frac{6.42 \times 10^4 \text{ cal}}{9 \times 10^{20} \text{ erg/g}} \times 4.19 \times 10^7 \text{ erg/cal}$$

$$= -2.98 \times 10^{-9} \text{ g} \tag{14}$$

The change in weight $= (2.98 \times 10^{-9})/20 \times 100 = 1.49 \times 10^{-8}\%$.

In nuclear reactions the energy change per atom is extremely large compared with chemical systems, usually in the range of keV to MeV per atom; for example, if, hypothetically, each atom of fluorine in 20 g of HF absorbed a 1 MeV gamma ray, the increase in weight of the HF would be

$$\Delta m = 1 \text{ MeV/atom} \times 6.023 \times 10^{23} \text{ atom/g-atom} \times 1.78 \times 10^{-27} \text{ g/MeV}$$

$$= 1.07 \times 10^{-3} \text{ g} \tag{15}$$

This gain in weight ($\sim\frac{1}{2}\%$) would be easily detected.

### 1.2.3   Binding Energy

Changes in nuclear composition $(Z + N)$ result in changes in energy because of nuclear binding forces. Observations of the stable nuclides in nature show that nuclear forces meet two requirements:

1. They must be strong to overcome the coulomb repulsion forces of the many positively charged protons in the nucleus.

2. They must be short-ranged; otherwise small stable nuclei would not exist.

Thus the forces between a free neutron and a free proton are not so great as the forces holding the neutron and proton together in a deuterium nucleus. Similarly, the forces between two free deuterium nuclei are smaller than the forces holding the helium nucleus together. In fact, such nuclear reactions as the "fusion" of two deuterons to form a larger nucleus in useful amounts are the goal of controlled thermonuclear power for generating electricity.

The energy released in the fusion of the appropriate number of free protons and neutrons in forming an atomic nucleus is called the binding energy (BE). From the Einstein equation of equivalence of mass and energy the release of binding energy is proportional to the loss in mass of the "fused" nuclides.

Thus the mass of $^4$He is smaller than the sum of the masses of two protons and two neutrons. The nuclide formation may be written as

$$2^1_1\text{H} + 2^1_0\text{n} \rightarrow {}^4_2\text{He} + \text{BE} \tag{16}$$

and the loss in mass is given by

$$\Delta M = 2M(^1\text{H}) + 2M(^1\text{n}) - M(^4\text{He}) \tag{17}$$

The mass of the neutral atoms is used instead of the mass of the nuclei alone to account for the mass of the orbital electrons. Thus

$$\Delta M = 2(1.0078252) + 2(1.0086654) - 4.0026036 = 0.0303776 \text{ amu} \tag{18}$$

and

$$\Delta E = \text{BE} = 0.0303776 \text{ amu} \times 931.4 \text{ MeV/amu} = 28.22 \text{ MeV} \tag{19}$$

An index of relative stability of nuclides is given by the average binding energy per nucleon in the nucleus; $^4$He, one of the most tightly bound stable light nuclides, has an average binding energy per nucleon of

$$\frac{\text{BE}}{A} = \frac{28.22}{4} = 7.05 \text{ MeV/nucleon} \tag{20}$$

In contrast, deuterium, $^2$H, one of the least-bound stable nuclides, has an average binding energy per nucleon of

$$\frac{\text{BE}}{A} = \frac{2.2}{2} = 1.1 \text{ MeV/nucleon} \tag{21}$$

The general relationship of average binding energy per nucleon with mass number is shown in Figure 1.2. Above the maximum value at $A \sim 60$ (the most-bound nuclides) the average value decreases slowly from 8.7 to 7.8 MeV/nucleon.

Another index of relative stability is the incremental binding energy required to add another neutron to a stable nuclide. Thus, if a neutron were added to $^{23}_{11}\text{Na}$, the change in mass would be

$$\Delta M = [11M(^1\text{H}) + 12M(^1\text{n}) - M(^{23}\text{Na})]$$
$$- [11M(^1\text{H}) + 13M(^1\text{n}) - M(^{24}\text{Na})]$$
$$= M(^{24}\text{Na}) - [M(^{23}\text{Na}) + M(^1\text{n})]$$
$$= 23.990967 - 23.998438 = -0.007471 \text{ amu} \tag{22}$$

and for the neutron addition

$$^{23}_{11}\text{Na} + \text{n} \rightarrow {}^{24}_{11}\text{Na} + \Delta\text{BE} \tag{23}$$

the change in binding energy is

$$\Delta\text{BE} = -0.007471 \text{ amu} \times 931.4 \text{ MeV/amu} = -7 \text{ MeV} \tag{24}$$

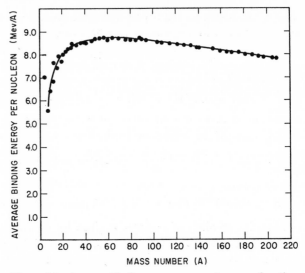

**Figure 1.2**   Average binding energy per nucleon as a function of the mass number of stable nuclides with isotopic abundance greater than 50%.

The minus sign signifies that $^{24}$Na has 7 MeV excess energy with respect to $^{23}$Na and an unbound neutron. Extra neutron binding energies range from about $-2$ to $-12$ MeV.

### 1.3   NUCLEAR STABILITY

The earth has been provided with the materials on it. Most of the existing atoms in this material are stable with respect to time; the remainder is radioactive. The earth is also bombarded continually with radiations from outer space, which convert some of the stable atoms to radioactive ones. Some radioactive atoms exist for very short times, ranging down to less than $10^{-20}$ sec; some exist for very long times, ranging up to more than $10^{18}$ yr. A particular nuclide may exist for so long a period of time that its rate of radioactive decay is unmeasurable. Such a nuclide is considered stable; for example, $^{209}_{83}$Bi, long considered the heaviest of the stable nuclides, is believed by some to be a radioactive nuclide. The lack of a measurable decay rate by any process of disintegration may be a sufficient criterion for defining nuclear stability. With present-day maximum-sensitivity measurement techniques nuclides whose lifetimes exceed $10^{18}$ yr may be considered stable.

Spontaneous radioactive decay can take place by some mode if the masses of the products are lighter than the mass of the original nuclide. The excess mass is released as radiation energy during the disintegration. A radioactive decay process can be written in the form

$$A \rightarrow B + b \tag{25}$$

in which the radioactive nuclide $A$ disintegrates to form nuclide $B$ with the release of some form of radiation $b$. Thus for the decay to occur spontaneously the change in mass

$$\Delta M = M_A - (M_B + M_b) \tag{26}$$

must be greater than zero. However, when a decay process is energetically possible because $\Delta M > 0$ for that process, it is determined only that the nuclide *can* decay spontaneously, not that it *will* decay or *when*. For $\Delta M < 0$ spontaneous decay cannot occur.

### 1.3.1 The $N/Z$ Ratio

The stable nuclides found in nature constitute specific combinations of neutrons and protons; other combinations result in radioactive nuclides.

For example, as shown in Figure 1.1, carbon nuclei ($Z = 6$) form stable isotopes only with six and seven neutrons, phosphorus ($Z = 15$) has only a monoisotopic stable isotope with 16 neutrons, whereas chlorine ($Z = 17$) forms stable isotopes with 18 and 20 neutrons but a radioactive isotope with 19 neutrons.

The systematics of nuclear stability remain the subject of significant study in nuclear physics research. About 300 stable nuclides are known. By examining the composition of these nuclides we note that the ratio of neutrons to protons ($N/Z$ ratio) follows a general relationship with increasing mass number. A plot of neutron numbers versus the proton numbers for the stable nuclides with greater than 10% natural abundance is shown in Figure 1.3. The $N/Z$ ratio is unity for many of the light nuclides through $^{40}_{20}$Ca and increases with atomic number to a value of about 1.5 for the heaviest stable nuclides. The line drawn through the nuclides represents a smoothed stability curve, approximated by the equation

$$Z_0 = \frac{A}{2 + 0.015 A^{2/3}} \tag{27}$$

Nuclides with $N/Z$ ratios differing from these represented by the solid curve in Figure 1.3 are normally unstable or radioactive. They tend to transform spontaneously to more stable states by nuclear transformations which change neutrons into protons or vice versa. Thus a radioactive nuclide of mass $A$ with an excess of neutrons can become more stable by converting neutrons into protons, and, conversely, a neutron deficient radioactive nuclide can become more stable by converting protons into neutrons. Such

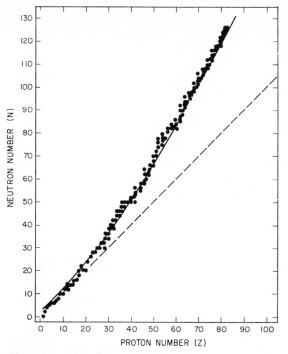

**Figure 1.3** The $N/Z$ ratio for the stable nuclides with isotopic abundance greater than $10\%$. The solid curve drawn
for $Z_0 = A/(2 + 0.015/A^{2/3})$, may be compared with the
dashed line, for which $Z = A/2$.

transformations in which the mass (nucleon) number is conserved in the
nucleus are manifested by the release of high-energy electrons and are known
as beta decay.

### 1.3.2  Instability of the Heaviest Elements

Figure 1.3 also shows that as the atomic number of the elements increase
an increasing excess of neutrons relative to protons is needed to maintain
stable combinations of nucleons in the nucleus. The extra neutrons may be
considered necessary to offset the rapid increase of coulomb repulsion forces
of the positively charged protons as the atomic number increases. However,
even though the binding energy per extra nucleon decreases slowly for larger
nuclei, the short-range nature of nuclear forces results in greater coulomb
instability as the atomic number increases. Thus all nuclides with $Z > 83$ are
unstable with respect to size because of the coulomb repulsion of the protons.
Since the helium nucleus is very stable, radioactive decay of the heavy

elements may take place by the emission of an alpha particle, a $^4$He nucleus, leaving a residual nucleus with two fewer protons and approximately 4 amu lighter. The residual nucleus may still be too large and repeated alpha emissions may occur until the atomic number becomes 83 or less. Although nuclides with $Z > 83$ are unstable with respect to alpha decay, some of them may also be more unstable with respect to other modes of radioactive disintegration, such as beta decay or spontaneous fission (the process of splitting into two large fragments).

### 1.3.3   De-excitation

Some radioactive nuclides have nucleon combinations that are stable with respect to alpha or beta decay, yet have an excess of internal energy. This excess energy may be released in several ways, the most common of which is the emission of electromagnetic radiation (photons) with discrete quanta of energy. These radiations are called gamma rays. If the de-excitation takes place within a lifetime that is easily measured (i.e., longer than a few milliseconds), such radioactive nuclides are known as metastable isotopes and the transition is considered a radioactive decay process called isomeric transition (IT). The designation for a metastable nuclide is given by adding an $m$ after the mass number; for example, the metastable state of $^{60}$Co is $^{60m}$Co.

The other common way for a nucleus to lose excess energy is by an electromagnetic interaction between the nucleus and orbital electrons which results in the emission of an electron whose kinetic energy is equal to the nuclear transition energy, less the binding energy of the emitted electron. This process called internal conversion (IC) competes with gamma-ray emission as a de-excitation process.

The emission of gamma rays or conversion electrons frequently occurs immediately after beta decay. The second process may occur so rapidly ($<10^{-10}$ sec) that the gamma rays or conversion electrons are emitted coincidentally with the beta particles. In activation analysis this coincident-radiation property may be exploited in the measurement of pertinent radionuclides.

### 1.3.4   Half-life

The decay of radioactive atoms, like unimolecular chemical reactions, is a first-order reaction process for which the rate of change is proportional to the amount present. Thus for a radioactive source of $N$ nuclei of the same nuclide the rate of decay is given by

$$-\frac{dN}{dt} = \lambda N \tag{28}$$

where $\lambda$, the constant of proportionality called the decay constant, has a characteristic value for each radioactive nuclide. The rate of change of $N$ is always negative for radioactive decay and is defined as the radioactivity or disintegration rate $(D)$ of the source.

$$D = -\frac{dN}{dt} \tag{29}$$

The solution to (28) may be obtained by integration

$$\int_{N_0}^{N} \frac{dN}{N} = -\int_{0}^{t} \lambda \, dt \tag{30}$$

between the limits of $N_0$ atoms at any reference time $t = 0$ and $N$ atoms at time $t$, later giving

$$\ln \frac{N}{N_0} = -\lambda t$$

$$N = N_0 e^{-\lambda t} \tag{31}$$

The radioactivity decay law follows from (28):

$$D = \lambda N = \lambda N_0 e^{-\lambda t}$$

$$D = D_0 e^{-\lambda t} \tag{32}$$

Figure 1.4$a$ shows the radioactivity decay law in its exponential form for a source of radioactive nuclei decaying initially at the rate of 100 disintegrations per second (dps) with a decay constant of 0.231/hr. It is more convenient, however, to plot radioactive decay in the linear form

$$\log D = \log D_0 - \lambda t \tag{33}$$

as shown in Figure 1.4$b$. On semilog paper the decay rate follows a straight line from the intercept $D_0 = 100$ dps at $t = 0$. It is also convenient to express the characteristic decay constant of a nuclide as a *half-life*, which is defined as the time required for any number of nuclei (or activity) to decay to half its initial value. Thus

$$\frac{D}{D_0} = \frac{N}{N_0} = \frac{1}{2} = e^{-\lambda T_{1/2}}$$

and

$$T_{1/2} = \frac{\ln 2}{\lambda} = \frac{0.693}{\lambda} \tag{34}$$

The half-lives of the nuclides are tabulated in convenient units, but note that since $\lambda t$ must be dimensionless $\lambda$ must be expressed in the *reciprocal* of the same unit used for $T_{1/2}$. In the sample above, in which the decay constant

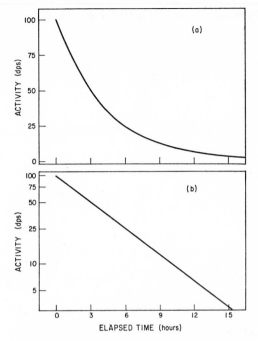

**Figure 1.4** The radioactivity decay law, shown (*a*) in exponential form on linear graph paper, and (*b*) in linear form on semilog graph paper for a radioactive source of 100 dps decaying with a half-life of 3 hr.

was given as $0.231/\text{hr}$, the half-life is $0.693/0.231 = 3$ hr. If the half-life were expressed as 180 min, $\lambda$ would be $0.693/180 = 3.85 \times 10^{-3}/\text{min}$.

## 1.4 NUCLEAR RADIATIONS

The disintegration of a radioactive isotope results in the release of its excess energy in the form of nuclear radiations. The disintegration product may itself be unstable with respect to some other product. Nuclear radiations were first studied in the decay series of the naturally occurring heavy elements; for example, the series beginning with $^{238}\text{U}$ and ending with stable $^{206}\text{Pb}$ involves 22 successive radioactive decays. By 1910 the radiations from these series had been classified into three types, given the names alpha, beta, and gamma rays. In 1932 the discovery of the neutron added not only another component of the nucleus but in the free state another important type of

**Table 1.4**  Characteristics of Nuclear Radiations

| Particle | Symbol | $Z$ (charge) | $A$ (mass number) | $M$ (mass, amu) |
|---|---|---|---|---|
| Alpha | $\alpha$ | +2 | 4 | 4.002604 |
| Helium-3 | $^3$He | +2 | 3 | 3.016030 |
| Triton | t | +1 | 3 | 3.016049 |
| Deuteron | d | +1 | 2 | 2.014102 |
| Proton | p | +1 | 1 | 1.007825 |
| Neutron | n | 0 | 1 | 1.008665 |
| Electron (beta) | $e^-, \beta^-$ | −1 | 1/1840 | 0.000549 |
| Positron | $e^+, \beta^+$ | +1 | 1/1840 | 0.000549 |
| Neutrino | $\nu$ | 0 | 0 | $\sim$0 |
| Gamma ray | $\gamma$ | 0 | 0 | 0 |
| x-ray (atomic) | x | 0 | 0 | 0 |

radiation. Some of the properties of the important radiations are given in Table 1.4.

### 1.4.1  Alpha Radiation

Alpha decay is the spontaneous emission of an alpha particle, the nucleus of the $^4$He atom. Although alpha decay is known in a few of the medium-weight nuclides (e.g., $^{147}_{62}$Sm), it is common to the heavy nuclides with $Z > 83$. Alpha decay proceeds according to the reaction

$$^A_Z X_N \rightarrow \, ^{A-4}_{Z-2} Y_{N-2} + \, ^4_2 He + Q \tag{35}$$

where $Q$, the energy equivalent to the excess mass, is converted into kinetic energy given to the product $Y$ and the alpha particle; alpha particles are emitted with discrete energies, ranging from about 3 to 9 MeV.

### 1.4.2  Beta Radiation

Beta decay is the spontaneous transformation within a nucleus of a neutron into a proton, or of a proton into a neutron, which results in a change of the $N/Z$ ratio toward a more stable configuration of the same mass number.

When the $N/Z$ ratio of the radioactive nucleus is greater than that for stable nuclei of the same mass number, a neutron is converted into a proton with the emission of a negatively charged high-speed electron and a neutrino:

$$n \rightarrow p^+ + e^- + \nu \tag{36}$$

The high-speed electron is a beta particle; the neutrino is an elusive partner (whose rest mass is about 0) required by the laws of physics to conserve momentum and to share the decay energy with the beta particle. Beta particles have a continuous energy spectrum, from 0 to a maximum value, with an average energy about one-third of the maximum beta decay energy. Beta decay for high $N/Z$ ratio nuclides proceeds by the reaction

$$\substack{A\\Z}X_N \rightarrow \substack{A\\Z+1}Y_{N-1} + \beta^- + \nu \tag{37}$$

When the $N/Z$ ratio of the radioactive nucleus is smaller than that for stable nuclei of the same mass number, a proton is converted into a neutron within the nucleus by one of two processes:

1. Positron emission         $p^+ \rightarrow n + e^+ + \nu$
2. Orbital electron capture   $p^+ + e^- \rightarrow n + \nu$

In either case the transformation results in a change of element

$$\substack{A\\Z}X_N \rightarrow \substack{A\\Z-1}Y_{N+1} \tag{38}$$

In the first case the positive electron, when it loses its kinetic energy, interacts with an electron in its environment and the positive and negative electrons annihilate each other to form two quanta of electromagnetic energy of 0.51 MeV each, equivalent to the rest mass of the electron. In the second case no nuclear radiation except the unmeasurable neutrino is emitted, but the radioactive decay event can still be observed by the characteristic x-rays emitted from the resulting atom as atomic electron transitions occur to fill in the hole left in the shell from which the electron was captured by the nucleus. Until the development of efficient gamma-radiation measurement systems beta particles were the radiations primarily used in activation analysis. Even with such systems now readily available beta particles are still often used for activation analysis measurements.

### 1.4.3 Gamma Radiation and Conversion Electrons

Gamma rays are electromagnetic radiation which is quantized into photons. The energy of a photon is given by

$$E_\gamma = h\nu \tag{39}$$

where $h$ = Planck's constant, $6.625 \times 10^{-27}$ erg-sec,
$\nu$ = the frequency of the radiation.

Gamma radiation associated with radioactive decay or nuclear reactions results from the de-excitation of product nuclei with excess energy. In this process the energy of the nuclear transition is emitted as a discrete quantum analogous to the x-rays emitted in orbital electron transitions. Thus for a given transition each gamma ray is emitted with the same energy. This property makes gamma radiation useful for identifying specific radionuclides, especially in the presence of many other radionuclides. The identification of several radionuclides by their characteristic gamma-ray energies has become an important method in radioactivation analysis.

It was noted in Section 1.3.3 that the internal conversion process competes with gamma-ray emission as a nuclear de-excitation process. In this process a monoenergetic electron is emitted in lieu of a gamma ray. For a given radionuclide the ratio of the number of internal conversion electrons to the number of gamma-ray photons emitted is called the internal conversion coefficient $\alpha$. Since the emitted electrons are reduced in energy by the orbital binding energies, separate coefficients exist for the $K, L, M, \ldots$, shells. Also, since internal conversion leaves a vacancy in one of the orbits, orbital electron transitions follow as they do after electron capture. The measurement of these electrons is often employed in activation analysis.

### 1.4.4 Neutrons

With the exception of six or seven delayed neutron emitters (described in Section 3.1.4), radionuclides with excess neutrons do not decay by neutron emission but, as stated in Section 1.42, by $\beta^-$ decay. Neutrons as unbound particles, however, are relatively easy to produce in the many devices discussed in Section 3.1, and it is appropriate to include neutrons as a type of radiation. The basic characteristics of neutrons are given in Table 1.4. The fact that the neutron has no charge gives it special properties: it cannot be accelerated in electric or magnetic fields, it cannot directly cause ionization of atoms and molecules, and it is not affected by coulomb interaction as it approaches the positive charge of a nucleus.

The neutron is not a fundamental particle. In the absence of other nuclear matter, neutrons disintegrate with a half-life of about 12.5 min into protons and electrons by the reaction given in (36). Free neutrons, however, do not generally exist long enough to decay, since they are absorbed by ambient material in times of less than 1 $\mu$sec.

The properties of neutrons are also a function of their kinetic energy, and it has become appropriate in examining the many ways that neutrons can interact with matter to classify them accordingly. Neutrons may be separated by energy ranged into groups that are pertinent to particular types of nuclear reaction. These classifications are not sharply defined; they overlap to a great extent, and different references to them may list significantly different nomenclature or ranges. However, they are generally convenient for separating the neutron interactions into the types discussed in Chapter 3.

1. *Slow neutrons: neutrons with energies up to about 1 keV*
   The most important subgroup of this classification is the "thermal" neutron, a neutron in thermal equilibrium with the atoms in its environment. Thermal neutrons have a distribution of velocity that approaches a Maxwell distribution (see Section 2.3.1) which, at room temperature, corresponds to a most probable energy of 0.025 eV. Other subgroups of slow neutrons include (a) epithermal neutrons, neutrons not in complete thermal equilibrium with their environment, and (b) resonance neutrons, neutrons in the energy range between 1 to 1000 eV. Many nuclides exhibit strong absorption of neutrons at discrete resonant energies in this region.

2. *Intermediate neutrons: neutrons with energies between about 1 and 500 keV*
   The relative lack of appropriate nuclear reactions for these energy neutrons has resulted in little information about this energy range. They are not considered further.

3. *Fast neutrons: neutrons with energy above 0.5 MeV*
   These neutrons are energetic enough to cause a variety of nuclear reactions which do not occur at lower neutron energies. They may be made in many sources and are of interest in activation analysis.

## 1.5 NUCLEAR REACTIONS

Nuclear reactions are changes induced in nuclei by interactions with projectile nuclei of sufficient kinetic energy. They are thus distinguished from radioactive decay processes, which are nuclear disintegration events that occur spontaneously. Nuclear reactions are somewhat analogous to chemical reactions, having in common a change in energy, a minimum energy requirement, and a reaction rate. The analogy fails in that changes in chemical identity can occur in nuclear reactions and that the energy changes per atom may be millions of times greater than those in chemical reactions. A comparison of reaction terminology is given below:

|  | Chemical | Nuclear |
| --- | --- | --- |
| Change in energy | $\Delta H$, heat of reaction | $Q = \Delta M c^2$ |
| Minimum energy requirement | activation energy | threshold |
| Reaction rate | $k$ | $R$ |

A chemical reaction written in the form

$$A + B \rightarrow C + D + \Delta H \tag{40}$$

where $A$ and $B$ are the reactants. $C$ and $D$ are the products, and $\Delta H$ is the energy change of the system, has as its nuclear counterpart

$$A + a \rightarrow B + b + Q \tag{41}$$

where $A$ is the target nuclide, $a$ is the colliding particle, $B$ is the product nuclide, $b$ is the resulting particle, and $Q$ is the change in energy of the system ($\Delta Mc^2$), given in units of MeV. Nuclear reactions are usually written by using an abbreviated convention

$$A(a,b)B \tag{42}$$

which is the same nuclear reaction represented by (41).

Irradiating projectiles may be any form of radiation, such as charged ions, neutrons, electrons, or photons. The resulting particles may also be any of these radiations. The usual ones are given below in order of increasing charge and mass:

| Irradiating Particle | Symbol of the (ionized) Projectile |
|---|---|
| Electromagnetic radiation (photons) | $\gamma$ |
| Electrons | e |
| Neutrons | n |
| Hydrogen, $^1$H (protons) | p |
| Deuterium, $^2$H (deuterons) | d |
| Tritium, $^3$H (tritons) | t |
| Helium-3, $^3$He | $^3$He |
| Helium-4, $^4$He (alpha particles) | $\alpha$ |
| Heavy ions | e.g., $^6$Li, $^{12}$C, etc. |

The interaction of a particular irradiating projectile with a particular nucleus may result in any of a number of possible nuclear reactions. For projectile energies used in radioactivations ($<50$ MeV) the nuclear reaction occurs in two independent steps:

1. The formation of a compound nucleus in which the energy of the captured projectile is shared among all of the nucleons.
2. The disintegration of the compound nucleus from which one or more nucleons are evaporated.

The probabilities for the several modes of disintegration of the compound nucleus are discussed in Section 2.2.5.

### 1.5.1 Conservation Laws

Nuclear reactions of interest in activation analysis obey four conservation laws:

1. Conservation of nucleons, $A$
2. Conservation of charge, $Z$
3. Conservation of mass-energy, $E$
4. Conservation of momentum, $p$

The first two conservation laws may be considered in our chemical reaction analogy as the "balancing" of the reaction. They are illustrated by the nuclear reaction

$$^{35}_{17}\text{Cl} + ^{1}_{0}\text{n} \rightarrow ^{32}_{15}\text{P} + ^{4}_{2}\text{He} + Q \tag{43}$$

or, in abbreviated form,

$$^{35}\text{Cl}(n,\alpha)^{32}\text{P} \tag{44}$$

The reaction is "balanced" with respect to nucleons and charge in that the reactants and products each have the same total number of nucleons (36) and protons (17).

### 1.5.2 Reaction Energy

The conservation of energy determines the reaction energy, which is equivalent to the difference in mass between the reactants and the products. Thus for the reaction

$$^{35}\text{Cl}(n,\alpha)^{32}\text{P} \tag{45}$$

$$\Delta M = M(^{35}\text{Cl}) + M(\text{n}) - M(^{32}\text{P}) - M(^{4}\text{He})$$
$$= 34.96885 + 1.00867 - 31.97391 - 4.00260$$
$$= +0.00101 \text{ amu} \tag{46}$$

$$Q = 931.4(\Delta M) = +0.94 \text{ MeV} \tag{47}$$

For this reaction $Q > 0$; that is, the mass of the products is lighter than the mass of the reactants. Such a reaction is exoergic. However, the reaction

$$^{32}\text{S}(n,p)^{32}\text{P} \tag{48}$$

has

$$\Delta M = M(^{32}\text{S}) + M(\text{n}) - M(^{32}\text{P}) - M(^{1}\text{H})$$
$$= 31.97207 + 1.00867 - 31.97391 - 1.00783$$
$$= -0.00100 \text{ amu} \tag{49}$$

$$Q = -0.93 \text{ MeV} \tag{50}$$

This reaction, for which $Q < 0$, that is, the mass of the products is heavier than the mass of the reactants, is an endoergic reaction. For such reactions to occur the projectile particle must have sufficient kinetic energy not only to supply the system with the energy $Q$ but also the necessary recoil energy required by the conservation of momentum criterion.

### 1.5.3 Threshold Energy

Part of the energy of the incident particle carried into a nuclear reaction must be used to conserve momentum of the colliding system. If we consider the target nucleus standing at rest and the incident particle approaching it with velocity $v$, we can picture the motion of the system before and after collision as follows:

| Before Collision | After Collision |
|:---:|:---:|

$$m \xrightarrow{v_a} \boxed{M} \qquad \boxed{m+M} \xrightarrow{v_{a+A}}$$

$$a \qquad\qquad A \qquad\qquad a+A$$

Conservation of momentum in the reaction system requires

$$mv_a = (m + M)v_{a+A} \tag{51}$$

Converting momentum to energy $(p^2/2m = E)$ yields

$$mE_a = (m + M)E_{a+A} \tag{52}$$

Thus the recoil energy is

$$E_{a+A} = \frac{m}{m + M} E_a \tag{53}$$

and the kinetic energy of particle $a$ available to the reaction is

$$E_Q = E_a - E_{a+A} = \frac{M}{m + M} E_a \tag{54}$$

Thus the minimum (threshold) energy of an incident particle required to produce a nuclear reaction, endoergic by $Q$, is

$$E_{\text{threshold}} = \frac{m + M}{M} Q \simeq \frac{a + A}{A} Q \tag{55}$$

The mass numbers are generally used in place of the actual masses. For an exoergic reaction that involves a neutral particle (such as the neutron) the threshold energy is zero; thus the reaction can take place theoretically with a colliding particle of zero kinetic energy. However, coulomb repulsion forces between charged particles may increase significantly the actual kinetic energy required to induce the reaction.

### 1.5.4 Types of Reaction

Nuclear reactions of interest in activation analysis are those in which the desired elements can be irradiated with "available" bombarding particles to produce radioactive nuclides whose radiations can be measured. Commonly "available" irradiating particles are discussed in Chapter 3. The nuclear reactions are conveniently grouped into three types:

#### Neutron Reactions

Since the neutron is an uncharged particle, it can approach a target nucleus unhampered by the coulomb forces of electrostatic interactions. Thus neutrons with essentially zero kinetic energy can collide with target nuclei. Neutrons in thermal equilibrium with their environment are called thermal neutrons and have an average kinetic energy of only 0.025 eV at 20°C. Neutron reactions can take place with thermal or higher energy neutrons in several ways. The four most common modes are given below.

NEUTRON CAPTURE. The most common activation reaction is neutron capture in which a low-energy neutron is absorbed by a nucleus with the prompt emission of a gamma ray. The reaction is illustrated by the neutron capture by sodium:

$$^{23}_{11}Na(n,\gamma)^{24}_{11}Na \tag{56}$$

Note that neutron capture results in an isotope of the same element, having increased the mass number from $A$ to $A + 1$. The neutron-to-proton ratio thus increases from $N/Z$ to $(N + 1)/Z$ and for most stable target nuclides this increase is sufficient to result in a product isotope which is unstable with respect to $\beta^-$ decay. In the example above

$$^{24}_{11}Na \xrightarrow{\beta^-} {}^{24}_{12}Mg \tag{57}$$

However, many elements have two or more successive $A$ isotopes that are stable; for example,

$$^{42}Ca, \,^{43}Ca, \,^{44}Ca \tag{58}$$

so that neutron capture by the lighter isotopes results in the production of stable isotopes unsuitable for radioactivity measurement; that is,

$$^{42}Ca(n,\gamma)^{43}Ca_{stable} \quad \text{and} \quad ^{43}Ca(n,\gamma)^{44}Ca_{stable} \tag{59}$$

The heaviest of the series does produce a radioactive isotope:

$$^{44}Ca(n,\gamma)^{45}Ca \xrightarrow{\beta^-} {}^{45}Sc \tag{60}$$

Almost every element in the periodic table has a stable isotope which on neutron capture leads to a beta-emitting radioisotope with a measurable half-life.

TRANSMUTATION.  The second most common activation reaction is the absorption of a neutron with the subsequent release of a charged particle, most often a proton. This reaction is the (n,p) reaction. Other reactions that occur usually with less probability are (n,d), (n,α), (n,t), and (n,$^3$He), in each of which the chemical identity of the target nucleus is changed. Transmutation reactions thus afford the possibility of removing the radioactive product by chemical means from the bulk of the target material. Chemical separation of such products is discussed in Chapter 6.

The (n,p) reaction, illustrated by

$$^{27}_{13}\text{Al}(n,p)^{27}_{12}\text{Mg} \tag{61}$$

results in both the addition of a neutron and the removal of a proton from the target nucleus. The corresponding change in neutron-to-proton ratio is $N/Z$ to $(N + 1)/(Z - 1)$. In almost every case the product nucleus is beta unstable and decays back to the original target nucleus. Thus

$$^{27}_{12}\text{Mg} \xrightarrow{\beta^-} {}^{27}_{13}\text{Al} \tag{62}$$

Many of the heavier elements ($Z > \sim 30$) have one or more light stable isotopes, which result in (n,p) products that are positron unstable; for example,

$$^{106}_{48}\text{Cd}(n,p)^{106}_{47}\text{Ag} \xrightarrow{\beta^+} {}^{106}_{46}\text{Pd} \tag{63}$$

A few nuclides form stable (n,p) reaction products; for example,

$$^{123}_{52}\text{Te}(n,p)^{123}_{53}\text{Sb}_{\text{stable}} \tag{64}$$

In general, the (n,α) reaction occurs less frequently with low-energy neutrons than the (n,p) reaction because of the greater internal energy required to remove an alpha particle from a nucleus. The (n,α) reaction is more common for the lighter elements. The products are generally on the neutron-excess side of stability, although stable nuclides are frequently formed; and with the lighter isotopes of the heavier elements products on the neutron-deficient side of stability may be formed. These three events are illustrated, respectively, with

$$^{27}_{13}\text{Al}(n,\alpha)^{24}_{11}\text{Na} \xrightarrow{\beta^-} {}^{24}_{12}\text{Mg} \tag{65}$$

$$^{32}_{16}\text{S}(n,\alpha)^{29}_{14}\text{Si}_{\text{stable}} \tag{66}$$

$$^{40}_{20}\text{Ca}(n,\alpha)^{37}_{18}\text{Ar} \xrightarrow{\text{EC}} {}^{37}_{17}\text{Cl} \tag{67}$$

FISSION. The fission process involves the absorption of a neutron into the very heaviest elements ($Z \geq 90$) and results in the splitting of the nucleus into two large fragments with the concomitant release of two to three neutrons. This process, which as a chain reaction leads to nuclear reactor sources of irradiation neutrons, is discussed in Chapter 3.

INELASTIC SCATTERING. Inelastic scattering differs from the three preceding absorption processes in that the neutron transfers only part of its kinetic energy to the target nucleus, escaping with only a degradation in its energy. Two processes of inelastic scattering are of interest for the production of radioactive isotopes: the (n,n′) and the (n,2n) reactions.

1. The (n,n′) reaction takes place when the scattered neutron imparts sufficient energy to the target nucleus to raise its energy to a metastable state (an isomer of the nuclide). No change in the $N/Z$ ratio occurs. The isomer decays back to the stable state with the emission of a gamma ray; for example,

$$^{103}_{45}\text{Rh}(n,n')^{103m}_{45}\text{Rh} \tag{68}$$

The prime on the emerging neutron denotes that the neutron is degraded in energy with respect to the original neutron and the $m$ of $^{103m}\text{Rh}$ denotes that the product is a metastable isomer of $^{103}\text{Rh}$. The decay process is

$$^{103m}_{45}\text{Rh} \xrightarrow{\gamma} {}^{103}_{45}\text{Rh} \tag{69}$$

2. The (n,2n) reaction may occur when the scattered neutron imparts enough energy to exceed the binding energy of the least bound neutron in the target nucleus. The net result is the removal of one neutron from the target nucleus; for example,

$$^{23}_{11}\text{Na}(n,2n)^{22}_{11}\text{Na} \tag{70}$$

Since the average binding energy for a neutron is about 6 to 8 MeV, only high-energy neutrons can be used for such reactions. The resulting nuclide, in all cases, is isotopic with the target nucleus, and, in most cases, since the $N/Z$ ratio has been reduced to $(N-1)/Z$, the product nucleus is unstable with respect to $\beta^+$ and/or EC decay. In the above example

$$^{22}_{11}\text{Na} \xrightarrow{\beta^+} {}^{22}_{10}\text{Ne} \tag{71}$$

### Charged Particle Reactions

Charged particle reactions differ from neutron reactions primarily in that a charged particle approaching a target nucleus experiences an electrostatic potential given by

$$V = \frac{Zze^2}{r} \tag{72}$$

where $Z$ = the atomic number of the target nucleus,
    $z$ = the atomic number of the charged particle,
    $e$ = the unit of electrostatic charge ($e^2 = 1.44 \times 10^{-13}$ for $V$ in MeV and $r$ in cm),
    $r$ = the distance of separation of centers of the two particles.

Charged particles, however, may be absorbed by nuclei, since at the outer boundary of the target nucleus the strong attractive nuclear forces exceed the coulomb repulsive forces. The magnitude of the potential when the two particles are just tangent is known as the potential barrier $V_c$ and is approximated by

$$V_c \simeq \frac{Zze^2}{R_0(A^{1/3} + a^{1/3})} \simeq 1.03 \frac{Zz}{(A^{1/3} + a^{1/3})} \tag{73}$$

where $A$ and $a$ are the mass numbers of the target and projectile nuclei, respectively, and $R_0$ is the unit nuclear radius ($\simeq 1.4 \times 10^{-13}$ cm). Thus the potential barrier for the reaction $^{197}_{79}\text{Au}(\alpha,\text{n})^{200}_{81}\text{Tl}$ is

$$V_c \simeq 1.03 \frac{79 \times 2}{(197^{1/3} + 4^{1/3})} = 22 \text{ MeV} \tag{74}$$

Although, according to the laws of quantum mechanics, the reaction can take place with alpha particles with kinetic energy just greater than the threshold energy of 12 MeV, the probability that reactions will occur does not become appreciable until the kinetic energy of the alpha particles used is greater than 22 MeV.

A schematic representation of the potential barrier is given in Figure 1.5. The potential energy at $R$ is the minimum kinetic energy, according to classical considerations, required by the charged particle to collide with the target nucleus. Quantum mechanics, however, show that in the nuclear domain a finite probability exists that a particle with less kinetic energy than the potential barrier will succeed in becoming absorbed into the nucleus. The resulting nucleus will have its internal energy raised by both the incoming kinetic energy ($V_E - V_0$) and the binding energy of the particle ($V_0 - V_B$), as illustrated in Figure 1.5. This excitation energy may be used to eject neutrons, charged particles, and/or photons. Thus many types of nuclear reaction can take place, each with some probability.

Charged particle accelerators which may be used for radioactivations are described in Chapter 3. These accelerators can provide proton-, deuteron-, and alpha-particle beams. Newer accelerators, which can produce beams of $^3$He ions and ions heavier than $^4$He, such as $^{12}$C, $^{14}$N, and $^{16}$O, are becoming available. These particles with high kinetic energy produce a greater variety of nuclear reaction. Because of the lower potential barrier associated with protons and deuterons, which allows irradiations to be made with lower

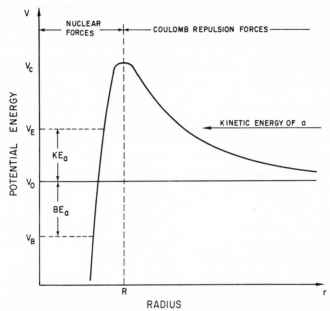

**Figure 1.5** A schematic representation of the nuclear potential energy of a target nucleus as a function of the separation distance from the irradiating particle $a$. Particles with less kinetic energy than $V_c$ may be absorbed by the nucleus according to quantum mechanics principles, leaving the nucleus with excitation energy equal to the incoming kinetic energy $V_E - V_0$, plus the binding energy $V_0 - V_B$.

energy accelerators, these two particles have been more widely used in activation analysis. Reactions with these two particles are examined further.

PROTON REACTIONS.  The most probable reaction with low-energy protons is the (p,n) reaction; for example,

$$^{45}_{21}\text{Sc}(p,n)^{45}_{22}\text{Ti} \tag{75}$$

The minimum "useful" kinetic energy required by the proton is given by the threshold energy for the reaction and by the potential barrier. For the above example (55) and (73) give the values

$$E_{\text{th}} = \frac{45 + 1}{45} Q = \frac{46}{45} \times 931.4 \, \Delta M = 2.9 \text{ MeV} \tag{76}$$

$$V_c \simeq 1.03 \times \frac{21 \times 1}{45^{\frac{1}{3}} + 1} \simeq 4.6 \text{ MeV} \tag{77}$$

These calculations show that the reaction may take place with protons with energies greater than 2.9 MeV but that the yield increases significantly at proton energies greater than 4.6 MeV. However, as the proton energy is further increased, other reactions such as (p,pn) or (p,2n) may occur with greater probability.

DEUTERON REACTIONS.   The low binding energy (2.2 MeV) of the deuteron makes it useful for transmutation reactions for several reasons.

1. Deuterons may be partly polarized in the field of the target nucleus; that is, the proton in being repelled turns the neutron toward the nucleus. The neutron may then be absorbed without complete capture of the deuteron. Such a reaction is called a "stripping" reaction; the neutron is stripped from the deuteron. The over-all reaction is shown as a charged particle reaction; for example,

$$^{55}_{25}\text{Mn}(d,p)^{56}_{25}\text{Mn} \tag{78}$$

The reaction is equivalent to the (n,$\gamma$) reaction and can occur with deuteron energies lower than those required for complete capture of the deuteron. For the above reaction the potential barrier is

$$V_c \simeq 1.03 \frac{25 \times 1}{55^{1/3} + 1} \simeq 5.4 \text{ MeV} \tag{79}$$

2. Most deuteron reactions involving single-particle emission, that is, (d,n), (d,p), and (d,$\alpha$) are exoergic; thus there is no threshold energy requirement for such reactions.

These two properties, which result from the small binding energy of the deuteron, produce relatively higher yields for deuteron irradiations than for proton or $\alpha$-particle irradiations. Deuteron beams available in cyclotrons are discussed in Chapter 3.

### Electron and Photon Reactions

The interaction of electrons with matter, as described in Chapter 4, results in the production of photons so that the two forms of radiation lead to the same types of nuclear reaction. The scattering or absorption of electromagnetic radiation by a nucleus leads to a deposition of energy in the nucleus with a subsequent emission of radiation, analogous to the scattering or absorption of neutrons. For electrons and photons with energies below the neutron binding energies inelastic scattering may result in the creation of metastable isomers such as $^{103m}$Rh or $^{115m}$In by the respective reactions

$$^{103}\text{Rh}(\gamma,\gamma')^{103m}\text{Rh} \quad \text{or} \quad ^{115}\text{In}(e,e'\gamma)^{115m}\text{In} \tag{80}$$

With energies above the neutron binding energies electrons and photons can induce photonuclear reactions with the emission of neutrons (and other particles as the kinetic energy of the photons are increased). Thus for

$E > 9.4$ MeV for $^{103}$Rh and $E > 9.0$ MeV for $^{115}$In the photonuclear reactions

$$^{103}\text{Rh}(\gamma,\text{n})^{102}\text{Rh} \quad \text{and} \quad ^{115}\text{In}(\text{e},\text{n})^{114}\text{In} \tag{81}$$

can occur.

Note that these two photon and electron reactions result in the same radioactivity production as the neutron reactions (n,n′) and (n,2n). Thus for $^{103}$Rh the corresponding reactions are

$$^{103}\text{Rh}(\gamma,\gamma')^{103\text{m}}\text{Rh}, \quad ^{103}\text{Rh}(\gamma,\text{n})^{102}\text{Rh} \tag{82}$$

$$^{103}\text{Rh}(\text{n},\text{n}')^{103\text{m}}\text{Rh}, \quad ^{103}\text{Rh}(\text{n},2\text{n})^{102}\text{Rh} \tag{83}$$

The number of electron linear accelerators becoming available in many places should increase the use of electron and photon beams as irradiation sources for activation analysis.

## 1.6  BIBLIOGRAPHY

### 1.6.1  General Texts of Nuclear Physics and Chemistry

W. E. Burcham, *Nuclear Physics* (McGraw-Hill, New York, 1963).

D. J. Carswell, *Introduction to Nuclear Chemistry* (Elsevier, Amsterdam, 1967).

J. C. Cunninghame, *Introduction to the Atomic Nucleus* (Elsevier, Amsterdam, 1964).

L. F. Curtiss, *Introduction to Neutron Physics* (Van Nostrand, Princeton, N.J., 1959).

R. D. Evans, *The Atomic Nucleus* (McGraw-Hill, New York, 1955).

G. Friedlander, J. W. Kennedy, and J. M. Miller, *Nuclear and Radiochemistry*, 2nd ed. (Wiley, New York, 1964).

M. Haissinsky, *Nuclear Chemistry and its Applications* (Addison-Wesley, Reading, Mass., 1964).

D. Halliday, *Introductory Nuclear Physics*, 2nd ed. (Wiley, New York, 1955).

B. G. Harvey, *Introduction to Nuclear Physics and Chemistry*, 2nd ed. (Prentice-Hall, Englewood Cliffs, N.J., 1969).

E. N. Jenkins, *Introduction to Radioactivity* (Butterworths, London, 1964).

N. F. Johnson, E. Eichler, and G. D. O'Kelly, *Nuclear Chemistry* (Wiley, New York, 1963).

I. Kaplan, *Nuclear Physics*, 2nd ed. (Addison-Wesley, Reading, Mass., 1963).

R. E. Lapp and H. L. Andrews, *Nuclear Radiation Physics*, 3rd ed. (Prentice-Hall, Englewood Cliffs, N.J., 1963).

M. Lefort, *Nuclear Chemistry* (Van Nostrand, Princeton, N.J., 1968).

W. E. Meyerhof, *Elements of Nuclear Physics* (McGraw-Hill, New York, 1967).

R. L. Sproull, *Modern Physics*, 2nd ed. (Wiley, New York, 1963).

L. Yaffe, Ed., *Nuclear Chemistry* (Academic, New York, 1968), 2 Vols.

### 1.6.2 Paperback Texts

I. Adler, *Inside the Nucleus* (Signet, New York, 1963).

G. R. Choppin, *Nuclei and Radioactivity* (Benjamin, New York, 1964).

B. L. Cohen, *The Heart of the Atom* (Anchor, New York, 1967).

C. S. Cook, *Structure of Atomic Nuclei* (Van Nostrand, Princeton, N.J., 1964).

B. G. Harvey, *Nuclear Chemistry* (Prentice-Hall, Englewood Cliffs, N.J., 1965).

R. T. Overman, *Basic Concepts of Nuclear Chemistry* (Reinhold, New York, 1963).

### 1.6.3 Sources of Nuclear Data

General Electric Company, *Chart of the Nuclides*, 10th ed., 1969, available from the General Electric Co., Schenectady, New York 12305.

R. C. Koch, *Activation Analysis Handbook* (Academic, New York, 1960).

C. M. Lederer, J. M. Hollander, and I. Perlman, *Table of Isotopes*, 6th ed. (Wiley, New York, 1967).

National Academy of Sciences—National Research Council, *Nuclear Data Tables*, available from Superintendent of Documents, U.S. Government Printing Office, Washington 25, D.C.

W. Seelmann-Eggebert, G. Pfennig, and H. Munzel, *Nuklidkarte*, 3rd ed. (Gersbach, Munich, 1968).

### 1.7 PROBLEMS

**1.** Nuclides have been made with the following compositions:

$$Z: \quad 8 \quad 21 \quad 33 \quad 43 \quad 90$$
$$+$$
$$N: \quad 8 \quad 19 \quad 42 \quad 55 \quad 142$$

Identify the nuclides and determine which of them are radioactive.

**2.** The chemical oxidation of carbon by the reaction

$$C + O_2 \rightarrow CO_2$$

liberates 94,000 calories of heat per gram of carbon. How many grams of carbon does this much heat represent? How many tons (1 ton = $9.07 \times 10^5$ g) of coal (as pure carbon) would be consumed in combustion to produce the energy equivalent to 1 g of carbon?

**3.** How much binding energy would be required (or released) for the nuclear decay of $^{16}O$ into four $^4He$ nuclei?

$$^{16}_{8}O \rightarrow 4\,^{4}_{2}He.$$

Suggest why this nuclear decay is not observed.

4. Experimentally, the coulomb barrier for helium ions approaching $^{209}_{83}$Bi nuclei has a height of 19.5 MeV. From these data calculate the radius parameter $R_0$ in

$$R = R_0 A^{\frac{1}{3}}.$$

5. Write in shorthand form a nuclear reaction

| | to produce | from |
|---|---|---|
| (a) | $^{24}$Na | sodium |
| (b) | $^{56}$Mn | cobalt |
| (c) | $^{58}$Co | manganese |
| (d) | $^{88}$Y | yttrium |
| (e) | $^{93m}$Nb | niobium |

6. Complete the following nuclear reactions:

(a) $\quad\quad\quad\quad {}^{18}_{8}O(^{3}He,\_\_){}^{19}_{9}F$

(b) $\quad\quad\quad\quad {}^{36}_{16}S(\_\_,n){}^{35}_{16}S$

(c) $\quad\quad\quad\quad {}^{40}_{18}Ar(p,4p5n)\_\_$

(d) $\quad\quad\quad\quad \_\_(d,n){}^{4}_{2}He$

(e) $\quad\quad\quad\quad {}^{239}_{93}Np(n,f){}^{94}_{38}Sr + \_\_ + 3\,{}^{1}_{0}n$

7. Thermonuclear reactions depend on the fusion of two light nuclei; for example, two deuterium nuclei may fuse by the reaction $^{2}$H(d,p)$^{3}$H, liberating 4.04 MeV of energy per event. If radiative capture of neutrons [i.e., the (n,$\gamma$) reaction] by deuterium liberates 6.25 MeV per event, compute the binding energy of the deuteron.

8. Are the reactions $^{27}_{13}$Al(n,$\alpha$)$^{24}_{11}$Na and $^{14}_{7}$N(n,p)$^{14}_{6}$C exoergic or endoergic and by how much? Compute the threshold energies for the two reactions.

9. Examine the GE Chart of the Nuclides or a similar nuclide chart and list the likely nuclear reactions that will take place with 14 MeV neutrons irradiating a silver foil and result in radioactive product nuclides.

10. Phosphorus in an organic material may be determined by neutron activation analysis from the reaction:

$$^{31}_{15}P(n,\gamma)^{32}_{15}P$$

Find two additional neutron reactions with other elements that may be present in organic materials and that will also lead to the production of $^{32}_{15}$P.

# Chapter 2

# Radioactivation

Activation analysis has been described in an operational sense as consisting of two major processes; (a) the production of a radioactive nuclide from the desired element by some nuclear reaction and (b) the measurement of the amount of the product radioactive nuclide that is related to the amount of the desired element present in the sample. The relationship is given by the radioactivity production equation, which commonly is written in the form:

$$D_i^0 = n_i \sigma_i \phi (1 - e^{-\lambda_i t}) \qquad (1)$$

$D_i^0$ is the radioactivity of the nuclide $i$ present at the end of the irradiation. Equation 2.1 shows that $D_i^0$ is a function of five parameters, three of which are determined by the identity of $i$, whereas two are general. These five parameters consist of the following:

$n_i$—the number of target nuclei in the sample available for the reaction to produce the nuclide $i$ (see Section 2.1).

34

$\sigma_i$—the cross section for the reaction producing the nuclide $i$ (see Section 2.2).

$\lambda_i$—the decay constant for the nuclide $i$ (see Section 1.3.4).

$\phi$—the irradiating particle flux (see Section 2.3).

$t$—the duration of the irradiation (see Section 2.4).

As an illustrative example, we examine the production of the radioactive isotope $^{64}Cu$ produced by the nuclear reaction $^{63}Cu(n,\gamma)^{64}Cu$ of the element copper from neutron irradiation. For this example

$D_i^0$  refers to the amount of $^{64}Cu$ present at the end of the irradiation.

$n_i$  refers to the number of $^{63}Cu$ target nuclei in the matrix.

$\sigma_i$  refers to the cross section of the $^{63}Cu(n,\gamma)^{64}Cu$ reaction.

$\lambda_i$  refers to the decay constant of the radioactive $^{64}Cu$ produced.

$\phi$  refers to the neutron flux to which the matrix was exposed.

$t$  refers to the irradiation time.

## 2.1  TARGET AND MATRIX MATERIALS

### 2.1.1  Targets

In the discussion of nuclear reactions we have often referred to the *target* nuclide. The target nuclide is the particular isotope of the element being determined which is involved in the particular nuclear activation reaction chosen to produce the particular radioactive product nuclide. Thus for the copper-activation illustration $^{63}Cu$ is the *target* nuclide. Since copper is the element we wish to determine, an appropriate nuclear reaction must be found which involves copper isotopes to produce some measurable radioactive nuclide (not necessarily a copper isotope). We chose the $^{63}Cu(n,\gamma)^{64}Cu$ reaction. We could also have chosen the $^{65}Cu(n,p)^{65}Ni$ or the $^{63}Cu(p,n)^{63}Zn$ reaction to measure copper by the amount of $^{65}Ni$ or $^{63}Zn$ produced, respectively. Note that the nuclear reaction is selected but that the *target* nuclide must be an isotope of the element being determined.

### 2.1.2  Matrices

In general, any sample being irradiated contains more than one nuclide. An exception is the activation to produce a pure radioisotope from a pure and monoisotopic element. Such elements are listed in Table 1.2. In the general case the desired element (or elements) may be a minor or trace constituent(s) of the sample. The total material of the sample being irradiated is called the *matrix*. Copper may be determined in a variety of matrix materials.

The matrix may be, among other things,

(a) organic or inorganic or both,
(b) solid, liquid, or gaseous,
(c) volatile or explosive,
(d) abundant or rare in available quantity for irradiation,
(e) all or part of the sample,
(f) rigid, plastic, or powder.

Each of these, and other, properties of the matrix determines, in part, the optimum method for achieving a successful irradiation.

### 2.1.3  Avogadro's Number

Generally, in activation analysis the desired element(s) is (are) a small fraction of the matrix material. Usually the *matrix* sample can be weighed before the irradiation. The weights of the *target* elements are obtained from the activation analysis, since the number of atoms of target nuclei in the matrix determined by the activation equation is related to the weight of the target nuclide by Avogadro's number N.

Avogadro's number expresses the fact that one gram-atomic weight of any element contains the same number of atoms. The value is generally accepted as $6.023 \times 10^{23}$ atoms/gram-atomic weight. Thus the weight $m_i$ (in grams) of our $n_i$ target nuclei is given by

$$m_i = n_i \times \frac{A_e}{N} \qquad (2)$$

where $A_e$ = atomic weight of the element (obtained from tables of atomic weights),

$N$ = Avogadro's number = $6.023 \times 10^{23}$ atoms/g-amu.

If the matrix contains $g$ grams of sample, the weight percent of the target nuclide $i$ in the matrix is given by

$$W_i(w/o) = \frac{m_i}{g} \times 100 \qquad (3)$$

### 2.14  Isotopic Abundance

Note that most elements consist of more than one stable isotope. Each isotope thus has a fractional isotopic abundance, abbreviated generally as $f$. For $^{23}$Na, which is the only stable isotope of sodium, $f = 1.0$. The other monoisotopic stable elements are listed in Table 1.2. Copper, however, consists in nature as 69.09 % $^{63}$Cu and 30.91 % $^{65}$Cu. Thus $^{63}$Cu has a fractional isotopic abundance of $f = 0.6909$. If we determine the amount of copper by the $^{63}$Cu(n,$\gamma$)$^{64}$Cu reaction, we determine the number of $^{63}$Cu atoms only. Thus the weight of the element $m_e$ is given by

$$m_e = n_i \times \frac{A_e}{f_i N} \tag{4}$$

$$m_e = \frac{m_i}{f_i} \tag{5}$$

## 2.2 CROSS SECTION

### 2.2.1 Definition

The rate of nuclear reactions in a given irradiation system is determined not only by the number of incident and target nuclei available for interactions but also by the probability that an incident particle *will* react with a target nucleus. This probability, expressed in terms of an area per incident particle, is called the cross section $\sigma$ of the particular reaction and is somewhat analogous to the rate constant $k$ used in chemical reaction kinetics.

A typical irradiation arrangement in an accelerator is shown in Figure 2.1 in which a thin target of uniform thickness $x$ is placed in a beam whose intensity is $I$ particles/sec. The beam intensity is measured with a Faraday cup, as given in Section 2.3.2. For a beam of particles striking a thin target (one in which the attenuation of incident particles is negligible), the cross

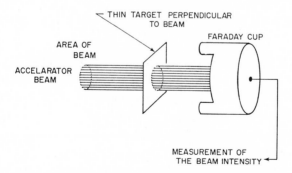

**Figure 2.1** In a typical accelerator irradiation a thin target is placed close to the exit port of the accelerator and a Faraday cup collects and integrates the total charge of the irradiating particles.

section $\sigma_i$ for the nuclear reaction $i$ is defined by the equation

$$\frac{dN_i}{dt} = R_i = I_0 n \sigma_i x \tag{6}$$

where $R_i$ = reaction rate for events $i$ (sec$^{-1}$),
   $I_0$ = number of incident particles per unit time (sec$^{-1}$),
   $n$ = number of target nuclei per unit volume (cm$^{-3}$),
   $x$ = target thickness (cm).

The cross section for a specified type of event may thus be defined as the ratio of the rate of those events to the rate of particle-target nuclei exposures per unit area:

$$\sigma_i = \frac{R_i}{I_0 n x}, \qquad \text{(cm}^2\text{/target nucleus)} \tag{7}$$

Thus the cross section has the dimensions of an area and expresses the probability that a particle will undergo the specified type of event by the ratio of the "effective" area to the total area of the target nucleus.

### 2.2.2   Total Cross Section

The limiting case for thin targets may be considered to be a target width of one nucleus per square centimeter. In practice, irradiation targets are generally thick such that on traversal the beam intensity may be significantly reduced. In this case the rate of reactions of all types occurring in the target is given by the loss in beam intensity. If $N_T$ is the total number of reactions removing a particle from the beam, then

$$\frac{dN_I}{dt} = -dI = I n \sigma_T \, dx \tag{8}$$

where $\sigma_T$ is the total cross section. Solution of (8) in the form

$$\frac{dI}{I} = -n \sigma_T \, dx \tag{9}$$

gives as the rate of interactions in target thickness $x$

$$\Delta I = I_0 - I = I_0 (1 - e^{-n \sigma_T x}) \tag{10}$$

### 2.2.3   Geometric Cross Section

Another concept of cross section as a probability of reaction between an incident particle and a target nucleus is expressed in terms of a geometrical cross section presented by the target nucleus to a point-size projectile. This cross-sectional area, $\sigma_{geo}$, is a good approximation to the reaction cross sections measured for fast neutron reactions but is not satisfactory for charged particles that must overcome coulomb barriers nor for slow neutrons that are in thermal equilibrium with their environment.

The geometrical cross section is given by

$$\sigma_{geo} = \pi R^2 \tag{11}$$

where from (1-1) in Section 1.1.2, $R = R_0 \times A^{\frac{1}{3}}$ cm

$$= 1.4 \times 10^{-13} A^{\frac{1}{3}} \text{ cm} \tag{12}$$

The geometric cross section gives us an idea of the magnitudes of reaction cross sections. Using the medium-weight nuclide $^{66}$Zn as an example,

$$\sigma_{geo} = 3.14 \times 1.4^2 \times 10^{-26} \times (66)^{\frac{2}{3}}$$
$$= 1.0 \times 10^{-24} \text{ cm}^2 \tag{13}$$

### 2.2.4 Units of Cross Section

Since most cross sections are of the order of $10^{-24}$ cm², it has become convenient to express cross sections in the units of the *barn*, where

$$1 \text{ barn} = 10^{-24} \text{ cm}^2 \tag{14}$$

For further convenience the barn is subdivided for low cross-section values in the metric system:

$$1 \text{ barn} = 10^3 \text{ millibarns } (1 \text{ mb} = 10^{-27} \text{ cm}^2)$$
$$= 10^6 \text{ microbarns } (1 \text{ } \mu\text{b} = 10^{-30} \text{ cm}^2) \tag{15}$$

Although the unit of barn is convenient for expressing or tabulating cross sections, it should be remembered that in activation calculations cross sections have the units of square centimeters.

### 2.2.5 Reaction Cross Section

The total cross section $\sigma_T$ was defined for a system in which the rate of nuclear events occurring was given by the rate of removal of the incident particles. Nuclear events involve either scattering or absorption processes. Each of these has a *partial* cross section, $\sigma_s$ and $\sigma_a$, respectively, in which

$$\sigma_T = \sigma_s + \sigma_a \tag{16}$$

Both general processes can in turn be further subdivided into individual processes; for example, scattering can occur as elastic collisions (in which the total kinetic energy of the system is conserved) or inelastic collisions (in which part of the kinetic energy of the incident particle results in the excitation of the target nucleus). When both processes occur in an irradiation,

$$\sigma_s = \sigma_{el} + \sigma_{inel} \tag{17}$$

Absorption of the incident particle leads to a nuclear transmutation; for example, irradiation of an aluminum foil with 14 MeV neutrons can result in several different nuclear reactions, including the more probable ones:

$$^{27}_{13}\text{Al}(n,\gamma)^{28}_{13}\text{Al} \quad \sigma_{n,\gamma}$$
$$^{27}_{13}\text{Al}(n,2n)^{26}_{13}\text{Al} \quad \sigma_{n,2n}$$
$$^{27}_{13}\text{Al}(n,p)^{27}_{12}\text{Mg} \quad \sigma_{n,p}$$
$$^{27}_{13}\text{Al}(n,d)^{26}_{12}\text{Mg} \quad \sigma_{n,d}$$
$$^{27}_{13}\text{Al}(n,\alpha)^{24}_{11}\text{Na} \quad \sigma_{n,\alpha}$$

Each of these reactions has its own partial cross section, called the reaction cross section. The sum of all partial cross sections for the reactions that are occurring with the target nuclei is equal to the partial cross section for absorption:

$$\sigma_a = \sigma_{n,\gamma} + \sigma_{n,2n} + \sigma_{n,p} + \cdots \tag{18}$$

Note that some absorption reactions lead to stable product nuclides; for example, $^{27}\text{Al}(n,d)^{26}\text{Mg}$. Such reactions, of course, are not useful for radioactivation. The part of the absorption cross section that leads to radioactive nuclides is generally called the activation cross section ($\sigma_{act}$). For low energy (thermal) neutrons the activation cross section is usually accounted for primarily by the radiative capture cross section $\sigma_{(n,\gamma)}$. The activation cross section has little meaning for high energy irradiating particles because of the multiplicity of nuclear reactions possible. Care should be exercised in obtaining appropriate cross-section values from sources in the literature to ensure that the cross section listed is for the desired reaction.

### 2.2.6 Excitation Functions

Cross sections vary not only with the nuclear reaction but, for a particular nuclear reaction, also with the energy of the irradiating particles. We have already seen (Section 1.5) that for endoergic and charged-particle reactions there is a threshold energy below which the reaction will not take place. The cross section for the reaction increases with energy above its threshold energy, generally reaches some maximum value, and then decreases with further increasing energy of the irradiating particles as other nuclear reactions, becoming energetically possible, increase in cross section. The dependence

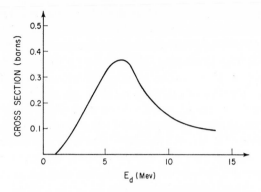

**Figure 2.2** A typical excitation function curve.

of the cross section on the energy of the irradiating particle is given by an *excitation function*. An excitation function for the $^{23}$Na(d,p)$^{24}$Na reaction is shown in Figure 2.2. The reaction shows a threshold at 1 MeV and a maximum cross section of 0.4 b at 6 MeV. For irradiations with fast ($E > 1$ MeV) monoenergetic particles the appropriate cross-section value must be used. If the irradiating particles have a spectrum of energies (e.g., in a nuclear reactor), the integrated cross section over the appropriate range of energies must be known. Measured cross sections for reactions with reactor neutrons are frequently refered to as pile neutron cross sections. Since the actual neutron energy spectrum varies from reactor to reactor, even in different locations within a given reactor, published pile neutron cross section values should be used with caution.

## 2.3 FLUXES AND BEAMS

Irradiations consist of exposing the matrix material to a source of irradiating particles. The sources of the more commonly used particles are described in Chapter 3. The production rate of the radionuclides from (2.1) involves the number of irradiating particles available for the activation which is expressed in terms of fluxes and beam intensities. The term *flux* is generally associated with slow or moderated neutron sources approaching the matrix nuclei from all directions, whereas the term *beam intensity* is generally associated with fast neutrons and charged particles which interact with the matrix from a single direction.

### 2.3.1 Fluxes

A thin matrix material presents a varying thickness to irradiating particles if the particles approach from varying directions. Since the total number of interactions depends on the thickness (i.e., the number of target nuclei) traversed, the total number of interactions will be greater for an irradiation intensity from all directions compared with one from a single direction only. The intensity of the irradiating particles in an irradiation in which the particles arrive from all directions is called the particle flux ($\phi$). For neutron irradiations the intensity of neutrons is, thus, the neutron flux and may be made more specific by classification into neutron fluxes of different neutron-energy groups (as given in Section 1.4.4). The neutron flux may be considered to be the number of neutrons crossing a unit area in unit time. Thus, if at steady state (i.e., equal numbers of neutrons are crossing a unit area in all directions), there is a density of $n$ neutrons/cm³ with an average velocity of $\bar{v}$ cm/sec, and the neutron flux is given by

$$\phi = n\bar{v} \quad \text{(in n/cm²-sec)} \tag{19}$$

for example, neutrons in thermal equilibrium with their environment are described as thermal (see Section 1.4.4). The neutron density-velocity relationship (i.e., the number of neutrons per unit volume in the velocity interval $dv$) is given by the kinetic theory of gases as

$$\frac{dn}{dv} = \frac{4n}{v_0^3\sqrt{\pi}} v^2 e^{-v^2/v_0^2} \tag{20}$$

where $n$ is the total number of neutrons per unit volume and $v_0 = (2kT/M_n)^{\frac{1}{2}}$ is the most probable velocity at an absolute temperature $T$; $M_n$ is the mass of the neutron and $k$ is the Boltzmann constant of kinetic energy $= 1.38 \times 10^{-16}$ erg/°C (or $8.56 \times 10^{-5}$ eV/°C). At an absolute temperature of 293°K (20°C) the most probable velocity is

$$v_0 = \left[\frac{2(1.38 \times 10^{-16})293}{1.67 \times 10^{-24}}\right]^{\frac{1}{2}}$$

$$v_0 = 2.2 \times 10^5 \text{ cm/sec}, \quad (E_0 = \tfrac{1}{40} \text{ eV}) \tag{21}$$

The average velocity is

$$\bar{v} = \frac{2v_0}{\sqrt{\pi}} = 2.5 \times 10^5 \text{ cm/sec} \tag{22}$$

and for a neutron density of $10^7$ n/cm³ the flux would be

$$\phi = n\bar{v} = 10^7 \times 2.5 \times 10^5 = 2.5 \times 10^{12} \text{ n/cm²-sec} \tag{23}$$

The total number of neutrons available during an irradiation lasting $t$ sec is called the integrated neutron flux and is given by

$$\Phi = n\bar{v}t, \quad \text{(in n/cm²)} \tag{24}$$

Thus a matrix exposed to the flux given in (23) for a period of one hour would have experienced an integrated neutron flux:

$$\Phi = 2.5 \times 10^{12} \times 3.6 \times 10^3 = 9.0 \times 10^{15} \text{ n/cm}^2 \tag{25}$$

## 2.3.2 Beams

A beam of particles, as contrasted to a flux, describes a generally collimated array of particles traveling in a single direction with intensity of $N$ particles/sec, expressed as a current. A sketch of a collimated-beam irradiation is given in Figure 2.1. However, some charged-particle accelerators may present an irradiation geometry in which a conically diverging beam of particles from the output tube of the machine traverses the sample being irradiated. The geometry is approximated by Figure 2.3, in which an object of volume $V$ is totally within the dimensions of the beam given by the variables $r$ and $\Omega$ and average intensity of $\bar{J}$ particles/cm² sec.

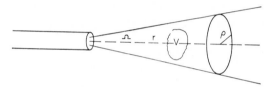

**Figure 2.3** Sample of volume $V$ irradiated in a conically diverging beam of accelerator particles. For a beam output of $I$ particles/sec the beam intensity at a distance $r$, of radius $\rho$, and diverging angle $\Omega$, is given by $J = I/\pi r^2 \tan^2 \Omega$ particles/cm²-sec.

For a sample in which the target element is not homogeneously distributed within the sample (e.g., concentrated at grain boundaries) the saturation activity (see Section 2.4) of the product radionuclide is given by the general equation

$$D^\infty = \int_V \int_V \int_{E\text{th}}^{E_{\max}} \left(\frac{\partial n}{\partial V}\right) dV \left(\frac{\partial J}{\partial V}\right) dV \left(\frac{\partial \sigma}{\partial E}\right) dE \tag{26}$$

in which the total number of target nuclei and beam particles is integrated over the total volume and the integrated cross section (in MeV-barns) of the reaction is given for the energy spectrum from the reaction threshold to the maximum particle energy.

For a homogeneous material (26) reduces to

$$D^\infty = n \int_V \int_{E\text{th}}^{E_{\max}} \left(\frac{\partial J}{\partial V}\right) dV \left(\frac{\partial \sigma}{\partial E}\right) dE \tag{27}$$

and for a monoenergetic beam of particles, passing through a sufficiently thin target in which the energy losses by collision and radiative processes are small, (27) reduces still further to

$$D^{\infty} = n\bar{J}\sigma \tag{28}$$

in which $\bar{J}$ is the average beam intensity passing through the target volume and $\sigma$ is the differential reaction cross section at the particular energy.

Charged-particle beams, generally given for accelerators as a current $I$ in microamperes ($\mu$A) or milliamperes (mA), are usually measured with a Faraday cup, an insulated electrode designed to stop the entire beam and measure the charge deposited. The total charge is measured by the voltage drop $\Delta V$ developed across a condenser of known capacity $C$ by

$$Q = C\,\Delta V \tag{29}$$

Since the charge per mole of singly charged ions is given by Faraday's constant,

$$\mathbf{F} = \mathbf{N} \times e = 9.65 \times 10^4 \text{ coulomb/mole}, \tag{30}$$

the current in microamperes is given by

$$1\,\mu\text{A} = \frac{6.023 \times 10^{23}}{9.65 \times 10^4} \times 10^{-6}$$

$$= 6.25 \times 10^{12} \text{ particles/sec} \tag{31}$$

and

$$I(\mu\text{A}) = \frac{C\,\Delta V \times 10^{-6}}{t} \tag{32}$$

is the average beam current for a total irradiation of $It$ $\mu$A-sec of total charge $C\,\Delta V$ coulomb. The average beam intensity $\bar{J}$ in particles per second may be expressed as

$$\bar{J} \text{ (particles/sec)} = 6.25 \times 10^{12}\, I\,(\mu\text{A}) \tag{33}$$

## 2.4 SATURATION ACTIVITY

The last three sections have examined the three factors that determine the reaction rate for a particular nuclear reaction. The reaction rate $R$ (in events per second) may be expressed as the number of product nuclei $N_p$ created per second by

$$R = \frac{dN_p}{dt} = n\sigma\phi \tag{34}$$

or

$$R = \theta\sigma J \tag{35}$$

where $\sigma$ = the reaction cross section, in cm²,
$n$ = the number of atoms in a flux $\phi$, in particles/cm²-sec,
$\theta$ = the thickness of the matrix, in number of atoms/cm², exposed to a beam $J$, in particles/sec.

If the product nuclei are stable, the total number created during an irradiation lasting $t$ sec is

$$N_p = Rt = (n\sigma\phi)t \tag{36}$$

However, we are interested in a radioactive product, which therefore decays during the irradiation with its characteristic half-life at a rate proportional to the amount present at any specified time [see (1-28)]. The rate of change of product nuclei during the irradiation is thus given by the difference between the production rate $R$ and the decay rate $\lambda N$, or

$$\frac{dN_p}{dt} = R - \lambda N_p \tag{37}$$

If the production rate is a constant, as it usually is, the solution to (37) is given by

$$N_p = \frac{R}{\lambda}(1 - e^{-\lambda t}) \tag{38}$$

or the disintegration rate $D_p$ of the nuclide resulting from the particular nuclear reaction at the end of the irradiation period $t$ is

$$D_p = \lambda N_p = R(1 - e^{-\lambda t}) \tag{39}$$

Equation 39 shows that for a constant reaction rate $R$ the activity produced in a given irradiation system will increase with irradiation time $t$, but as the irradiation time becomes large in relation to the half-life, so that $\lambda t \to \infty$, $e^{-\lambda t} \to 0$, and the decay rate of the sample approaches the production rate as a limiting value. This relationship can also be seen from (37), for, when the decay rate $\lambda N_p$ becomes equal to the production rate $R$, $dN_p/dt = 0$, and the number of radioactive product nuclei has reached its maximum value

$$N_{p(\text{max})} = \frac{R}{\lambda} \tag{40}$$

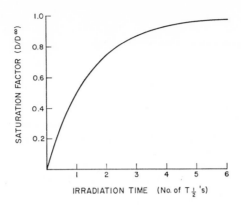

**Figure 2.4** Growth of a radioactive nuclide at a constant production rate, $D = D^\infty(1 - e^{-\lambda t})$, for which $D^\infty$, the saturation activity, is equal to the production rate. The curve shows the saturation factor as a function of irradiation time, expressed in number of half-lives. The value for any number of half-lives $n$, is $D/D^\infty = (1 - 1/2^n)$

and

$$D_{\max} = R \tag{41}$$

This activity is called the *saturation activity* and is designated by the symbol $D^\infty$. Thus the general equation of the radioactivity produced in a given irradiation of time $t$ is shown as

$$D(t) = D^\infty(1 - e^{-\lambda t}) \tag{42}$$

Figure 2.4 shows the growth curve of a radioactive product nuclide in units of fraction of saturation $(D/D^\infty)$ as a function of the irradiation time in units of its half-life. Note that the radioactivity production curve is the reciprocal of the radioactive decay curve. Note further that the increase in radioactivity reaches the "economic law of diminishing returns" and, when irradiation time is costly, that significantly increased production is not obtained with irradiation times much longer than two or three half-lives.

In concluding this section we refer to our illustration of the activation of copper by the $^{63}Cu(n,\gamma)^{64}Cu$ reaction. The irradiation consists of the exposure of a 1 g thin-sheet of material containing 1% copper to the thermal neutron flux of (23) for a period of 3 hr. Referring to the nomenclature in Section 2.0,

$D^0 =$ the unknown activity of the $^{64}$Cu product at the end of the irradiation,

$n =$ the number of $^{63}$Cu target nuclei,

$$n = \frac{m_e f N}{A_e} = \frac{0.01(0.6909)6.023 \times 10^{23}}{63.54}$$

$$= 6.55 \times 10^{19} \text{ atoms,}$$

$\sigma =$ the reaction cross section for the $(n_{th}, \gamma)$ reaction

$= 4.5$ b $= 4.5 \times 10^{-24}$ cm$^2$,

$\phi =$ the thermal neutron flux

$= 2.5 \times 10^{12}$ n/cm$^2$ sec,

$(1 - e^{-\lambda t}) =$ the saturation factor,

$T_{1/2} = 12.8$ hr for $^{64}$Cu,

$\lambda = \ln 2/T_{1/2} = 0.693/12.8 = 0.054/$hr

$t = 3$ hr,

$\lambda t = 0.162$ (dimensionless),

$(1 - e^{-\lambda t}) = 0.150$.

Thus the activity of the $^{64}$Cu in the matrix at the end of the irradiation is calculated to be

$$D^0 = n\sigma\phi(1 - e^{-\lambda t})$$

$$= 6.55 \times 10^{19} \times 4.5 \times 10^{-24} \times 2.5 \times 10^{12} \times 0.150$$

$$= 1.10 \times 10^8 \text{ dps.}$$

## 2.5 BIBLIOGRAPHY

Discussions of the production of radioactive nuclides by nuclear reactions are contained in most textbooks concerned with nuclear science, and thus the references to some of the textbooks which cover this subject in greater detail appear in the bibliographies of other chapters.

The general principles of radioactivation are covered in the nuclear chemistry and physics textbooks listed in the bibliography in Section 1.6. The experimental details of activation are given in the textbooks listed in the bibliography in Chapter 6, and the applications to activation analysis are covered in the books listed in Chapter 7.

## 2.5.1  Activation Nuclear Reactions

References that are more specific to the production of radioactive products by nuclear reactions are listed below.

F. Ajzenberg-Selove, Determination of Nuclear Reactions, in *Methods of Experimental Physics 5B (Nuclear Physics)* L. C. L. Yuan and C. S. Wu, Eds. (Academic, New York, 1963), pp. 339–366.

O. Chamberlain, Determination of Flux of Charged Particles, in *Methods of Experimental Physics 5B (Nuclear Physics)*, L. C. L. Yuan and C. S. Wu, Eds. (Academic, New York, 1963), pp. 485–507.

J. B. Cumming, Monitor Reactions for High-Energy Proton Beams, *Ann. Rev. Nucl. Sci.* **13**, 261–286 (1963).

P. M. Endt and M. Demeur, *Nuclear Reactions* (North-Holland, Amsterdam, 1959, 1962), 2 Vols.

W. M. Garrison and J. G. Hamilton, Production and Isolation of Carrier-Free Radioisotopes, *Chem. Rev.* **49**, 237–272 (1951).

I. J. Gruverman and P. Kruger, Cyclotron-Produced Carrier-Free Radioisotopes, *Intern. J. Appl. Radiation Isotopes* **5**, 21–31 (1959).

J. Hoste, D. DeSoete, and R. Gijbels, Neutron, Photon, and Charged Particle Reactions for Activation Analysis, in J. R. DeVoe, Ed., *Modern Trends in Activation Analysis*, Nation Bureau of Standards Special Publication 312, 699–750 (1969), 2 vols.

J. W. Meadows, Excitation Functions for Proton-Induced Reactions with Copper, *Phys. Rev.* **91**, 885–889 (1953).

J. M. Miller and J. Hudis, High Energy Nuclear Reactions, *Ann. Rev. Nucl. Sci.* **9**, 159–202 (1959).

L. Rosen and D. W. Miller, Total Interaction Cross Sections, in *Methods of Experimental Physics 5B (Nuclear Physics)*, L. C. L. Yuan and C. S. Wu, Eds. (Academic, New York, 1963), pp. 366–485.

W. Rubinson, The Equations of Radioactive Transformation in a Neutron Flux, *J. Chem. Phys.* **17**, 542–547 (1959).

E. Segrè, *Experimental Nuclear Physics*, Vol. II (Wiley, New York, 1953).

D. H. Wilkinson, Nuclear Photodisintegration, *Ann. Rev. Nucl. Sci.* **9**, 1–28 (1959).

L. Yaffe, Preparation of Thin Films, Sources, and Targets, *Ann. Rev. Nucl. Sci.*, **12**, 153–188 (1962).

## 2.5.2  Tabulation of Cross Sections (see also Chapter 7)

O. U. Anders and W. W. Meinke, Excitation Functions and Cross Sections, ADI-4999 ADI Auxiliary Publications Project, Photoduplication Service, Library of Congress, Washington 25, D.C.

N. Baron and B. L. Cohen, Activation Cross-Section Survey of Deutron-Induced Reactions, *Phys. Rev.* **129**, 2636 (1963).

D. J. Hughes, *Neutron Cross Sections* (Pergamon, New York, 1957) pp. 182.

D. J. Hughes and R. B. Schwartz, Neutron Cross Sections, BNL-325, 2nd ed. (Superintendent of Documents U.S. Government Printing Office, Washington, D.C., 1958), revised periodically.

R. J. Howerton, Neutron Cross Sections, in *Nuclear Data Tables*, K. Way, Ed. National Academy of Sciences—National Research Council, 1959.

N. Jarmie, J. D. Seagrave, and H. V. Argo, Charged-Particle Cross Sections; Hydrogen to Fluorine, LA-2014, U.S. Atomic Energy Commission, 1956.

D. B. Smith, N. Jarmie, and J. D. Seagrave, Charged Particle Cross Sections, Neon to Chromium, LA-2424, U.S. Atomic Energy Commission, 1960.

## 2.6  PROBLEMS

1.  Determine the number of atoms of the following nuclides in 1 g each of the respective material:

| Nuclide | Material |
|---------|----------|
| $^{27}Al$ | aluminum |
| $^{62}Ni$ | nickel |
| $^{31}P$ | $P_2O_5$ |
| $^{44}Ca$ | $Ca(NO_3)_2$ |
| $^{54}Cr$ | $(NH_4)_2CrO_4$ |

2.  A number of $1 \times 1$ in. thin foils are prepared for activation. For each foil material of the weight listed below determine the target thickness in atoms/cm²:

| Foil material | Density (g/cm³) | Foil weight (mg) |
|---------------|-----------------|------------------|
| aluminum | 2.70 | 40 |
| copper | 8.95 | 40 |
| silver | 10.5 | 80 |
| gold | 19.3 | 100 |

3.  Calculate the geometric cross sections for the nuclides

$$^{27}_{13}Al \quad ^{88}_{38}Sr \quad ^{113}_{48}Cd \quad ^{135}_{54}Xe \quad ^{179}_{79}Au$$

Find their respective neutron capture cross-section values (e.g., GE Chart of the Nuclides) and estimate whether there is any order in the neutron capture cross section as a function of size, $Z$, or $A$.

4.  Determine the neutron flux for the source given in (23) if the ambient temperature were 100°C.

5.  Calculate the average beam intensity of an electron linear accelerator if the Faraday cup showed a total integrated charge of $4.8 \times 10^{-6}$ coulomb in an irradiation of 600 sec.

6. Determine the saturation factor for the production of 12.8-hr $^{64}$Cu during a 1-hr irradiation and the saturation activity if, at the end of the 1-hr irradiation, $3.5 \times 10^4$ dps $^{64}$Cu were present.

7. One gram of copper is irradiated in a reactor flux of $10^{13}$ n/cm²-sec for 2 hr. On removal from the reactor, what is the activity (in curies) of each copper (n,$\gamma$) radioisotope?

8. A 10-mg gold foil (100% $^{197}$Au) was irradiated in a thermal neutron flux of $10^{10}$ n/cm² sec for 1 hr. One day after the end of the irradiation the foil contained $2.53 \times 10^5$ dps of 64.8-hr $^{198}$Au. From these data calculate the cross section for thermal neutron radiative capture of gold. Compare the thermal neutron cross section of $^{197}$Au to the geometric cross section, using $R_0 = 1.4 \times 10^{-13}$ cm.

9. One gram of a hydrocarbon is irradiated in a nuclear reactor with a thermal neutron flux of $10^{12}$ n/cm²-sec. After a 2.58-hr irradiation, followed by 5.16 hr of sample preparation time, the measured activity of 2.58-hr $^{56}$Mn is $6.0 \times 10^4$ dpm. If the cross section for the reaction $^{55}$Mn(n,$\gamma$)$^{56}$Mn is 13.3 b, calculate the percentage of manganese impurity in the hydrocarbon.

10. A thin aluminum foil (10 mg/cm²) is exposed to a beam of 15-MeV protons. A Faraday cup measures the beam intensity at 100 $\mu$A. If the activity of 9.5-min $^{27}$Mg is $6.8 \times 10^3$ dps after one half-life following a 5-min irradiation, calculate the cross section for the $^{27}$Al(p,n)$^{27}$Mg reaction at this proton energy.

# Chapter 3

# Irradiation Sources

The degree to which radioactivation methods are adapted for analysis and tracing, especially when short-lived radionuclides are involved, rests to a considerable extent on the availability of irradiation facilities. The choice of an irradiation source implies the choice of the nuclear reaction for producing the desired radionuclide. In the choice of an irradiation facility several factors must be considered; the three principal ones are the following:

1. The type of irradiating particle.
2. The energy of the irradiating particle.
3. The intensity of the beam or flux.

Other properties, such as maximum sample size that can be accommodated or uniformly irradiated, sample-cooling limitations, transportation delays, cost, and availability, are sometimes important.

The first of the principal factors, the type of irradiating particle, is generally chosen by the nuclear reaction which specifies the activation product sought from the target nuclide; for example, if $^{63}$Cu is the target nucleus, the activation reaction can be chosen to give any of several activation products shown:

| Activation product | Irradiating particle | Nuclear reaction |
|---|---|---|
| $^{64}$Cu | slow neutrons | $(n,\gamma)$ |
| $^{64}$Cu | deuterons | $(d,p)$ |
| $^{62}$Cu | fast neutrons | $(n,2n)$ |
| $^{62}$Cu | photons | $(\gamma,n)$ |
| $^{63}$Ni | fast neutrons | $(n,p)$ |
| $^{63}$Zn | protons | $(p,n)$ |
| $^{60}$Co | fast neutrons | $(n,\alpha)$ |

Thus for a given target nuclide any of several radionuclides can be chosen as the activation product and the actual choice determines the irradiation source required. However, in many locations only a particular irradiation facility is locally available, and the choice of activation product may thus be dictated. At an installation in which only an electron accelerator is available copper would be determined conveniently by the measurement of $^{62}$Cu produced by the $(\gamma,n)$ reaction. When an accelerator and a reactor are both available, the choice of reaction would most likely be $^{63}$Cu$(n,\gamma)^{64}$Cu with neutrons in the reactor. A fast neutron generator allows any of the nuclides $^{64}$Cu, $^{62}$Cu, $^{63}$Ni, and $^{60}$Co to be the activation product.

Because of the convenience of irradiating in nuclear reactors, with their large irradiation volumes, large neutron fluxes, and the generally large cross sections for radiative capture reactions with low-energy neutrons, reactor neutrons are most often considered as the irradiation source. Increasing use is being made of fast neutron generators, charged-particle accelerators, and photo- and electronuclear machines. Each of the irradiation sources has characteristic features which make it suitable for particular types of irradiation. Familiarity with all of these irradiation sources is desirable, so that a greater range for selection of a suitable activation reaction is available.

This section reviews some of the more useful irradiation sources. For convenience the sources are grouped by particle type into (a) neutrons, (b) charged particles, and (c) electrons and photons. By nuclear reactions with suitable targets each of these primary sources can be used to produce secondary sources of any of the other types of radiation. The most important example is the neutron generator in which fast neutron beams are obtained from charged-particle nuclear reactions.

## 3.1 NEUTRONS

Neutrons are produced only by nuclear reactions. Therefore a source of neutrons can exist only in a device in which nuclear reactions (which release neutrons as a product particle) can take place. Such devices cover a wide range of mechanisms. Neutron-producing nuclear reactions can be induced by the following:

1. Alpha and gamma radiations from radioactive nuclides.
2. Charged particles and gamma radiation from accelerating devices.
3. Nuclear fission (and fusion) under controlled or uncontrolled conditions.

These devices yield neutron sources which vary both in energy spectrum and flux. They also vary considerably in simplicity of construction, size, location, and cost from simple, low-cost, low-flux radioisotope sources to complex, expensive high-flux nuclear reactors.

Neutrons from any of these sources have in common the ability to be thermalized, in which energy state they have large probabilities for capture by most elements to form measurable radioactive nuclides. With fast neutrons, obtainable in most of these devices, other neutron reactions can be chosen if appropriate.

### 3.1.1 Radioisotope Sources

The simplest of the neutron sources is the radioisotope source. Such sources may consist of alpha- or gamma-emitting radionuclides intimately mixed with a target material with a low neutron binding energy. Another type of radioisotope neutron source consists of the very heavy elements that disintegrate wholly or in part by spontaneous fission. Radioisotope neutron sources have the advantages of being small, compact, portable, of having no moving parts or high voltages, of being easily calibrated, and of having a neutron output that is either constant or decays at a known rate. Their major disadvantage is the relatively low neutron output obtainable even with multicurie amounts of radioactivity. A minor disadvantage is that they cannot be turned off conveniently, although "shutter" systems can provide adequate shielding against radiation hazards. The dissipation of the radiation heat from radioisotope sources must also be considered.

#### Radioactive ($\alpha,n$) Sources

The nuclear reaction most used for the ($\alpha,n$) production of neutrons is the very reaction that led to their discovery:

$$^{9}_{4}\text{Be} + ^{4}_{2}\text{He} \rightarrow ^{12}_{6}\text{C} + ^{1}_{0}\text{n} + 5.7 \text{ MeV} \qquad (1)$$

in which beryllium was irradiated with alpha particles from the naturally occurring radioactive elements. Until recently radioisotope neutron sources used the 5.3 MeV alpha particles from 138-d $^{210}$Po. Other naturally occurring alpha-particle emitters used include 1620-y $^{226}$Ra (in equilibrium with its daughters) and 21.2-y $^{227}$Ac. Modern sources include the transuranium elements plutonium, americium, and curium. Light elements other than beryllium can be used as the target, but since beryllium yields the largest output of neutrons per alpha particle, compared with any other element under similar conditions, it is almost exclusively used as the target element in large sources.

Equation 1 shows a reaction change of energy of 5.7 MeV. For $^{210}$Po-$\alpha$-Be sources, with $E_\alpha = 5.3$ MeV, the neutron energy would be expected to vary between 11 MeV (with incoming alpha-particle and outgoing neutron in the same direction) to about 6.7 MeV (with the two particles in the opposite direction). For actual sources the spread in neutron energy from 11 MeV downward is even greater because of alpha-particle energy losses in the target and the production of excited states of the $^{12}$C product nuclei. $^{226}$Ra-$\alpha$-Be sources also have a considerable spread in neutron energy. Although $^{226}$Ra has a long half-life compared with $^{210}$Po, radium sources are not extensively used for neutron production because of the intense gamma radiation emitted by $^{226}$Ra. Precautions against this radiation must always be observed when using $^{226}$Ra sources.

The production of kilogram quantities of the transuranium elements has resulted in newer forms of $(\alpha,n)$ neutron sources which contain the nuclides $^{239}$Pu, $^{241}$Am, or $^{242}$Cm. These radionuclides are commercially available as

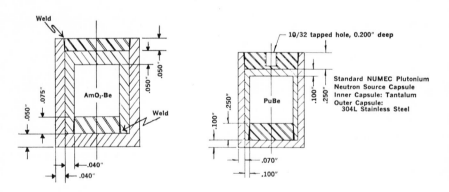

Capsule Material: Type 304L stainless steel

**Figure 3.1** Commercially available encapsulated sources of Am-$\alpha$-Be and Pu-$\alpha$-Be neutrons yield neutron outputs of about $2.1 \times 10^6$ and $1.8 \times 10^6$ neutrons/sec-curie, respectively. (*Courtesy of the Nuclear Materials and Equipment Corporation.*)

**Table 3.1** Alpha-Particle Neutron Sources

| Nuclide | Half-Life | $E_\alpha$ | Form | Neutron Yield (nps/Ci) |
|---------|-----------|------------|------|------------------------|
| $^{210}$Po | 138 d | 5.3 | Be-mixed | $\sim$2.5 × $10^6$ |
| $^{226}$Ra | 1,620 y | 4.48 | Be-mixed | $\sim$1.1 × $10^7$ |
| $^{239}$Pu | 24,400 y | 5.1 | $PuBe_{13}$ | $\sim$1.6 × $10^6$ |
| $^{241}$Am | 458 y | 5.4 | $AmO_2$-Be | $\sim$2.2 × $10^6$ |
| $^{242}$Cm | 163 d | 6.1 | | |

intermetallic compounds with beryllium (e.g., $PuBe_{13}$) which have a density of 3.7 g/cm$^3$ and an output of about 1.8 × $10^6$ neutrons per second per curie (nps/Ci) of plutonium. Somewhat greater specific intensity may be obtained with $^{241}$AmBe$_{13}$ which yields about 2.1 × $10^6$ nps/Ci $^{241}$Am. These materials can be procured in encapsulated cylindrical forms, shown in Figure 3.1.

A comparison of the neutron yields from ($\alpha$,n) radioisotope sources is given in Table 3.1. Although the $^{226}$Ra-$\alpha$ sources offer a greater neutron yield

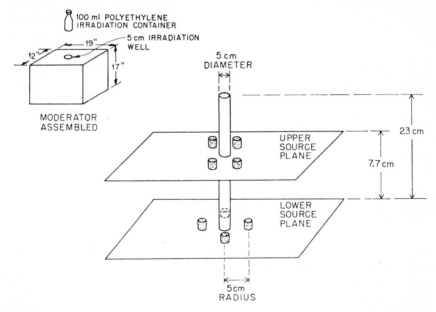

**Figure 3.2** An irradiation facility of eight 2-Ci Am-$\alpha$-Be sources placed about a central irradiation hole in paraffin. The thermal neutron flux in the irradiation position is $\sim$8 × $10^4$ n/cm$^2$-sec. [From J. E. Strain and W. S. Lyon, The Use of Isotopic Neutron Sources for Chemical Analysis, in *Radiochemical Methods of Analysis*, Vol. 1 (International Atomic Energy Agency, Vienna, 1965).]

per curie, the lack of intensive gamma radiation and the decreasing cost of plutonium may make Pu-α-Be or Am-α-Be sources useful as radioisotope neutron sources.

For radioactivations with thermal neutrons, in which ultimate sensitivity is not an important criterion, a useful irradiation facility can be constructed by placing eight radioisotope sources in a cubic array around a central sample-irradiation hole in a paraffin-moderated container. A sketch of an array is given in Figure 3.2. With eight 2-Ci Am-α-Be neutron sources, a uniform thermal neutron flux of $8 \times 10^4$ n/cm²-sec is obtainable over a 2-cm cubic sample. Such a device is useful for rapid activation analysis of many elements in milligram quantities whose activation products have short half-lives; for example, by measuring the activity of 3.77-m $^{51}$V the element vanadium can be measured in amounts as little as 1 mg.

### Radioactive (γ,n) Sources

The (γ,n) reaction with gamma rays emitted by commonly available radio-isotopes are possible with only two nuclides, namely beryllium and deuterium. The (γ,n) threshold for beryllium is 1.67 MeV and for deuterium, 2.23 MeV. All other nuclides have (γ,n) thresholds above 6 MeV. Several available radionuclides emit gamma radiation with energy in excess of 1.67 MeV. The energy of the emitted neutron $(E_n)$ from a monoenergetic gamma ray $(E_\gamma)$ interacting with a nucleus of beryllium or deuterium with neutron binding energy $(Q)$ is given* by the angle $(\theta)$ between the incident photon and the emitted neutron by

$$E_n = \frac{A-1}{A} E_\gamma - Q - \frac{E_\gamma^{\,2}}{1862(A-1)} + \delta \cos \theta \tag{2}$$

with the correction term

$$\delta \simeq E_\gamma \frac{2(A-1)(E_\gamma - Q)}{931 A^3} \tag{3}$$

The spread in neutron energy is given by

$$\Delta E_n = 2\delta \tag{4}$$

which is increased considerably as the amount of beryllium and deuterium is increased to obtain usable neutron intensities. The spread in neutron energy is prevalent in large sources because of Compton scattering of the photons before the photoneutron reaction and the scattering of the neutrons before emerging from the source.

A summary of neutron yields from radionuclide photoneutron sources is given in Table 3.2. The yields are given as standard yields, that is, the number

---

* A. Wattenberg, Photo-neutron Sources and the Energy of the Photo-neutrons, *Phys. Rev.* **71**, 497–507 (1947).

**Table 3.2** Neutron Yields from Radioisotope Photoneutron Sources

| Nuclide | $T_{1/2}$ | $E_\gamma$ (MeV) | with | $E_n$ (MeV) | Approximate Yield ($10^4$ n/sec) |
|---|---|---|---|---|---|
| $^{24}$Na | 15.0 h | 2.75 | Be | 0.2 | 14 |
| | | 2.75 | $D_2O$ | 0.8 | 29 |
| $^{56}$Mn | 2.6 h | Several | Be | 0.2 | 3 |
| | | 2.5, 2.7, 3.0 | $D_2O$ | 0.2 | 0.3 |
| $^{72}$Ga | 14.1 h | Several | Be | 0.2 | 6 |
| | | 2.2, 2.5 | $D_2O$ | 0.13 | 7 |
| $^{88}$Y | 108 d | 0.9, 1.8 | Be | 0.16 | 10 |
| | | 2.8 | $D_2O$ | 0.3 | 0.3 |
| $^{124}$Sb | 60.4 d | 1.7 | Be | 0.02 | 19 |
| $^{140}$La | 40.2 h | 1.6, 2.5 | Be | 0.6 | 0.2 |
| | | 2.5 | $D_2O$ | 0.15 | 0.7 |

of neutrons per second from 1 Ci of radioisotope in 1 g of beryllium or heavy water at a distance of 1 cm.

Comparison of Table 3.2 with Table 3.1 shows that the neutron yield per curie is much lower for photoneutron than for alpha-neutron sources. With the further disadvantages of shorter half-lives and added hazard due to penetrating gamma radiation, photoneutron sources are not widely used for neutron-activation purposes, even when low neutron fluxes are acceptable.

### Spontaneous Fission Sources

Those transuranic elements that disintegrate by spontaneous fission, releasing several neutrons in the process, represent a potential source of radioisotopes for neutron generation. A practical example is 2.65-y $^{252}_{98}$Cf, for which about 3.2% of its decays are spontaneous fission, releasing 3.76 neutrons per fission.* The neutron yield $I_n$ from a radionuclide source undergoing spontaneous fission can be calculated with the aid of (1-28), since the neutron production is determined by the rate of decay of the radionuclide. With the above data for $^{252}$Cf each disintegration yields $0.032 \times 3.76 = 0.12$ neutron on the average. Thus 1 g of $^{252}$Cf would produce a neutron yield of $2.34 \times 10^{12}$ n/sec. The nuclide $^{254}$Cf, which decays primarily by spontaneous fission, with a half-life of 56 d, would have a specific neutron yield of $\sim 10^{16}$ n/sec-g.

A large-scale effort has been undertaken to examine the potential use of $^{252}$Cf as neutron sources.* It is the only heavy radionuclide that has its own

---

* Californium-252, Its Use and Market Potential, U.S. Atomic Energy Commission brochure, May 1969.

**Table 3.3**    Spontaneous Fission Neutron Sources[a]

| Nuclide | Alpha Decay $T_{1/2}{}^{b}$ (years) | Fission $T_{1/2}{}^{b}$ (years) | Fissions per $10^6$ $\alpha$ | Neutrons per Fission[c] |
|---|---|---|---|---|
| $^{238}$U | $4.5 \times 10^9$ | $6.5 \times 10^{15}$ | $5.6 \times 10^{-1}$ | — |
| $^{236}$Pu | 2.85 | $3.5 \times 10^9$ | $8.1 \times 10^{-4}$ | $1.89 \pm 0.20$ |
| $^{238}$Pu | 86.4 | $4.9 \times 10^{10}$ | $2.3 \times 10^{-3}$ | $2.04 \pm 0.13$ |
| $^{240}$Pu | 6580 | $1.3 \times 10^{11}$ | $5.4 \times 10^{-2}$ | $2.09 \pm 0.11$ |
| $^{242}$Pu | $3.8 \times 10^5$ | $7.1 \times 10^{10}$ | 5.3 | $2.32 \pm 0.16$ |
| $^{244}$Pu | $\sim 7.6 \times 10^7$ | $2.5 \times 10^{10}$ | $3.0 \times 10^{-3}$ | — |
| $^{242}$Cm | 0.45 | $7.2 \times 10^6$ | $6.2 \times 10^{-2}$ | $2.33 \pm 0.11$ |
| $^{244}$Cm | 17.6 | $1.3 \times 10^7$ | 1.3 | $2.61 \pm 0.13$ |
| $^{252}$Cf | 2.6 | 85 | $3.3 \times 10^4$ | $3.51 \pm 0.16$ |
| $^{254}$Cf | Long | 0.17 | Large | — |

[a] G. D. O'Kelley, Radioactive Sources, in L. Yuan and C. Wu, Eds., *Methods of Experimental Physics*, Vol. 5B (Academic, New York, 1963), p. 562.
[b] C. M. Lederer, J. M. Hollander, and I. Perlman, *Table of Isotopes*, 6th ed., (Wiley, New York, 1967).
[c] W. W. T. Crane, G. H. Higgins, and H. R. Bowman, Average Number of Neutrons per Fission for Several Heavy Nuclides, *Phys. Rev.* **101**, 1804–1805, (1956).

quarterly journal* which describes the feasibility studies underway with microgram to milligram sources of encapsulated $^{252}$Cf$_2$O$_3$. Such sources are available on loan to qualified researchers. One of the early proposed applications is as a neutron source for radioactivation analysis.

The data for these and other potentially useful spontaneous fission neutron sources are given in Table 3.3. These data indicate that as significant quantities of these very heavy elements become available (see Section 3.1.5) they will probably be a useful type of radioisotope-neutron source. However, the engineering aspects of removing the heat dissipated by such a source requires considerable attention (see Problem 3.1).

### 3.1.2 Neutron Generators

The neutron generator in its present commercial forms is a compact charged-particle accelerator (see Section 3.2.1) designed to produce efficiently a beam of neutrons by some appropriate nuclear reaction. The reactions most

* *Californium-252 Progress*, available on request to Savannah River Operations Office, U.S. Atomic Energy Commission, P.O. Box A, Aiken, South Carolina, 29801.

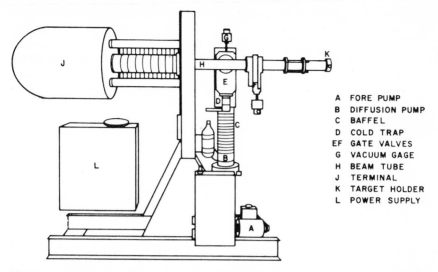

A  FORE PUMP
B  DIFFUSION PUMP
C  BAFFEL
D  COLD TRAP
EF GATE VALVES
G  VACUUM GAGE
H  BEAM TUBE
J  TERMINAL
K  TARGET HOLDER
L  POWER SUPPLY

**Figure 3.3**  A schematic diagram of a neutron generator. [From W. W. Meinke and R. W. Shideler, Activation Analysis: New Generators and Techniques Make It Routine, *Nucleonics* **20**, No. 3, 60–65 (1962).]

commonly used which have high yields of fast neutrons are

$$^2\text{H(d,n)}^3\text{He} \qquad Q = 3.25 \text{ MeV} \tag{5}$$

$$^3\text{H(d,n)}^4\text{He} \qquad Q = 17.6 \text{ MeV} \tag{6}$$

$$^9\text{Be(d,n)}^{10}\text{B} \qquad Q = 3.79 \text{ MeV} \tag{7}$$

A schematic diagram of a neutron generator is shown in Figure 3.3. In common with the dc generators described in Section 3.2.1, the neutron generator consists of an ion source that delivers deuterium ions (deuterons) to an accelerating tube in which the deuterons are accelerated through a potential of about 150 kV. The accelerated ions strike a suitable target from which the neutrons are emitted isotropically. The fast neutrons may be moderated in some suitable material (e.g., water or paraffin) to produce thermal neutrons.

Many types of commercial neutron generator suitable for radioactivation analysis are available. A comprehensive survey of these instruments has been published in *Nucleonics*,* data from which are reproduced in part as Table 3.4.

The deuteron-tritium neutron generator is generally designed to operate at voltages in the order of 150 kV and currents up to about 2.5 mA. These

* A Nucleonics Survey—Commercially Available Neutron Sources for Activation Analysis, *Nucleonics* **23**, No. 4, 60J–K (1965).

**Table 3.4**    Types of Commercially Available Accelerator Neutron Source

| Type | Approximate Cost ($ 1000) | Beam Particles | Targets | Target Life-time (hr) | Maximum Voltage (kV) | Maximum Average Neutron Yield (n/sec) | Maximum Average Thermal n-Flux (n/cm²-sec) |
|---|---|---|---|---|---|---|---|
| Sealed tube | 7 | d, t | Ti-t | 1–1000 | ~125 | ~$10^8$ | ~$10^5$ |
| Cockcroft-Walton | 20 | d, p, t | Ti-t | 5 | ~200 | ~$10^{11}$ | ~$10^8$ |
| Accelerators | 4–250 | d, p, He, e | Ti-t, Be | 2–∞ | 150–2000 | ~$10^{12}$ | ~$10^9$ |

generators produce a neutron yield of about $2 \times 10^{11}$ 14-MeV neutrons/sec distributed essentially uniformly over $4\pi$ sr. Thus at a distance of 3.16 cm from the target the fast neutron flux would be

$$\phi \simeq \frac{\text{yield}}{4\pi r^2} \simeq \frac{2.5 \times 10^{11}}{4\pi 10} \simeq 2 \times 10^9 \, \text{n/cm}^2\text{-sec} \tag{8}$$

The usable thermal neutron flux at this distance would be determined by the type and amount of moderator surrounding the target. Table 3.4 shows that for most of the available machines the thermal neutron flux is about a factor of 10 or so lower than the fast neutron flux.

The obvious advantage of the neutron generator is its moderate initial and facility cost compared with the costs for even a small nuclear reactor. They also allow elements to be determined by fast neutron reactions when the thermal neutron activation products lack convenient properties, for example, in the determination of fluorine:

$$^{19}\text{F}(n,\gamma) \, 11\text{-s} \, ^{20}\text{F} \tag{9}$$

$$^{19}\text{F}(n,2n) \, 1.87\text{-h} \, ^{18}\text{F} \tag{10}$$

Neutron generators are also available in installations coupled to computer-integrated radioactivity measurement instruments which may make automatic assembly-line radioactivation analysis available as a routine service for selected applications involving large numbers of samples. Analyses involving activation, radioactivity measurement, and computer data calculations may be completed in minutes.

The major disadvantage of the neutron generator, especially for the d-t reaction systems, is the need for large sources of radioactive tritium which is incorporated into the target in the form of a metal tritide. The heat dissipated in the target results in the loss of tritium from the target, which in the higher current generators results in a decay of the neutron output with half-life values of the order of a few hours. Thus, with the further disadvantage of lower neutron fluxes compared with reactors, neutron generators are not useful for maximum sensitivity determinations of nuclides resulting in long half-life products. Considerable development, of course, is being made to lengthen the target lifetimes of tritium targets (e.g., by rotating the target) and to increase the neutron output of the generators. Neutron generators appear to be attractive in many commercial situations in which large numbers of rapid analyses are required nondestructively and in which sufficient activity of a short-lived product of a desired element can be produced in a short irradiation.

### 3.1.3 Cyclotrons and Accelerators

The cyclotrons and accelerators discussed in Section 3.2 as charged-particle irradiation sources are useful devices for the production of secondary neutron beams. The neutron generator, designed specifically for this purpose and described in Section 3.1.2, accelerates deuterons for interactions with deuterium and tritium to produce neutrons by the $^2H(d,n)^3He$ and $^3H(d,n)^4He$ reactions, respectively. Cyclotrons and accelerators can accelerate deuterons, as well as protons, alpha particles, and electrons, to produce high-energy neutrons in a variety of nuclear reactions and in high yield.

Alpha-particle accelerating devices have been used to produce neutrons by the $(\alpha,n)$ and $(\alpha,2n)$ reactions. Alpha particles with energies of about 20 MeV can produce the $(\alpha,n)$ reaction in every stable nuclide, but even at 30 MeV the neutron yield is only about $2 \times 10^9$ n/sec-$\mu$A-sr for beryllium and decreases rapidly with increasing $Z$ of the target element. Thus alpha-particle accelerators are not used extensively for neutron-radioactivation purposes.

The deuteron reaction, as used in the neutron generator, is a prolific source of neutrons. Thus the availability of deuteron-accelerating devices, such as the cyclotrons and accelerators of universities and national laboratories, offers the opportunity for neutron irradiations. However, the cross section for $^3H(d,n)^4He$ reaches a peak value at a deuteron energy of about 150 kV, as optimized in the neutron generator, and falls off rapidly with increasing deuteron energy. Other useful (d,n) reactions are $^7Li(d,n)^8Be$ and $^9Be(d,n)^{10}B$. With high-energy deuterons, neutrons are produced in good yield by the deuteron stripping reaction [see (78), Section 1.5.4].

Proton accelerators are also used, when available, as neutron sources. They include cyclotrons, Van de Graaff generators, and proton linear accelerators. Neutrons are most often made in these machines by the irradiation of lithium by the reaction

$$^7Li(p,n)^7Be \quad Q = -1.63 \text{ MeV} \tag{11}$$

**Table 3.5**   Neutron Production and Fluxes Available
from Electron Irradiation of Beryllium Targets[a]

| Electron Energy (MeV) | Total Neutron Production (n/sec-kW) | Thermal Neutron Flux in a Water-Moderated Source (n/cm²-sec-kW) |
|---|---|---|
| 2 | $5.0 \times 10^8$ | $1.2 \times 10^7$ |
| 3 | $6.7 \times 10^9$ | $1.0 \times 10^8$ |
| 4 | $1.8 \times 10^{10}$ | $2.1 \times 10^8$ |
| 6 | $3.0 \times 10^{10}$ | $3.0 \times 10^8$ |
| 20 | $7.0 \times 10^{11}$ | $5.0 \times 10^9$ |

[a] Adapted from data collected by Malcolm H. MacGregor, Applied Radiation Corporation *News* (July 1958) for the neutron yields from the electron irradiation of a beryllium target located behind a thin, high $Z$ converter.

with protons of about 2.3-MeV energy. This reaction produces intermediate-energy neutrons with maximum yield. Thick target yields in excess of $10^9$ n/sec-$\mu$A have been obtained with a 3-MeV Van de Graaff accelerator.

Electron accelerators may be used conveniently as a neutron source. The electron beam is usually allowed to interact with a high atomic number element, such as tantalum or tungsten, to convert part of the electron beam into Bremsstrahlung (photons with a continuous energy spectrum) which then interact with targets such as beryllium to produce neutrons by the (e,e′n) and ($\gamma$,n) reactions. The neutrons can be moderated to thermal energy with suitable water or paraffin assemblies around the accelerator target. Table 3.5 shows the neutron production and water-moderated thermal neutron fluxes available from a 20-MeV, 12-kW electron linear accelerator operated at full power. The thermal neutron flux of $6 \times 10^{10}$ n/cm²-sec at full power is about equivalent to the thermal neutron flux available in a 2-kW nuclear reactor and is sufficient for the activation of many elements in the microgram range; for example, in comparison to the value of 1 mg given in Section 3.1.1, vanadium can be measured by neutron activation in such a machine in amounts as little as $5 \times 10^{-9}$ g. The use of electron linear accelerators for direct activation of samples by electro- and photonuclear reactions is described in Section 3.3.

### 3.1.4   Nuclear Reactors

The most copious and controllable sources of neutrons for radioactivation are found in nuclear reactors in which neutrons interact with "fissionable" nuclei and fission into two products with the concomitant release of more than 1 neutron. The net gain in available neutrons with each fission event results in a nuclear chain reaction.

The fission process has as its basis the fact that all nuclei with mass $A >$ ~110 are potentially unstable with respect to division into two smaller fragments by as much as 200 MeV. However, the "surface tension" forces of such nuclei allow stable nuclides with $A >$ ~110 to exist in nature. Fission occurs when the internal energy is increased or distributed within the nucleus to distort it sufficiently so that the increased coulomb repulsion force exceeds the surface-tension binding force. Nuclear fission reactions occur spontaneously with the very heaviest transuranium nuclides, for example, $^{254}$Cf. Irradiation of all heavy elements with sufficiently energetic charged particles, neutrons, and photons produces nuclear fission reactions. In general these reactions all require considerably more energy to occur than they produce. So far the only reactions of practical importance are the neutron fission of uranium and plutonium. The three *fissionable nuclides* with thermal neutrons are $^{233}$U, $^{235}$U, and $^{239}$Pu, of which only $^{235}$U (0.71% of natural uranium) occurs in nature. The other two can be produced from the *fertile materials* $^{232}$Th and $^{238}$U by the neutron reactions

$$^{232}\text{Th}(n,\gamma)^{233}\text{Th} \xrightarrow{\beta^-} {}^{233}\text{Pa} \xrightarrow{\beta^-} {}^{233}\text{U} \tag{12}$$

$$^{238}\text{U}(n,\gamma)^{239}\text{U} \xrightarrow{\beta^-} {}^{239}\text{Np} \xrightarrow{\beta^-} {}^{239}\text{Pu} \tag{13}$$

Nuclear reactors can be operated for fission with thermal or fast neutrons. Fast-neutron reactors are used to convert the fertile materials into fissile materials. The number of neutrons produced per fission, however, is about the same. The fission reaction proceeds schematically, for neutron capture by $^{235}$U, for example, by the creation of a compound nucleus, which fissions into two heavy nuclei with the release of one to several ($\nu$) neutrons;

$$^{235}_{92}\text{U} + {}^{1}_{0}\text{n} \rightarrow [^{236}\text{U}]^* \rightarrow {}^{A_1}Z_1 + {}^{A_2}Z_2 + \nu^{1}_{0}\text{n} + Q \tag{14}$$

where $Q$ is ~200 MeV and $\bar{\nu} \simeq 2.5$ neutrons per fission. The conservation of nucleons requires

$$Z_1 + Z_2 = 92 \tag{15}$$

$$A_1 + A_2 + \nu = 236 \tag{16}$$

The fission fragments $Z_1$ and $Z_2$ cover a range of about 35 elements from about $Z = 30$ (zinc) to about $Z = 65$ (terbium). The masses of the primary fission fragments are those that have neutron-to-proton ratios about equal to that of the fissioning nucleus ($N/Z \sim 1.56$), and thus the primary fragments have an excess of neutrons compared with stable nuclei with $Z$ ranging from 30 to 65. The fission fragments are therefore all beta unstable and decay by several successive beta emissions with generally increasing half-lives to form stable nuclides; for example, two such $\beta^-$ decay chains are

$$^{91}_{36}\text{Kr} \xrightarrow[10 \text{ sec}]{\beta^-} {}^{91}_{37}\text{Rb} \xrightarrow[72 \text{ sec}]{\beta^-} {}^{91}_{38}\text{Sr} \xrightarrow[9.7 \text{ hr}]{\beta^-} {}^{91}_{39}\text{Y} \xrightarrow[59d]{\beta^-} {}^{91}_{40}\text{Zr}_{\text{stable}} \tag{17}$$

$$^{143}_{56}\text{Ba} \xrightarrow[12 \text{ sec}]{\beta^-} {}^{143}_{57}\text{La} \xrightarrow[14 \text{ min}]{\beta^-} {}^{143}_{58}\text{Ce} \xrightarrow[33 \text{ hr}]{\beta^-} {}^{143}_{59}\text{Pr} \xrightarrow[13.7d]{\beta^-} {}^{143}_{60}\text{Nd}_{\text{stable}} \tag{18}$$

The distribution of the various nuclides varies considerably with the most probable fission event resulting in the masses $A = \sim95$ and $A = \sim140$. The distributions vary also with neutron energy and are shown for the fission of $^{235}\text{U}$ with thermal neutrons and with 14-MeV neutrons in Figure 3.4. The distributions show that the symmetric fission event with 14-MeV neutrons is about 100 times more probable than with thermal neutrons.

In all but about 0.64% of $^{235}\text{U}$ fissions the neutrons are released during the fission event (within $10^{-12}$ sec) and are called *prompt neutrons*. A few ($\sim9$) radioactive fission products with half-lives of less than 1 to 55 sec decay by beta emission, some of which leave the product nucleus with excitation energy greater than its neutron separation energy. Neutrons are released from such nuclei before further beta decay or gamma-ray emission can occur and appear to occur with a half-life identical to that of the preceding $\beta^-$ decay. These neutrons are called *delayed neutrons*. The decay scheme (see Section 4.1) for the longest lived delayed neutron emitter, 55.6-sec $^{87}\text{Br}$, is shown in Figure 3.5. A list of known delayed-neutron emitters is given in Table 3.6.

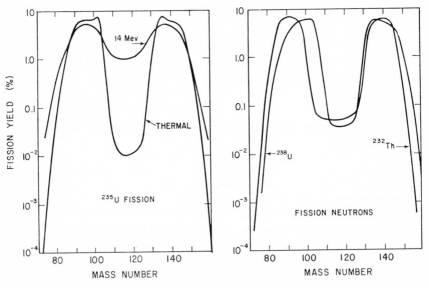

**Figure 3.4** Mass distribution of $^{235}\text{U}$, $^{238}\text{U}$, and $^{232}\text{Th}$ fission products. [Adapted from S. Katcoff, *Nucleonics* **16** (4), 78 (1958).]

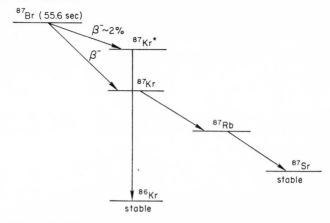

**Figure 3.5** The decay scheme for delayed neutron emission from 55.6-s $^{87}$Br. The excitation energy of $^{87*}$Kr is greater than its neutron separation energy and neutron emission occurs almost instantaneously (before a beta particle or gamma ray can be released) and thus appears to occur with a half-life identical to that of the preceding $\beta^-$ decay.

Although they represent only a small fraction of the total neutron yield, their occurrence is of great importance in the control of nuclear reactors.

The energy distribution for the neutrons released from fission of $^{235}$U is shown in Figure 3.6a and given by the empirical relationship

$$N(E)\, dE = \left(\frac{2}{\pi e}\right)^{1/2} \sinh \sqrt{2E}\ e^{-E}\, dE \tag{19}$$

**Table 3.6** Delayed Neutron Emitters[a]

| Nuclide | $T_{1/2}$ (sec) | Product | Neutron Energy (MeV) |
|---------|-----------------|---------|----------------------|
| $^{85}$As | 0.43 | 3-m $^{84}$Se | |
| $^{87}$Br | 55 | Stable $^{86}$Kr | 0.3 |
| $^{88}$Br | 16 | 76-m $^{87}$Kr | |
| $^{89}$Br | 4.5 | 2.8-h $^{88}$Kr | 0.5 |
| $^{90}$Br | 1.6 | 3.2-m $^{89}$Kr | |
| $^{135}$Sb | <24 | 42-m $^{134}$Te | |
| $^{137}$I | 24 | Stable $^{136}$Xe | 0.6 |
| $^{138}$I | 6.3 | 3.9-m $^{137}$Xe | |
| $^{139}$I | 2.0 | 14-m $^{138}$Xe | |

[a] Adapted from D. T. Goldman, Chart of the Nuclides, General Electric Co., Schenectady, New York, 8th ed., June 1965.

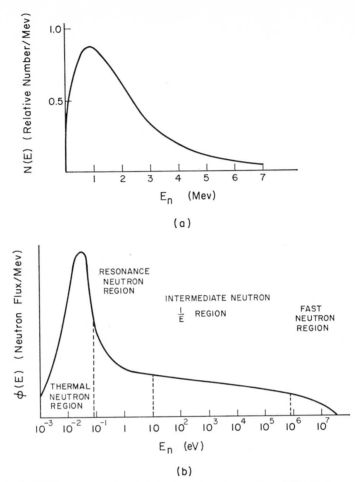

**Figure 3.6** (a) Energy distribution of neutrons from $^{235}U$ fission; (b) typical distribution of neutron energies in a thermal reactor. [Adapted from S. E. Liverhant, *Elementary Introduction to Nuclear Reactor Physics* (Wiley, New York, 1960), pp. 120, 135.]

where $N(E)\, dE$ is the number of neutrons emitted with energy in the interval $E$ to $E + dE$. The most probable neutron energy is just under 1 MeV and the average neutron energy is about 2 MeV. Neutrons with energy above 15 MeV have been detected for $^{235}U$ fission.

Fission-produced neutrons quickly become thermalized by the dissipation of their energy in collisions with the moderator nuclei. For activation purposes the ambient neutron energy distribution of the neutrons in the irradiation position is important. This distribution will vary considerably in relation to the type of reactor and the position relative to the reactor core. Near the core the ratio of fast-to-thermal neutrons will be large, decreasing with distance away from the core and becoming extremely small in the *thermal column* of the reactor. In contrast to the description of neutrons in Section 1.4.4, which were classified by energy according to the nuclear reactions they cause, reactor neutrons are more loosely classified by the manner in which they are obtained in the reactor: that is, *fast, resonance*, and *thermal* neutrons. Fast neutrons are fission neutrons not yet moderated by collisions, resonance neutrons constitute the $1/E$ spectrum produced by moderation (ranging from $\sim$1 MeV to $\sim$1 eV), and thermal neutrons are those that have reached thermal equilibrium with their moderator and have a Maxwellian distribution of velocities. The distribution of the neutron flux per unit energy increment is shown as a function of neutron energy in a thermal reactor in Figure 3.6*b*.

The thermal reactor, designed to maximize thermal neutron fission, is the type generally used for radioactivation services, although any reactor available is suitable for neutron irradiations. Most research or testing reactors are designed to make neutron irradiations easily available, having many irradiation positions near the core, in port holes, in the "thermal column," etc., methods for getting samples quickly into and out of the reactor, temperature controls, and other features. A typical, small, research-type reactor used for radioactivation analysis is the Triga,* shown schematically in Figure 3.7. In this reactor the fast neutrons produced by fissions are moderated rapidly to thermal energies to replenish the supply of thermal neutrons for further fissions. The core consists of an array of fuel-moderator elements surrounded by a graphite reflector cooled and shielded by water. Irradiations can be made either in the core-grid position through a pneumatic transfer system or in spaces provided in the reflector. A rotary specimen rack in which samples can be rotated around the core at the top of the reflector provides a means of irradiation of many samples at one time with a uniform total neutron flux. This feature is of particular value for the comparitor method of activation analysis described in Chapter 7.

The ratio of neutrons with energy above 0.4 eV to neutrons of energy below 0.4 eV can be measured in terms of the cadmium ratio (CR). This ratio allows the calculation of the thermal neutron flux at any irradiation position. The particular property of cadmium that makes it useful for measuring thermal neutron fluxes is evident from the cross-section curve in Figure 3.8 for the neutron capture of cadmium. The curve shows that cadmium absorbs neutrons with energy below about 0.4 eV almost exclusively, compared with neutrons with energy greater than 0.4 eV. Thus the activation of a suitable material may be made with all neutrons present in the facility and then, by

---

* Produced by the Gulf General Atomic Corp.

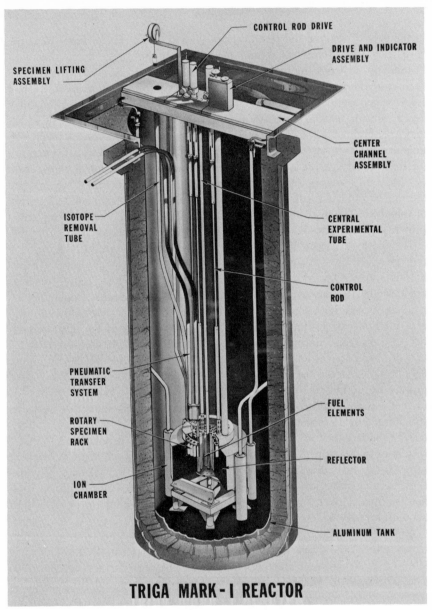

**CONTROL ROD DRIVE**

**DRIVE AND INDICATOR ASSEMBLY**

**SPECIMEN LIFTING ASSEMBLY**

**CENTER CHANNEL ASSEMBLY**

**ISOTOPE REMOVAL TUBE**

**CENTRAL EXPERIMENTAL TUBE**

**CONTROL ROD**

**PNEUMATIC TRANSFER SYSTEM**

**FUEL ELEMENTS**

**ROTARY SPECIMEN RACK**

**REFLECTOR**

**ION CHAMBER**

**ALUMINUM TANK**

# TRIGA MARK - I REACTOR

**Figure 3.7** A cutaway scale diagram of a Triga nuclear reactor suitable for neutron activation analysis. (*Courtesy of Gulf General Atomic Corp.*)

surrounding the material with cadmium, only with neutrons with energy greater than 0.4 eV. The difference in activation intensity is, of course, that due to the thermal neutrons absorbed in the cadmium shield. If $\phi_{\mathrm{th}}$ is the thermal neutron flux and $\phi_r$ is the resonance neutron flux (neutrons with

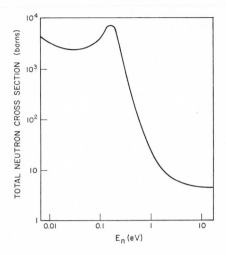

**Figure 3.8** Total neutron cross section of cadmium as a function of neutron energy.

energy greater than $\sim 0.4$ eV), with a detector of known sensitivity, for example, indium or gold, the measurement of the cadmium ratio is given by the ratio of the specific saturation activity, $D^\infty$, of a bare irradiated foil to the specific saturation activity of a cadmium-shielded foil irradiated in the same position. The specific saturation activity of the bare foil is given from (2-42) as

$$D_b^\infty = \frac{D_b^0}{n(1 - e^{-\lambda t})} = \sigma_{th}\phi_{th} + \sigma_r\phi_r \qquad (20)$$

and the specific saturation activity of the cadmium-shielded foil is given by

$$D_c^\infty = \frac{D_c^0}{n(1 - e^{-\lambda t})} = \sigma_r\phi_r \qquad (21)$$

The cadmium ratio is thus given by

$$CR = \frac{D_b^\infty}{D_c^\infty} = \frac{\sigma_{th}\phi_{th} + \sigma_r\phi_r}{\sigma_r\phi_r} \qquad (22)$$

and the thermal neutron flux can be calculated from the measured cadmium ratio by

$$\sigma_{th}\phi_{th} = D_b^\infty - D_c^\infty = D_b^\infty - \frac{D_b^\infty}{CR} = D_b^\infty\left(1 - \frac{1}{CR}\right) \qquad (23)$$

Thus

$$\phi_{th} = \frac{D_b^0}{n\sigma_{th}(1 - e^{-\lambda t})}\left(1 - \frac{1}{CR}\right) \qquad (24)$$

### 3.15  Thermonuclear Explosions

The detonation of a thermonuclear device with its inherent and awesome release of explosive energy can hardly be considered as an irradiation source for activation analysis, yet one of the stated objectives of the U.S. Atomic Energy Commission Plowshare Program for the peaceful uses of nuclear explosions is the production of radioisotopes of the very heaviest elements $(Z > 92)$.* Such nuclides are now made in our highest flux reactors in only minute quantities in stepwise increase of $Z$ and only after years of irradiation time. The feature of "contained" thermonuclear explosions is the tremendous number of neutron interactions that can take place instantaneously in a relatively confined volume. Approximately $1.5 \times 10^{23}$ neutrons are created in a 1-kiloton explosion (1 kiloton is the power equivalent to the detonation of one thousand tons of TNT). This large number of neutrons, available in small volume for only a microsecond or so, yields an equivalent instantaneous "neutron flux" that can activate all the immediate device and geological materials. Such explosions have been shown to produce radionuclides that result from reactions involving the successive capture of 16 or more neutrons,

$$^{238}_{92}\text{U} + 16\,^{1}_{0}\text{n} \rightarrow [^{254}_{92}\text{U}] \xrightarrow{6\beta^-} {}^{254}_{98}\text{Cf} \qquad (25)$$

The hypothetical product nucleus $^{254}_{92}\text{U}$ would be so $\beta^-$-unstable that six successive $\beta^-$ decays would be required to return the $N/Z$ ratio back to the $\beta^-$-stability curve of (1-27). The resulting nuclide would be $^{254}_{98}\text{Cf}$, which decays by *spontaneous* fission with a half-life of 56 days. Even heavier nuclides are being sought in the Plowshare Program. Thus a long-term potential exists for the economic production of high-intensity neutron sources of transuranic elements created as a by-product of engineering applications of nuclear explosions.

Obviously not a practical means for activation analysis, this research, nevertheless, is of interest in the study of nuclear reactions and "high-flux" activation phenomena.

### 3.2  CHARGED PARTICLES

Charged-particle accelerators are useful for radioactivations for two prominent reasons: (a) in many places they are conveniently available and (b) some elements do not yield convenient neutron activation products or they require neutron energies not attainable in nuclear reactors or neutron

* Engineering with Nuclear Explosives, *Proc. Third Plowshare Conference*, U.S.A.E.C. Report TID-7695, April 1964.

generators. The first reason is generally a compelling one and the second is often comparative; for example, the radioactivation of carbon is considered from this second reason. Neutron-radiative capture in carbon by the reaction

$$^{13}C(n,\gamma)^{14}C \tag{26}$$

takes place with the 1.11 % abundant isotope $^{13}C$ and results in the product nuclide $^{14}C$ whose decay constant is only $3.81 \times 10^{-12}$/sec. These data indicate that even with long irradiation times at high flux levels the sensitivity for carbon is low enough to make neutron activation analysis unattractive compared with other analytical techniques. However, the reaction of the abundant isotope $^{12}C$ with energetic particles in which a neutron is removed results in the easily detected radionuclide 20.5-m $^{11}C$ with high sensitivity for analysis. Because of the large neutron binding energy of $^{12}C$ (18 MeV), particles with energies well in excess of 18 MeV are required for adequate yields. Other possible nuclear reactions for determining carbon include $^{12}C(p,\gamma)^{13}N$ and $^{13}C(p,n)^{13}N$ leading to the product nuclide 10-m $^{13}N$. Thus charged-particle or photon accelerators are suitable for the activation of carbon.

A given radionuclide one neutron removed from stability can be produced by several nuclear reactions involving neutrons, charged particles, or photons; for example, the (n,2n), (p,d), and (γ,n) reactions. The choice is generally based on machine availability, but it can also be based on considerations of sample size, beam homogeneity, intensity, and energetics, relative reaction cross sections, and other factors.

These reactions are possible, of course, only with irradiation beams with energies at least equal to the neutron binding energy of the neutron in the target nucleus. With the exception of deuterium and beryllium, all nuclei have neutron binding energies in excess of about 6 MeV. Thus, in general, nuclear reactors are not suitable for the (n,2n) reaction. Neutron generators producing 14-MeV neutrons from the (d,t) reaction may be used for (n,2n) reactions with many elements, but because of rapid target burnup over a few hours of full beam they are used primarily for activation products which have very short half-lives. Heavy charged-particle accelerators are excellent devices for (p,d) and other activation reactions and should be considered when available for activations when reactor irradiations are either not feasible or give an excess of (n,γ) activation interferences. Photon and electron devices offer the same general advantages as the heavier charged-particle accelerators over neutron sources.

An accelerator is any device that imparts kinetic energy to ions or electrons. Many types of accelerator device have been developed to give sufficient kinetic energy to charged particles to induce nuclear reactions in high yield. Before the development of accelerator devices natural alpha-emitting radioisotopes were available as sources of $\sim$5 MeV charged particles. The intensities were small. A 1-Ci source emits $3.7 \times 10^{10}$ alpha particles/sec isotropically. Thus the "beam" intensity decreases as $1/r^2$ and, at a distance of only 10 cm from a "point" source, is

$$J = \frac{3.7 \times 10^{10}}{4\pi(10)^2} = 2.94 \times 10^7 \ \alpha/cm^2\text{-sec} \tag{27}$$

Modern accelerators provide beams of charged particles with energies ranging from keV to BeV and of considerably greater intensities. A small proton accelerator of only 1-$\mu$A current, whose beam can easily be collimated to an area of 1 cm$^2$, has a beam intensity of

$$J = 6.25 \times 10^{12} \ p/cm^2\text{-sec} \tag{28}$$

which decreases only slowly with distance along the beam direction.

The simplest way to accelerate a particle of charge $e$ is to move it through a potential charge $V$, in which the particle will acquire kinetic energy

$$E = eV \tag{29}$$

The devices that achieve such acceleration to produce beams of particles with sufficient kinetic energy to induce nuclear reactions generally can be classified in three types:

1. Direct-current (dc) generators.
2. Pulsed radio-frequency (rf) accelerators.
3. Pulsed non-rf accelerators.

Modern dc generators include the Cockcroft-Walton cascade transformer-rectifier generator and the Van de Graaff electrostatic generator which provide a steady high voltage to an accelerating tube. The pulsed rf accelerators include the orbital types such as the fixed-frequency cyclotron and the linear types, such as the proton linear accelerator. The only pulsed non-rf accelerator in use today is the betatron, and since these accelerate only electrons they are described in Section 3.3 with the electron linear accelerator.

### 3.2.1 Direct-Current Generators

The dc generator consists of an ion source, an accelerating tube, a method of generating high voltage to the accelerating tube, and an analyzing system for selecting particular ions or their energies. The more important types of such accelerators are (a) the cascade generator and (b) the electrostatic generator.

#### Cascade Generators

The cascade generator is based on ac-voltage multiplication through transformers and rectification into dc voltage, using high-voltage rectifiers

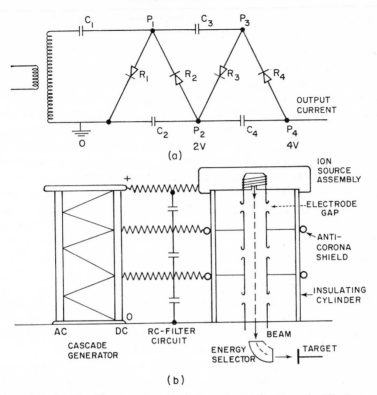

**Figure 3.9** (a) Cascade generator circuit (two stages of voltage doubling); (b) schematic of the generator and accelerator tube showing the ion source, energy selector, and target positions. [Adapted from W. E. Burcham, *Nuclear Physics* (McGraw-Hill, New York, 1963), pp. 302–303.]

such as the selenium rectifier. Rectifiers charge capacitors during one half-cycle and other rectifiers transfer charge during the other half-cycle to produce a steady doubled dc voltage. With $N$ similar capacitors and $N$ rectifiers, a voltage multiplication of $N$ is achieved. With no current drain the output voltage is just $N$ times the input voltage.

A schematic of a circuit for two stages of voltage doubling is shown in Figure 3.9a. As current is drained from the generator, a ripple voltage is obtained which has been shown to be proportional to $\sim N^2$ and inversely proportional to the frequency and condenser capacity. Thus modern generators, to reduce the ripple voltage, use high frequency ac, large capacitors, and as few stages of $N$ as possible.

The output voltage is used to accelerate a beam of ions produced by the ion source through the accelerating tube, and the ions emerge from the tube with a velocity corresponding to the accelerating voltage. The beam of ions can be bent in an energy-selecting magnet to strike a target at the exit tube. A Cockcroft-Walton-type accelerator is shown schematically in Figure 3.9b. An advantage of this type of generator over the electrostatic generator is its lack of moving parts.

Cascade generators generally have large beam currents; up to 10 mA of positive ions can be generated with energies to ∼1 MeV and up to 3 MeV under special conditions. Cascade generators in the 200 to 300 keV range are in common use as neutron generators, as described in Section 3.1.2.

Conventional Cockcroft-Walton accelerators are manufactured by the Philips Company of Eindhoven, Holland, and by Haefely of Basel, Switzerland. In the United States they are made by Radiation Dynamics of Westbury, N.Y., whose "Dynamitron" operates at a high frequency of about 300 kc with voltage available to about 3 MeV and by High Voltage Engineering Corporation of Burlington, Mass., whose "Insulating-Core Transformer," which consists of a stack of 30-kV transformer-rectifiers, yields an output voltage as high as ∼600 keV.

### Electrostatic Generators

The electrostatic generator is the most widely used of the dc generators. It was described by R. J. Van de Graaff in 1931, and, although called the Van de Graaff generator since then, this name is now a registered trademark of the High Voltage Engineering Corporation.

The electrostatic generator consists of a large-radius terminal cap on an insulated column which receives charge from a motor-driven fabric belt. Positive charge is sprayed onto the moving belt by means of corona discharges from sharp needles (corona points) fed by a high voltage source (∼20 to 50 kV). The belt carries the charge inside the high voltage terminal, where it is removed by a second set of corona points. On the return trip of the belt cycle a high voltage source may spray negative charge on the belt, thereby doubling the charge-carrying capacity of the belt. The terminal cap is thus continuously charged until leakage from the terminal is in equilibrium with the supply. The terminal then maintains a steady high-voltage potential. This potential is applied to a regulated acceleration tube which accelerates ions produced by an ion source located inside the high voltage terminal. A schematic diagram of the electrostatic generator is shown in Figure 3.10a. By control of the spray-charging source potential, the high voltage terminal potential can be controlled, and thus the kinetic energy of the accelerated particles can be regulated. Beam-bending magnets at the exit of the accelerating tube and before the target location may be used to control the charge-spraying source automatically and thus allow self-regulation of the beam particles' kinetic energy. An important advantage of the electrostatic generator is the nearly monoenergetic character of the beam; the energy spread at 6 MeV is as little as 20 keV (a resolution of ∼0.3%). Modern devices can yield beam

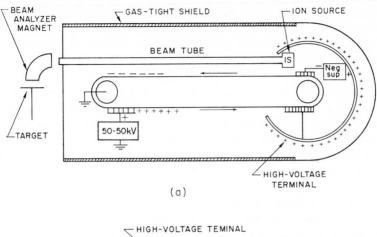

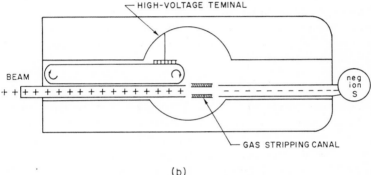

**Figure 3.10** (a) Schematic of the Van de Graaff electrostatic generator in a horizontal mode; (b) schematic of the tandem electrostatic generator.

currents of ~10 μA of p, d, t, ³He, ⁴He, and heavier ions which are monoenergetic and energy adjustable. Beams up to 10-MeV protons are readily available at this current and beams of ~5 MeV protons are available at currents of 0.5 mA.

The newer devices based on the tandem accelerator allow doubling of the kinetic energy of the accelerated particles. A schematic of the tandem electrostatic generator is shown in Figure 3.10b. In these devices the high voltage terminal is in the center of the accelerator and the ion source at one end produces *negatively* charged ions (e.g., H⁻ for protons) which are *accelerated* toward the high voltage terminal. Inside the terminal they collide with gas molecules in a "stripping" canal, where electrons are stripped off the negative ions and converted into *positive* ions which are now *repulsed* down the other side of the accelerating tube away from the high voltage terminal. At the exit port they have twice the kinetic energy than when starting at rest at the high voltage terminal. Beams of a few $\mu$A of 20-MeV p, d, and t are now available. Heavier ions, such as $^{12}$C and $^{16}$O, have also been accelerated.

### 3.2.2  Pulsed Radio-Frequency Accelerators

Pulsed rf accelerators have in common the ability to accelerate positive ions to high energy without the use of the corresponding high voltage potentials required. Thus they avoid the limitations of insulation breakdown affecting dc generators. The pulsed (or resonance) accelerators also have in common the general feature of applying a high-frequency voltage to a series of gaps through which the ion beam passes. The devices must match the accelerating frequency to the arrival of the ions at each gap so that the field is always in the accelerating direction as the particle crosses each gap. Radio frequency oscillators provide the accelerating potentials. The ions may move in (a) a circle under the influence of a magnetic field passing through the same gaps (the cyclotron) or (b) in a straight line of gapped tubes (the linear accelerator). Such devices are generally available at university and government research centers.

### *The Cyclotron*

The cyclotron, developed by E. O. Lawrence in 1930, is the most widely used charged-particle accelerator for proton and deuteron beams with energies between 5 and 20 MeV. A schematic diagram of a cyclotron is given in Figure 3.11. Positive ions produced near the center of the "dees" are accelerated toward the negative dee. On arrival at the gap the rf oscillator reverses the polarity and the positive ion is accelerated toward the other dee which is now negatively charged. The magnetic field accelerates the ion in a circle given by the cyclotron equation. The force on an ion of mass $m$, charge

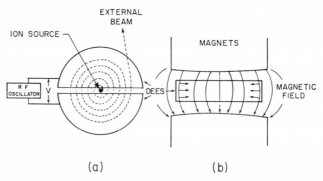

(a)                                  (b)

**Figure 3.11**   Schematic drawing of a cyclotron: (*a*) cyclic acceleration through the "dees"; (*b*) axial (vertical) focusing.

$e$, and velocity $v$ moving perpendicular to a uniform magnetic field $B$ is $evB$ which causes the ion to travel in a circle of radius $r$ given by the balanced centrifugal force

$$\frac{mv^2}{r} = evB \tag{30}$$

The frequency of revolution is

$$f = \frac{v}{2\pi r} = \frac{eB}{2\pi m} \tag{31}$$

and if the particle's mass does not appreciably increase relativistically the frequency is constant for a uniform magnetic field. The energy of the particle becomes [from (31)]

$$E = \tfrac{1}{2}mv^2 = \frac{e^2 B^2 r^2}{2m} \tag{32}$$

The constant rf to the "dees" continuously increases the radius of the ions (and their energy) until they spiral into the deflector, which has a negative potential, pulling the ions out of the "dee" system. The energy of the particles for the final radius $R$ is

$$E = \frac{e^2 B^2 R^2}{2m} \tag{33}$$

Typical cyclotrons require electromagnets with fields of the order of 10 to 20 kilogauss, an ion source (for $H^+$, $D^+$, $^4He^{2+}$, etc.) near the center of the accelerator, and an rf power supply of some tens of kilowatts at frequencies that are usually between 10 and 20 Mc/sec. Limitations of simple cyclotrons operating at fixed frequency result in maximum useful energies of about 25 to 30 MeV for protons and about 30 to 50 MeV for deuterons. Beam intensities generally range from microamperes to milliamperes.

### The Proton Linear Accelerator

The growth of radar technology during the early 1940's resulted in the development of high-power high-frequency sources and microwave techniques which were applied to the construction of linear accelerators that can accelerate protons and other heavy ions to energies in the tens of MeV range and electrons to energies in the tens of BeV range. The heavy ion and electron linear accelerators have much in common, for they give high-energy beams well collimated in the forward direction with small energy spreads, from pulsed accelerations down a series of waveguides by an electromagnetic wave. However, because the heavier ions move more slowly compared with electrons moving at highly relativistic speeds, a series of drift tubes is required to shield the heavier ions from the rf wave during the nonaccelerating part of the pulse cycle.

The proton linear accelerator can deliver currents of the order of 1 mA in a sharply focused monoenergetic beam of energy up to about 70 MeV. These values made this device attractively competitive with the fixed-frequency cyclotron, but the high cost of operation (which requires power of about 60 kW/MeV of acceleration) has inhibited its wide-spread construction. The few that exist are used primarily for studies in nuclear physics. However, these machines do result in prolific outputs of secondary neutrons, and neutron irradiations can be made as by-product irradiations whenever they are located.

## 3.3  PHOTONS AND ELECTRONS

The two machines most commonly available for photon or electron activation are the betatron and the electron linear accelerator. As a rather large expensive machine, the betatrons are not abundant, but when available they offer an adequate source of photonuclear radioactivation. On the other hand, electron linear accelerators are commercially produced by several manufacturers as a relatively versatile, low-cost source of high energy (2 to $\sim$30 MeV) ionizing radiation, and their present and future availability makes them attractive as general sources of irradiations for producing radionuclides by the $(\gamma,n)$ and $(e,e'n)$ reactions

In general, photon activation cross sections are higher than the corresponding electron activation cross sections. The deflection of irradiating electrons by the coulomb field of atomic nuclei produces a continuous spectrum of gamma (x-) rays known as Bremsstrahlung. Since this radiative process involves the coupling of the electron with the electromagnetic field of the emitted photon, the cross sections for radiations are of the order of the fine structure constant ($\alpha = \frac{1}{137}$) times the cross sections for elastic scattering. Thus for comparable irradiations with photons and electrons the ratio of the photonuclear reaction cross sections is given theoretically by

$$\frac{\sigma_{(e,e'n)}}{\sigma_{(\gamma,n)}} \simeq \alpha = \frac{1}{137} \tag{34}$$

Since in any given irradiation an electron beam is only partially converted into Bremsstrahlung, the actual ratio observed is much smaller. In practice, the electron beam is converted into a Bremsstrahlung photon beam with a distribution of energies between 0 and $E_e$ MeV, where $E_e$ is the maximum energy of the electrons.

A high atomic weight material, such as tantalum, platinum, or tungsten, which acts as a photon radiator, is interposed between the beam output source and the sample. Under these conditions the given activation product is produced in greater yield per electron compared with direct electron irradiation. The cross section for the reaction with the Bremsstrahlung beam is difficult to evaluate and is usually determined experimentally as

$$\int_0^E \sigma(E) \, dE \quad \text{(in MeV-b)} \tag{35}$$

for the irradiation conditions.

### 3.3.1 Betatrons

The cross-sectional view of a betatron shown in Figure 3.12 is similar to that of a cyclotron, but in a betatron electrons are accelerated by varying the magnetic flux $\phi$. The cyclotron equation, given in Section 3.2.2, shows that an electron moving in a fixed orbit $r_0$ with a changing magnetic flux, induces a tangential electric field $E_t$ by

$$2\pi r_0 E_t = \frac{d\phi}{dt} = \pi r_0{}^2 \frac{d\bar{B}}{dt} \tag{36}$$

where $\bar{B}$ is the average magnetic flux density within the orbit; but, since the momentum of an electron $p$, traveling in an orbit of radius $r_0$, is

$$E_t e = \frac{dp}{dt} = \frac{d}{dt}(B_0 r_0) \tag{37}$$

where $B_0$ is the value of $B$ at the orbit $r_0$, it follows that to accelerate an electron at a fixed radius

$$\frac{d\bar{B}}{dt} = 2\frac{dB_0}{dt} \tag{38}$$

This condition is obtained in betatrons by adjusting the magnet gap shown in Figure 3.12.

The betatron has the advantages of relative simplicity of construction and low cost compared with other charged-particle accelerators. Although it is limited to energies below 100 to 300 MeV because of radiation losses, it operates well for radioactivation in the energy range of 15 to 30 MeV. Most betatrons built have electron energies in the range of about 20 MeV and electron beam currents greater than 0.01 $\mu$A can be extracted. The application of a betatron for activation analysis has been reviewed by Brune et al.*

* D. Brune, S. Mattson, and K. Liden, Application of a Betatron in Photonuclear Activation Analysis, in *Modern Trends in Activation Analysis* (NBS, 1968) pp. 949–952.

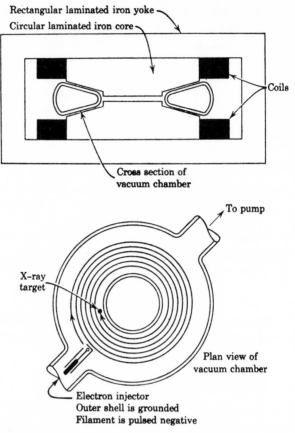

**Figure 3.12**   Schematic drawing of a betatron. [From E. M. McMillan, Particle Accelerators in *Experimental Nuclear Physics*, Vol. III (Wiley, New York, 1959), p. 681.]

### 3.3.2   Electron Linear Accelerators

The electron linear accelerator uses an oscillating electric field to accelerate electrons in a straight line at relativistic speeds. Most modern electron linear accelerators are powered by klystrons with outputs of more than 20-MW power of traveling rf waves at frequencies of about 3000 Mc (10-cm wavelength). Since the electron is traveling with the phase velocity of the wave at almost the speed of light, the drift tubes of the heavy-particle accelerators are not required, but to maintain an exactly correct phase velocity below the speed of light the waveguide is loaded with a series of hollow disks.

A typical electron linear accelerator of interest for radioactivation is the Stanford University Mark II shown schematically in Figure 3.13. This

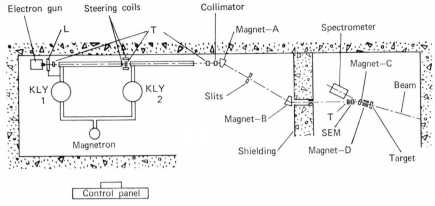

Figure 3.13 Schematic arrangement of the Stanford University Mark II electron accelerator and the target position for irradiations.

device produces a monoenergetic beam of electrons with energies up to 75 MeV and an average electron beam intensity of about $10^{12}$ electrons/sec over a beam area at the irradiation target position of about 1.4 cm².

The intensity-areal distribution of this beam at an energy of 57 MeV has been measured by the activation of a thin copper foil and subsequent counting of the 1.34 MeV $\gamma$-radiation of 12.9-h $^{64}$Cu produced by the

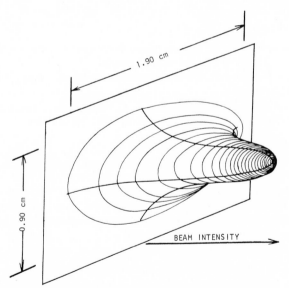

Figure 3.14 Abstract representation of the Stanford University Mark II Electron Linear Accelerator beam-intensity distribution plotted in the beam direction.

$^{65}$Cu(e,e'n)$^{64}$Cu reaction. The foil was sectioned into 2.4 × 2.4 mm squares which were weighed and counted individually. The activities corrected to a common weight and time represent the intensity distribution. These data are shown in Figure 3.14, where the beam-intensity distribution is plotted in the beam direction. More than 50% of the beam is concentrated in a circle of 0.32-cm radius concentric to the beam axis. Positioning of samples for irradiation within this beam area may be made by exposing a glass slide at the irradiation position and centering the sample holder on the radiation-blackened picture of the beam.

Several manufacturers produce these accelerators and some of them also offer commercial activation irradiations. Two of the larger electron accelerators at Stanford University include the Mark III, which can produce a beam of about 1 $\mu$A of 1-BeV electrons, and the two-mile SLAC accelerator which will produce a beam of about 30 $\mu$A of 20-BeV electrons. These linear accelerators are not useful for activation analysis because of the multiplicity of nuclear reactions caused by such high energy particles and large number of different elements in a sample which can produce the same radionuclide.

## 3.4  BIBLIOGRAPHY

### 3.4.1  Neutrons

RADIOISOTOPE SOURCES

L. F. Curtiss, *Introduction to Neutron Physics* (Van Nostrand, Princeton, N.J. 1959). Chapter III.

B. T. Feld, The Neutron in E. Segrè, Ed. *Experimental Nuclear Physics*, **2** (Wiley, New York, 1953), pp. 357–460.

G. D. O'Kelley, Radioactive Sources in L. C. L. Yuan and C. S. Wu, Ed. *Methods of Experimental Physics* **5B** (Academic, New York, 1963), pp. 555–580.

NEUTRON GENERATORS

E. A. Burrill and M. H. MacGregor, Using Accelerator Neutrons, *Nucleonics* **18**, No. 12, 64–68 (1960).

J. H. Cherubeni, Fast Activation Analysis with Van de Graaff Neutron Sources, Bulletin FA, High Voltage Engineering Co., Burlington, Mass., 1962, 24 pp.

A. S. Gillespie and W. W. Hill, Sensitivities for Activation Analysis with 14-MeV Neutrons, *Nucleonics* **19**, No. 11, 170–173 (1961).

W. W. Meinke and R. W. Shideler, Activation Analysis: New Generators and Techniques Make it Routine, *Nucleonics* **20**, No. 3, 60–65 (1962).

REACTORS

C. F. Bonilla, *Nuclear Engineering* (McGraw-Hill, New York, 1957).

L. F. Curtiss, *Introduction to Neutron Physics* (Van Nostrand, Princeton, N.J., 1959).

S. Glasstone and M. C. Edlung, *The Elements of Nuclear Reactor Theory* (Van Nostrand, New York, 1952).

W. R. Harper, *Basic Principles of Fission Reactors* (Wiley-Interscience, New York, 1961).

D. J. Hughes, *Pile Neutron Research* (Addison-Wesley, Cambridge, Mass., 1953).

S. E. Liverhant, *Elementary Introduction to Nuclear Reactor Physics* (Wiley, New York, 1960).

G. Murphy, *Elements of Nuclear Engineering* (Wiley, New York, 1961).

R. Stevenson, *Introduction to Nuclear Engineering* (McGraw-Hill, New York, 1954).

### 3.4.2  Charge-particle Accelerators

W. E. Burcham, *Nuclear Physics* (McGraw-Hill, New York, 1963), Chapter 8.

G. Friedlander, J. W. Kennedy, and J. M. Miller, *Nuclear and Radiochemistry*, 2nd ed. (Wiley, New York, 1964), Chapter 11.

D. Halliday, *Introductory Nuclear Physics* (Wiley, New York, 1955), Chapter 12.

B. G. Harvey, *Introduction to Nuclear Physics and Chemistry* (Prentice-Hall, Englewood Cliffs, N.J., 1962), Chapter 13.

R. E. Lapp and H. L. Andrews, *Nuclear Radiation Physics*, 3rd ed. (Prentice-Hall, Englewood Cliffs, N.J., 1863), Chapter 10.

M. S. Livingston and J. P. Blewett, *Particle Accelerators* (McGraw-Hill, New York, 1962).

E. M. McMillan, Particle Accelerators in E. Segrè, Ed., *Experimental Nuclear Physics*, Vol. 3 (Wiley, New York, 1959), Part XII.

L. C. L. Yuan and C. S. Wu, Eds., *Methods of Experimental Physics* **5B**, Nuclear Physics (Academic, New York, 1963), Chapter 3.

## 3.5  PROBLEMS

1.  Calculate the neutron intensity from 1 g of 56-d $^{254}$Cf if it disintegrates solely by spontaneous fission, releasing three neutrons per fission. Calculate the heat generated in 1 g of $^{254}$Cf if 170 MeV of heat per fission is deposited in the source. Assume that the heat liberated by alpha decay is negligible.

2.  Determine the minimum number of curies of $^{241}$Am required per neutron source in the eight-source irradiation facility in Figure 3.2 if a routine analysis procedure for 0.1 mg of sodium at a detection limit of 1 dps sets the maximum irradiation time at 10 min for each sample.

3.  Identical thin 1-mg gold foils are irradiated in the same position in a reactor for an equal length of time but one of the foils is covered with cadmium. The bare irradiated foil had $6 \times 10^4$ dpm of $^{198}$Au saturation activity after the irradiation, whereas the Cd-covered foil had $1 \times 10^4$

dpm. Compute the thermal neutron flux at the irradiation position in the reactor.

4. A 1-cm² colimated beam of 6.7-MeV protons from a cyclotron impinges on a 0.1-mm thick silver foil. If the proton beam current is 0.1 mA, how much 6.5-h $^{107}$Cd will be produced after a 1-hr irradiation? The cross section for the reaction at 6.7 MeV is 0.14 b.

5. Choose for each irradiation facility a convenient activation product for the determination of strontium in pure CaO. Which is best?
   (a) A thermal neutron reactor.
   (b) A fast reactor with beam ports and thermal columns.
   (c) A d-t neutron generator.
   (d) A cyclotron set up to accelerate deuterons.
   (e) An electron linear accelerator.

Chapter 4

# Radionuclides

The first part of a radioactivation analysis results in the production of radio-active nuclides. The completion of the analysis (or of a radiotracer experiment) involves the detection and measurement of the desired nuclides from among all those produced in the irradiated matrix. The means of isolation, both electronic and chemical, are presented in Chapter 6. The detection is achieved by observing interactions of the radiations released by the disintegrating nuclei, and quantitative measurement is made by observing these interactions in suitably calibrated detection systems. The more useful systems are discussed in Chapter 5. The principles of radioactivity measurement are based on two general considerations: (a) the modes by which radioactive nuclei disintegrate (i.e., the types and energies of the radiations released) and (b) the mechanisms by which these radiations may be observed.

Radionuclides are characterized by both their decay schemes and their decay kinetics. Each radionuclide has a unique combination of values for these two sets of parameters. The decay scheme summarizes the modes of disintegration that show the types and energies of the radiations released. The kinetics describe the rates of disintegration of the radioactive nuclei and subsequent product nuclei in those cases in which the decay product is also radioactive. The four general types of radiation emitted by radioactive decay were described in Section 1.4. Of these, neutron emission by nuclei is important only for the few short-lived delayed-neutron emitters present only in nuclear reactors and thus are of little concern in radioactivation analysis. Further, since alpha particle emitting radionuclides are produced by radioactivation only with the very heaviest nuclides ($Z > 83$) and are seldom used as radioactive tracers for several reasons, including the nontechnical ones of hazard, rarity, and high cost, the description of $\alpha$-radiation given in Section 1.4.1 will suffice for our needs. Thus by this process of elimination beta-ray and gamma-ray emission become the two major decay processes used in radioactivation analysis measurement.

We have seen in Section 1.4.2 that beta decay occurs for nuclides with neutron-to-proton ratios different from those of stable nuclides. Figure 1.3 showed that the $N/Z$ ratio for the stable nuclides followed along a "smoothed" curve with values ranging from $\sim 1.0$ for the light nuclides to $\sim 1.5$ for the heavy ones. For a given mass number nuclei with larger values of $N/Z$ than those that are stable decay to stable nuclei by converting neutrons to protons within the nucleus and emitting negatively charged electrons (beta emission) according to the equation

$$n \rightarrow p + \beta^- + \nu \tag{1}$$

Nuclei with smaller values of $N/Z$ can become stable by converting protons into neutrons within the nucleus by either or both of two processes: the emission of a positively charged electron (positron emission) or the capture of an atomic orbital electron (electron capture) by

$$p \rightarrow n + \beta^+ + \nu \tag{2}$$

and/or

$$p + e^- \rightarrow n + \nu \tag{3}$$

The product nucleus may or may not be left in an excited state relative to its minimum energy state (ground state). If it is, the excess excitation energy is released by the emission of one or more photons of electromagnetic radiation (gamma emission), by orbital electrons and x-rays (internal conversion), or both. In addition, decay by positron emission culminates in the eventual annihilation of the positron and a negatively charged electron and results in the liberation of two quanta of electromagnetic radiation, each having a kinetic energy equal to the rest mass of the electron (0.51 MeV).

Thus the radiations resulting from the disintegration of nuclei made radioactive because of irradiation-altered $N/Z$ ratios are beta particles (either negatively or positively charged) and gamma radiations in those cases in which the decay product nucleus is left in an excited state. For positron emission 0.51-MeV gamma rays are also available. Orbital electron capture, on the other hand, results in no measurable nuclear radiations, since the neutrino is, for all practical purposes, undetectable. However, subsequent atomic processes (x-ray and conversion electron emission) do allow this disintegration process to be measured.

## 4.1 DECAY SCHEMES

A decay scheme is a representation of nuclear energy levels of a radionuclide and the modes of de-excitation. The decay scheme shows each mode of decay, its abundance, the energy of the radiations, the sequence of emissions, the half-lives involved, and the product nuclide. The representation is made by depicting relative energy levels (as line bars in the vertical dimension) for the nuclides involved (by atomic number in the horizontal dimension). The lowest bar for each nuclide shows the nuclide in its lowest energy state (the ground state) even if the nuclide is radioactive. Decay schemes vary from simple, involving only one mode of decay, to complex, with two or more modes of decay or by one mode to many energy states. The latter are usually accompanied by several gamma-ray transitions. The decay scheme is important in radioactivity measurement since it relates the number of radiations of a given kind and energy to the actual number of disintegrations of the particular radionuclide.

### 4.1.1 Decay Modes

Figure 4.1 illustrates the decay schemes of some useful radionuclides that show the simpler modes of decay. The two top rows of decay schemes are of $\beta^-$-emitting nuclides, the $\beta^-$-decay arrows going to the right to indicate the increase in atomic number from $Z$ to $Z + 1$; for example, 14.3-d $^{32}$P is representative of those radionuclides that decay solely by beta emission directly to the ground state of the product nucleus. This is shown in the decay scheme by placing a zero to the right of the product nucleus energy level. In this example $^{32}_{15}$P is shown to decay directly to the ground state of $^{32}_{16}$S, emitting beta particles with maximum energy of 1.71 MeV. Other such beta emitters, formed by radiative neutron capture, include 12.3-y $^3$H, 5770-y $^{14}$C, 86.7-d $^{35}$S, 165-d $^{45}$Ca, 3.5-m $^{55}$Cr, 50.4-d $^{89}$Sr, 4.20-m $^{206}$Tl, and 3.3-h $^{209}$Pb. Their decay schemes not only show that these nuclides can be measured only by beta-particle detectors but that for each 100 disintegrations of the nuclide 100 beta particles of varying energy up to the maximum will be released.

The second nuclide in Figure 4.1, 2.3-m $^{28}$Al, is typical of the group of single-transition beta emitters that decay to an excited state of the product nucleus, in this case $^{28}$Si, with 1.78 MeV of excitation. The energy state is

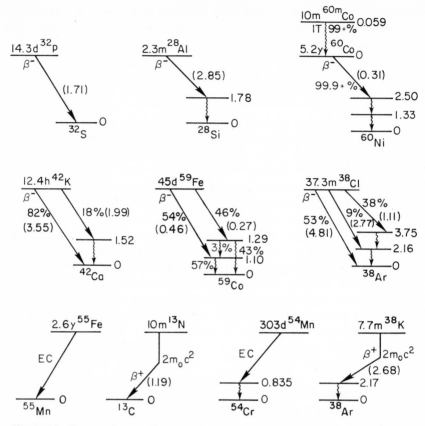

**Figure 4.1** Decay schemes of some useful radionuclides showing the simpler modes of decay and the decay energies. Decay schemes of some nuclides with more complex modes of disintegration are shown in Figure 4.2. (Adapted from C. M. Lederer, J. M. Hollander, and I. Perlman, *Table of Isotopes* 6th ed. (Wiley, New York, 1967).]

shown as a bar in the decay scheme labeled at the right 1.78 (MeV above the ground state). Unless a half-life value is given for the state, showing it to be a metastable isotope of the product nuclide, it may be assumed that de-excitation occurs promptly after or in coincidence with the beta decay, with the emission of a gamma ray of the total excitation energy, in this case, 1.78 MeV. Thus [28]Al could be measured by detecting either the 2.86-MeV maximum-energy beta particles or the 1.78-MeV gamma rays or both in coincidence. For each 100 disintegrations of [28]Al 100 beta particles *and* 100 gamma rays are available for detection. Other nuclides formed by (n,γ) reactions which decay in this manner include 10.7-s [20]F, 3.8-m [52]V, 39.5-m [123]Sn, 5.3-d [133]Xe, and 47-d [203]Hg.

The irradiation of $^{59}$Co with neutrons leads to two isomeric states of $^{60}$Co, the 10-min metastable state $^{60m}$Co, which is 0.059 MeV above the 5.26 y $^{60}$Co ground state. A few tenths of 1% of the $^{60m}$Co disintegrations have been shown to decay by beta emission to $^{60}$Ni, but for clarity this mode is not shown in the decay scheme in Figure 4.1. For all practical purposes the 10-min activity would be measured by the isomeric transition (IT) gamma rays of 0.059 MeV released as $^{60m}$Co decays to the $^{60}$Co ground state. The decay scheme as shown also notes that a few hundredths of 1% of the $^{60}$Co beta decays do not proceed through the decay sequence shown (they go directly to the ground state); but again, for all practical purposes, all of the beta decays (with $E_{\beta^-max} = 0.31$ MeV) result in the 2.50-MeV excited state of $^{60}$Ni from which de-excitation occurs promptly by *two successive* gamma-ray transitions, the first released with $E_\gamma = 1.17$ MeV, followed by the second with $E_\gamma = 1.33$ MeV. Thus $^{60}$Co can also be measured by its beta or gamma radiations, but the result of each 100 disintegrations is not only 100 (i.e., 99.9+) beta particles of 0.31 MeV maximum energy but also 200 gamma rays, 100 with 1.17 MeV energy and 100 with 1.33 MeV energy. Another radio-nuclide with the same general decay scheme, but with its own characteristic values of half-life and radiation energies, is 20-s $^{46m}$Sc $\xrightarrow{\text{IT}}$ 84-d $^{46}$Sc, which is the product of neutron irradiation of the monoisotopic element scandium ($^{45}$Sc). Several radionuclides, for example, 1.86-h $^{83m}$Kr, 2.8-h $^{87m}$Sr, 16-s $^{89m}$Y, 57-m $^{103m}$Rh, and 40-s $^{109m}$Ag, are metastable isotopes of stable nuclides which may be formed by the (n,n') or ($\gamma,\gamma'$) reactions. These radioisotopes decay by isomeric transition and release gamma rays of the transition energy. The decay scheme for these nuclides are similar to the $^{60m}$Co transition to $^{60}$Co, except that in these cases the ground-state nucleus is stable toward further radioactive decay.

The second row of decay schemes in Figure 4.1 contains nuclides that show beta-decay branching, that is, decay by beta emission to more than one energy level of the product nuclide. The radionuclide 12.4-h $^{42}$K decays by beta emission, with 82% of the disintegrations resulting in the ground state of $^{42}$Ca and 18% ending in the 1.52 MeV excited state of $^{42}$Ca. Prompt de-excitation yields a gamma ray with 1.52 MeV energy. Thus 100 disintegrations of $^{42}$K would yield 82 beta particles with energies up to the maximum of 3.55 MeV, 18 beta particles with energies up to the maximum of 1.99 MeV, and 18 gamma rays of 1.52 MeV energy. Other nuclides with similar decay schemes include 5.0-m $^{37}$S with 90% gamma rays, 33-d $^{141}$Ce with 30% gamma rays, and 9.4-d $^{169}$Er with 15% gamma rays.

The radionuclide $^{59}$Fe also decays with two beta branches, but both result in excited states of $^{59}$Co. The decay scheme shows that the 0.46-MeV maximum-energy beta decay is followed by a 1.10-MeV gamma ray to reach the ground state of $^{59}$Co. It also shows the 0.27-MeV maximum-energy beta decay results in the 1.29-MeV state of $^{59}$Co. This state exhibits gamma-ray branching, with about 6% of the gamma-ray transitions going to the 1.10-MeV state and about 94% going directly to the ground state of $^{59}$Co. Gamma-radiation measurement by energy would show for each 100 disintegrations of $^{59}$Fe, three gamma rays with $E_\gamma = 0.19$ MeV, 57 gamma rays with $E_\gamma = 1.10$ MeV, and 43 gamma rays with $E_\gamma = 1.29$ MeV. Another radionuclide, 5.80-m $^{51}$Ti, decays by two beta branches, both to excited states of the product nucleus $^{51}$V.

The decay scheme for 37.3-m $^{38}$Cl shows three beta-decay branches, two resulting in excited states and the third in the ground state of the product nuclide, $^{38}$Ar. Since each of the excited states decays to the one below it, each 100 disintegrations of $^{38}$Cl would yield, in addition to the 31, 16, and 53 beta particles with energies up to their respective maxima, 31 gamma rays with $E_\gamma = 1.59$ MeV and 47 gamma rays with $E_\gamma = 2.17$ MeV. The radionuclide 2.58-hr $^{56}$Mn also decays with three beta-decay branches, but each results in an excited state of the product $^{56}$Fe. Since the two upper levels decay with branching of the gamma-ray transitions, there are six possible gamma-ray energies available for measurement (see Problem 4.2).

The simpler modes of decay by electron capture and positron emission are shown in the bottom row of Figure 4.1. Several nuclides, such as 2.6-y $^{55}$Fe, decay entirely by orbital electron capture to the ground state of the product nuclide, in this case $^{55}$Mn. Other such nuclides include 35-d $^{37}$Ar, 330-d $^{49}$V, and 11-d $^{71}$Ge. Since orbital electron capture results in the emission of only the neutrino, these disintegrations are not measurable by nuclear radiations. Fortunately, the events may be determined from the secondary atomic processes that occur, namely the characteristic x-rays and conversion electrons that are emitted as electrons from the outer shells fall into the vacancies left by the captured electrons. In other cases of decay solely by electron capture, such as 303-d $^{54}$Mn, the product nucleus is left in an excited state, and the gamma-ray transition (or transitions) to the ground state allows the radionuclide to be measured by the gamma rays or by the characteristic x-rays of the product atom.

Positron emission is illustrated by the decay of 10-m $^{13}$N to the ground state of $^{13}$C. It is customary, when distinguishing between positron emission and electron capture, to show a vertical line from the radioactive nuclide that represents an energy drop of 1.02 MeV for positron emission to allow for the two 0.51-MeV photons that accompany the annihilation of the positron. Thus it should be noted that radionuclides which decay by positron emission directly to the ground state of the product nuclide may still be measured by the two 0.51-MeV annihilation photons. Other nuclides, such as 7.7-m $^{38}$K, decay by positron emission to an excited state of the product nuclide. The decay scheme for $^{38}$K shows that for every 100 disintegrations there are 100 positrons with energy up to the maximum energy of 2.68 MeV, 100 gamma rays of 2.17-MeV energy, and 200 photons of 0.51-MeV energy.

### 4.1.2 Complex Decay Schemes

Some additional decay schemes of activation products are illustrated in Figure 4.2. The nuclides involved decay by more than one mode; for example, 245-d $^{65}$Zn, formed by neutron irradiation of $^{64}$Zn, decays by both positron emission (1.7% of the disintegrations) and electron capture (98.3% of the disintegrations), of which the 44% branch of electron captures results in the

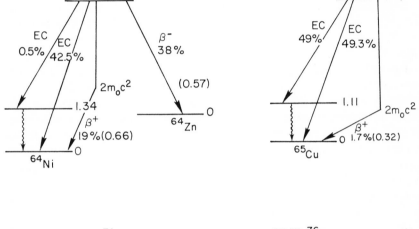

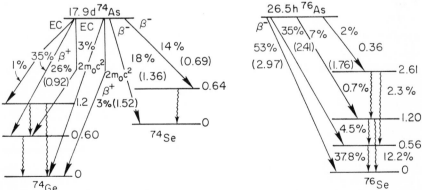

**Figure 4.2** Decay schemes of some nuclides with complex modes of disintegration. Quantitative measurement of these radionuclides may be made by selecting the type and energy of a particular radiation and correcting the disintegration rate for the fraction of decays by that radiation.

1.11-MeV excited state of $^{65}$Cu. Thus for each 100 disintegrations of $^{65}$Zn there would be available for measurement 1.7 positrons with maximum energy of 0.32 MeV, three annihilation photons of 0.51 MeV, and 44 gamma rays of 1.11 MeV; $^{65}$Zn is easily measured by gamma-ray detection. An x-ray counter could be used to detect the $^{65}$Cu x-rays resulting from the electron capture decays.

The decay scheme of $^{64}$Cu shows that it decays by three modes, electron capture and positron emission to $^{64}$Ni and beta emission to the ground state of $^{64}$Zn. The decay scheme for 62% of the decays leading to $^{64}$Ni is similar to the decay scheme of $^{65}$Zn but with the difference that only $\sim$0.5% leads to a prompt gamma ray from an excited state. Thus gamma-ray measurement would be difficult for $^{64}$Cu except for the 38 0.51-MeV annihilation photons of each 100 disintegrations. However, with maximum energies of 0.57 and 0.66 MeV for the $\beta^-$ and $\beta^+$ transitions, respectively, the 57 beta particles per 100 disintegrations make $^{64}$Cu relatively easy to measure in a beta counter.

The (n,$\gamma$) activation product of $^{75}$As (100% abundant) is 26.5-h $^{76}$As, whose decay scheme is similar to that of $^{56}$Mn, except that, in addition, 50% of the beta decays result in the ground state of the product nucleus $^{76}$Se. There are five gamma-ray transitions, the most prominent of which is the 0.56-MeV transition which gives on the average 38 photons per 100 disintegrations. The (n,2n) and ($\gamma$,n) products of $^{75}$As are even more complicated, with six decay branches, four leading to $^{74}$Ge and two to $^{74}$Se. Each 100 disintegrations would yield gamma radiation consisting of 58 0.51-MeV, 63 0.60-MeV, and 14 0.64-MeV photons. This isotope could be measured by any or all of the three photon groups, by the four beta branches, and even by the characteristic x-rays of germanium.

It should be apparent that the ability to make absolute measurement of the disintegration rate of a radionuclide depends strongly on the knowledge of the exact decay scheme of the nuclide. Even on a relative basis, the decay scheme aids in choosing the optimum radiations for detection and measurement.

## 4.2  DECAY RATES

### 4.2.1  Half-Life Considerations

In addition to the *modes* of disintegration, which are characterized for each radionuclide by the types and energies of the radiations released, radionuclides also have characteristic *rates* of disintegration. Both parameters assist in identifying and measuring radionuclides. We have seen in Section

1.3.4 that the decay of a radioactive nuclide is a first-order reaction process for which the rate of decay $(-dN/dt)$ is proportional to the number of radio-active nuclei $(N)$ present. The disintegration rate of a source of a radioactive nuclide was defined as the radioactivity $D$ of the source and given in (1-28) as $\lambda N$, where $\lambda$ is the decay constant whose value is a characteristic of the radionuclide. The decay constant is also expressed in terms of half-life, $T_{1/2}$, the time required for the decay of any initial number of nuclei to one-half that number. Equation 1-34 is repeated here

$$T_{1/2} = \frac{\ln 2}{\lambda} = \frac{0.693}{\lambda} \qquad (4)$$

to emphasize that although radionuclides are characterized by half-lives in the literature it is the decay constant that appears in the decay rate equation:

$$D = D^0 e^{-\lambda t} \qquad (5)$$

The decay constant of each radionuclide is involved in the activation analysis determination in two ways:

1. It determines for a given irradiation time the saturation factor $(1 - e^{-\lambda t_1})$ for that nuclide (see Section 2.4) where $t_1$ is the irradiation time.
2. It determines by (5) the amount of radioactivity still present at the time of measurement for a delay time $t$ from the end of the irradiation.

Thus the decay rate (as expressed by half-life) may be a more important factor than the decay scheme in many instances in activation analysis. A nuclide with a very short half-life may be irradiated to its saturation value with only a short irradiation, but it may also decay to a negligible level following the irradiation before it can be prepared for measurement. On the other hand, a nuclide with a very long half-life may decay negligibly from the end of irradiation to the time of measurement, but the small fraction of saturation radioactivity obtainable from a limited quantity of sample even with a long irradiation time may not be sufficient for an accurate radioactivity measurement.

As an example of the importance of considering the half-life we examine the determination of beryllium and fluorine (e.g., in the pure compound $BeF_2$) following thermal neutron irradiation. This compound was chosen because both elements are monoisotopic; $^9Be$, which forms the long-lived nuclide $2.7 \times 10^6$-y $^{10}Be$ with a cross section of 0.010 b, and $^{19}F$, which forms the short-lived nuclide 11-s $^{20}F$ with an approximately equal cross section of 0.009 b. If 1 g of $BeF_2$ were exposed to a flux of $10^{12}$ n/cm²-sec for 1 hr, $^{20}F$ would have reached its saturation activity, since

$$(1 - e^{-\lambda t}) = 1 - e^{-(0.693/11) \times 60 \times 60} = 1 - e^{-227} \simeq 1$$

whereas for $^{10}$Be the saturation factor would be

$$1 - \exp\left(-\frac{0.693 \times 1}{2.7 \times 10^6 \times 365 \times 24}\right) = 1 - e^{-2.92 \times 10^{-11}} \simeq 2.92 \times 10^{-11}.$$

The initial activities of $^{20}$F and $^{10}$Be, respectively, would be

$$D^0(^{20}\text{F}) = \frac{2 \times 10^{-3}}{47} \times 6.023 \times 10^{23} \times 9 \times 10^{-27} \times 10^{12} \times 1$$

$$= 2.31 \times 10^8 \text{ dps}$$

$$D^0(^{10}\text{Be}) = \frac{1 \times 10^{-3}}{47} \times 6.023 \times 10^{23} \times 10 \times 10^{-27} \times 10^{12} \times 2.92 \times 10^{-11}$$

$$= 3.75 \times 10^{-3} \text{ dps}$$

The $^{20}$F activity would be readily measurable after irradiation with even the crudest types of counting equipment, whereas the $^{10}$Be might be barely detectable with even the most sensitive detection equipment available today. However, if detectable, the $^{10}$Be activity would be measurable for centuries, whereas the $^{20}$F activity would be unmeasurable in about 5 min. The maximum delay time before measurement of the $^{20}$F when it has decayed to the initial $^{10}$Be activity value can be calculated from

$$\log_{10} \frac{D^0}{D} = \frac{\lambda t}{2.303} \tag{6}$$

which is the decay equation (5) in logarithmic (base 10) form. Thus

$$t = \frac{2.303}{0.693/11} \log_{10} \frac{2.31 \times 10^8}{3.75 \times 10^{-3}}$$

$$= 36.6 \log_{10} 6.15 \times 10^{10}$$

$$= 288 \text{ sec} = 4.8 \text{ min.}$$

With modern activation analysis systems, provided no transportation delays were encountered, 4.8 min would be more than enough time to determine the $^{20}$F radioactivity, and with a somewhat longer irradiation at this flux (e.g., 1 week) or an equal irradiation in a larger flux (e.g., $10^{13}$ n/cm$^2$-sec) the $^{10}$Be in 1 g of BeF$_2$ would be readily determined. Since almost all other nuclides produced by (n,$\gamma$) reactions have half-lives much shorter than $2.7 \times 10^6$ y $^{10}$Be, minimum detectability for such quantities is not a problem if the amount of sample is not severely limited.

### 4.2.2  Radioactive Equilibria

With the exception of pure monoisotopic elements activated with thermal neutrons, irradiation of a matrix will result in the production of more than one radioactive nuclide. The measurement of the several nuclides in the sample may be accomplished by the measurement of their radiation types and energies or of their half-lives or both. Since the half-life value is a characteristic of each radioactive nuclide, if several radionuclides are present in a sample, the total activity $D_T(t)$ at any time after irradiation $t$ will be the sum of the activity of each radionuclide present:

$$D_T(t) = D_1{}^0 e^{-\lambda_1 t} + D_2{}^0 e^{-\lambda_2 t} + \cdots + D_n{}^0 e^{-\lambda_n t} \tag{7}$$

The initial activities $D_i{}^0$ can be determined experimentally by graphical analysis of the decay curve if the mixture contains not more than three or four nuclides with half-lives that differ from one another by at least a factor of two (see Problem 4.6).

An important exception to (7) occurs when a radionuclide decays to a product nuclide which is itself radioactive. Such successive transformations, called a decay chain, may be written as

$$P \xrightarrow{T_{1/2}P} D \xrightarrow{T_{1/2}D} S \tag{8}$$

where $P$ is the parent nuclide, $D$ is the daughter nuclide, and $S$ is the daughter's product nuclide, which may itself be a radioactive nuclide. In activation analysis practice, $S$ is almost always a stable nuclide. Several exceptions exist, especially in the cases in which radionuclides whose granddaughter nuclide is the metastable state of the stable nuclide. The decay chains of $^{111}Pd$, $^{117}Cd$, and $^{124}Sn$ are examples of the latter exception.

Table 4.1 lists for several elements the decay chain resulting from neutron activation. Included in the table are several fission products which are sometimes used to determine uranium or as radioisotopic tracers.

**Table 4.1** Decay Chains of Some Product Nuclides Following Neutron Activation

| Target Nuclide | Reaction | Activation Product ($P$) | Parent Half-Life | Daughter Nuclide ($D$) | Daughter Half-Life | Stable Nuclide ($S$) |
|---|---|---|---|---|---|---|
| $^{46}Ca$ | $(n,\gamma)$ | $^{47}Ca$ | 4.5 d | $^{47}Sc$ | 3.4 d | $^{47}Ti$ |
| $^{48}Ca$ | $(n,\gamma)$ | $^{49}Ca$ | 8.7 m | $^{49}Sc$ | 57.5 m | $^{49}Ti$ |
| $^{94}Zr$ | $(n,\gamma)$ | $^{95}Zr$ | 65 d | $^{95}Nb$ | 35 d | $^{95}Mo$ |
| $^{96}Zr$ | $(n,\gamma)$ | $^{97}Zr$ | 17 h | $^{97}Nb$ | 72 m | $^{97}Mo$ |
| $^{100}Mo$ | $(n,\gamma)$ | $^{101}Mo$ | 14.6 m | $^{101}Tc$ | 14 m | $^{101}Ru$ |
| $^{235}U$ | $(n,f)$ | $^{90}Sr$ | 28 y | $^{90}Y$ | 64 h | $^{90}Zr$ |
| $^{235}U$ | $(n,f)$ | $^{137}Cs$ | 30 y | $^{137m}Ba$ | 2.6 m | $^{137}Ba$ |
| $^{235}U$ | $(n,f)$ | $^{140}Ba$ | 12.8 d | $^{140}La$ | 40 h | $^{140}Ce$ |

In each of these chains the parent nuclide decays according to the decay equation

$$-\frac{dN_P}{dt} = \lambda_P N_P \tag{9}$$

which has the familiar solution

$$N_P = N_P{}^0 e^{-\lambda_P t} \tag{10}$$

However, the net change in the number of daughter nuclei with time is a function not only of its own decay constant but also of the rate with which it is being produced by the decay of its parent. The rate equation for the daughter nuclide is given by

$$\frac{dN_D}{dt} = \lambda_P N_P - \lambda_D N_D \tag{11}$$

in which $\lambda_P N_P$ is the decay rate of the parent (which is the production rate of the daughter) and $\lambda_D N_D$ is the decay rate of the daughter. Equation 11 may be rearranged into the form of a linear differential equation of the first order

$$\frac{dN_D}{dt} + \lambda_D N_D - \lambda_P N_P{}^0 e^{-\lambda_P t} = 0 \tag{12}$$

The solution to this differential equation is derived in several textbooks* and is given as

$$N_D = \frac{\lambda_P}{\lambda_D - \lambda_P} N_P{}^0 (e^{-\lambda_P t} - e^{-\lambda_D t}) + N_D{}^0 e^{-\lambda_D t} \tag{13}$$

In terms of radioactivities, and with no initial daughter present (i.e., $N_D{}^0 = 0$), (13) becomes

$$D_D = \frac{\lambda_D}{\lambda_D - \lambda_P} D_P{}^0 (e^{-\lambda_P t} - e^{-\lambda_D t}) \tag{14}$$

Two interesting types of radioactive equilibrium occur when the half-life of the parent nuclide is greater than the half-life of its daughter (i.e., $\lambda_P < \lambda_D$). Equation 14 shows that if $\lambda_P < \lambda_D$, as $t$ increases, $e^{-\lambda_D t}$ becomes negligible much faster than $e^{-\lambda_P t}$, and when $e^{-\lambda_D t} \ll e^{-\lambda_P t}$

$$D_D = \frac{\lambda_D}{\lambda_D - \lambda_P} D_P{}^0 e^{-\lambda_P t} \tag{15}$$

---

* R. T. Overman and H. M. Clark, *Radioisotope Techniques* (McGraw-Hill, New York, 1960), Appendix H, p. 460, or E. Segrè, Radioactive Decay, in *Experimental Nuclear Physics*, Vol. III (Wiley, New York, 1959), pp. 2–8.

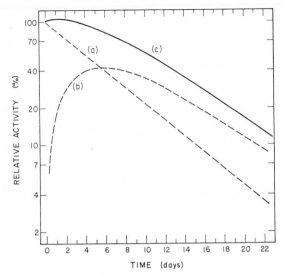

**Figure 4.3** Growth and decay curves for the radio-nuclides 4.5-d $^{47}$Ca $\rightarrow$ 3.4-d $^{47}$Sc: (a) the decay curve for an initially pure source of $^{47}$Ca, (b) the growth curve for $^{47}$Sc in the source; (c) the observed total activity of the source.

or

$$\frac{D_D}{D_P} = \frac{\lambda_D}{\lambda_D - \lambda_P}$$ (16)

This condition is known as *transient equilibrium*. For an initially pure parent the total activity of the unseparated parent and daughter activities reaches a maximum value before equilibrium is attained. At equilibrium the daughter activity decays with the half-life of the parent.

Neutron irradiation of $^{46}$Ca leads to the product nuclide 4.5-d $^{47}$Ca which decays to 3.4-d $^{47}$Sc. Figure 4.3 shows the activity of an initially pure source of $^{47}$Ca as a function of time. If $^{47}$Sc were a stable isotope, the decay of $^{47}$Ca would be given by line *a*. The growth of the 3.4-d $^{47}$Sc is shown as line *b*. The sum of the two activities in the unseparated source is given as line *c*. Extrapolation of line *b* back to $t = 0$ gives the hypothetical value

$$D_D^{\;0} = \frac{\lambda_D}{\lambda_D - \lambda_P} D_P^{\;0} = \frac{0.204}{0.204 - 0.154} D_P^{\;0} = 4.08 D_P^{\;0}$$ (17)

This value of the amount of $^{47}$Sc which would be present in the source if the $^{47}$Ca were already in equilibrium with the $^{47}$Sc at $t = 0$. If this equilibrium

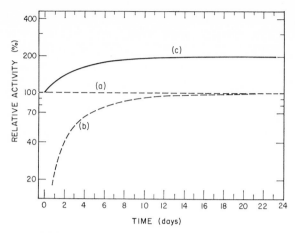

**Figure 4.4** Growth and decay curves for the radio-nuclides 27-y $^{90}$Sr → 64-h $^{90}$Y: (*a*) the decay curve for an initially pure source of $^{90}$Sr; (*b*) the growth curve for $^{90}$Y in the source; (*c*) the observed total activity of the source.

amount of $^{47}$Sc were then chemically separated from the $^{47}$Ca, the ratio of the separated activities would be 4.08.

The limiting case of radioactive equilibrium occurs when $\lambda_P \ll \lambda_D$ and the parent does not decay appreciably during many daughter half-lives. This condition is called *secular equilibrium*. It follows from $\lambda_P \ll \lambda_D$ that (15) reduces still further to

$$D_D = D_P{}^0 = D_P \qquad (18)$$

Secular equilibrium is shown in Figure 4.4 by the well-known radionuclide 27-y $^{90}$Sr, which decays to 64-hr $^{90}$Y. Line *a* shows the activity $D_P{}^0$ (constant within 0.1% over a period of 350 hr) of a freshly purified $^{90}$Sr source. Line *b* shows the growth of the $^{90}$Y daughter in the sample and line *c* shows the total activity. It may be noted from (14) that if $\lambda_P \ll \lambda_D$ and $e^{-\lambda_P t} \simeq 1$, then, since $\lambda_P N_P{}^0 = R$, the rate of production of the daughter nuclide

$$N_D = \frac{R}{\lambda_D}(1 - e^{-\lambda_D t}) \qquad (19)$$

which gives the production equation in the same form as that of a radio-nuclide produced by a nuclear reaction at a constant production rate ($R = n\sigma\phi$)

$$D_D{}^0 = R(1 - e^{-\lambda_D t}) \qquad (20)$$

One feature of secular equilibrium is the ability to have an almost constant source of a short-lived radiotracer available in the laboratory (e.g., 68-m $^{68}$Ga, 64-h $^{90}$Y, 30-s $^{106}$Rh, 2.6-m $^{137m}$Ba, 40-h $^{140}$La, 17.3-m $^{144}$Pr, from their respective long-lived parents) by repeated separations of the short-lived daughters after equilibrium or near-equilibrium is attained. A convenient way to prepare a laboratory source of the short-lived nuclide is to absorb the parent nuclide on an ion-exchange resin or alumina column (see Section 6.2.2) and elute ("milk") the daughter nuclide from the column as needed. Such a device is colloquially called a "cow" and the analogy is well deserved.

It is apparent that if the daughter nuclide is longer-lived than the parent equilibrium is *not* attained at any time. This decay system is the most frequently encountered situation for successive beta decays and is illustrated in Figure 4.5 for the pair 8.7-m $^{49}$Ca decaying to 57.5-m $^{49}$Sc. Line *a* is the decay of $D_P{}^0$ initially pure $^{49}$Ca decaying as if $^{49}$Sc were stable. Line *b* is the growth of $^{49}$Sc in the $^{49}$Ca source and line *c* is the total activity of the sample.

It is sometimes desirable to calculate, for the systems of transient equilibrium and no equilibrium, the time after separation when the daughter activity in the purified parent source reaches its maximum value. This time, $t_m$, can be obtained from the differentiation of (13) to give (for initially pure

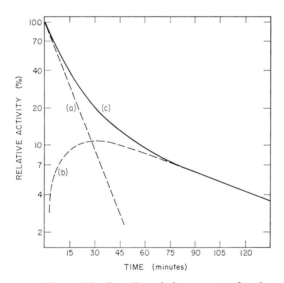

**Figure 4.5** Growth and decay curves for the radionuclides 8.7-m $^{49}$Ca → 57.5-m $^{49}$Sc: (*a*) the decay curve for an initially pure source of $^{49}$Ca; (*b*) the growth curve for $^{49}$Sc in the source; (*c*) the observed total activity of the source.

parent)

$$\frac{dN_D}{dt} = \frac{\lambda_P^2}{\lambda_D - \lambda_P} N_P^0 e^{-\lambda_P t} - \frac{\lambda_P \lambda_D}{\lambda_D - \lambda_P} N_P^0 e^{-\lambda_D t} \tag{21}$$

Setting $dN_D/dt = 0$ at $t = t_m$ gives

$$\frac{\lambda_P}{\lambda_D - \lambda_P} e^{-\lambda_P t_m} = \frac{\lambda_D}{\lambda_D - \lambda_P} e^{-\lambda_D t_m}$$

$$t_m = \frac{1}{\lambda_D - \lambda_P} \ln \frac{\lambda_D}{\lambda_P} \tag{22}$$

Equation 22 shows that the time $t_m$ for secular equilibrium approaches infinity as the half-life of a parent increases to large values.

A practical example of the use of (22) is noted for the determination of zirconium in the presence of hafnium,* following neutron irradiation from which the 65-d $^{95}$Zr activity is separated from the 70-d $^{175}$Hf and the 43-d $^{181}$Hf radioactivities only with great difficulty, since the two elements are chemically similar. The daughter, 35-d $^{95}$Nb, however, can be readily separated chemically from both zirconium and hafnium. The time when the $^{95}$Nb activity reaches its maximum value is

$$t_m = \left( \frac{0.693}{35} - \frac{0.693}{65} \right)^{-1} \ln \frac{65}{35} = 68 \text{ d.} \tag{23}$$

## 4.3  INTERACTION OF RADIATION WITH MATTER

Radioactivity measurement processes depend on the interactions of the radiations emitted by the disintegrating nuclei with some material in their environment. The radiations passing through matter in gaseous, liquid, or solid state are affected by individual atoms of the matter causing the radiations to lose some of their kinetic energy with each interaction. In turn, the radiations have a pronounced effect on the atoms involved.

There are four sets of parameters with which the interactions of radiation with matter can be examined. These sets, which are summarized for electrons and photons in Table 4.2, are the following:

1. *The type of radiation.* Nuclear radiations may be classified as heavy charged particles (e.g., protons, deuterons, and alpha particles), electrons

---

* J. E. Hudgens, Jr., and H. J. Dabagian, Radioactivation Determination of Zirconium in Zr-Hf Mixtures, *Nucleonics* **10**, No. 5, 25–27 (1952).

(positive or negative charged), photons, and neutrons. Different types of radiation affect matter in different ways.

2. *The kind of matter.* Such properties of matter as the physical and chemical state, the density, and the atomic number affect the mechanisms and rates with which radiations are stopped.

3. *The part of individual atoms affected.* The interaction mechanism for individual encounters for a given radiation depends on the part of the atom involved. Radiations may interact with the nucleus, with individual orbital electrons, or even with the electric field of the nucleus or the orbital electrons.

4. *The type of interaction.* The interaction can take place in three ways:

i. Elastic scattering, which changes the energy and direction of the radiation but causes no change in the internal energy of the scattering medium.

ii. Inelastic scattering, which changes the energy and direction of the radiation but also causes a change in the energy of the scattering medium.

iii. Absorption, in which the radiation becomes part of a system with the medium and some other process releases the excess energy of the new system.

**Table 4.2**   Summary of the Interactions of Electrons and Photons with Matter[a]

| Particle | Matter | Interaction | | |
|---|---|---|---|---|
| | | Elastic Collision | Inelastic Collision | Absorption |
| Electron $(\beta^-, \beta^+)$ | Nucleus | Rutherford scattering | Bremsstrahlung | Electron capture |
| | Orbital electrons | Negligible | Ionization and excitation | Annihilation of $\beta^+$ |
| Photon $(\gamma)$ | Nucleus | Negligible | Nuclear resonance (e.g., Mossbauer effect) | Photodisintegration |
| | Orbital electrons | Negligible | Compton scattering | Photoelectric effect |
| | Field | Negligible | Negligible | Pair production for $E_\gamma > 1.02$ MeV |

[a] Heavy-lined interactions are the major interactions for radiation detection and measurement.

### 4.3.1 Heavy Particles

In the realm of activation analysis heavy charged particles and neutrons are not radiations encountered from the radioactivation of stable nuclides and thus are not included in Table 4.2. Their interactions with matter are of considerable importance in other respects and detailed discussions of their interaction properties may be found in the several references listed in the bibliography. In general, charged particles (e.g., alpha particles) interact with matter as do electrons, that is, primarily by ionization and excitation of the atoms and molecules along the path of the particles. The major difference, however, is that charged particles, "heavy" compared with the electron, do not suffer large deflections in ionizing interactions and thus are stopped in reasonably straight paths. A head-on elastic collision of an alpha particle with an electron would result in an energy loss by the alpha particle of

$$\frac{\Delta E}{E} = \frac{4M_e}{M_\alpha} \tag{24}$$

Thus a collision of a 5-MeV alpha particle with an orbital electron would free the electron and impart to it a kinetic energy of about 2700 eV. Such secondary electrons can produce secondary ionizations, since on the average the energy required to ionize an atom into an *ion pair* (ip) is about 35 eV/ip. A 5-MeV alpha particle would produce approximately $5 \times 10^6/35 \simeq 1.5 \times 10^5$ ip in a very short distance (about 3.5 cm of air or about 0.004 cm of aluminum). Since the energy loss per ionization is essentially constant and alpha particles travel in straight paths, it is possible to obtain an empirical relationship between the range of monoenergetic alpha particles and their energy. For radionuclide-emitted alpha particles with energies in the region of 4 to 7 MeV the range in air (in centimeters) is given by

$$R = 0.309 E_\alpha^{3/2} \tag{25}$$

where $E_\alpha$ is in MeV.

The range in solid materials may be estimated by the approximate relationship

$$R_s = \frac{3.2 \times 10^{-4} R A^{1/2}}{\rho} \tag{26}$$

where $R_s$ = range in the solid material of density $\rho$ and mass number $A$,
$\quad\quad R$ = range in air.

Neutrons have no charge and therefore exert a negligible effect on atomic electrons. Their interactions are primarily with the nuclei of the atoms in their path. High-energy neutrons lose their energy by elastic collisions (neutron moderation, as given in Section 3.1.4). The loss per collision is greater with lower $Z$ materials, since a greater fraction of the kinetic energy can be imparted to a smaller struck nucleus. Neutrons may also undergo resonance scattering (inelastic collisions) during the process of moderation. A neutron, on reaching thermal velocity, is eventually absorbed in some nucleus by the $(n,\gamma)$ process. Although the neutron produces essentially no ionization while losing its kinetic energy, two events which permit the detection of neutrons by ionization processes do occur; the elastic scattering of neutrons by hydrogen atoms results in recoil protons which, as heavy-charged particles, can cause ionizations, and the release of a prompt gamma ray following the neutron absorption allows the detection of the ionization it produces.

### 4.3.2 Electrons

Electrons with energy in the range of those emitted as beta particles lose energy primarily by ionization and excitation processes. In this regard electrons behave similarly to the heavier charged particles. They also lose approximately the same amount of energy per ion pair formed, about 34 eV.* Since, however, the mass of the electron is so much smaller than that of an alpha particle, the *specific ionization* (i.e., the number of ion pairs produced per centimeter of path length) is much smaller. As a result a beta particle travels a longer path length than an alpha particle of equal energy. Further, an electron can lose up to half its energy in a single encounter if it is scattered through large angles. Thus the path of a beta particle deviates increasingly from a straight line as it loses its energy. With the added fact that beta particles have continuous distribution of energy (as contrasted to monoenergetic alpha particles) it becomes difficult to define the range of a beta particle as easily as that of an alpha particle. Even a beam of initially monoenergetic electrons become continuous in energy as they traverse an absorber.

The combination of a continuous energy spectrum and of continuous scattering angle spectrum leads to an absorption rate in matter that can be approximated by an exponential absorption law of the form

$$I = I_0 e^{-\mu_l x} = I_0 e^{-\mu_m \theta} \tag{27}$$

where $I/I_0$ is the fraction of electrons remaining after traversing an absorber, for which $\mu_l$ = the linear absorption coefficient (in cm$^{-1}$),

$x$ = the thickness traversed (in cm),

---

* A recommendation has been made by the International Commission on Radiological Units and Measurements that the energy loss per ion pair produced in air by an electron with energy greater than 0.3 MeV be taken as 34 eV. (From U.S. Department of Commerce, *National Bureau of Standards Handbook*, No. 62, 1957.)

or

$$\mu_m = \text{the mass absorption coefficient (in cm}^2/\text{g)},$$
$$\theta = \text{the thickness traversed (in g/cm}^2).$$

Since the loss of energy per ionization is nearly constant, the loss of energy of a beta particle ($\Delta E$) is proportional to the atomic number (the number of electrons per atom) and to the atom density ($N\rho/A$) (the number of atoms per cubic centimeter). Thus

$$\Delta E = \frac{N\rho Z}{A} \tag{28}$$

where $N$ = Avogadro's number,

$A$ = the atomic weight of the absorber,

$\rho$ = the absorber density.

Since the variation of $Z/A$ with $Z$ is small (decreasing from about 0.5 for aluminum to about 0.4 for lead), absorption depends essentially on the density of the absorber. Thus, if the absorption length is given in units of g/cm$^2$ (i.e., density $\times$ thickness), the fractional absorption for a given energy beam of electrons is almost independent of the atomic number of the absorber material. The mass absorption coefficient for given beta maximum energies may be calculated from the empirical relation

$$\mu_m = \frac{22}{E_{\max}^{1.33}} \quad \text{(in cm}^2/\text{g)} \tag{29}$$

Equation 27 [analogous somewhat to (1-32), the radioactive decay law] indicates that there is a half-thickness value $\theta_{1/2}$ for which the initial intensity is reduced to half its value (usually for only one or two half thicknesses):

$$\frac{I}{I_0} = \tfrac{1}{2} = e^{-\mu_m \theta_{1/2}} \tag{30}$$

$$\theta_{1/2} = \frac{\ln 2}{\mu_m} = \frac{0.693}{\mu_m} \tag{31}$$

Figure 4.6 gives an empirical curve relating the maximum beta-particle energy, $E_\beta(\max)$, with half-thickness value $\theta_{1/2}$. The curve is useful for estimation of maximum beta energies by simple absorption measurements.

A better estimate of the maximum beta-particle energy may be made from a determination of the beta range by a total absorption measurement. A typical absorption curve for a pure radionuclide is seen in Figure 4.7, in which the beta absorption curve for 55-m $^{69}$Zn which decays by emission of beta-particles with maximum energy of 0.91 MeV, shows an extrapolated range of 0.36 g/cm$^2$. Such determinations, however, are seldom made for activated radionuclides unless qualitative identifications are being attempted.

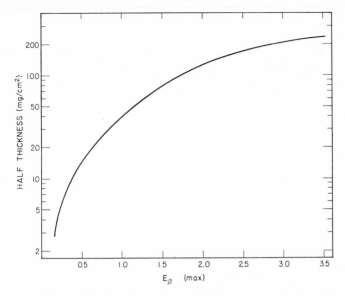

**Figure 4.6** The absorption half-thickness as a function of maximum beta-particle energy. [Adapted from G. I. Gleason, D. L. Tabern, and I. D. Taylor, Absolute Beta Counting at Defined Geometrics, *Nucleonics* **8**, No. 5, 12–21 (1951).]

Prior studies have resulted in several empirical energy-range relationships; a useful set, taken from Glendenin and Coryell,* is given by the equations

$$R = 0.542E_{\max} - 0.133 \quad \text{for} \quad E_{\max} > 0.8 \text{ MeV} \tag{32}$$

$$R = 0.407E_{\max}^{1.38} \quad \text{for} \quad 0.15 < E_{\max} < 0.8 \text{ MeV} \tag{33}$$

Figure 4.8 shows a curve prepared by Glendenin* which relates the range in aluminum with maximum beta energies. The curve is especially useful for energies below 0.8 MeV, when (33) is less exact.

The only other important interaction of electrons with matter is the inelastic scattering of high-speed electrons by the coulomb attraction of the positively charged nucleus. This coulomb-induced scattering results in the emission of continuous-energy photons, or Bremsstrahlung (braking rays). The gamma-ray background of the "pure" beta emitting radionuclide $^{69}$Zn shown in Figure 4.7 is due to the Bremsstrahlung produced in the absorber by the $^{69}$Zn beta particles. Advantage may be taken in some instances of the

---

* See L. E. Glendenin, Determination of the Energy of Beta Particles and Photons by Absorption, *Nucleonics* **2**, No. 1, 12 (1948).

Bremsstrahlung from beta emitters for their measurement with gamma-ray detection equipment.

The total loss of energy of an electron traversing an absorber, $\Delta E_T$, is the sum of the loss by the ionization collisions, $\Delta E_c$, and the loss by radiation, $\Delta E_r$. The loss of energy by radiation has been shown to depend on $Z^2$ of the absorber and the electron energy

$$\Delta E_r = c_1 Z^2 E \tag{34}$$

and from (28) the loss of energy by collisions in the same material is

$$\Delta E_c = c_2 Z \tag{35}$$

The ratio of losses by radiation to collision is given by

$$\frac{\Delta E_r}{\Delta E_c} \simeq \frac{ZE}{700} \tag{36}$$

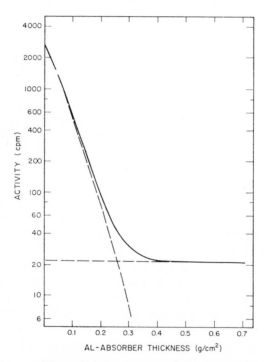

**Figure 4.7** Aluminum-absorption curve for 55-m $^{69}$Zn, decaying by emission of beta particles with 0.91-Mev maximum energy. The extrapolated range is about 0.36 g/cm$^2$.

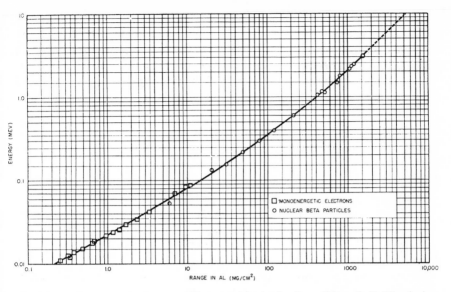

**Figure 4.8**   The range of electrons and beta particles in aluminum. [From L. E. Glendenin, Determination of the Energy of Beta Particles and Photons by Absorption, *Nucleonics* **2**, No. 1, 16, 1948).]

Thus for absorption in aluminum $(Z = 13)$ the loss by radiation is unimportant for the energy range of beta particles emitted by all usual radionuclides. In lead $(Z = 82)$, however, the loss by radiation is $\sim 12\%$ of the loss by collision for a 1 MeV beta particle. The production of Bremsstrahlung for higher energy beta-emitting radionuclides should be considered in absorption measurements.

### 4.3.3   Gamma Rays

Table 4.2 shows that gamma rays interact with matter not only differently than do electrons but also that they interact in three significant ways.

1. The photoelectric effect, an absorption process in which a gamma ray ejects an electron from an atomic orbit and disappears by transferring all its energy in the process.

2. The Compton effect, an inelastic scattering process in which the photon ejects an electron but escapes with degraded energy.

3. Pair production, an absorption process in which the photon vanishes in creating a positively and negatively charged pair of electrons.

The differences in interactions between gamma rays and charged particles result from the fact that charged particles lose their energy in small discrete steps ($\sim$34 eV per ion pair produced), whereas a gamma ray can lose most or all of its energy in a single event. For even a collimated beam of mono-energetic gamma rays a single collision by any one photon in traversing an absorber of thickness $dx$ removes it from the beam. The rate of photons removed from the beam ($-dI$) by any of the three processes is proportional to the incident rate ($I$), the linear absorption coefficient ($\mu_l$), and the thickness ($dx$):

$$-dI = \mu_l I dx \tag{37}$$

which on integration gives

$$I = I_0 e^{-\mu_l x} \tag{38}$$

Thus it is seen that a beam of monoenergetic gamma rays is absorbed exponentially in matter and without a definite maximum range. This is in contrast to the absorption of electrons which approximately follows an exponential absorption [see (4-27)] but as their energy diminishes slowly reaches a maximum range.

Gamma-ray absorption coefficients are more conveniently expressed as mass absorption coefficients $\mu_m$ (in $cm^2/g$) for which

$$\mu_m = \frac{\mu_l}{\rho} \tag{39}$$

where $\rho$ is the density of the absorber.

Since gamma rays of a given energy have no definite range and are absorbed exponentially, the definition of a half-thickness value $X$ becomes more significant for a given photon energy than for electrons. Thus

$$X = \frac{\ln 2}{\mu_l} = \frac{0.693}{\rho \mu_m} \tag{40}$$

The half-thickness value has the same usefulness in resolving gamma-ray energy absorption curves as the half-life value $T_{1/2}$ in resolving decay curves. Some typical mass absorption coefficients for gamma rays are given in Table 4.3.

Since there are three mechanisms for the absorption (or scattering) of photons, the absorption coefficient reflects the sum of all three processes; namely,

$$\mu = \mu_{pe} + \mu_c + \mu_{pp} \tag{41}$$

where the partial coefficients are for the photoelectric effect, the Compton effect, and pair production, respectively. The absorption coefficient $\mu$ is a complicated function of gamma-ray energy, since each of the three processes is itself a function of gamma-ray energy. Figure 4.9 shows the total and partial

**Table 4.3**   Some Mass Absorption Coefficients

| $E_\gamma$ (MeV) | Absorber Material | | | |
|---|---|---|---|---|
| | Water (cm²/g) | Aluminum (cm²/g) | Iron (cm²/g) | Lead (cm²/g) |
| 0.1 | 0.171 | 0.169 | 0.370 | 5.460 |
| 0.2 | 0.137 | 0.122 | 0.146 | 0.942 |
| 0.3 | 0.119 | 0.104 | 0.110 | 0.378 |
| 0.5 | 0.097 | 0.084 | 0.084 | 0.152 |
| 1.0 | 0.071 | 0.061 | 0.060 | 0.070 |
| 2.0 | 0.049 | 0.043 | 0.042 | 0.046 |
| 3.0 | 0.040 | 0.035 | 0.036 | 0.041 |
| 5.0 | 0.030 | 0.028 | 0.031 | 0.043 |

linear absorption coefficients for lead ($Z = 82$) as a function of energy. Figure 4.10 shows these functions for aluminum ($Z = 13$). It is apparent that each of the three gamma-ray interaction processes becomes the dominant one over certain energy ranges.

If an amount of radionuclide produced or acquired constitutes a radiation hazard, the absorption coefficient may be useful in the estimation of the shielding requirements to reduce the external radiation intensity to an acceptable value.

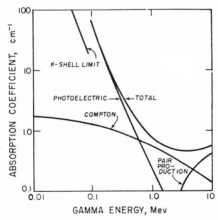

**Figure 4.9** Total and partial linear absorption coefficients for gamma rays in lead. [From G. R. White, National Bureau of Standards Circular 585, 1957].

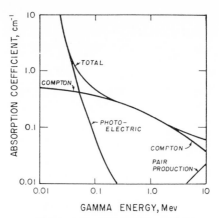

GAMMA ENERGY, Mev

**Figure 4.10** Total and partial linear absorption coefficients for gamma rays in aluminum. (From G. R. White, National Bureau of Standards Circular 583, 1957).

## *The Photoelectric Effect*

The photoelectric effect cannot take place with free electrons (since the conservation of momentum requires a recoiling partner, namely the residual atom). Thus it is reasonable that the most tightly bound electron, the $K$ shell electron, has the greatest probability of absorbing the photon incident on the atom. For incident photon energies above the $K$-shell binding energy, about 80 % of the photoelectric absorption processes take place with $K$-shell electrons. Most of the remainder take place with $L$-shell electrons. The process is shown schematically in Figure 4.11. The result is the ejection of an electron with kinetic energy

$$E_c = E_\gamma - B_c \tag{42}$$

where $B_c$ is the binding energy of the ejected electron in its orbital shell. For nuclear gamma rays the photoelectron energy is essentially equal to the incident photon energy.

Absorption by the photoelectric effect decreases rapidly with increasing gamma-ray energy (as is shown in Figure 4.9) but also increases rapidly with increasing absorber atomic number. It has been shown that the probability of photoelectric absorption for gamma-rays with energies in the range between the binding energy of the $K$-shell and about 0.5 MeV is approximated by

$$\mu_{pe} = k \frac{Z^5}{E_\gamma^{3.5}} \tag{43}$$

making photoelectric absorption more important for lower energy gamma rays in the heavier elements.

*The photoelectric effect.* The incident photon interacts with the atom to eject a photoelectron (primarily from the $K$-shell) whose energy is $E_e = E_\gamma - B_e$. The recoil of the parent atom conserves momentum. Atomic processes, that is characteristic x-ray and Auger electron emissions in filling the inner-shell vacancy, complete the event.

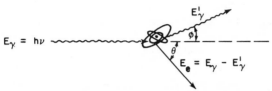

*The Compton effect.* The incident photon ejects an electron from the atom and is scattered through angle $\phi$ with degraded energy $E'_\gamma$. The Compton electron is ejected at angle $\theta$ with energy $E_e = E_\gamma - E'_\gamma$, neglecting the relatively small binding energy of the electron.

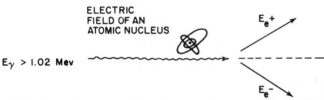

*Pair production.* The photon, having energy greater than 1.02 MeV, may interact with the electric field of the nucleus. The photon is absorbed to create a positive and negative pair of electrons whose total kinetic energy is $E_\gamma - 1.02$ MeV. The positron, after coming to rest, annihilates with an electron, creating two photons of 0.51 MeV energy each.

**Figure 4.11**  Schematic diagrams of the three principle iteration mechanisms of gamma rays with matter.

### The Compton Effect

The Compton effect is shown schematically in Figure 4.11. From Figure 4.9 it is seen that the effect in lead becomes the dominant process for photon energies between 0.6 and 5 MeV. By comparison the range in aluminum is between 0.05 and 15 MeV.

In Compton scattering the photon transfers only part of its energy to an electron, emerging as a photon with lower energy. It can then undergo another Compton scattering or a photoelectric absorption. From the conservation of energy it follows that

$$E_e = E_\gamma - E'_\gamma = h\nu - h\nu' \tag{44}$$

where the primes refer to the scattered photons. From the conservation of momentum it follows that the change in wavelength $\Delta\lambda$ for the scattered photon is

$$\lambda' - \lambda = \frac{h}{mc}(1 - \cos\phi) \tag{45}$$

where $\phi$ is angle of the scattered photon with respect to the incident photon. Equation 45 states that the change in wavelength of the scattered photon is independent of the wavelength itself. Therefore the fractional energy lost by the photon in a Compton scattering increases with increasing photon energy. Table 4.4 shows some typical values for $\phi = 90°$ Compton scattering.

Equation 45 can be converted to give the equation for the energy of the degraded photon as

$$E'_\gamma = \frac{E_\gamma}{1 + (E_\gamma/mc^2)(1 - \cos\phi)} \tag{46}$$

which for $E_\gamma \gg mc^2$ reduces to

$$E'_\gamma \simeq \frac{mc^2}{(1 - \cos\phi)} \tag{47}$$

An interesting observation from (47) is that for 90° scattering, the escaping photon approaches a limiting energy value of 0.51 MeV ($= mc^2$) and that for a back-scattered photon ($\phi = 180°$) the limiting energy value approaches $\sim 0.26$ MeV ($= mc^2/2$). This back-scattered photon will be seen in Chapter 5 to play a role in gamma-ray spectrometry (see also Problem 4.10).

**Table 4.4**  Compton Effect for $\phi = 90°$[a]

| $E_\gamma$ MeV | $E'_\gamma$ MeV | $\dfrac{E_\gamma - E'_\gamma}{E_\gamma}$ (%) |
|:---:|:---:|:---:|
| 0.01 | 0.010 | 1 |
| 0.10 | 0.0837 | 16 |
| 1.00 | 0.337 | 66 |
| 10.00 | 0.486 | 95 |

[a] Adapted from D. Halliday, *Introductory Nuclear Physics*, 2nd ed. (Wiley, New York, 1955) p. 171.

### Pair Production

Photons of energy greater than twice the rest mass of the electron ($2m_e c^2 = 1.02$ MeV) may be absorbed by the process of pair production. The production of the positron-negatron pair must take place in the electrostatic field of a nucleus to conserve the momentum of the incident photon. The excess energy of the photon is converted into kinetic energy of the pair

$$KE = E_\gamma - 1.02 \text{ MeV} \tag{48}$$

with the positron having slightly more than half the available kinetic energy, since it is accelerated slightly by the nucleus, whereas the negatron is slightly retarded. The process is shown schematically in Figure 4.11.

The partial absorption coefficient $\mu_{pp}$ is obviously 0 for $E_\gamma \leq 1.02$ MeV. It has been shown empirically that $\mu_{pp}$ increases as the square of the atomic number of the absorber and linearly at low energies just above 1.02 MeV:

$$\mu_{pp} \simeq kNZ^2(E_\gamma - 1.02) \tag{49}$$

and for high energy gamma rays (as $E_\gamma$ becomes large compared with 1.02 MeV)

$$\mu_{pp} \simeq kNZ^2 \ln E_\gamma \tag{50}$$

Figures 4.9 and 4.10 show that absorption by pair production becomes the dominant mode for $E_\gamma > 5$ MeV in lead and $E_\gamma > 15$ MeV in aluminum. Since almost no radioactive isotope emits gamma rays with energy in excess of 5 MeV, pair production is seldom a significant mechanism in radioactivity measurement. However, many radionuclides emit gamma rays with energy in excess of 1.02 MeV, and pair production may make a significant contribution to the absorption of the gamma radiation. It should be recalled that when the positron loses its kinetic energy the annihilation process occurs in which the rest mass energy reappears as two 0.51 MeV photons. These photons are indistinguishable from the two 0.51 MeV photons resulting from the annihilation of positrons from beta disintegrating nuclei.

### 4.4 BIBLIOGRAPHY

The general discussion of the properties of radioactive nuclides is given in the several texts listed under Section 1.6.1 in the bibliography for Chapter 1. Most of these texts separate the various modes of decay into individual chapters, and several of them contain detailed discussions of the rates of radioactive decay; for example, Chapter 3 in Friedlander, Kennedy, and Miller, and Chapter 8 in Overman and Clark. Many texts contain discussion of the interaction of radiation with matter as separate chapters which range in scope from descriptive to mathematical theory. Some of the more useful as references are listed below.

H. A. Bethe and J. Ashkin, Passage of Radiations Through Matter in *Experimental Nuclear Physics*, E. Segrè, Ed. (Wiley, New York, 1963), Vol. 1, Part 2.

R. D. Evans, *The Atomic Nucleus* (McGraw-Hill, New York, 1955), Chapters 23–25.

G. Friedlander, J. W. Kennedy, and J. M. Miller, *Nuclear and Radiochemistry*, 2nd ed. (Wiley, New York, 1964), Chapter 4.

M. Haïssinsky, *Nuclear Chemistry and its Applications* (Addison-Wesley, Reading, Mass., 1964), Chapter 2.

D. Halliday, *Introductory Nuclear Physics* (Wiley, New York, 1955), Chapter 7.

B. G. Harvey, *Introduction to Nuclear Physics and Chemistry* 2nd ed. (Prentice-Hall, Englewood Cliffs, N.J., 1969), Chapter 11.

R. T. Overman and H. M. Clark, *Radioisotope Techniques* (McGraw-Hill, New York, 1960), Chapter 1.

W. J. Price, *Nuclear Radiation Detection*, 2nd ed. (McGraw-Hill, New York, 1964), Chapter 1.

### 4.5  PROBLEMS

1.  The nuclide $^{236}$U decays by emission of $\alpha$-particles with energy of 4.5 MeV, as measured with an ionization instrument. Coincident measurements show about 26% of the alpha decays are accompanied by conversion electrons of 0.050 MeV. Draw the decay scheme of $^{236}$U.

2.  From the data given below for the radiations from 2.58-h $^{56}$Mn draw the decay scheme and give the number of each radiation per 100 disintegrations of the radionuclide:

$\beta^-$:    2.81 MeV (50%),    1.04 MeV (30%),    0.65 MeV (20%),

$\gamma$:    $\gamma_1 = 0.845$ MeV,    $\gamma_2 = 1.81$ MeV,    $\gamma_3 = 2.13$ MeV,

$\gamma_4 = 2.65$ MeV,    $\gamma_5 = 2.98$ MeV

relative yields of gamma rays:

$$\gamma_1(100):\gamma_2(29):\gamma_3(15):\gamma_4(1.8):\gamma_5(0.4).$$

3.  Look up the decay schemes for 53.6-d $^7$Be, 9.5-m $^{27}$Mg, 2.62-h $^{31}$Si, and 4.4-h $^{80m}$Br in a reference [such as C. M. Lederer, J. M. Hollander, and I. Perlman, *Table of Isotopes*, 6th ed. (Wiley, New York, 1967)], and suggest the radiations (with their quantity per disintegration) to be detected in measuring the respective radionuclides.

4.  One milligram of pure arsenic was irradiated in a neutron flux of $10^{13}$ thermal neutrons/cm²-sec and $2 \times 10^{12}$ fast neutrons/cm²-sec for a period of 24 hr. The cross section for thermal neutron radiative capture

by $^{75}$As is 4.3 b; the cross section for the (n,2n) inelastic scattering reaction is 0.5 b. The decay schemes for the product nuclides $^{76}$As and $^{74}$As are given in Figure 4.2. Calculate the $D^0$ activity (in dps) of the two arsenic radionuclides. How many gamma rays (in $\gamma$ps) from each of the transitions of 0.56, 0.60, and 0.64 MeV would be emitted by the sample 24 hr after the end of the irradiation?

5. How long an irradiation of pure beryllium would be required in a neutron flux of $10^{14}$ n/cm²-sec to obtain a specific activity of 3.7 × $10^4$ dps $^{10}$Be per gram of beryllium?

6. The following data were obtained by repeated counting of the gross beta activity of a sample following a neutron irradiation. Plot the data on semilog paper and by the graphical analysis determine the half-lives and $D^0$ activities of the radionuclides present.

| Activity (dps) | Time After Irradiation (min) | Activity (dps) | Time After Irradiation (min) |
|---|---|---|---|
| 1113 | 2 | 197 | 40 |
| 915 | 4 | 156 | 50 |
| 768 | 6 | 126 | 60 |
| 657 | 8 | 75 | 90 |
| 573 | 10 | 52 | 120 |
| 434 | 15 | 37 | 150 |
| 352 | 20 | 29 | 180 |
| 298 | 25 | 23 | 210 |
| 255 | 30 | 19 | 240 |

7. If $10^5$ dpm of pure $^{47}$Ca were present at the end of an irradiation, what would be the total activity present one week later? At what time after the irradiation would the $^{47}$Sc have reached its maximum disintegration rate?

8. Plot the total activity decay curve for 100 dps of $^{101}$Mo at time zero from the activities of the two nuclides at 2, 4, 7, 10, 15, 20, 40, and 60 min.

9. A 10-ml aqueous solution of 100-mCi $^{32}$P is kept in a 1-cm diameter thin plastic vial. Estimate the beta- and gamma-ray flux intensities (in number/cm²-sec) at a distance of one meter from the source. Assume the absorption in the plastic is negligible.

10. The kinetic energy of a Compton-scattered electron has its maximum when $\cos \phi = -1$ ($\phi = 180°$), for which

$$E_{e^-}(\text{max}) = \frac{E_\gamma^0}{1 + m_0 c^2/2E_\gamma^0}.$$

Determine the maximum electron energy for Compton-scattering interactions with the gamma rays of $^{28}$Al, $^{60}$Co, $^{137}$Cs, and $^{170}$Tm.

11. A gamma-ray energy analysis was made by determining the absorption through increasing thicknesses of lead foils placed between the sample and a gamma-ray detector. From the data given below determine the energy of the gamma rays emitted by the sample.

| Thickness of Pb Absorber (cm) | Intensity (cps) | Thickness of Pb Absorber (cm) | Intensity (cps) |
|---|---|---|---|
| 0.2 | 1601 | 3.0 | 63 |
| 0.4 | 1187 | 3.5 | 46 |
| 0.6 | 891 | 3.9 | 37 |
| 0.9 | 580 | 4.2 | 29 |
| 1.2 | 386 | 4.6 | 24 |
| 1.5 | 268 | 5.0 | 20 |
| 2.0 | 151 | 5.3 | 17 |
| 2.5 | 93 | 5.7 | 14 |

12. If all of the radiations from 1 $\mu$Ci $^{22}$Na were stopped in a conducting material and the total number of electrons produced were collected, compute the current of electricity that could be measured by a sufficiently sensitive ammeter.

Chapter 5

# Radiation Detectors

Radioactive nuclides are determined qualitatively and quantitatively by the interactions of their emitted radiations with materials used as radiation detectors. A radioactivity measurement system generally consists of two parts; the detector in which the radiations interact and the associated measuring device that presents the information from the detector in a suitable form.

Several types of radiation interaction mechanism were described in Section 4.3. Of these, only two, excitation and ionization, are widely used as the

**117**

basis for radiation detectors. These two processes are the primary means by which charged particles lose their energy in matter. Uncharged particles, such as gamma rays, are detected by the ionization interactions of secondary charged particles whose formation mechanisms were described in Section 4.3.3.

A third process, molecular dissociation, is involved in chemical and photographic methods of radiation detection. This process, although important in many aspects of radiation utilization and protection, is not generally used for quantitative measurements of radioactivated nuclides.

The second half of any general radiation measurement system is the electronic components that convert the signal from the detector into a usable form for data presentation. For detectors based on ionization of gases or semiconductors the signal consists of a charge of electricity. For scintillation detectors the signal is a quantum of fluorescent light that can be converted into a charge of electricity. Associated electronic instruments amplify the charges of electricity into easily measured current or voltage outputs. The output may be used to count the number of pulses in a given time period in a counter system, to be sorted by pulse amplitude in a radiation energy spectrometer, or to serve as a feedback signal in a process control system.

This chapter includes a discussion of the random errors in radiation measurement which are due both to statistical processes in the transfer and amplification of radiation energy in any detection system and to the statistical nature of the radioactivity decay process itself.

## 5.1  DETECTION METHODS

The two major radiation detection systems in modern practice employ detectors that are based on ionization of gases (e.g., proportional and G-M detectors), excitation of crystals which result in luminescence (e.g., scintillation detectors), or ionization of solids (e.g., semiconductor detectors). The principle of ionization in the semiconductor is similar to that in a gas, except that the charge is carried by electrons and electron vacancies (holes) in the crystal instead of electrons and positive ions in the gaseous atoms. Semiconductors are becoming the most widely used form of radiation detector as the related technology continues to develop.

Radioactivity measurement systems are generally of the pulse type; that is, the output of the detector is given as a series of electrical pulses separated in time. Each pulse represents the interaction of a unit of radiation with the detector. The nonpulse-type system, referred to as a mean-level detection system, is generally employed in survey-type radiation measurement devices and is not widely used for radionuclide measurement.

Modern radioactivity measurement systems not only determine the number of radiations detected per unit time (i.e., the counting rate) but they also allow, to varying degrees, the separation of the radiations by type and energy. Systems with which energy resolution can be made are called spectrometers and are available for alpha, beta, and gamma radiations, the gamma-ray spectrometer being especially useful for radionuclide analysis. In gamma-ray spectrometers the amplitude of each electrical pulse is proportional to the gamma-ray energy deposited in the detector. Electronic sorting devices (pulse-height analyzers) can separate these pulses by amplitude and frequency.

A multichannel pulse-height analyzer, operated in conjunction with an analog-to-digital converter, an appropriate memory device, and supporting data-presentation equipment is commonly used to perform analyses of complex mixtures of gamma-ray emitting radionuclides without the need for chemical separation of the mixture by element. When many routine analyses of the same type are required, the memory readout may be coupled to a computer for complete data processing. Electronic identification of specific radiations from several specific radionuclides may also be made by the use of coincidence measurements. For an event to be recorded as a disintegration of a specific radionuclide the interactions of two simultaneous or sequential radiations must be detected within a predetermined time period. Normally two detectors are required. The two radiations may be a beta ray and its subsequent gamma ray, as in 2.3-m $^{28}$Al, two successive gamma rays, as in $^{60}$Co, or the two annihilation photons that occur in any positron emitter. Thus the radionuclides that are most suitable for measurement by coincident means are generally those with the more complex decay schemes. These systems are discussed in Chapter 6.

## 5.2 GAS-FILLED DETECTORS

Gas-filled radiation detectors are among the oldest types of radiation detector available and are still widely used. The detector types include the ionization chamber, the proportional chamber, and the Geiger-Muller (G-M) tube. These types have in common a gas-filled chamber with a central electrode insulated from the chamber walls. A voltage applied to the central electrode creates an electrostatic field across the chamber so that ion pairs resulting from the ionizing radiations are accelerated toward the collection electrodes. Schematic diagrams of typical end-window proportional and G-M detectors are shown in Figure 5.1.

Collection of the ions produced by the ionization of the gaseous atoms in the chamber results in an electric current pulse. The current can consist of the free electrons or the ion pairs, that is, the electrons and positive ions. The behavior of the free electrons and positive ions depends on the properties of

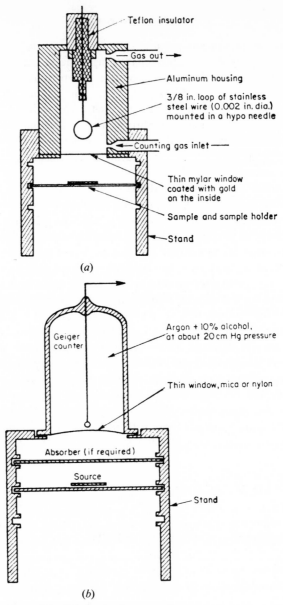

Figure 5.1 Schematic diagrams of gas-ionization detectors: (*a*) a gas-flow proportional chamber; (*b*) an end-window Geiger-Muller tube. [From Bernard G. Harvey, *Introduction to Nuclear Physics and Chemistry*, 2nd Ed. (Prentice-Hall Inc., Englewood Cliffs, N.J., © 1969). By permission of the publisher.]

the gas in the chamber and the electric field (voltage) applied. Since ionic mobilities, even in gases, are relatively slow compared with those of electrons, pulse ion-collection detectors are designed for rapid collection of the electrons.

It was noted in Section 4.3.2 that the average loss in energy of a charged particle in ionizing a gaseous atom was about 34 eV/ip. Thus, if all of the energy of a beta particle of energy $E_\beta$ were expended in causing ionizations in a gas-filled detector, the number of electrons (ion pairs) produced would be

$$n = \frac{E_\beta(\text{eV})}{34} \qquad (1)$$

If all of these electrons were collected at the central electrode, the charge would be

$$Q = en = 1.6 \times 10^{-19}\, n \qquad (\text{in C}) \qquad (2)$$

If $N$ such beta particles per second were completely stopped in the detector, the resulting current would be

$$I = QN = 1.6 \times 10^{-19}\, nN \qquad (\text{in A}) \qquad (3)$$

for example, a source of $N = 100$ betas per second entering the detector, expending an average energy of 1 MeV in the gas, would produce a current of

$$I = 1.6 \times 10^{-19} \times \frac{10^6}{34} \times 10^2 = 4.7 \times 10^{-13}\, \text{A}. \qquad (4)$$

Such a current is easily measured with an electrometer. Electrometers of various types can measure currents in the range $10^{-8}$ to $10^{-14}$ A and are the usual reading devices for mean-level current-type ionization chambers.

Ideally, the number of ions collected (current) in an ionization chamber is proportional to the number of like particles stopped in the chamber per unit time. However, the actual number of electrons (ion pairs) collected varies with the voltage applied to the chamber. At low voltages the collection of charge at the electrodes is in competition with the loss of ion pairs by recombination into neutral atoms. At high voltages the electrons freed in the primary ionizations caused by the radiation acquire sufficient kinetic energy during their acceleration toward the collecting electrode to cause additional ionizations (secondary electrons) which add to the collected charge. This process is called gas multiplication.

The relationship between the relative numbers of electrons (ion pairs) collected per event or the pulse height (i.e., the emf of the signal produced by the current of ions as it passes through a resistance in the circuit) and the applied voltage in a typical cylindrical chamber is illustrated in Figure 5.2 for

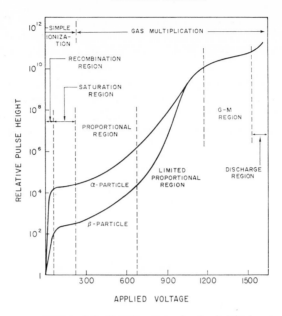

**Figure 5.2**   Relative size of pulse height in an
ionization chamber shown as a typical function
of the applied voltage for alpha and beta particles.

alpha and beta particles. The voltage range is conveniently divided into five
regions as marked. The first two regions constitute the range of primary or
simple ionization, whereas the last three regions cover the range of gas
multiplication in which the secondary electrons add to the collected charge.

In the recombination region the average velocity of the ions accelerated
toward the electrodes increases with voltage, the time available for recombination decreases, and the efficiency of charge collection increases. Therefore
the output pulse height increases with applied voltage.

In the saturation region the recombination losses become negligible
because the time required for collection of all ions in the chamber becomes
very short. The collected charge is given by (2), and the pulse height is
proportional to the energy of the radiation expended in the chamber. For a
given energy deposition the same number of ion pairs is collected. Thus the
pulse height is independent of the voltage applied. This region is also called
the ionization chamber region, since at these voltages the saturation current
is proportional to the mean energy deposited in an ionization chamber. The
ionization chamber is useful when operated as a mean-level radiation detection system. Also, since the path of alpha particles is relatively short, even in
gases, the ionization chamber may be used in conjunction with a pulse-height

analyzer as an alpha-qarticle spectrometer. Since neither of these operations involves measurement of radioactivated nuclides, the ionization chamber is not often used in activation analysis.

These first two regions shown in Figure 5.2 are the regions of simple ionization. When the applied voltage produces a field in excess of about 200 V/cm, the collected charge is increased by the process of gas multiplication in which the accelerated electrons acquire sufficient kinetic energy to produce additional ionizations by collisions. The multiplication factor in the proportional region increases rapidly with increasing applied voltage. Since it is independent of the initial ionization, the pulse size remains proportional to the intensity of the initial energy deposited. However, as the voltage is continually increased into the limited proportional region, the density of secondary charges interferes with the multiplication process. The differences in output pulse heights are no longer proportional to the initial ionizations. The relationship between pulse height and energy deposition slowly vanishes.

In the G-M region the detector produces a pulse of essentially constant size because the magnitude of the charge collected becomes independent of the primary ionization. A beta particle and an alpha particle produce the same final pulse size, regardless of the number of primary ionizations produced in the G-M tube. Thus the G-M tube is not useful as a detector for pulse-height analysis, but, because of its relative simplicity and large-amplitude output pulse, it is still a useful detector for radionuclide analysis of chemically separated elements.

As the voltage across a gas-filled detector is increased even further, the chamber operates as a continuous discharging tube unsuitable for radiation detection. Prolonged operation of a gas-filled chamber in the discharge region may seriously affect the chamber's usefulness as a radiation detection device.

### 5.2.1 The Proportional Chamber

The proportional chamber has become the most general type of detector for the measurement of beta radioactivity in solid or gaseous samples. The proportional chamber combines the advantages of the ionization chamber in maintaining proportionality between output pulse and primary ionization and the G-M tube in obtaining a sufficiently large gas-multiplication pulse for each event detected. Many commercial proportional counters are available.* They generally contain a bellshaped detector tube, illustrated in Figure 5.1a, through which a counting gas flows. A common counting gas is a mixture of 10% methane and 90% argon (P-10 gas). This particular gas mixture has

* See, for example, *Nucleonics*, **23**, No. 10, 84–89 (1965).

been found to be a good compromise between pure argon and pure methane. Pure argon is unsuitable as a counting gas because of the existence of a long-lived excited state of the ions which gives rise to after-pulsing, and higher concentrations of methane require operating voltages up to 4000 V, well above those shown in Figure 5.2. Other gases are equally good but also more expensive. Samples to be counted are generally placed on some form of disk or planchet and placed underneath the thin-window of the counting tube. Detectors are routinely available with end-window thicknesses as thin as as 80 $\mu$g/cm$^2$. Such ultra-thin windows, according to Figure 4.6, allow beta particles of $\sim$0.1 MeV energy to enter with a reduction in intensity of only 50%. For very low energy beta emitters, such as $^{14}$C ($E_\beta$ max = 0.156 MeV) and $^3$H ($E_\beta$ max = 0.018 MeV), windowless proportional chambers (i.e., the sample is placed directly into the counting chamber) may be used. Many of the problems associated with the end-window counting of solid samples on planchets are reviewed in Section 6.3.2. The proportional counter is used for the measurement of very low levels of radioactivity, those found in environmental samples or in radioactivated samples near the limit of sensitivity.

The term "low-background" counters generally refers to systems in which the normal levels of counter backgrounds are reduced substantially. The sources of counter backgrounds are due primarily to cosmic radiation, consisting of charged particles and secondary gamma and neutron radiations, and to gamma and beta radiations from radioactive materials present in the laboratory and detector and shield materials. These sources are reduced in low-background proportional counting systems by surrounding the detector with "thick" shielding of radioactivity-free materials and surrounding the detector within the shield with "guard" detectors that operate in "anticoincidence" with the sample detector. A high-energy cosmic ray which penetrates the shield and produces ionization in the sample detector would also produce ionization in a "guard" detector. The pulses produced by the two detectors cancel each other and the event in the main detector is not recorded. Only those events that occur solely with the main detector are counted.

. One additional advantage of a proportional chamber is its ability to measure alpha and beta radiations separately, even when the radiations are coming from the same sample. This capability results from the difference in specific ionization of alpha and beta particles, as shown in Figure 5.2. In the proportional region the pulse size depends markedly on applied voltage. By electronic discrimination of the pulse sizes the counting rate of a given type of radiation may be made independent of the applied voltage over a small ($\sim$200–300 V) voltage range. This range of voltage (occurring at about 1000 V for alpha radiation and at about 2000 V for beta radiation) is called "the plateau" of the detector. The operating voltage for the counter is set

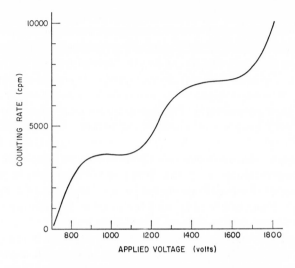

**Figure 5.3** Plateau curves for alpha and beta particles in a gas flow proportional counter.

at an appropriate value within this range. With voltage regulation, the counting rate of a radioactive sample is thus made independent of small changes in the line voltage. A typical plateau curve for a flow-type proportional counter is seen in Figure 5.3. To count the alpha radiation this counter would be operated at a voltage of about 1000 V. To count the beta radiation the counter would be operated at about 1600 V. The total counting rate at 1600 V includes the contribution from both the beta and alpha radiation. To obtain the counting rate of the beta activity alone the counting rate of the alpha radiation (as determined by extending the alpha plateau to 1600 V with the use of a pure alpha emitter) would be subtracted from the total count.

### 5.2.2 The Geiger-Muller Tube

The G-M tube is by far the simplest of the radiation detectors. Few adjustments of the counter system are needed for steady operation. However, the G-M tube has been gradually replaced by other types of detector because its intrinsic limitations frequently outweigh the advantage of simplicity of operation, particularly for radioactivation analysis applications. Nevertheless, G-M counters continue to exist in many laboratories, and in spite of the counting corrections required of them will continue to provide adequate measurements for many radioactivation requirements, especially when chemical separations are part of the procedure.

The more apparent advantages of the G-M tube are its high sensitivity, response to many different types of radiation, wide variety in size and shape, large output pulse signal, and low cost of supporting electronics. The form most widely used in radiochemical and radioactivation laboratories is the commercially available end-window G-M tube used with a shelf-arrangement to hold samples at fixed distances from the window, as shown in Figure 5.1b. They are normally placed inside a lead or steel shield with a door for access of samples to the shelf arrangement. The electronic circuitry consists of a high-voltage source, a discriminator, a scaler, and a timer. This total system is called a G-M counter.

The counting rate characteristics of a G-M counter are simpler than those of the proportional counter, since the G-M pulse is essentially independent of the primary ionization event causing the pulse. The plateau of a good G-M tube is generally about 300 V long, and the increase in counting rate with applied voltage is generally less than 3%/100-V increase.

One of the general limiting characteristics of G-M tubes is the relatively long recovery time required by the counter to distinguish between two successive discharges in the tube. This long "dead" time (around 250 $\mu$sec compared with about 0.5 $\mu$sec for a proportional counter) results in significant overlaps of ionizing events (coincidences) which, in turn, leads to significant count-rate losses for more intense radioactive samples. This effect becomes more important as the counting rate exceeds $10^4$ cpm. A correction for these "dead-time" losses may be determined from the resolving time $\tau$. If $n$ is the true counting rate for $\tau = 0$ and $m$ is the observed counting rate, then

$$n = \frac{m}{1 - m\tau} \tag{5}$$

for example, for a counter whose resolving time is 250 $\mu$sec a sample with observed counting rate $m = 525$ cps would actually have a true counting rate

$$n = \frac{5.25 \times 10^2}{1 - (5.25 \times 10^2 \times 250 \times 10^{-6})} = 605 \text{ cps} \tag{6}$$

the observed counting rate being low by more than 13%. In a proportional counter the corresponding loss would be less than 0.03%. G-M counting has in common with proportional counting the need to determine the over-all efficiencies for counting samples under end-window counters to convert the measured counting rate into absolute disintegration rates for the desired radionuclides. The over-all efficiency of the end-window detector system is dependent on many factors, some of which involve the detector, some the counting arrangement, and others the source itself.

## 5.3 SCINTILLATION DETECTORS

It is interesting to note that the scintillation detector, which contributed an important part in the development of radioactivation analysis, is one of the oldest types of detector used in radiation measurement. The *spinthariscope*, developed by 1908, was based on the fluorescence emitted when alpha particles interacted with thin films of zinc sulfide crystals. The interactions were observed with a microscope in a darkened room by "human counters" whose counting rate was limited, physiologically, to about 60 scintillations per minute. The development of electronic counters or scalers in the 1930's made the visual counter obsolete. The photomultiplier tube, emerging in the 1940's, initiated the advance of modern scintillation counting. The scintillation detector became important for gamma-ray spectrometry, especially in radioactivation analysis, with the discovery of specific high-density scintillating crystals sensitive to gamma radiations and the concurrent development of electronic instrumentation which can separate and sort the resulting electrical pulses by amplitude (pulse height).

### 5.3.1 Scintillation Detection Principles

The scintillation detector depends on the property of certain solid crystals which can dissipate energy resulting from ionization and excitation in the form of luminescence emission. These emissions of visible or ultraviolet light are generally in the form of fluorescence (with lifetimes of the order of $\sim 10^{-8}$ sec) or phosphorescence (longer wavelengths with lifetimes of the order of $\sim 10^{-4}$ sec). Scintillating crystals may consist of organic or inorganic compounds. The luminescence of organic compounds (e.g., anthracene) is an inherent property of the organic molecule, whereas in inorganic compounds the luminescence is a property of the crystalline state. Since almost all scintillation detectors used for radiation spectroscopy are made of inorganic crystals, our discussion is limited to such crystals. The band theory of solids formulated since 1928 has been applied to the conduction properties of metals, semiconductors, and crystalline insulators. These classifications are generally made by the difference in their electrical resistance. Approximate values are given in Table 5.1.

The band theory of solids considers the quantum-mechanic separation in a crystal lattice of the outer and inner electronic energy levels of the bound atoms. Although the inner electrons are tightly held to the nucleus of the atom, the outer electrons (the "valence" electrons) are sufficiently affected by the neighboring atoms to form a series of continuous "allowed" energy bands separated by ranges of "quantum-mechanically forbidden" energy values. A

**Table 5.1** Electrical Resistance of Solids

| Material | Resistivity ($\Omega$-cm) |
|---|---|
| Conductors | $\sim 10^{-5}$ |
| Semiconductors | $10^{-2}$ to $10^{9}$ |
| Insulators | $10^{14}$ to $10^{22}$ |

schematic diagram of the electronic energy-band system in an ionic crystal insulator is shown in Figure 5.4. In the ground state of the crystal the valence band is completely filled with electrons, whereas the conduction band is empty. The quantum separation in energy between these two bands is called the energy gap $E_G$ and is the minimum energy required theoretically for ionization of an electron from the valence band to the conduction band. In crystals of insulator materials the energy gap is sufficiently large that the number of electrons in the conduction band at room temperature is negligible. The transition of an electron from the valence band to the conduction band creates an electron vacancy or "hole" in the valence band and results in an electron-hole pair. Since the electron and the hole are no longer bound to the

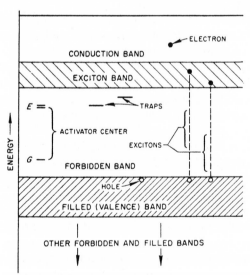

**Figure 5.4** The electronic energy band system in an ionic crystal insulator. [From R. B. Murray, Scintillation Counters, Chapter 2 in A. H. Snell, Ed, *Nuclear Instruments and Their Uses* (Wiley, New York, 1962).]

atom, both may migrate independently through the crystal lattice and contribute to the electrical conductivity in the crystal.

An electron may also be excited to an energy state lower than the conduction band in which it is still electrostatically bound to the hole in the valence band. This process of *excitation* results in an electron-hole pair (exciton) which has no net charge but can migrate through the crystal lattice. Figure 5.4 shows an excitation band of discrete excitation energies with a maximum at the bottom of the conduction band and a minimum corresponding to the excitation ground state. Radiation absorbed by an insulating crystal may result in both ionization and excitation. Recombination of electrons from the conduction band with holes in the valence band may result in further excitation.

The presence of lattice defects and impurities (activators) in ionic crystals produces local energy levels (centers and traps) in the forbidden region of the energy-band diagram, below the conduction band. An activator ion can exist in both the ground state and excited states, shown by $G -$ and $E =$ in Figure 5.4. The annihilation of an exciton may occur in three ways:

1. Luminescence, a transition to the ground state by light emission.
2. Quenching, a transition to the ground state by radiationless thermal dissipation of the excitation energy.
3. Traps, a transition to the conduction band by acquiring thermal energy or to the ground state by radiationless thermal dissipation.

The conditions for luminescence of a center are given by the potential energy of the ground and excited states of the molecule as a function of some configuration coordinate $r$, as shown in Figure 5.5. An impurity center may be raised to the excited state by absorption of radiation (or by capture of an exciton). The absorption process is shown by the transition $A \to B$, which, according to the Franck-Condon principle occurs in a time short compared with atom movements. This excited state seeks minimum potential energy, migrating to position $C$ by thermal dissipation of the excess energy. The luminescent photon is emitted as transition $C \to D$, following which the center returns to minimum energy in the ground state by further thermal dissipation of the excess energy. The luminescence emission spectrum is a band rather than a sharp line because of the thermal fluctuations about the potential energy minimum at $C$. De-excitation along the path $C \to E \to A$ in which the energy $C \to E$ is provided by thermal excitation is also possible. In this case, internal quenching, no radiation occurs. The probability of this process $K_i$ is dependent on temperature and activation energy $E = \Delta E(C \to A)$ by

$$K_i = ae^{-E/kT} \tag{7}$$

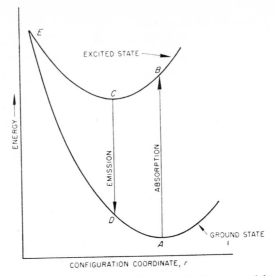

**Figure 5.5** Schematic diagram of the potential energy levels of an activator center showing one of many possible excited states. [From R. B. Murray, Scintillation Counters, Chapter 2 in A. H. Snell, Ed., *Nuclear Instruments and Their Uses* (Wiley, New York, 1962).]

and the luminescence quantum efficiency $q_0$ may be defined as

$$q_0 = \frac{k_f}{k_f + k_i} = \frac{1}{1 + be^{-E/kT}} \tag{8}$$

where $k_f$ = probability for luminescence emission,

$k_i$ = probability for the competing internal quenching,

$k$ = Boltzmann constant,

$a$ and $b$ = constants.

The luminescence emitted by a scintillating crystal following excitation by absorbed radiation generally follows an exponential decay with time:

$$I = I_0 e^{-t/\tau} \tag{9}$$

where $\tau$ is the decay time.

Inorganic crystals that exhibit useful scintillating properties are termed phosphors, some of which occur as pure crystals (e.g., diamonds, alkali halides, uranyl salts). Others are self-activating; that is, heat treatment produces an excess of one of the ions in interstitial positions in the crystal lattice which act as luminescent activators. Such crystals include ZnS with excess Zn, CdS with excess Cd, and ZnO with excess Zn.

Most scintillation crystals used for radiation detection are impurity-activated and are generally alkali halides activated by heavy metals such as thallium, europium, and lead. Although scintillation detection has become the primary measurement method for gamma radiation, scintillation-detection methods are also useful for counting alpha particles (using the original scintillating material zinc sulfide) or beta particles (especially in the absence of coincident gamma radiation) by liquid or thin-plastic scintillating organic compounds. The gas-scintillation method has been developed for counting heavy charged particles. The noble gases Xe and Kr are generally used as the scintillating gas, but these detectors have not yet proved useful for the routine measurement of radioactive nuclides.

The basic functions of the inorganic-crystal scintillation detector (consisting of a scintillation crystal optically coupled to a photomultiplier tube) are shown in Figure 5.6. The operation of the detection system may be considered to proceed in five stages:

1. Absorption of the gamma-ray energy of the scintillator.
2. Conversion of the absorbed energy into photons of luminescence.
3. Migration of the direct and reflected photons to the photocathode of the photomultiplier tube.
4. Emission of several photoelectrons which are accelerated to the first dynode.
5. Electron multiplication to produce a measurable pulse at the anode of the photomultiplier tube.

The energy of the incident photon deposited in the crystal is converted into photons of light of specific wavelength. These photons are reflected from the walls of the crystal container, generally made of $Al_2O_3$ or $MgO$, until they pass through the optical coupling and strike the photocathode, where they produce a number of photoelectrons. The photocathode is made of semiconductor materials such as $SbCs_3$ or $BiCs_3$ which have high photoelectric quantum efficiency of the order of 0.2 electrons per photon. The photomultiplier tube is designed to focus and accelerate the primary photoelectrons through a series of multiplier electrodes (dynodes). Each dynode is maintained at a voltage of 75 to 150 V greater than the preceding one, and for each electron from the photocathode the average multiplication per dynode is about a factor of 4. Thus for a 10-dynode photomultiplier tube the amplification of the current of photoelectrons from the photocathode is about $(4)^{10}$ or about $10^6$. The output current at the anode flows through a load resistor $R_L$ which gives the negative voltage drop pulse shown in Figure 5.6. This pulse is coupled through the blocking capacitor $C$ to the supporting electronics.

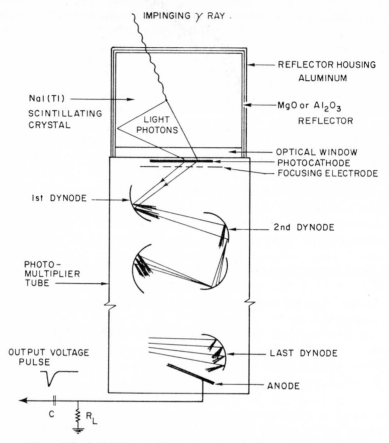

**Figure 5.6** A NaI(Tl) scintillation crystal optically coupled to a photo-multiplier tube. The absorption of an impinging gamma-ray in the detector results in an output voltage pulse proportional to the incident energy absorbed.

### 5.3.2   The NaI (Tl) Scintillation Detector

Most inorganic materials used as scintillators for measuring x- and gamma radiations are alkali halides. The most common compound, available commercially in many sizes and shapes, is the NaI crystal activated with about 0.1 % thallium. The thallium is added in the form of thallous ions (Tl⁺) to increase the scintillation efficiency at room temperatures and to shift the wavelength of the fluorescence photons to about 4200 Å, compatible with the response of standard photomultiplier tubes. CsI is sometimes used for special purposes. NaI(Tl) has several attractive properties. The crystal has a high density (3.67 g/cm³) for efficient absorption of the gamma radiation.

Iodine provides a high atomic number for efficient output of light per unit of gamma radiation absorbed. The gamma-ray attenuation coefficients for NaI are shown in Figure 5.7. The attenuation coefficients for the photoelectric effect and Compton scattering become equal at about 0.3-MeV energy, and pair production is not important for gamma rays with energy less than 2 MeV. NaI(Tl) can be grown into large single crystals (>8-in. diameter) and is very transparent to its own fluorescent light. These crystals have the added advantage of a very short fluorescence decay time (~0.25 $\mu$sec), which allows counting of high activity samples with small dead-time losses.

NaI(Tl) is hygroscopic (it absorbs water vapor from the air), and therefore the crystal (and the light-reflector surface) must be hermetically sealed in a closed container with an optical couple to the photomultiplier tube. Another important restriction in the preparation of crystal-phototube assemblies is the avoidance of potassium, which contains the naturally occurring radio-isotope $^{40}$K, and of other radioactive elements in any of the construction materials to reduce the detector background.

The interactions of gamma rays with matter were shown in Section 4.3.3 to involve some combination of photoelectric absorption, Compton scattering, and (for gamma rays with $E_\gamma > 1.02$ MeV) pair production. The

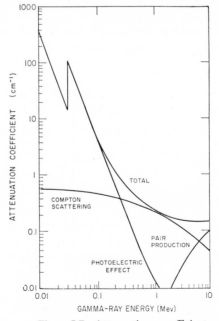

**Figure 5.7** Attenuation coefficients for NaI as a function of gamma-ray energy.

attenuation coefficients for these interactions in NaI(Tl) scintillation crystals are shown in Figure 5.7. A source of monoenergetic gamma rays produces in a NaI crystal a spectrum of energy depositions according to the individual interactions of the gamma rays and the crystal; for example, a source of $^{85}$Sr, with $E_\gamma = 0.514$ MeV, would produce photoelectrons and Compton electrons with a ratio of about 1 to 6 in an "ideal" NaI(Tl) scintillation detector system, according to Figure 5.7. The theoretical gamma-ray spectrum obtained in such an "ideal" detector is shown in Figure 5.8; the delta function at $E_0 = 0.514$ MeV corresponds to the total absorption of the gamma ray, and the continuous function of energies corresponds to Compton scatterings from the Compton edge [given by (4.46)] to the minimum detectable. The actual spectrum obtained in a "real" NaI(Tl) scintillation detector is also shown in Figure 5.8. The broadening of the photopeak line is due to both the fluctuations in light output from the scintillator and the number of photoelectrons formed at each dynode in the photomultiplier tube. The amplitude of the photopeak is also increased somewhat by the probability that a Compton-scattered gamma ray may undergo additional Compton scatterings, resulting in the absorption of the total gamma-ray energy. Thus the total peak (photopeak) amplitude is expected to increase as the size of the detector increases because of the increased probability for multiple events. Other features of gamma-ray spectra are examined in Section 6.3.3.

The quality of a scintillation detection system is determined by several

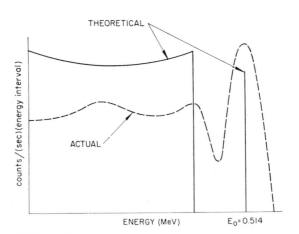

**Figure 5.8** Theoretical and actual gamma-ray spectra of $^{85}$Sr ($E_\gamma = 0.514$ MeV) resulting from Compton and photoelectric interactions in a NaI detector. [from R. L. Heath, Scintillation Spectrometry, USAEC Report IDO-16880-1, 2nd ed., 1964.]

parameters:

1. The detection efficiency.
2. The linearity of pulse height with incident gamma-ray energy.
3. The resolution of the detector-phototube assembly.
4. The background of the system.

The scintillation efficiency (i.e., the ratio of emitted photon energy to incident radiation energy) of NaI(Tl) is the largest of the alkali halide phosphors, about 12%, and about twice as efficient as anthracene, a good organic scintillator. The detection efficiency of a scintillation detector (i.e., the ratio of counting rate to disintegration rate of the radionuclide source) is dependent on the size and shape of the crystal and the geometrical relationship of the source. Commercial NaI(Tl) detectors come in a variety of sizes, mostly cylindrical, coupled to matching photomultiplier tubes. Many systems used for counting gamma-ray emitting radionuclides have well-type crystals, into which vials containing the active sample in the form of a solution or precipitate can be placed. A standard scintillation detector, a 3-in. diameter × 3-in. deep flatfaced NaI(Tl) detector, used to prepare standard gamma-ray spectra* is shown in Figure 5.9. The well-type detector is sometimes more convenient in activation analysis, especially when used in conjunction with radiochemical separations.

The resolution of a detection system is a measure of the ability of the detector system to produce a single pulse-height value for monoenergetic radiation totally absorbed in the detector. The resolution or line width in percent, $W_{1/2}$, is defined as the width of the total energy peak at half of the peak maximum.

$$W_{1/2} = \frac{\Delta E}{E_\phi} \times 100\% \tag{10}$$

where $E_\phi$ = the total gamma-ray energy (maximum pulse height) value
$\Delta E$ = the energy (pulse height) interval for the full width at half the maximum value of the peak (FWHM).

The values of $\Delta E$ and $E_\phi$ may both be expressed in energy or volts. Since the fraction of incident gamma rays that is totally absorbed in a given size detector decreases with gamma-ray energy, the resolution also changes with gamma-ray energy. Thus the resolution value of a detector must be given for a particular gamma-ray energy. The 0.662-MeV gamma ray of 30-y $^{137}$Cs is generally used to measure NaI(Tl) scintillation detector resolution. Resolution values for $^{137}$Cs generally range from 7 to 9% for flat-crystal detectors and from 10 to 12% for well-crystal detectors.

* R. L. Heath, Scintillation Spectrometry, USAEC Report IDO-16880, 2nd ed., 1964.

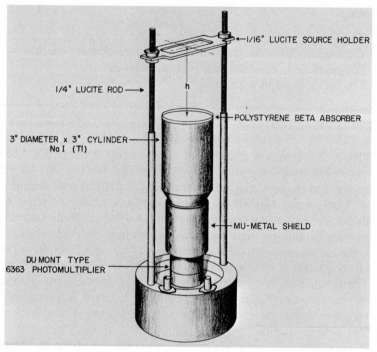

**Figure 5.9** A 3 in. dia. × 3 in. deep flat-faced NaI(Tl) scintillation detector used to obtain standard gamma-ray spectra. (Courtesy of R. L. Heath, Idaho Nuclear Corp.)

### 5.3.3 Liquid Scintillation Detectors

Another means of scintillation counting is sometimes used in activation analysis. Organic compounds that scintillate in the liquid phase are useful for counting radionuclides that decay solely by low-energy beta emission. Two such examples are $^{35}$S ($E_\beta$ max = 0.168 MeV) and $^{45}$Ca ($E_\beta$ max = 0.25 MeV). In liquid scintillation counting the sample is incorporated directly into the scintillator solution by one of three common methods: (a) directly, if it is soluble in the organic scintillator, (b) in miscible alcohol solution if it is soluble in water, or (c) by grinding to achieve a fine suspension if it is insoluble in either one.

Common organic-liquid scintillators include toluene, xylene, terphenyl, and other aromatic hydrocarbons. These organic compounds have the advantage of fast response and ease of use. The sample must be incorporated into the scintillator in a chemical form that does not quench the fluorescence. Since the fluorescence spectra for these compounds are in the far violet, modern scintillator solutions include a secondary phosphor which shifts the wavelength to those detected more efficiently by photomultiplier tubes.

When quenching occurs to a significant extent, internal standards are sometimes introduced for absolute counting. The original samples cannot then be recounted. To reduce the noise level of the photomultiplier due to thermionic emission the liquid scintillation counter is generally shielded in a refrigerator, and for additional improvement in efficiency two photomultiplier tubes in coincidence circuits are sometimes used for counting one solution. Both phototubes must receive light from the fluorescence event simultaneously (within 1 $\mu$sec) to be recorded. Since the thermal noise pulses are random in either phototube, many of them are rejected by the coincidence circuit.

Another method for scintillation counting of low-energy beta emitters is the use of thin plastic scintillating wafers in place of a liquid scintillator. The use of plastic scintillators offers several novel ways for counting; for example, the liquid sample can be intermixed with plastic beads for intimate contact, poured into a plastic scintillator dish mounted on the phototube, or run through a plastic scintillator tube spiraled around the phototube for flow counting. Such techniques may be of interest for activated nuclides in special-design problems.

## 5.4  SEMICONDUCTOR DETECTORS

The discovery of the transistor in 1948 led not only to the rapid development of solid-state electronic devices for the electronic counting of pulses from conventional radiation detectors but also to a new type of radiation detector. The semiconductor radiation detector has in common with gas detectors the absorption of incident radiation by direct ionization. The difference lies primarily in the kind of ionization produced; in the gas counter electron-ion pairs are created, whereas in the semiconductor electron-electron hole pairs are created. The semiconductor detector has become quite useful in activation analysis, especially for gamma-ray and conversion-electron energy spectroscopy.

### 5.4.1  Semiconductor Principles

A semiconductor is generally defined as a material whose resistance to electricity is between that of insulators and conductors. Approximate values of their resistivity are given in Table 5.1.

The properties of semiconductors are explained by the quantum mechanical considerations of the band theory of solids described in Section 5.3.1. The difference in conductivity of semiconductors is determined by the number density of electron transitions from the *valence* band (the band of electrons most loosely bound to individual atoms in the solid) to the *conduction* band (the band of electrons free to move within the solid). In semiconductors the valence band is almost filled with electrons, whereas the conduction band is nearly empty. A schematic diagram of the energy bands of an intrinsic semiconductor is shown in Figure 5.10a. Some typical energy gap values at 300°K are 0.66 eV for germanium, 1.08 eV for silicon, and 1.50 eV for selenium. However, the minimum energies of radiation particles required to produce such transitions are about 3.6 eV for silicon and about 2.8 eV for germanium, the only two semiconductors commonly used as radiation

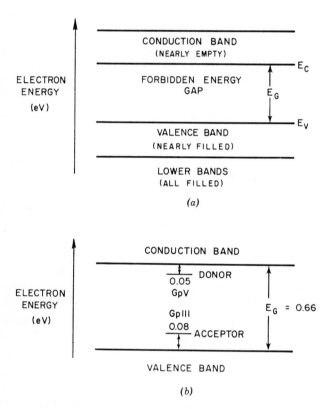

**Figure 5.10** (a) Energy band diagram of an intrinsic semiconductor material. The energy gap $E_G = E_C - E_V$ is 0.66 eV for germanium and 1.05 eV for silicon, at 300°K. (b) Impurity energy levels in germanium showing the donor and acceptor impurity levels relative to the energy gap for group-V and group-III elements, respectively.

**Table 5.2** General Properties of Silicon and Germanium Semiconductors

|  | Silicon | Germanium |
|---|---|---|
| Atomic number, $Z$ | 14 | 32 |
| Atomic weight, $M$ | 28.09 | 72.59 |
| Density, g/cm$^3$ | 2.33 | 5.32 |
| Dielectric constant, $k$ | 12 | 16 |
| Energy gap, at 300°K, eV | 1.08 | 0.66 |
| Energy required per e$^-$ + hole pair, eV | 3.6 | 2.8 |
| Electron drift mobility at 300°K, $\mu$(cm$^2$/V-sec) | 1,450 | 3,800 |
| Electron drift mobility at 77°K, $\mu$ | 21,000 | 36,000 |
| Hole drift mobility at 300°K, $\mu$ | 480 | 1,800 |
| Hole drift mobility at 77°K, $\mu$ | 11,000 | 42,000 |
| Intrinsic resistivity at 300°K, $\Omega$-cm | $2.3 \times 10^5$ | 47 |
| Intrinsic carrier concentration at 300°K, cm$^{-3}$ | $1.5 \times 10^{10}$ | $2.35 \times 10^{13}$ |

detectors. The general properties of these two chemical group IVb elements are listed in Table 5.2.

The transition of an electron from the valence band to the conduction band creates an electron vacancy or "hole" in the valence band and thus the number of electrons in the conduction band and the number of holes in the valence band of an intrinsic semiconductor are equal. At a given temperature the rate of electron transitions and the rate of "recombination" of electrons and holes reach an equilibrium; the concentration of electrons, n(cm$^{-3}$), and holes, p(cm$^{-3}$) is given by

$$n = p = N(E)e^{-E_G/2kT} \qquad (11)$$

where $k$ is the Boltzmann constant and $T$ is the absolute temperature. At 300°K, $kT = 0.026$ eV; $N(E)$, the density of states, is the number of allowed energy states per unit energy interval, which is a property of the band structure of each semiconductor and is approximately equal to $10^{19}$. Thus at 300°K n and p are both about $1.5 \times 10^{10}$ cm$^{-3}$ for silicon and about $2.35 \times 10^{13}$ cm$^{-3}$ for germanium.

The electrons in the conduction band and the holes in the valence band both contribute to the electrical conductivity of the semiconductor and are called *carriers*. If an electric field $E$ is applied across the crystal, the electrons and holes are both accelerated and the total current density $J$ is given by

$$J = Ee(n\mu_n + p\mu_p) \qquad (12)$$

where $e$ is the electron charge, $\mu_n$ is the mobility of the electrons in the conduction band, and $\mu_p$ is the mobility of the holes in the valence band. Some typical values for electron and hole-drift mobilities are given in Table 5.2.

Intrinsic semiconductors are difficult to produce as chemically pure crystals. Impurities or imperfections in the crystal lattice structure introduce another source of carrier excitations, in which the number of electrons and holes need not be the same. The major semiconductors are the group IV elements, such as silicon and germanium, which crystalize in the diamond structure with chemical valence of 4. Atoms of the group V elements may substitute as impurities in the lattice structure. Four of the valence electrons of each group V element can enter the valence band of the crystal, but the fifth valence electron will be loosely bound to its parent atom. In semiconductors the energy level of such electrons lies in the energy gap close to the level of the conduction band. Thus a small excitation energy (small compared with the energy gap) can result in the transition of the electron from the impurity atom to the conduction band. In this case, however, the hole left by the electron is not mobile and not a current carrier. Such impurities are called *donors*, since they donate an excess of negative electron carriers, and the resulting crystal is called a n-type semiconductor.

Impurities from the group III elements introduce the inverse effect; that is, they create holes in the valence band, for they lack valence electrons to complete the lattice bonds. These holes may be filled by electrons with energy excitation just above the valence-band level, thus leaving holes in the valence band without corresponding mobile electrons in the conduction band. Such impurities are called *acceptors*, since they accept electrons that leave an excess of positive hole carriers. The resulting crystal is called a p-type semiconductor. The energy level diagram of these donor and acceptor impurities is shown in Figure 5.10*b*.

Defects in a pure crystal, which result from vacant sites or interstitial atoms, create the same type of donor and acceptor. Real crystals always have both donor and acceptor impurities or defects in them, and since each cancels the effect of the other the type of semiconductor is determined by the one present in excess. The occurrence of these impurities and crystal defects is important to the mobility of the electrons and holes in the semiconductor.

## 5.4.2 Semiconductor Radiation Detectors

The use of semiconductors as detectors for ionizing radiations is based on the production of electron-hole pairs which are free-charge carriers and the collection of the liberated charge by application of an electric field. Two general features of semiconductors important in their use as radiation detectors are the charge-collection time and the electrical capacity. The charge-collection time $\tau$ is given approximately by

$$\tau = \frac{w^2}{\mu V} \tag{13}$$

where $w$ = the sensitive depth of the detector,
$\mu$ = the carrier mobility,
$V$ = the applied voltage.

The electrical capacity must be low to give a good signal-to-noise ratio. The capacity (in picofarads) is about

$$C = 1.1 \frac{kA}{4\pi w} \tag{14}$$

where $k$ = the dielectric constant for the semiconductor,
$A$ = the detector area (cm²).

Since the charge-collection time is proportional to $w^2$ and the capacitance to $1/w$, these two features are in conflict. The values of these parameters are usually optimumized for detection of specific types of nuclear particles.

The semiconductor has several advantages over the gas-filled detector, two major ones being the larger number of electron-hole pairs produced in the solid compared with the electron-ion pairs produced in the gas and the higher density of the solid. Both advantages result in potentially higher resolution in the measurement of the incident particle energy. The collection of the liberated charge in the solid is more difficult, however, than in the gas. In addition to the losses of electron holes by recombination, the impurities and crystal defects may act as centers for trapping the moving carriers before their collection by the electric field. Further, the use of an electric field for charge collection requires the detector to be an insulator (i.e., with a very high resistivity); otherwise the resulting high currents would not only lead to excessive electronic noise but the fluctuations in the current might be greater than the small current pulses of the collected electron-hole pair charges. Thus the major consideration in the use of semiconductors as radiation detectors is the application of a high electric field without excessive leakage currents. Several methods have been developed to achieve this condition for semiconductor crystals; the three major ones are the p-n junction detector, the surface barrier detector, and the lithium ion-drift detector.

### 5.4.3 The p-n Junction Detector

A semiconductor diode used as a radiation detector is illustrated in Figure 5.11. This diode is called a diffused n⁺p junction, the n⁺ signifying a semiconductor "doped" with a donor impurity. A typical detector consists of a

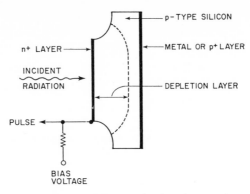

**Figure 5.11** A diffused n⁺p junction semiconductor diode used as a radiation detector.

thin ($\sim 1\mu$) layer of n⁺-type silicon (doped with phosphorous, which gives a
high concentration of free electrons with almost no free holes) in a slab of
p-type silicon (with an excess of holes and very few electrons). The application
of the reverse bias voltage (one that puts the + charge on the n-type and the
− charge on the p-type side) sweeps the free electrons and holes out of a layer
between the n and p regions to form a depletion layer. The width of the
depletion layer $w$ is given by

$$w = K(\rho V)^{\frac{1}{2}} \tag{15}$$

where $K$ = a parameter of the semiconductor,
$\rho$ = the resistivity ($\Omega$-cm),
$V$ = the applied voltage (V).

For p-type silicon $K \simeq 3.2 \times 10^{-5}$, and with $\rho = 10^4 \, \Omega$-cm and $V \simeq 300$ V
depletion layers of the order of 0.1 cm are achieved.

This depletion layer becomes the sensitive region of the radiation detector.
The number of electron-hole pairs produced in the depletion region is proportional to the energy expended by the incident radiation in the region. Since
the reverse bias voltage has removed the excess electrons and holes, the
collection of the electron-holes produced by the ionizing radiation results in
a useful output pulse. The measurement of this pulse by an external circuit
can serve as the basis for a radiation energy spectrometer.

An alternate method to the reverse-bias voltage detector is the surface
barrier detector in which a n-type silicon or germanium wafer is oxidized at
the surface to form a thin oxide layer. The junction is close enough to the
surface so that with the use of a metal coating as an electrode the wafer acts
as a surface barrier detector in a manner similar to the diffused junction
detector, except the polarity of the applied voltage is reversed. Figure 5.12

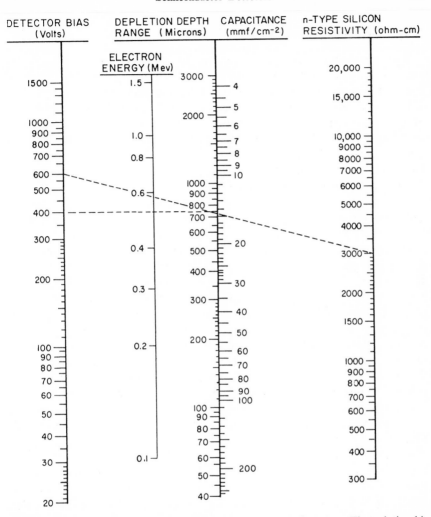

**Figure 5.12** Nomograph for n-type surface-barrier type detectors. The relationship between depletion depth and capacitance is given for detector resistivity and applied voltage. The horizontal intercept on the energy scale gives the electron energy that could be completely stopped in the depletion depth. [Adapted from J. Blankenship and C. Borkowski, *IRE Trans.* **NS-7,** 192 (1960).]

shows a nomograph for n-type surface-barrier detectors which gives the relationship between depletion depth and capacitance as functions of detector resistivity and applied voltage. The electron energy that can be completely stopped in a given depletion depth can also be determined from the nomograph.

### 5.4.4   The Lithium Ion-Drift p-i-n Junction Detector

With the increased emphasis in semiconductor development for electron and gamma-ray spectroscopy, the lithium ion-drift p-i-n junction semiconductor is gaining popularity for such radiation measurements. The advantage of the lithium ion-drift detector results from the high mobility of lithium ions in silicon and germanium. The depletion layer may be increased significantly in depth (to more than 1 cm), since the resistivity of the wafer can be increased without loss in charge collection. The increased thickness of the depletion layers also gives a significantly increased efficiency for the absorption of electrons and gamma rays.

Lithium, as an interstitial impurity, acts as a *donor* in silicon or germanium, and as an ionized impurity drifts in an electric field. At elevated temperatures (100 to 400°C) the drift velocity is greater than the diffusion velocity.

The preparation of a lithium ion-drift detector is illustrated in Figure 5.13. A p-type crystal of silicon or germanium with uniform concentration of acceptors $N_A$ is coated on one side with lithium which diffuses into the crystal forming a p-n junction inside the crystal at the interface, where $N_D = N_A$ (Fig. 5.13a). Now if the crystal is reverse-biased and the temperature is raised so that the ion drift in the depletion region is large compared with the diffusion, the lithium ions drift from the donor rich region, balancing the concentrations $N_A = N_D$ for some distance into the crystal and producing a layer of an "intrinsic" semiconductor between the p-type and $n^+$-type regions (Figure 5.13b). At room temperature, and under reverse bias, the residual carriers are swept from the intrinsic region, and the crystal is operable as a radiation detector (Figure 5.13c). Such techniques can produce crystals with intrinsic layers more than 1 cm thick and yield detectors useful for electron and gamma-ray spectroscopy. Germanium, with its higher efficiency for the gamma-ray photoelectric effect, is useful for gamma-ray spectroscopy. However, because at room temperature lithium ion diffusion is still appreciable in germanium, such detectors are usually cooled to liquid nitrogen temperatures, ~77°K.

The advantage of the semiconductor detector is its potential for energy resolution. The line width of a full-energy peak, expressed in terms of full width at half maximum (FWHM) is

$$\omega_i = 2.36(E\epsilon)^{1/2} \tag{16}$$

where $E$ = radiation energy (eV),

$\epsilon$ = energy in eV to produce an electron-hole pair;

for example, a gamma-ray from $^{137}$Cs fully absorbed in germanium would have a line width of

$$\omega_i = 2.36 \times (0.662 \times 10^6 \times 2.9)^{1/2} = 3.3 \text{ keV}$$

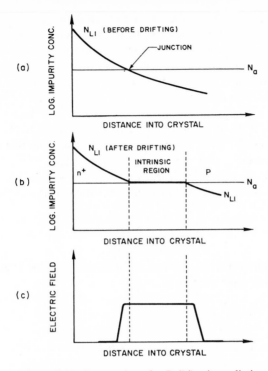

**Figure 5.13**  Preparation of a Ge(Li) n-i-p radiation detector with profiles for acceptor impurities $(N_a)$ and lithium donor impurities $(N_{Li})$: (a) before the drifting process; (b) after the drifting process; (c) electric field distribution with applied reverse bias. [Adapted from W. J. Price, *Nuclear Radiation Detection*, 2nd Ed. (McGraw-Hill, New York, 1964), p. 231.]

The line width of a scintillation crystal with an excellent resolution of $7\%$ is 46 keV. The energy per ionizing event of 2.9 eV for germanium may be compared with 30 eV required to produce an ion pair in a gas ionization chamber and with 500 eV required to produce a photoelectron at the photocathode of a scintillation detector.

## 5.5  STATISTICS OF RADIOACTIVITY MEASUREMENT

Radioactive decay, like other nuclear processes, was shown at an early date to be a random process and therefore subject to statistical fluctuations. In fact, the exponential decay law, derived as a first-order kinetic process in

Section 1.3.4, was initially derived from probability considerations and does not require any apriori knowledge of the decay process itself.

The derivation is based solely on the assumption that for a population $N^0$ of a given radionuclide the probability of any one atom decaying in a small time interval $\Delta t$ is a constant $\lambda$ and that $\lambda$ is the same for all successive time intervals $\Delta t$. For the time interval $\Delta t$ (small compared with $1/\lambda$) the probability that a particular atom will decay is $\lambda\,\Delta t$. Its chance of surviving during $\Delta t$ is $1 - \lambda\,\Delta t$. If it survives that period, its chance of surviving the next period $\Delta t$ is also $1 - \lambda\,\Delta t$, and by the law of compounding probabilities its chance of surviving two periods $\Delta t$ is $(1 - \lambda\,\Delta t)^2$. For a finite period $t = n\,\Delta t$ the total probability for survival is $(1 - \lambda\,\Delta t)^n = (1 - \lambda t/n)^n$. This expression can be expanded in a series in which the limit as $n$ approaches infinity is

$$\lim_{n \to \infty} \left(1 - \frac{\lambda t}{n}\right)^n = e^{-\lambda t} \tag{17}$$

Since the probability for survival is $e^{-\lambda t}$ for each atom, the average fraction $N/N^0$ that survives after a time $t$ is

$$\frac{N}{N^0} = e^{-\lambda t} \tag{18}$$

which is the same equation derived from kinetic theory.

### 5.5.1  Errors in Radioactivity Measurement

In physical measurements two types of error (other than gross mistakes) determine the quality or reliability of the data.

1. Systematic errors, introduced by experimental conditions that cause disagreement between the individual measured values and the "true" value. An example of a systematic error in radioactivity measurement, which can occur frequently, is an incorrect calibration of the efficiency of the counter as a function of the radiation energy, sample thickness, or other counting condition. The accuracy of a radioactivity measurement is determined primarily by the success with which systematic errors are reduced or eliminated.

2. Random errors, introduced by either variabilities in the experimental conditions or inherent fluctuations in the process being measured. These errors cause repeated measurements under apparently fixed conditions to differ from one another. An example of a random experimental error is the change in measurement caused by surges in the line voltage or changes in the background of a counter while a sample is being counted. Such errors can be controlled to a considerable degree by improving the counting system. The

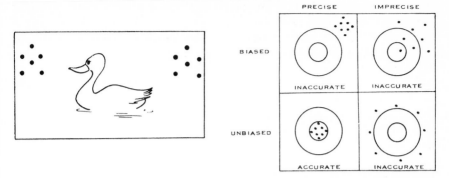

**Figure 5.14** (*a*) "On an average the duck was dead." A hunter fired both barrels of a shotgun at a duck. The first hit 2 ft in front, the second hit 2 ft behind. On an average the duck was dead. What the hunter really wanted was meat on the table. In duck hunting one must keep trying until a single shot hits the mark, but in estimating the activity of a radioactive source the best estimate is usually the average. [Adapted from G. D. Chase and J. L. Rabinowitz, *Principles of Radioisotope Methodology* (Burgess, Minneapolis, 1962).]

random errors due to the statistical nature of radioactive decay are fixed by probability distributions of the process. The precision of a radioactivity measurement is determined primarily by the success with which the random experimental errors are controlled; the ultimate precision is determined by the statistics of the radioactivity decay process itself.

An interesting cartoon illustrating the relation between the terms *bias*, *precision*, and *accuracy* is shown in Figure 5.14. It should be noted that the term *inaccurate* is somewhat ambiguous in that it implies the presence of systematic and/or random errors. The term *accurate* is not ambiguous, since it implies that both systematic and random errors are small.

### 5.5.2 The Statistical Distribution for Radioactive Decay

Since the process of radioactive decay occurs at random intervals of time, the number of decays $N$ occurring in any finite time interval $t$ is subject to statistical fluctuations, which give rise to an error in the observed counting rate $R = N/t$. The probability law that governs the statistical fluctuations of radioactive decay is the Poisson distribution. The Poisson distribution is derived from the binomial distribution, which is the fundamental statistical law that governs random events. For large sampling (counts) the Poisson distribution becomes equivalent to the Gaussian (normal) distribution.

#### The Binomial Distribution

The binomial law, in the general case, considers a large population of two kinds, in which the probability of selecting a member of the first kind from the population is $p$. Then the probability of selecting a member of the second kind is $1 - p$ for each selection event. The probability $W(n)$ that exactly $n$ members of the first kind will be selected from a total population $N$ of both kinds is given by the binomial law as

$$W(n) = \frac{N!}{(N - n)! \, n!} \, p^n (1 - p)^{N-n} \tag{19}$$

The true average value $\bar{x}_t$ of the binomial distribution is given by

$$\bar{x}_t = \sum_{n=0}^{n=N} n \, W(n) = pN \tag{20}$$

and the variance by

$$\sigma^2 = Np(1 - p) = \bar{x}_t(1 - p) \tag{21}$$

The standard deviation is

$$\sigma = \pm [\bar{x}_t(1 - p)]^{1/2} \tag{22}$$

For radioactive decay we can consider that $N$ is the total number of radioactive atoms in the radionuclide source and that $n$ is the number of disintegrations that occur in counting time $t$. From Section 5.5 the probability of an atom not decaying in time $t$ is $1 - p = e^{-\lambda t}$ and therefore the probability for decay is $p = 1 - e^{-\lambda t}$. Equation 19 for radioactive decay is thus

$$W(n) = \frac{N!}{(N - n)! \, n!} \, (1 - e^{-\lambda t})^n (e^{-\lambda t})^{N-n} \tag{23}$$

The true average number decaying in time $t$ is

$$\bar{x}_t = N(1 - e^{-\lambda t}) \tag{24}$$

and the variance from (21) is

$$\sigma^2 = N(1 - e^{-\lambda t})e^{-\lambda t} = \bar{x}_t e^{-\lambda t} \tag{25}$$

The standard deviation is

$$\sigma = \pm [\bar{x}_t e^{-\lambda t}]^{1/2} \tag{26}$$

If the counting time $t$ is short compared with the half-life, as is usually the case, then $e^{-\lambda t} \simeq 1$ and the standard deviation is

$$\sigma = \pm \sqrt{\bar{x}_t} \tag{27}$$

Further, if a sufficiently large number of counts $C$ is obtained, then $\bar{x}_t$ may be replaced by $C$ and the standard deviation is

$$\sigma = \pm \sqrt{C} \tag{28}$$

### The Poisson Distribution

For $p \ll 1$ (i.e., $\lambda t \ll 1$, $N \ll 1$, and $C \ll N$), the binomial distribution can be transposed into a more convenient form, the Poisson distribution

$$W(n) \simeq \frac{C^n e^{-C}}{n!} \tag{29}$$

Equation 29 shows that the Poisson distribution is characterized by a single parameter $C$ (the estimate of $\bar{x}_t$, the true average number of counts). The standard deviation of $C$ is given by (28):

$$\sigma_c = \pm\sqrt{\bar{x}_t} \simeq \pm\sqrt{C} \tag{30}$$

### The Gaussian (Normal) Distribution

The Poisson distribution, which applies to the phenomena of radioactive decay, is correct only for integer values of $n$ (i.e., the *number* of particles counted). However, the statistical theory of errors is generally based on the Gaussian distribution of a *continuous* variable. This distribution is often called the normal distribution. The Poisson distribution can be approximated by the Gaussian distribution $w(n)$ if $C$ is large and $|\bar{x}_t - C| \ll \bar{x}_t$:

$$w(n) = \frac{1}{\sqrt{2\pi\sigma^2}}\, e^{(C-\bar{x}_t)^2/2\sigma} \tag{31}$$

The standard deviation of the Gaussian distribution is also

$$\sigma = \pm\sqrt{\bar{x}_t} \simeq \pm\sqrt{C} \tag{32}$$

### 5.5.3 Counting Statistics

The statistical nature of radioactive decay is easily seen from the variation observed in the data obtained from the repetitive counting of a long-lived radionuclide in any radiation counting system. The data given in Table 5.3 are 12 successive 5-min counts of a $^{90}$Sr source in an end-window gas-flow proportional counter. These data are used to examine counting statistics.

### Average Counting Rate

The average (mean) value for a set of $N$ similar measurements is given by

$$\bar{x} = \frac{1}{N}\sum_{i=1}^{N} x_i \tag{33}$$

Thus the mean number of total counts per 5-min counting period is $1670/12 = 139.16$ counts. The average counting rate is $R = 139.16/5 = 27.83$ cpm.

**Table 5.3**  Variations Observed During
Repetitive 5-min Counts[a]

| Count Number | Total Counts |
|:---:|:---:|
| 1 | 139 |
| 2 | 128 |
| 3 | 110 |
| 4 | 144 |
| 5 | 142 |
| 6 | 140 |
| 7 | 137 |
| 8 | 142 |
| 9 | 144 |
| 10 | 138 |
| 11 | 156 |
| 12 | 150 |

[a] Background count: 270 counts in 60 min.

### Standard Deviation of an Individual Measurement

The precision of radioactivity measurement reflects the extent of scatter in the data of repetitive measurements due to the random nature of the radioactive process itself. A measure of the extent of the statistical fluctuations about the mean value of a set of data is given by the variance, defined as

$$\sigma_x^2 = \frac{1}{N-1} \sum_{i=1}^{N} (\bar{x} - x_i)^2 \qquad (34)$$

and, more conveniently, by the standard deviation $\sigma_x$. The use of $N-1$ for $N$ results from the need to use up one of the independent measurements for the estimation of $\bar{x}$, leaving $(N-1)$ independent measurements left to estimate the variance.

For the data in Table 5.3 the variance of a single 5-min count is

$$\sigma_c^2 = \tfrac{1}{11}(1446) = 131.45$$

and the standard deviation is

$$\sigma_c = \pm 11.5 \text{ counts}$$

The standard deviation of the counting rate is

$$\sigma_R = \pm 2.3 \text{ cpm}$$

## Standard Deviation of the Average Counting Rate

The longer or more frequently the counting rate of a sample is measured, the closer the measured average counting rate will be to the counting rate that would be determined by counting the sample until all of the atoms had decayed. This, however, is generally impractical to accomplish, especially for a long-lived radionuclide. There are two methods with which the reliability of $\bar{x}$ as an estimate of the "true" average value can be estimated.

1. The first would be the repetition of the $N$ measurements a large number of times. Each set of $N$ measurements would yield a measured average counting rate, the distribution of which would yield a mean value of the average counting rate. However, the variance of the average counting rate may be estimated from the variance of just a single set of $N$ measurements by

$$\sigma_{\bar{x}}^2 = \frac{\sigma_x^2}{N} = \frac{1}{N(N-1)} \sum_{i=1}^N (\bar{x} - x_i)^2 \tag{35}$$

The standard deviation of the mean counting rate from the data in Table 5.3 is

$$\sigma_{\bar{R}} = \left(\frac{26.29}{12}\right)^{1/2} = 1.48 \text{ cpm} \tag{36}$$

2. The second method of evaluating the reliability of $\bar{x}$ and also a method of checking the operation of a radiation detection system to determine whether the observed variations in the counting rate of a given source are purely random in nature is the Chi-square ($\chi^2$) test. Chi square is defined for $N$ equal counts as

$$\chi^2 = \frac{1}{\bar{x}} \sum_{i=1}^N (\bar{x} - x_1)^2 \tag{37}$$

Chi square is related to the number of degrees of freedom $F$, which for the Poisson distribution is $N - 1$, and the probability $P$ that repeated counts will give larger deviations than those expected from the Poisson distribution. Figure 5.15 gives the values of $\chi^2$ for this relationship. In general, values of $\chi^2$ yielding values of $P$ between 0.1 and 0.9 indicate an acceptable dispersion in the counting data, whereas values of $P$ smaller than 0.02 or greater than 0.98 would indicate that the counter system has introduced errors other than those of normal counting statistics.

The data in Table 5.3 give a $\chi^2$ value of

$$\chi^2 = \frac{1}{139.2} (1446) = 10.39 \tag{38}$$

Figure 5.15 shows for $F = 11$ that $P \simeq 0.5$ and that these data show an acceptable dispersion.

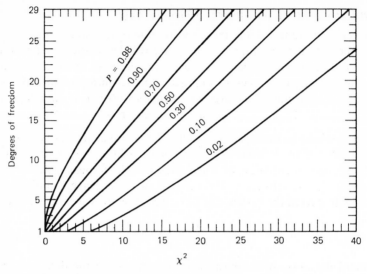

**Figure 5.15**   The chi-square test. [From R. D. Evans, *The Atomic Nucleus* (McGraw-Hill, 1955), p. 776.]

### *Standard Deviation from the Poisson Distribution*

Equation 30 gave the standard deviation of the number of radioactive atoms decaying in a counting time $t$, where $t \ll T_{1/2}$ and for a sufficiently large number of counts ($C > 100$) as $\sigma_c \simeq \pm\sqrt{C}$. The standard deviation of the counting rate is

$$\sigma_R = \pm \frac{\sqrt{C}}{t} \tag{39}$$

If the sample whose counting data are given in Table 5.3 had been acquired in an uninterrupted measurement for 60 min, the counting rate for $C = 1670$ counts and $t = 60$ min would have been

$$R = \frac{1670}{60} = 27.83 \text{ cpm} \tag{40}$$

and the standard deviation,

$$\sigma_R = \pm \frac{\sqrt{1670}}{60} = 0.68 \text{ cpm} \tag{41}$$

It is interesting to compare the magnitudes of the standard deviation calculated from the general definition of the standard deviation of a mean [see (36)] and from the Poisson distribution, specific for the statistics of radioactive decay [see (39)]:

$$\bar{R} \pm \sigma_{\bar{R}} \text{ (general definition)} = 27.8 \pm 1.5 \text{ cpm}$$

$$\bar{R} \pm \sigma_R \text{ (Poisson distribution)} = 27.8 \pm 0.7 \text{ cpm}$$

The smaller standard deviation for the Poisson distribution for the same set of data reflects the increase in precision in estimating the mean value by knowing the specific statistics for the process involved (radioactive decay). The counting rate of the sample used to compile the data in Table 5.3 is correctly given as $27.8 \pm 0.7$ cpm.

### Propagation of Errors

It was noted in Table 5.3 that the counter used to measure the sample had a background measurement of 270 counts in 60 min, or

$$B \pm \sigma_B = \frac{270}{60} \pm \frac{\sqrt{270}}{60} = 4.50 \pm 0.27 \text{ cpm}$$

To obtain the net counting rate (activity) of the sample this background value must be subtracted from the gross counting rate. The standard deviation of the net counting rate is obtained from the general rule for the propagation of errors of values that are added or subtracted. For the relationship

$$A = B \pm C \pm D \pm \cdots \tag{42}$$

the standard deviation of $A$ is given by

$$\sigma_A = [\sigma_B{}^2 + \sigma_C{}^2 + \sigma_D{}^2 + \cdots]^{1/2} \tag{43}$$

Thus the net counting rate of the sample is

$$A = R - B = 27.8 - 4.5 = 23.3 \text{ cpm}$$

and its standard deviation is

$$\sigma_A = [(0.68)^2 + (0.27)^2]^{1/2} = \pm 0.73 \text{ cpm}$$

The propagation of errors of values that are multiplied or divided is related to fractional errors of the respective values. For the relationship

$$W = \frac{XY}{Z} \tag{44}$$

the standard deviation of $W$ is given by

$$\frac{\sigma_W}{W} = \pm \left[ \left( \frac{\sigma_X}{X} \right)^2 + \left( \frac{\sigma_Y}{Y} \right)^2 + \left( \frac{\sigma_Z}{Z} \right)^2 \right]^{1/2} \tag{45}$$

Thus, if the $^{90}$Sr in the source were present in a sample weighing $0.012 \pm 0.001$ g, the specific activity of the sample would be

$$S \pm \sigma_s = \frac{23.3}{0.012} \pm \frac{23.3}{0.012}\left[\left(\frac{0.73}{23.3}\right)^2 + \left(\frac{0.001}{0.012}\right)^2\right]^{1/2}$$

$$= (1.94 \pm 0.17) \times 10^3 \text{ cpm/g}.$$

## 5.6  BIBLIOGRAPHY

### 5.6.1  General References

N. F. Johnson, E. Eichler, and G. D. O'Kelley, *Nuclear Chemistry* (Wiley, New York, 1963), Chapter VI.

W. F. Mann and S. B. Garfinkel, *Radioactivity and its Measurement* (Van Nostrand, Princeton, N.J., 1966), paperback,

G. D. O'Kelley, Detection and Measurement of Nuclear Radiation, Nuclear Science Series on Radiochemical Techniques, NAS-NS-3105 (1962) 138 pp. Available from Office of Technical Services, Department of Commerce, Washington, D.C., $1.50.

W. J. Price, *Nuclear Radiation Detection*, 2nd ed. (McGraw-Hill, New York, 1964).

K. Siegbahn, *Alpha, Beta, and Gamma-Ray Spectroscopy* (North-Holland, Amsterdam, 1955), 2 vol.

A. H. Snell, *Nuclear Instruments and Their Uses* (Wiley, New York, 1962).

### 5.6.2  Gas Counters

C. S. Curran and J. D. Craggs, *Counting Tubes* (Academic, New York, 1949).

B. Rossi and H. H. Staub, *Ionization Chambers and Counters* (McGraw-Hill, New York, 1949).

D. H. Wilkinson, *Ionization Chambers and Counters* (Cambridge University Press, Cambridge, 1950).

### 5.6.3  Scintillation Detectors

J. B. Birks, *The Theory and Practice of Scintillation Counting* (Macmillan, New York, 1964),

C. E. Crouthamel, *Applied Gamma-Ray Spectrometry* (Pergamon, New York, 1960).

### 5.6.4  Semiconductor Detectors

G. Dearnaley and D. C. Northrop, *Semiconductor Counters for Nuclear Radiation*, 2nd ed. (Wiley, New York, 1966).

G. T. Ewan and A. J. Tavendale, High-Resolution Studies of Gamma-Ray Spectra Using Lithium-Drift Germanium Gamma-Ray Spectrometers, *Can. J. Phys.*, **42**, 2286–2331, 1964.

F. S. Goulding, Semiconductor Detectors—Their Properties and Applications, *Nucleonics*, **22**, No. 5, 54–61 (1964).
J. M. Taylor, *Semiconductor Particle Detectors* (Butterworth, London, 1963).

### 5.6.5  Counting Statistics

Y. Beers, *Introduction to the Theory of Error* (Addison-Wesley, Reading, Mass. 1957).
R. D. Evans, *The Atomic Nucleus* (McGraw-Hill, New York, 1955), Chapters 26–28.

## 5.7  PROBLEMS

1. The following elements are to be activated as given. Suggest for each the optimum type of radiation detector with which the activation product could be counted:

   (a) carbon $(\gamma,n)$  (b) nitrogen $(n,2n)$
   (c) sodium $(n,\gamma)$  (d) aluminum $(n,\gamma)$
   (e) calcium $(n,\gamma)$  (f) manganese $(p,n)$

2. Determine the plateau and optimum operating voltage of a proportional counter in which a RaDEF ($^{210}$Pb, $^{210}$Bi, $^{210}$Po) source gave the following counting rates as the applied voltage was increased:

   | volts | cpm | volts | cpm | volts | cpm |
   |---|---|---|---|---|---|
   | 1400 | 10 | 1900 | 3240 | 2400 | 8185 |
   | 1500 | 1850 | 2000 | 5825 | 2500 | 8205 |
   | 1600 | 1935 | 2100 | 7955 | 2600 | 8295 |
   | 1700 | 2010 | 2200 | 8135 | 2700 | 8415 |
   | 1800 | 2025 | 2300 | 8165 | 2800 | 9705 |

3. The resolving time of a NaI(Tl) scintillation detector is limited by the decay of the luminescence and approximately equal to the decay time of the scintillator. Determine the counting rate at which the counting losses would exceed 3 % in a scintillator with a 0.25 $\mu$sec resolving time.

4. The intrinsic efficiency $\epsilon$ for a beam of gamma radiation, collimated normal to an absorber of thickness $d$ is $\epsilon = 1 - e^{-\mu d}$. What is the maximum intrinsic efficiency for such counting of the 1.37 MeV gamma rays of $^{24}$Na with

   (a) a 1-in. thick NaI(Tl) scintillation crystal,
   (b) a 1-in. depletion depth Ge semiconductor?

5. Determine the charge-collection time and the electrical capacity of a 2.5-cm dia × 3.5-mm deep Ge(Li) semiconductor detector, operated at 77°K with an applied voltage of 400 V.

6. A surface-barrier n-type silicon detector with resistivity of $1.8 \times 10^4$ $\Omega$-cm has a depletion depth of 2.4 mm. Estimate the detector bias voltage and the maximum beta-particle energy that could be stopped in the depletion depth.

7. Compare the theoretical line widths of the full-energy peak for the $^{24}$Na 1.37-MeV gamma in a germanium semiconductor to a NaI(Tl) scintillation crystal.

8. Determine from the Poisson distribution the number of counts that should be taken to reduce the standard deviation of the count
   (a) to less than 2%,
   (b) to less than 0.2%.

9. The Gaussian distribution approximates the observed spread in pulse heights from monoenergetic radiation in radiation detection equipment. Express the usual term for the resolution of a NaI(Tl) scintillation spectrometer ($W_{1/2}$, the full width at half maximum) in terms of the standard deviation $\tau$ of the distribution of pulse heights about the average value.

10. A technician transferred five 10-$\mu$l aliquots of a radioactive solution with a micropipet to vials for scintillation counting. One of the aliquots was counted five times and gave the following net counting rates for 1-min counts: 5423, 5506, 5416, 5478, and 5453 cpm. The other four aliquots were counted once each for 1 min and gave the counting rates of 5327, 5372, 5555, and 5450 cpm:
   (a) From these data determine the standard deviation (in $\mu$l) with which this technician can pipet 10 $\mu$l aliquots;
   (b) Determine from a $\chi^2$ test whether the five counts of the one aliquot indicate a properly functioning scintillation counter with respect to Poisson statistics.

Chapter 6

# Radiochemistry and Radioactivity Measurement

Radiochemistry in radioactivation analysis has been a subject of considerable controversy. Radiochemistry may be defined as the application of radioactivity to the study of chemical systems. Thus a nuclear chemist might consider activation analysis as an application of radiochemistry. On the other hand, a nonchemist who performs an analysis by radioactivation and radiation measurement might consider activation analysis as a form of instrumental analysis. In this latter method the identity of a radioelement is made solely by the characteristic radiations of its radioisotopes.

Many successful activation analyses are readily and accurately performed without any chemical treatment of the irradiated sample. In many other cases isolation of the desired radioelements may be required. The radiochemical activation analysis generally includes an isolation of the sought radioelements by chemical separations. In many instances, however, even with chemical isolation, radiation spectrometry is required to measure the individual radio-isotopes. Thus the question of radiochemistry in activation analysis is not one of whether radiochemistry should be used or avoided but what is the best combination of sample treatment, both chemical and instrumental, that will give the most convenient determination at the level of accuracy and precision required. The two methods complement each other; chemical separations make the radiation measurement simpler (in many cases eliminating the need for radiation spectroscopy); conversely, radiation spectroscopy makes the need for chemical purity less stringent (in many cases eliminating the need for some or any chemical treatment).

It is apparent, for example, that the measurement of a thermal neutron flux by the activation of pure gold foils does not require treatment of the foils to measure the activity of the sole product $^{198}$Au. On the other hand, activation analysis for elements at the very limits of sensitivity present in matrices that are easily activated and obscure the desired radiations in even the most sensitive of radiation spectrometers require efficient chemical separations with high degrees of purification, and possibly measurement in low-background radiation counters. Between these extremes are the activation analysis that can be performed by simple chemical separations and routine radiation measurements. The "simple" chemical separation might be a one-step removal of the desired element or perhaps the removal of a major interfering radionuclide. In general, for any given problem, the basic considerations in estimating the relative importance between radiochemical and instrumental means are the level of sensitivity required and the desired precision (or resolution of ambiguities) of radionuclide identification and calibration.

## 6.1  GENERAL  CONSIDERATIONS

Radiochemical analysis may be considered as the qualitative and quantitative determination of radionuclides by analytical chemistry and radiation measurement. It differs from classical chemical analysis in that the mass of the radionuclide sought is generally unweighable and must be separated not only from a large amount of matrix material but also from many other interfering radionuclides.

The determination of a radionuclide in a source free from other interfering radionuclides is generally achieved by choosing an efficient method of radiation measurement. Such measurements can generally be made with any desired degree of accuracy. However, except for thermal-neutron irradiation of a pure monoisotopic stable element, radioactivation results in samples that contain mixtures of radionuclides. The determination of a desired radio-nuclide at a given precision requires qualitative identification and quantitative measurement by specific or selective means. Radiation-energy spectroscopy is one such means.

If the desired radionuclide is present in low-activity levels in a matrix with many activated elements, the determination is usually made in two general parts:

1. Separation of the chemical element of the desired radionuclide.

2. Determination of the radioisotopes of the element by selective radiation measurement.

The former might be a painstaking high-decontamination purification of the element or a simple, rapid group-separation of chemically similar elements, The latter might be a single radioactivity measurement in a simple counter, a lengthy low-level radioactivity measurement in a low-background counter or an analysis of a complex gamma-ray energy spectrum. In each case the ob-jective is the measurement of the radiations from a desired radionuclide in the presence of radiations from other radionuclides and counter background.

Radiochemistry, as it applies to radioactivation analysis, utilizes all of the separation methods of analytical chemistry. However, the low mass of the radionuclides in the sample introduces a number of changes. Not only are direct gravimetric procedures precluded but also the integrity of the sample is endangered by possible losses of the radionuclide to the walls of the container or other materials. The latter is especially true for those irradiations in which the activation product is not isotopic with the target element. For these reasons radiochemistry in activation analysis generally utilizes the *carrier* method, the deliberate addition of a weighable amount of the element isotopic with the radionuclide of interest after the irradiation. The carrier method of radiochemistry generally consists of the following procedures:

1. Sampling
2. Preirradiation treatment (if required)
3. Preparation for irradiation
4. Dissolution of the irradiated sample
5. Addition of carriers
6. Chemical separations
7. Preparation for counting

### 6.1.1 Preirradiation Treatment

#### Sampling

One of the fundamental requirements in a chemical analysis is that the sample be representative of the material in which the sought elements are contained. This requirement is especially important in applications in which the material is a mixture of solids, in which there is a distribution of particle sizes, or in which a crystalline material contains grain boundaries. A "grab" sample, taken with the assumption that the material is homogeneous, may lead to a determination of concentration of a trace element that is precise but not representative.

The "art" or technique of sampling has been developed in considerable detail. Standard sampling procedures have been developed for almost all materials.* The procurement of a sample for analysis varies for different materials and even for different states of the same material but usually includes three considerations:

1. Taking a laboratory sample from the source material.
2. Taking the actual samples for analysis from the laboratory sample.
3. Treating the actual samples before weighing.

The homogeneity of the source material is the basic factor in the adequacy of the laboratory sample. If the source material is not perfectly homogeneous, extensive care must be taken in selecting the laboratory sample. Even greater care must then be taken in aliquoting the laboratory sample into test samples for analysis. Grinding, dissolution, and other means of increasing homogeneity may result in contamination of the sample by the sought elements.

The weight of the test samples also requires a consideration of the treatment given the samples. Many samples can be weighed "as is." Those that can contain variable amounts of hygroscopic water may be dried at $105-110°C$ before weighing or a determination of the moisture can be made to convert the measured concentrations from an "as is" basis to a "dry basis." If chemically combined water is removed on drying, the samples may require ignition at a higher temperature to obtain samples with definite composition. The absorption of $CO_2$ and other gases by finely ground samples may also be of concern.

The number of replicate samples to be taken depends considerably on the levels of concentration and precision desired. In some practices duplicate tests are made on two samples; in other practices single tests are made on three samples. In general, the level of errors in sampling should be consistent with the level of errors in the over-all analysis.

### Preirradiation Chemical Treatment

It is almost axiomatic in activation analysis that chemical treatment of a small sample before irradiation should be expressly avoided. The problem of

---

* See, for example, I. M. Koltoff, P. J. Elving, and E. C. Sandell, *Treatise on Analytical Chemistry*, Part 1, Vol. 1 (Wiley-Interscience, New York, 1959), pp. 93–96.

potential contamination by adding reagents that may have trace concentrations of the desired elements in excess of those in the sample is readily apparent. Yet there are occasions when preirradiation chemical treatment is necessary.

1. Samples too bulky for irradiation in which the concentration in smaller aliquots of the sample would contain too little of the desired elements. Reduction in bulk or volume is required. The sample might be ashed or dissolved and concentrated by evaporation or chemical means.

2. Samples that are too inhomogeneous to obtain representative samples for irradiation. The treatment to obtain homogeneous samples may be mechanical or chemical.

3. Samples that contain major constituents whose activation would lead to radioactive sources containing hazardous levels of radiation. Chemical separation of these constituents may be required before activation.

It is also almost axiomatic that in such cases considerable testing and quality control of the procedure are required to ensure that the samples are *not* contaminated by the specific materials and operations required for the treatment.

### Preparation for Irradiation

Treated or untreated samples are prepared for irradiation by the choice of an encapsulation method. The method considers the size and form of the sample, its geometry with respect to the irradiation beam or flux, and the presence of suitable standards, flux monitors, and blanks. The sample is generally condensed into the smallest volume consistent with minimum treatment.

### Dissolution

Following the irradiation solid samples are dissolved in the most convenient method that ensures no losses of the sought elements; for example, iodine and other such volatile elements may require dissolution of the sample in a closed system. Organic liquid samples may be mixed with miscible aqueous solutions.

Organic liquids or solids may require ashing to liberate inorganic trace elements. The ashing may be done "dry" or "wet." Dry ashing is simple but entails possible losses of trace elements by volatization or possible conversion to refractory compounds. Wet ashing, usually with $HNO_3$, is more tedious but avoids the problems of dry ashing.

Nonorganic materials may be dissolved in appropriate aqueous acids; in some cases the sought elements may be quantitatively leached from the matrix to provide initial separation from the bulk material. Refractory materials can be dissolved by chemical fusion. Specific acid, basic, and oxidizing fluxes are available for each material.

## 6.1.2  Carriers

The carrier method of radiochemistry is similar to the radiotracer method, except that the role of the tracer is reversed. In the radiotracer method an unweighable amount of a radioactive isotope is added to a system in which some particular component is to be followed. In the carrier method a weighable amount of inert material is added to a system in which some particular radionuclide is to be separated. In both cases the fundamental assumption is made that the carrier material and the radionuclide behave identically throughout any subsequent process. This assumption is almost completely valid for chemical separations when the carrier and radionuclide are isotopic to each other. It may be sufficiently valid for nonisotopic tracer and carrier when the chemical behavior of the two are similar in the procedures to be used. For example, the proton irradiation of a mixture of rubidium and cesium might produce the alkaline-earth-element radionuclides $^{85}$Sr and $^{133}$Ba. Calcium, as another member of the alkaline earth series, could be used to remove the $^{85}$Sr *and* the $^{133}$Ba together, provided chemical reactions common to all three elements were used. Specific reactions, however, can separate the three elements from one another. To ensure an efficient separation of the $^{85}$Sr *from* the $^{133}$Ba addition of strontium *and* barium carriers would be desired.

### *Carrier-Free Chemistry*

A radionuclide source that contains no weighable amounts (usually $<1$ $\mu$g) of stable isotopes of the same element is called *carrier-free*. A solution of such a source sometimes exhibits unusual behavior; for example, a carrier-free solution of 64-d $^{85}$Sr containing an activity concentration of 1 $\mu$Ci would have a molarity of Sr$^{2+}$ given by

$$C = \frac{N}{V \times \mathbf{N}} = \frac{D/\lambda}{V \times \mathbf{N}} = \frac{2.22 \times 10^{6}\,\mathrm{dpm} \times \dfrac{64 \times 24 \times 60}{0.693}\,\mathrm{min}}{10^{-3}l \times 6.023 \times 10^{23}}$$

$$C \simeq 5 \times 10^{-10}\ \mathrm{M}.$$

At such extremely low concentration, ions can adsorb on almost any solid material such as dust or container walls. The presence of even a microgram of carrier is generally sufficient to prevent such losses by adsorption. High specific-activity chemistry is sometimes desirable (e.g., to reduce self-absorption losses in a sample with only weak radiations), and processes that are not mass-dependent such as ion exchange or solvent extraction are frequently used for carrier-free radiochemistry.

## Properties of Carriers

To be suitable for a quantitative procedure two general properties are required of a carrier:

1. The desired radionuclide must follow the carrier quantitatively in all of the separation steps.
2. The carrier must mix completely with the radionuclide.

The first property is an obvious one. For isotopic carriers only the "isotope effect" (the separation of isotopes due to the differences in atomic weight) could be of concern. The enrichment of the 0.7% abundant $^{235}U$ in natural uranium to more than 90% is an example of the deliberate application of the isotope effect. In all normal laboratory practices, however, the isotope effect is of no importance.

The second property can be of considerable importance, especially if the radionuclide is not in ionic solution when the carrier is added or if the element can exist in several oxidation states; for example, the precipitation of carrier iodine, added as $I^-$, can leave all the radioactive iodine behind if it is present in the solution as $IO_3^-$. In cases in which the radionuclide can exist in several oxidation states the element (carrier and radioisotope) is usually oxidized and reduced through all of its oxidation states by suitable reagents before initiating any separation steps.

## Holdback Carriers

Carriers are also useful for coping with interfering radionuclides that have chemical properties similar to those of the desired radionuclide. Such carriers are called *holdback carriers*, since they prevent the low-mass interfering radionuclides from being adsorbed or carried by the sought element. Holdback carriers are generally used for those radionuclides that are expected to have been produced in quantity much larger than the sought radionuclide or to separate homologous elements in procedures in which the separation efficiency is not too great.

## Chemical Yield

The carrier method of radiochemistry involves a *chemical yield* determination. If a carrier is added in known amount (generally in milligram quantities), the separation of the radionuclide need not be quantitative. In the interest of rapid separations for short-lived radionuclides or of extreme-purity separations for maximum sensitivity, partial loss of the radionuclide can be tolerated. Since the specific activity of the radionuclide is constant after mixing with the carrier, the fractional recovery of the carrier (the chemical yield) is also the fractional recovery of the radionuclide.

Obviously the amount of the carrier element present in the original sample must be small compared with the added amount. This is generally the case when activation analysis is used to measure trace elements in microgram or smaller amounts.

Following the chemical separations the chemical yield is determined by measuring the amount of the remaining carrier. This determination may be made by several methods:

1. *Gravimetric.* Although this method is the easiest, it is also nonspecific; that is, other chemicals may separate with the carrier. In some cases the compound weighed may not be stoichiometric, that is, of indefinite or unknown composition, especially with respect to water of crystalization.

2. *Volumetric or Photometric and Other Instrumental.* Such methods are usually more specific for individual elements. They also require more extensive calibrations, standards, and blanks.

3. *Radiometric.* The use of a known amount of another radioactive isotope as a tracer of the carrier. This method is convenient but not always possible. Besides the usual properties of a suitable tracer, namely, similar chemical behavior, compatible half-life, and easily measured radiations, the tracer must not interfere with the radioactivity measurement of the desired radionuclide. An example of this technique is noted for vanadium activation analysis. The carrier for the 3.7-m $^{52}$V produced by the $^{51}$V $(n,\gamma)$ reaction is vanadium tagged with 330-d $^{49}$V, which decays solely by electron capture and therefore does not contribute to the photopeak measurement of $^{52}$V. After the 3.7-m $^{52}$V has decayed completely the $^{49}$V tracer is measured by liquid scintillation counting.

4. *Reactivation Analysis.* This method is useful for short-lived activation products. After the measured radionuclide decays, the counting sample which contains the recovered carrier is irradiated in the same manner as the original sample. Since the carrier amount is much larger than the original trace element amount, the radioactivity produced is now a measure of the amount of carrier present in the sample counted. This technique has also been used for vanadium chemical-yield determination.

### 6.1.3  Chemical Separations

Chemical separations are the part of a radiochemical procedure that results in the isolation of the activation product radionuclide from the other nonisotopic radionuclides produced in the irradiated matrix. In the carrier method this is achieved by the isolation of the carrier element in a form suitable for radiation measurement and chemical-yield determination.

The literature of analytical chemistry abounds with chemical procedures that will adequately separate any given element from all others. Section 6.4

lists a number of general references that offer a wide variety of radiochemical procedures for all of the elements. At first glance the basis for a choice of a specific procedure by an inexperienced analyst must indeed appear to be greatly confusing. Yet the selection of specific steps in a radiochemical procedure generally follows a common set of goals that may be considered to be the *strategy* of the separation. These goals will generally include the following:

1. Reduction of the mass of the sample to be optimum or consistent with the radiation measurement conditions to be used. These conditions and the chemical yield anticipated will determine the amount of carrier to be added; for example, the final sample may be a thin precipitate on a filter paper for beta counting, a small volume of solution for gamma-ray counting in a well-type scintillation counter, or a plated sample on a metal disk for alpha-particle spectroscopy.

2. Isolation of the carrier element with sufficient chemical purity and definite composition for chemical-yield determination. Several different types of separation method may be used to obtain separations from several groups of impurities. The element separated from all others may still require a further step to convert it into a chemical form for quantitative yield measurement.

3. Sufficient chemical separation of the product radionuclide from the other nonisotopic radionuclides, especially those pressnt in carrier-free or high specific-activity concentrations. The radioactivity measurement must be sufficiently above the background activities to be compatible with the desired precision for the analysis. Radiation spectroscopy may still be required to measure the desired radionuclide in the presence of its other radioisotopes.

### Reduction of Mass

Following the sample postirradiation dissolution and equilibration of the desired radionuclides with their carriers the initial steps of the separation procedures usually deal with the gross separation of the carriers from the matrix solution. Several factors influence the choice of these steps, which include the half-life and the total activity of the individual radionuclides. For short-lived radionuclides rapid procedures are required. For high activity radionuclides individual aliquots of the matrix solution may be taken for each element; for low activity radionuclides sequential separation of each of the carriers is required.

In all of these cases it is usually the strategy to remove the desired element from the bulk sample by some mass- or volume-reducing process such as precipitation, solvent extraction, ion exchange, or distillation. The removal of the bulk matrix element is generally avoided, since it may result in the partial or complete co-removal of the desired radionuclide even in the presence of its carrier. A good separation of the carrier element may carry small amounts of the matrix elements, but they are generally removed in subsequent separations.

Besides the partial purification of the carrier element from the matrix solution, the reduction of mass to the milligram quantities of carrier facilitates the remaining chemical operations, since they can be carried out in small volumes.

In many specific activation analyses the initial separation steps may be sufficient for the required degree of chemical separation for radiation measurement. For short-lived activation products the total chemical separations must be achieved in time to measure the radiations before their decay. Extensive development of rapid radiochemical separation procedures have evolved; for example, the chemical separation of 8-s $^{14}N$ produced from the $^{16}O$ (n,p) reaction has been achieved by a distillation process in 7 sec. The general procedure for such rapid radiochemical separations is to prepare the target material in a chemical and physical form suitable for rapid dissolution and carrier exchange and to decrease the transfer time from end of irradiation to beginning of separations as much as possible. Pneumatic transfer systems are generally used for rapid delivery of irradiated samples to the radiochemical laboratory.

### Carrier Isolation

Following the removal of the carrier element from the matrix solution a great number of separation techniques are available. The choice is generally determined by the specific interfering elements to be removed. In many cases several equally good methods are suitable, and the final choice is governed by the experiences of the analyst. Some of the more common type of separation include precipitation, ion exchange, solvent extraction, distillation, and electrodeposition. The principles of these methods are summarized in Section 6.2.

Limitations in the choice of separation methods are generally experienced only when carriers of similarly behaving elements have been added to the matrix; for example, the removal of barium and strontium carriers together from a matrix solution will require a subsequent gross separation of the barium and strontium carriers before each is individually purified.

Following the gross separation of each carrier subsequent chemical operations may be carried out by using the techniques of "semimicro chemical analysis," that is, procedures involving only milligram quantities of reagents. Such procedures are generally performed in small volumes (<40 ml), and the solutions and precipitations are manipulated in centrifuge tubes, the smallest sizes of ion exchange columns, solvent extraction funnels, etc.

Such techniques offer the added advantage of rapid completion of the operations.

When the carrier is sufficiently pure, it is generally converted into some chemical form suitable for chemical-yield determination or mounting for a radiation measurement. In the latter case the chemical yield may be determined after the radiation measurement. Some of the several methods for determining the chemical yield are listed in Section 6.1.2.

### Radionuclide Isolation

The complete isolation of a "pure" carrier element is not a sufficient requirement for a "pure" radionuclide. A chemically pure barium compound may carry carrier-free strontium radionuclides to the extent that they could completely mask the radiation measurement of the desired barium radioisotopes. The general evaluation of a separation, both for carriers and radionuclides, is determined by the change in composition resulting from the chemical separation procedure.

For carrier quantities of chemicals the effectiveness of a particular separation method for any two elements is given by the separation factor SF, which is defined as the ratio of concentrations of the two elements before and after separation:

$$SF = \frac{[A]_i/[B]_i}{[A]_f/[B]_f} \qquad (1)$$

where $[\ ]_i$ = the concentration before separation,

$[\ ]_f$ = the concentration after separation of the two elements A and B being separated.

A separation factor of unity implies no chemical separation.

In radiochemical separations the degree of separation of two nonisotopic radionuclides is given by the decontamination factor DF, defined as the ratio of radioactivities of the two nuclides before and after separation:

$$DF = \frac{C_i/D_i}{C_f/D_f} \qquad (2)$$

where $C_i$ = the initial activity of the interfering radionuclide,

$D_i$ = the initial activity of the desired radionuclide,

and the $f$ subscripts refer to the final activities of the two after separation.

In general, decontamination factors of $10^4$ or greater are sought for radionuclide separation procedures. However, if the trace element has produced a radionuclide whose activity is just measurable, radiochemical methods that yield much higher decontamination factors are required.

In addition to the specific separation methods that remove the desired radionuclide from the interfering ones, separation methods are available to remove specific interfering radionuclides from the carrier solution. These are called *scavenging processes*, and they generally involve the addition of a suitable carrier to the solution and the precipitation of the added carrier. The precipitate carrying the interfering radionuclide is discarded. Another separation technique is *masking*, the complexing of an interfering radionuclide such that it does not follow the separation of the desired radionuclide in procedures in which it would do so in simple ionic solution. This technique is often used in solvent extraction and ion exchange separations.

For activation analysis of trace elements at levels near maximum sensitivity more stringent practices in radiochemistry have been developed for low-level radioactivity measurement. If matrix or major component elements yield interfering radionuclides, decontamination factors of $10^9$ or higher may be required.

Low-level radiochemistry may be defined as the separation of radionuclides whose counting rates are of the order of the counting rates of blank samples plus counter backgrounds. Low-background counters used for low-level radioactivity measurement have been described in Section 5.2.1. Such backgrounds generally range from about 0.2 to 1 cpm.

A blank in a chemical analysis is a sample that is treated identically to the sample being analyzed. Reagent blanks are used in activation analysis when preirradiation treatment of samples is required. The blank sample starts with the same amount of reagent or deionized water used to dissolve the actual sample. The blank thus measures the contribution of the element being determined from the reagents used in the preirradiation treatment of the sample and any radioactivity introduced into the sample from the reagents and carriers used in the total treatment before radiation measurement. Since almost all manufactured chemicals contain some radioactivity from either natural- or nuclear-weapons testing sources, the blank sample generally contains some radioactivity. In some cases the counting rate of the blank is considerably greater than the low-level counter background. These two contributions to the counting rate of the unknown sample must be subtracted to obtain the net counting rate of the sample. The activity levels at which low-level radiochemistry is required may be defined generally as those for which the ratio of net activity to blank plus background activity is approximately unity.

The major considerations for low-level radiochemistry (which are also pertinent to higher activity level radiochemistry) are the following:

1. Selection of reagents and procedures that reduce the blank to the level of the counter background or lower.
2. Selection of procedures that have high chemical yield and high radiochemical purity.

3. Selection of counting conditions to maximize over-all counting efficiency.

The chemical separation procedures are common to those for higher level activity samples. The difference between the two in the factors listed above is the greater care needed in the selection of reagents and carrier chemicals and the more repetitions required of the specific separation steps to achieve the necessary decontamination factors.

Following the achievement of radiochemical purity all of these procedures require the preparation of the carrier into a form suitable for radiation measurement. Techniques for the collection and mounting of samples for counting are also varied. Consideration must be given to such factors as sample area and mass, type and energy of the radiations, and the types of sample mount suitable for the available detectors.

Solutions can be evaporated on dish or plate planchets. Usually great care is required to avoid splattering of the samples and to obtain uniformly thick deposits. For gamma-ray measurements the solutions can be transferred directly to a small vial in which a standard volume of solution is counted. Another common method is to precipitate the carrier and filter it onto a weighed filter paper. The dried precipitate can then be weighed for chemical-yield determination before mounting for radiation measurement. In many cases radionuclides can be electrodeposited, either carrier-free or in the presence of microgram amounts of carriers, onto thin metal disk electrodes.

## 6.2  SEPARATION METHODS

The principles employed in the choice of separation methods in radio-chemistry have included consideration of several general objectives. Yet in most applications the nature of the interfering radionuclides or macro constituents require consideration of their specific chemical properties with respect to the desired elements. Although a great number of techniques has been used in radiochemical separations, the number actually used in most separations is small. They involve the major processes for treating low-concentration inorganic ions. These are precipitation (or coprecipitation), ion-exchange chromatography, and solvent extraction. For a number of pertinent elements the techniques of distillation and electrodeposition are frequently used by radiochemists.

### 6.2.1  Precipitation

Precipitation is the most classical of the chemical techniques available for chemical separations. It is the basis for gravimetric analysis, the determination of the amount of a material of known composition by weighing its mass.

If the precipitate is dissolved, the determination may be made by volumetric analysis, the measurement of the quantity by its titration with a stoichiometric amount of reagent. Precipitation separations have been the primary means of radiochemical separations in methods adapted from conventional chemical analysis.

Precipitation may be defined as the deposition of a solid from a solution by the action of an added agent, such as a chemical reagent, heat, electricity, or sonic energy. The mechanism of precipitation is complex. The visible formation of the solid phase follows after the process of nucleation, which is itself not fully understood. Solutions are known to contain molecular aggregates (crystal embryos) of the solute distributed randomly throughout the solvent. These aggregates undergo continuous formation and dissolution in equilibrium with the surrounding solvent. If molecular aggregates of a critical size are exceeded, the aggregate becomes a crystal nucleus and crystal growth continues. The process of nucleation requires a supersaturated solution; that is, one in which the solubility product amount of the solute is exceeded. The rate at which precipitation occurs depends partly on the degree of supersaturation. Many other factors affect the purity, completeness, and rate of precipitation. Among them are impurity concentrations, the type and excess of precipitating agent used, temperature, and stirring rate.

The effect of the rate of precipitation may be divided into two types; slow and fast precipitation. Precipitation at slow rates is desired for analytical separations, since they generally result in the formation of large well-shaped crystals. Such crystals occupy a smaller volume, are less contaminated with the solution, and are more easily filtered and washed. Slow precipitation may be accomplished in two ways:

1. By the addition of a very dilute solution of the precipitant, gradually with stirring, to a warmed or hot solution.

2. By the addition of the precipitant indirectly to the solution; for example, as the result of the hydrolysis of the reagent.

The first method increases the amount of recrystallization during digestion of the precipitate, whereas the second method results in the nucleation of a smaller number of crystals, both processes leading to the formation of large crystals. The latter method is known as precipitation from homogeneous solution and several examples are listed in Table 6.1.

Precipitates that occur in crystals of very small size, that is, less than 1 $\mu$ in diameter, are called *collodial*. Collodial precipitates are generally avoided in gravimetric analysis because they are difficult to process. They can pass directly through ordinary filter papers and require such techniques as ultrafiltration and ultracentrifugation to be collected efficiently. In radiochemical

**Table 6.1**  Summary of Methods that Utilize Precipitation from Homogeneous Solution[a]

| Precipitant | Reagent | Element Precipitated |
|---|---|---|
| Hydroxide | Urea | Al, Ga, Th, Fe(III), Sn, Zr |
|  | Acetamide | Ti |
|  | Hexamethylenetetramine | Th |
|  | Metal chelate and $H_2O_2$ | Fe(III) |
| Phosphate | Triethyl phosphate | Zr, Hf |
|  | Trimethyl phosphate | Zr |
|  | Metaphosphoric acid | Zr |
|  | Urea | Mg |
| Oxalate | Dimethyl oxalate | Th, Ca, Am, Ac, rare earths |
|  | Diethyl oxalate | Mg, Zn, Ca |
|  | Urea and an oxalate | Ca |
| Sulfate | Dimethyl sulfate | Ba, Ca, Sr, Pb |
|  | Sulfamic acid | Ba, Pb, Ra |
|  | Potassium methyl sulfate | Ba |
|  | Ammonium persulfate | Ba |
|  | Metal chelate and persulfate | Ba |
| Sulfide | Thioacetamide | Pb, Sb, Bi, Mo, Cu, As, Cd, Sn, Hg, Mn |
| Iodate | Iodine and chlorate | Th, Zr |
|  | Periodate and ethylene diacetate (or $\beta$-hydroxyethyl acetate) | Th, Fe(III) |
|  | Ce(III) and bromate | Ce(IV) |
| Carbonate | Trichloroacetate | Rare earths, Ba, Ra |
| Chromate | Urea and dichromate | Ba, Ra |
|  | Potassium cyanate and dichromate | Ba, Ra |
|  | Cr(III) and bromate | Pb |
| Periodate | Acetamide | Fe(III) |
| Chloride | Ag ammonia complex and $\beta$-hydroxyethyl acetate | Ag |
| Arsenate | Arsenite and nitrate | Zr |
| Tetrachlorophthalate | Tetrachlorophthalic acid | Th |
| Dimethylglyoxime | Urea and metal chelate | Ni |
| 8-Hydroxyquinoline | Urea and metal chelate | Al |
| Fluoride | Fluoboric acid | La |

[a] From I. M. Kolthoff, P. J. Elving, and E. B. Sandell, *Treatise on Analytical Chemistry*, Part I, Vol. 1 (Wiley-Interscience, New York, 1959), p. 741. Description of the methods is given by L. Gordon, M. L. Salutsby, and H. H. Willard, *Precipitation from Homogeneous Solution* (Wiley, New York, 1959).

analysis, however, colloidal precipitates may be of interest in two ways:

1. The isolation of carrier-free radionuclides from solutions in which the solubility product of a compound, which would be considered insoluble in carrier concentrations, is not exceeded.

2. The adsorption of interfering radionuclides present in carrier-free concentrations.

The latter property, in which a precipitating compound carries with it compounds of other solutes from the solution, is called *coprecipitation*. It is of general importance in radiochemistry. The incorporation of the coprecipitated impurity may occur through several mechanisms such as mixed crystals, surface adsorption, occlusion (and inclusion), and postprecipitation.

In mixed crystal coprecipitation the contaminants are contained as a solid solution. Isomorphous compounds that form mixed crystals, such as the rare earths, can coprecipitate in the form of a crystalline solution. Mixed crystals generally form from components whose lattice constants do not differ more than $\pm 5\%$ from one another. This requirement need not be met if the coprecipitated radionuclide is present in carrier-free or trace concentrations. If the concentration of the impurity is homogeneous throughout the precipitate, the distribution of the impurity between the precipitate and the solution is given by the Berthelot-Nernst distribution law as

$$\frac{[T]_s}{[C]_s} = D \frac{[T]_l}{[C]_l} \tag{3}$$

where $D$ is the homogeneous distribution coefficient, the brackets refer to the concentration of the trace radionuclide T and the carrier compound C, and the subscripts refer to the solid $s$ and the solution $l$ phases.

Under slow precipitation conditions the tracer radionuclide does not become uniformly distributed throughout the precipitate. The growth of larger crystals tends toward a nonuniform distribution, since it allows only the surface of the crystal to come into equilibrium with the solution. Under these conditions the distribution of the trace radionuclide is given by the Doerner-Hoskins distribution law as

$$\ln \frac{[T]_i}{[T]_f} = \lambda \ln \frac{[C]_i}{[C]_f} \tag{4}$$

where $\lambda$ is the logarithmic distribution coefficient, the brackets refer to the initial ($i$) and final ($f$) concentrations of the trace radionuclide (T) and the carrier compound (C) in the solution. In most radiochemical precipitations in which some recrystallization occurs, the actual distribution of a coprecipitated trace radionuclide will lie somewhere between the distribution given by these two laws. Experimentally determined distribution coefficients are useful in comparing decontamination factors achievable by alternate precipitation methods.

In surface absorption the incompletely coordinated ions at the surface of a crystal results in a net attraction for other ions in the solution. Ions homologous to the precipitate are strongly attracted. If one of the ions of a precipitate is in excess in the solution, impurity ions of the opposite charge may be attached to the surface of the crystals, continuing the growth of the crystals. The two properties of the precipitate that affect the extent of surface adsorption are the specific surface area (i.e., the surface area per unit weight) and the electric charge at the crystal-solution interface. These properties are especially important for collodial precipitates, which exhibit a high efficiency for surface adsorption.

The mechanism for surface adsorption may be explained by the diffuse double-layer theory in which one layer is fixed at the surface and the outer adjacent layer is free to move (the diffuse layer). The charged layers owe their existence to adsorbed ions which may be quite different from ions already present in the inner portion of the precipitate. This theory helps to explain why the electrolyte content in a solution affects the magnitude and sign of the charge of a colloidal precipitate. The ions present in the diffuse outer layer of the colloid extend into the external solution, with no sharp boundary between the ions in the diffuse outer layer and those in the equilibrium external solution. The concentration of the ions constituting the diffuse layer varies and is dependent on the concentration and pH of the external solution. With the addition of an electrolyte, flocculation of a colloid generally results in the surface adsorption of the electrolyte.

The extent of adsorption has been described by several empirical rules; for example, the *Paneth-Fajans-Hahn adsorption rule:* A precipitate will efficiently adsorb those ions whose compounds with the opposite-charged ion of the crystal have low solubility. The extent of adsorption of a particular impurity is also dependent on its concentration. The empirical relationship is given by the *Freundlich adsorption isotherm:*

$$A = \alpha C^{1/n} \tag{5}$$

where $A$ is the amount of impurity adsorbed per unit weight of adsorbent, $C$ is the concentration of the impurity after absorption, and $\alpha$ and $n$ are constants. This equation indicates that the efficiency for adsorption is large for low concentrations, and the extent of adsorption reaches a saturation value as the impurity concentration increases. Thus surface adsorption is an efficient process for the coprecipitation of carrier-free radionuclides. Coagulating agents such as $Al_2(SO_4)_3$ and $Fe(OH)_3$ are commonly used to coprecipitate (scavenge) many impurity carrier-free radionuclides. According to the Paneth-Fajans-Hahn adsorption rules, cations that form insoluble sulfates and hydroxides, respectively, would be efficiently coprecipitated. According to the Freundlich adsorption isotherm, the presence of an added carrier (a holdback carrier) reduces the extent of coprecipitation if the carrier opposite-charged ion salt is nominally soluble.

Occlusion of an impurity within a precipitate is generally defined as the mechanical entrainment or trapping of the impurity by the rapidly growing crystal layers. In fast-precipitation processes, in which the precipitating reagent is added to the solution, high concentrations of the precipitant occur locally in the solution, and both solutes and solvent can be mechanically occluded as the precipitate forms under these concentrated conditions. Coprecipitation by occlusion occurs more commonly with collodial precipitates such as freshly precipitated hydroxides and sulfides. Aging of such precipitates may result in the release of some of the occluded impurities. Precipitations in radiochemical practice is generally made under nonequilibrium conditions in which occlusion of carrier-free impurity radionuclides may occur. The remedies include the use of holdback carriers for known pertinent interfering radionuclides, aging of the precipitates, and substitution of some slow-precipitation process.

Postprecipitation is the formation of a second (impurity) solid phase on the precipitate of the primary carrier compound. It can occur when the impurity tends to form a stable supersaturated solution that is induced to precipitate by the primary substance. An example of this process in analytical chemistry is the postprecipitation of magnesium oxalate on calcium oxalate. Magnesium oxalate forms supersaturated solutions that do not precipitate in pure solution. It does not coprecipitate extensively with freshly precipitated calcium oxalate. However, if the calcium oxalate is not filtered immediately, a second solid phase of magnesium oxalate will precipitate on the calcium oxalate crystals.

The principles of these precipitation processes indicate that there are no fixed precipitation rules to avoid the coprecipitation of all types of interfering radionuclides. Many practices have developed in radiochemistry to reduce the extent of coprecipitation. Several are summarized in Table 6.2. The general procedure is to precipitate from hot dilute solution, adding the precipitant slowly, with agitation, and allowing the precipitate to age before filtering and washing with a suitable wash solution.

Table 6.2 also shows that reprecipitation is an efficient process in radiochemistry to improve the purity of the carrier precipitate, regardless of the process by which the interfering radionuclides are coprecipitated. In reprecipitation the precipitate is dissolved in a suitable solvent, diluted, and the conditions for precipitation re-established, sometimes with a different precipitant or precipitating condition. Reprecipitations generally allow any

Table 6.2   Influence of Precipitation Conditions on the Purity of Precipitates[a]

| Condition | Form of Impurity[b] | | | |
| | Mixed Crystals | Surface Adsorption | Occlusion | Post-precipitation |
| --- | --- | --- | --- | --- |
| Dilute solutions | 0 | + | + | 0 |
| Slow precipitation | + | + | + | − |
| Prolonged digestion | − | + | + | − |
| High temperature | − | + | + | 0 |
| Agitation | + | + | + | 0 |
| Washing the precipitate | 0 | + | 0 | 0 |
| Reprecipitation | + | + | + | + |

[a] Adapted from I. M. Kolthoff, P. J. Elving and E. B. Sandell, *Treatise on Analytical Chemistry*, Part 1, Vol. 1 (Interscience, New York, 1959).
[b] + = increased purity,
− = decreased purity,
0 = little or no change in purity.

required decontamination factor to be achieved. If particular interfering radionuclides are known to be present, holdback carriers can be added to increase the decontamination factor for those radionuclides.

A further advantage of precipitations in radiochemistry results from the common practice of separating precipitates from the supernatant solutions by centrification and decanting rather than by filtration. This practice also makes it easier to wash the precipitate with small aliquots of dilute precipitating reagent solutions or water. The precipitate can then be dissolved in the same centrifuge tube with small volumes of solvent for subsequent processing. Filtering on small filter papers (<1-in. dia) is usually made for the final preparation of the counting source in a compound of stoichiometric or reproducible composition for chemical-yield determination.

### 6.2.2   Ion Exchange Chromatography

Since the description of the first chromatogram in 1906, chromatography has been defined in many ways; but it is basically a separation process in which ionic or molecular mixtures are separated by their differential distribution between two phases, generally one moving and one stationary. If the stationary phase is any of a large number of ion-exchange resins, the separation process is termed ion-exchange chromatography. Other common methods of separation that involve a moving and a stationary phase include partition chromatography and gas chromatography.

Partition chromatography, named in 1942, is primarily a liquid-liquid extraction process in which one of the liquid solvents si maintained in a stationary phase, such as a column of silica gel. Silica gel can adsorb more than 50% its weight of water without becoming appreciably wet and is therefore useful for immobilizing an aqueous phase. Other materials used as the supporting material include alumina, asbestos, barium sulfate, calcium sulfate, cellulose, charcoal, magnesia, and titanium dioxide. Although partition chromatography is useful for the separation of many inorganic materials and is used in many radiochemical analyses, its successes have been particularly noteworthy in the separation of organic substances. A recent development in adsorption chromatography has been the use of strips of filter paper as a support for the stationary phase. This process termed paper chromatography, has become popular in radiochemical separations of high specific activity-materials, since the paper strip can be subsequently cut into fractions for radioactivity measurement.

Gas chromatography, introduced in 1952, is a separation method for mixtures of gases in a moving carrier gas phase passing through a stationary phase, either a solid adsorbent or a liquid supported on an inert material, as in partition chromatography. The carrier gas is usually an inert gas, such as helium, hydrogen or nitrogen. The solid adsorbents include alumina, silica gel, synthetic zeolites, and activated charcoal in particle sizes ranging from 20 to 200 mesh. Liquid adsorbents include such organic solvents as liquid paraffin, benzylbiphenyl, or acetonylacetone. The use of a flowing gas rather than a flowing liquid has introduced several advantages in chromatographic techniques; for example, gas chromatography is necessary for separations of radioactive nuclides of gases, such as krypton and xenon, and is of great use in radiochemical separations in which the desired radionuclide can be easily volatilized. The method is generally useful for substances with vapor pressure greater than about 10 mm Hg at the operating temperature of the gas chromatographic column. General references on partition and gas chromatography are given in Section 6.4.2.

An ion exchange resin, as used in radiochemistry, is generally a polymerized, insoluble hydrocarbon with repetitive sites for ionic groups which are easily dissociated. Such resins can exchange either cations or anions, according to the charge of the ionizable group; for example, cation exchange resins are usually made of resins that have fixed anion sites in such forms as

$$\text{sulfonic acids:} \quad R_s\text{—}SO_3\text{—}H$$

$$\text{carboxylic acids:} \quad R_s\text{—}COOH$$

where $R_s$ is the polymerized hydrocarbon and, in the acid form, hydrogen is the exchangeable cation. Anion exchange resins are usually high-molecular-weight insoluble bases with the fixed ion in the form of a cation, such as

amino groups:            $R_s - NH_2, R_s - NHR, R_2 - NR_2$

quaternary ammonium groups:       $R_s - NR_3^+$

The latter type is characteristic of the strongly basic resin, which is useful for inorganic-anion separations. Two important properties of ion exchange resins are insolubility and chemical stability. Since most linear polymers containing sulfonic acid or quaternary ammonium groups are soluble in water, network polymers, that is, those having a crosslinked molecular structure, are most useful for ion exchange resins. Such resin hydrocarbons may be obtained by polymerization of phenol and formaldehyde or of styrene and devinylbenzene. Two popular resins for radiochemical separations are the sulfonated phenol-formaldehyde resins for cation exchange and quaternary ammonium base polystyrene-divinylbenzene resins for anion exchange. The crosslinkage unit of these two ion-exchange resins is shown in Figure 6.1. Hydration, which tends to swell the lattice structure, facilitates exchange of ions with internal lattice sites.

The number of ion exchanging groups introduced into the network polymer determines the exchange capacity of the resin on a unit weight basis. Exchange capacities are usually given in units of milliequivalents per gram of dry resin or milliequivalents per milliliter of wet resin. Typical capacities are about 4.6 meq/g and 2.0 meq/ml, respectively, for the sulfonated cation resin and about 3.5 and 1.1 for the quaternary strong base anion exchange resin.

Ion-exchange chromatography in radiochemistry is generally achieved by preparing a column of the desired ion-exchange resin and introducing a solution of the mixed radionuclides at the top of the column under conditions in which the desired radionuclide will exchange onto the stationary phase. After the original solution has passed through the column the radionuclides are *eluted* from the column with suitable solvents; the rate with which they move down the column is determined by the distribution coefficient of each radionuclide for the resin-solvent system. If two chemically similar radionuclides move down the column at different rates, an elution chromatogram of the two radionuclides might look like the two Gaussian-shaped chromatographic bands illustrated in Figure 6.2. The basis for this shape of elution curve in ion exchange chromatography has been the subject of considerable study. Theories of ion-exchange chromatography, evolved from considerations of the mechanisms of ion-exchange, include the crystal lattice theory, the double layer theory, and the Donnan membrane theory.

The crystal lattice theory draws an analogy between the exchange of crystal lattice ions and the mixing of two soluble electrolytes. Extensive studies with the natural zeolites, materials of highly porous, chainlike silicate structures,

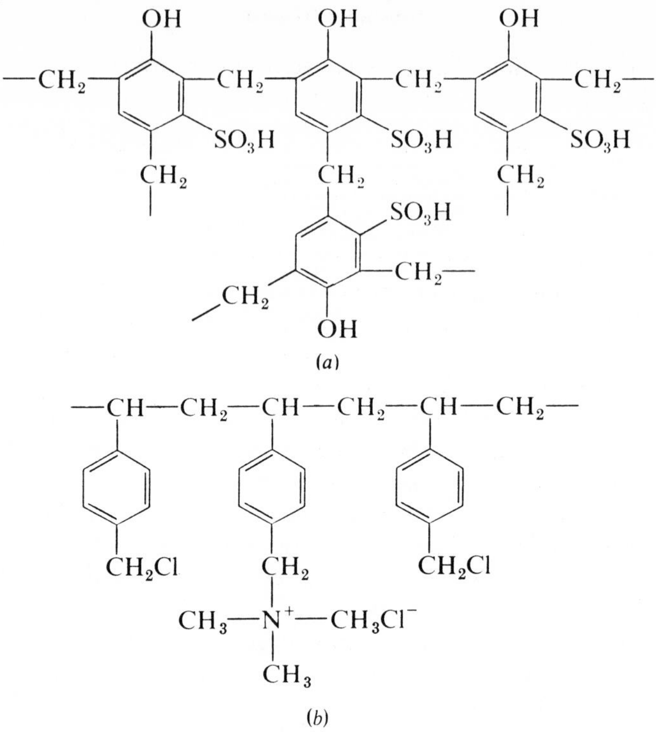

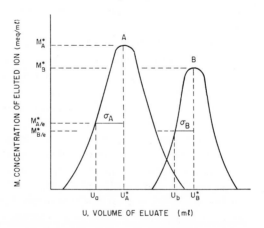

(a)

(b)

**Figure 6.1** Unit structures of the crosslinked ion-exchange resins; (a) sulfonated phenolformaldehyde cation exchange resin; (b) quaternary ammonium polystryrene-divinylbenzene anion exchange resin.

**Figure 6.2** An elution curve showing the separation of the two ions A and B through an ion-exchange resin column.

have shown that the sodium and calcium ions which occupy essential lattice sites can be reversibly exchanged. In crystalline silicate clay minerals such as kaolinite the exchange of anions has been demonstrated, whereas in montmorillinite, ions of both signs are exchangeable.

The double-layer theory of ion exchange stems from the electrokinetic properties of colloids, discussed in Section 6.2.1. With the ion exchange resin taken as a colloidal precipitate in equilibrium with the external solution, the concentration of exchangeable ions in the diffuse layer may be considered to vary continuously in relation to the concentration and pH of the external solution. The addition of another ion to the solution results in a new equilibrium system in which some of the added ions enter the diffuse outer layer and replace, stoichiometrically, ions previously held in that layer. This model differs from the crystal lattice model primarily in that the number of crystal exchange sites is determined independently of pH or concentration, whereas the double-layer exchange is dependent on these two parameters. In many systems both types of exchange may occur simultaneously.

The Donnan membrane theory relates the distribution of ions on two sides of a membrane in which one side contains an electrolyte, one of whose ions is not able to diffuse through the membrane. The analogy to ion exchange assumes that the unit resin structure holding the exchangeable ion is the nondiffusable ion; for example, in a strongly acidic cation-exchange resin, the sulfonate, entirely retained in the resin phase, is the nondiffusible ion. With a solution of HCl the Donnan theory relates the concentrations at equilibrium as

$$[H^+]_e[Cl^-]_e \cong [H^+]_r[Cl^-]_r \qquad (6)$$

where the brackets refer to the ion concentrations in the external ($e$) and internal ($r$) solutions. Electroneutrality requires

$$[H^+]_r = [-SO_3^-]_r + [Cl^-]_r \qquad (7)$$

where $[-SO_3^-]$ refers to the sulfonate ion concentration in the internal solution. Substitution for $[H^+]_r$ gives the equation

$$[[-SO_3^-]_r + [Cl^-]_r][Cl^-]_r \cong [H^+][Cl^-] \qquad (8)$$

The Donnan theory thus offers a means of calculating the concentration of an ion (in this example, Cl⁻) if the concentration of the ion in the external solution is known. This theory is also consistent with the crystal lattice theory. In fact the only major difference between all three theories is in the position and origin of the exchange sites. Since any of these theories must satisfy the law of neutrality, they are analogous to the theories of solutions of electrolytes.

Several relationships other than the Donnan theory for ion-exchange equilibria have been proposed. Two such formulations for column ion-exchange processes have been based on the mass action law and the distillation plate theory.

The use of the mass action law considers the resin as an ideal solution in which the exchange process is completely reversible. Thus the exchange of strontium ions on a sodium-form cation exchanger may be written in the form

$$2NaR + Sr^{2+} \rightleftharpoons SrR_2 + 2Na^+ \tag{9}$$

The mass action law defines a thermodynamic equilibrium constant $K$ as the ratio of the products to reactant activities or, in dilute solutions (as the activity coefficients approach unity), the ratio of the product to reactant concentrations:

$$K = \frac{[SrR][Na^+]^2}{[Sr^{2+}][NaR]^2} \tag{10}$$

Although $K$ has been shown to be reasonably constant within specific concentration ranges, and mostly for exchanges involving ions of the same valence, large variations in $K$ are frequently encountered. In practice, an empirical distribution coefficient $K_d$ is defined for ion-exchange equilibria in which the resin is considered as a solid phase and the distribution of a cation is measured by the ratio of concentrations in the two phases as

$$K_d = \frac{E_r}{E_s} \cdot \frac{V}{W} \tag{11}$$

where $E_r$ and $E_s$ are the number of equivalents $E$ of the cation in the resin $r$ and solution $s$, respectively, $W$ is the weight (in grams) of resin, and $V$ is the volume (in ml) of solution in the system. This distribution coefficient $K_d$ may also be derived from the Donnan theory.

The plate theory of ion-exchange column operation is derived from the theory of distillation columns in which a true equilibrium is never reached. A plate is defined as a length of column in which the average concentration of the ion in the two phases is the same as that for true equilibrium for the two phases in that length of column. The physical length of the column in which this equilibrium occurs is called the "height equivalent to a theoretical plate," HETP. On the basis of the assumptions that the ion exchange resin column is a series of equal theoretical plates, each of which is in equilibrium, and that the number of ions in each plate is small compared to the total quantity of the ion in the eluate, the elution curves shown in Figure 6.2 are given by the elution equation

$$U^* = V(C + 1) \tag{12}$$

where $U^* =$ the volume of eluant passed through the column at any point of elution (ml),

$C =$ the distribution ratio, the amount of the ion in the resin in any plate to the amount of the ion in the interstitial solution of the same plate; $C$ may be considered to be the distribution coefficient $K_d$ [defined in (11)] for each plate and may be expressed as

$$C = \frac{KQW}{V[El^{\pm}]^z} \tag{13}$$

where $K =$ the equilibrium constant,

$Q =$ the exchange capacity of the resin (meq/g),

$W =$ weight of the resin in column (g),

$[El]^{\pm} =$ concentration of the eluant ion,

$z =$ valence of the ion,

Equation 13 indicates the variables that can alter the elutability (and therefore the separability) of ions. These variables are the resin form and capacity, the amount of resin (column length), and the ionic form, ionic charge, and concentration of the eluant ion. Thus ion exchange separability is also affected by pH (e.g., acidity or buffering agents) and complexing agents that can alter the charge of the exchanged or eluant ions.

The elution curves in Figure 6.2 are further simplified by assuming that they have the shape of the Gaussian normal equation [see (5.31)]. This assumption allows the eluant concentrations to be related to the volume of eluant:

$$M = M^* e^{-\sigma^2(U-U^*)^2} \tag{14}$$

where the maximum $M = M^*$ occurs at the maximum $U = U^*$ and $\sigma$ is the width of the Gaussian curve at $M = (1/e)M^*$. The width $\sigma$ is given by

$$\sigma = \frac{1}{V}\left[\frac{p}{2C(C+1)}\right]^{1/2} \tag{15}$$

where $p$ is the number of theoretical plates in the column. The elution curve is then described, from a combination of (12), (14), and (15), as

$$M = M^* \exp\left[-\frac{p}{2}\left(\frac{C+1}{C}\right)\left(\frac{U^*-U}{U}\right)^2\right] \tag{16}$$

and in logarithmic form as

$$\log M = \log M^* - 0.217p\left(\frac{C+1}{C}\right)\left(\frac{U^*-U}{U}\right)^2 \tag{17}$$

Finally, the number of theoretical plates $p$ may be calculated from

$$p = \left(\frac{2C}{C+1}\right)\left(\frac{U^*}{U_a-U^*}\right)^2 \tag{18}$$

In the development of an ion-exchange separation method for a carrier element and interfering radionuclides the parameters of (16) are determined experimentally and adjusted until a satisfactory decontamination factor is achieved.

The plate theory of elution can be used to estimate the minimum length of an ion exchange column required to achieve a quantitative separation set at some acceptable degree of cross-contamination; for example, 0.01 %. It also allows the estimation of the minimum volume of eluant required to remove a particular ion from the column. When interfering radionuclides have been adequately eluted, the eluant may be changed in concentration or kind to remove the carrier element or other interfering radionuclides. With proper selection of resins, ionic form of the desired radionuclide, and eluants, adequate radiochemical purity may be obtained from nonisotopic interfering radionuclides solely by ion-exchange chromatography. Ion exchange is, of course, a useful substitute for stills in providing distilled water in many radiochemistry laboratories. It is also of use in the removal of trace quantities of interfering impurities from reagent-grade chemicals to be used in pre-irradiation treatment of activation analysis samples.

### 6.2.3   Solvent Extraction

Solvent extraction is useful in radiochemical separations for several reasons; notable among them are simplicity, specificity, and speed. The property of speed is especially important for the separation of short-lived activation products. Solvent extraction, as it applies to radiochemistry, consists generally of converting a carrier element from ionic form in aqueous solution into a nonpolar form in which it can be transferred (extracted) into an organic solvent.

Most carrier element salts are strong electrolytes. Although they are generally insoluble in organic solvents, they are chosen as carriers because they are soluble in water. As a solvent, water is a highly polar material that can solvate the ions of a strong electrolyte. For solvent extraction the hydrated ion must be converted into some uncharged form or complex that is more soluble in some organic solvent. This generally requires, for almost all metal ions, that the coordinated water molecules be removed from the ion before a complex molecule which is soluble in an organic solvent can be formed. The large variety of both *specific* complex forms of cations and ions and of *specific* organic solvents in which they are soluble affords an even larger variety of *specific* extraction methods for many elements with a high degree of separation from all other elements.

Another advantage of solvent extraction in radiochemistry is obtained for separations from carrier-free interfering radionuclides. The phenomenon of coextraction (i.e., the extraction of a trace constituent with the carrier element, in which the trace constituent element would not extract in macro quantity), is almost unknown in solvent extraction.

There are two general forms in which nonpolar complexes can be made from metallic ions. The first is the form of chemical coordination-bonded complexes, of which the chelate is the most important for solvent extraction. The second is the form of electrostatic-bonded complexes that produce ion association compounds.

The coordination-bond complexes of interest for extraction consist of three general types:

1. Simple coordination complexes.
2. Central complex ion complexes.
3. Chelate complexes.

The simple coordination complex consists of the metal ion combined with the appropriate number of monofunctional ligands. An example of such a complex is $GeCl_4$, which is extractable into carbon tetrachloride.

The central complex ion complex consists, as its name implies, of a complex ion in place of a monatomic ion as the center of the extractable complex. An example is the heteropoly acids in which such oxygen-containing acids as boric and phosphoric, can incorporate a heteroatom such as Mo, Fe, and Cr. An example of the latter might be a zirconium-phosphoric acid complex with the formula $H_w P_x Zr_y O_z \cdot nH_2O$.

The most common form of coordination complex is the chelate in which the complexing agent occupies two or more coordination positions of the metal ion to form a cyclic compound. Some of the types of organic reagent that form multicovalent bonds with metals are listed in Table 6.3. Most of these reagents form stable chelates with five- or six-member rings.

Some chelating reagents form anionic instead of neutral complexes which are especially useful as "masking" agents. They prevent the extraction of metals with which they form strong anionic complexes. Thus the combination of an extraction agent for the desired carrier element and a masking agent for an important interfering radioelement can result in a markedly increased decontamination factor.

Metal complexes that form as cations or anions may still be extractable if they can be transformed into uncharged compounds by electrostatic association with oppositely charged ions. Examples of such complex systems are also given in Table 6.3. The metal may be contained in either the cationic or ionic member of the ion-pair. As an example the ion association complex $[(C_6H_5)_4As^+, MnO_4^-]$ could be used for the extraction of arsenic or manganese carriers.

**Table 6.3**    Metal Solvent-Extraction Systems[a]

|  | Reactive Grouping |
|---|---|
| I. Chelate Systems | |
|   A. Four-membered ring systems | |
|     1. Dialkyldithiocarbamates | $—N—C—S^{(-)}—$ |
|     2. Xanthates | $—S—C—S^{(-)}—$ |
|   B. Five-membered ring systems | |
|     1. N-Benzoylphenylhydroxylamine | $—O{=}C—N—O^{(-)}—$ |
|     2. Cupferron | $—O{=}N—N—O^{(-)}—$ |
|     3. α-Dioximes | $—N{=}C—C{=}N^{(-)}—$ |
|     4. Dithizone | $—N—N{=}C—S^{(-)}—$ |
|     5. 8-Quinolinols | $—N{=}C—C—O^{(-)}—$ |
|     6. Toluene-3,4-dithiol | $—^{(-)}S—C{=}C—S^{(-)}—$ |
|     7. Catechol | $—^{(-)}O—C{=}C—O^{(-)}—$ |
|   C. Six-membered ring systems | |
|     1. β-Diketones and hydroxycarbonyls | $—O{=}C—C{=}C—O^{(-)}—$ |
|       a. Acetylacetone | |
|       b. Thenoyltrifluoracetone (TTA) | |
|       c. Morin | |
|       d. Quinalizarin | |
|     2. Nitrosonaphthols | $—O{=}N—C{=}C—O^{(-)}—$ |
|     3. Salicylaldoxime | $—N{=}C—C{=}C—O^{(-)}—$ |
|   D. Polydentate systems | |
|     1. Pyridyl-azo-naphthol (PAN) | $—N{=}C—N{=}N—C{=}C—O^{(-)}—$ |
| II. Ion Association Systems | |
|   A. Metal contained in cationic member of ion-pair | |
|     1. Alkylphosphoric acids | |
|     2. Carboxylic acids | |
|     3. Cationic chelates | |
|       a. Phenanthrolines | |
|       b. Polypyridyls | |
|     4. Nitrate | |
|     5. Trialkylphosphine oxides | |
|   B. Metal contained in anionic member of ion-pair[b] | |
|     1. Halides ($GaCl_4^-$) | |
|     2. Thiocyanates ($Co(CNS)_4^-$) | |
|     3. Oxyanions ($MnO_4^-$) | |
|     4. Anionic chelates ($Co(Nitroso\ R\ salt)_3^{3-}$) | |

[a] From G. H. Morrison and H. Freiser, Extraction, *Anal. Chem.* **30,** 632–640 (1958), by permission of the American Chemical Society.

[b] The cation member associated with these metal-containing anions is usually of an "onium" type such as oxonium, e.g., $ROH_2+$, $R_2OH+$, $R_2COH+$; ammonium, e.g., $RNH_3+,\ldots, R_4N+$; arsonium, $R_4As+$; phosphonium, $R_4P+$; stibonium, $R_4Sb+$; sulfonium, $R_3S+$.

The second part of the extraction procedure is the transfer of the complexed element into the organic solvent. This is generally accomplished in radiochemical procedures by the batch-extraction method in which the solution containing the extractable complex is shaken with the organic solvent in a separatory funnel, and after the two immiscible solvents have separated the heavier one is removed by running it out through the stopcock.

The degree of transfer from one solvent to the other is given by the *Nernst partition isotherm*, which defines the distribution of a solute shared between two immiscible solvents. For a system at constant temperature, and in which the solute has the same molecular weight in both solvents, the ratio of concentrations at equilibrium is given by a distribution coefficient $K_d$ defined as

$$K_d = \frac{[M]_2}{[M]_1} \tag{19}$$

where the brackets denote the concentration of the solute $M$ in the two solvents 1 and 2. This distribution law, although not thermodynamically rigorous, gives a reasonable approximation to the distribution of a carrier-free radioelement separated in laboratory-scale volumes but may involve serious deviations for carrier concentrations, especially if the carrier element is involved in any chemical association or dissociation with either of the solvents.

A rigorous derivation of the Nernst partition isotherm may be made from thermodynamic considerations in which the distribution of the solute is at equilibrium when the partial molal free energy $\bar{F}$ of the solute in each solvent is equal. The partial molal free energy is given by

$$\bar{F} = \bar{F}^0 + RT \ln \{A\} \tag{20}$$

where $\bar{F}^0 =$ a constant value for a standard solution, such as 1 molal,

  $R =$ the gas constant,

  $T =$ the absolute temperature,

  $\{A\} =$ the molal activity.

The molal activity is defined as

$$\{A\} = \gamma[A] \tag{21}$$

where $[A] =$ the solute concentration, in molality,

  $\gamma =$ the molal activity coefficient.

Thus at equilibrium

$$\bar{F}_1^0 + RT \ln \{A\}_1 = \bar{F}_2^0 + RT \ln \{A\}_2 \tag{22}$$

and

$$\frac{\{A\}_2}{\{A\}_1} = e^{-[(F_1{}^0 - F_2{}^0)/RT]} \tag{23}$$

Since inorganic solutes do not generally affect the immiscibility of the two solvents, the values of $F_1^0$ and $F_2^0$ are constant, and the exponential term may be defined as a constant partition coefficient $p$ given by

$$p = \frac{\{A\}_2}{\{A\}_1} = \frac{\gamma_2}{\gamma_1} K_a \tag{24}$$

In the general case of solvent extraction, in which the extent of chemical effects between the solute and the solvents is unknown, the over-all distribution of the carrier element between the two solvents is given by a stoichiometric distribution ratio $D$ defined as

$$D = \frac{[C]_0}{[C]_a} \tag{25}$$

where the brackets denote the total carrier element concentration in the organic and aqueous solvent phases. In an ideal system with no interactions the distribution ratio $D$ would be equal to the partition coefficient $p$. In practice, however, $D$ can vary greatly with such changes in the extraction conditions as pH and total salt concentrations. The distribution ratio can also be expressed as an efficiency term, the percent extracted, $\%E$, by

$$\%E = \frac{100D}{D + V_w/V_0} \tag{26}$$

where $V_w$ and $V_0$ are the volumes of the aqueous and organic phases, respectively.

In the batch-extraction method a separation of the desired radionuclide from interfering radionuclides is made by choosing suitable extraction forms that give large differences in extractability. The effectiveness of a separation may be expressed as a separation factor $\beta$, which is related to the distribution ratios of the two radionuclides in the two solvents:

$$\beta = \frac{(C_1)_0/(C_1)_w}{(C_2)_0/(C_2)_w} = \frac{D_1}{D_2} \tag{27}$$

where $C_1$ is the concentration of the desired radionuclide in each phase and $C_2$ is the concentration of the interfering radionuclide in each respective phase.

The most important consideration in a solvent extraction separation procedure is the choice of a complex-solvent system that makes $\beta$ large enough to achieve a satisfactory separation. Several techniques have been developed for radiochemical solvent extraction procedures that maximize $\beta$ for particular radionuclide separations, among which are included

(a) mixing of solvents,
(b) stripping or back-extraction,
(c) back-washing,
(d) oxidation and reduction,
(e) use of masking agents,
(f) use of salting-out agents.

The mixing of solvents is used to increase the separation factors for specific radionuclides or to dilute certain organic compounds to increase their extraction ability. Stripping is the removal of the extracted carrier from the organic solvent for further radiochemical separations. Stripping may be achieved by volatilizing the organic solvent in the presence of a small amount of water to hold the solute or by wet-ashing to fumes in the presence of strong acids such as sulfuric, nitric, or perchloric. Back-extraction is a chemical means of stripping in which the solute is extracted back into the aqueous phase with a solution containing acids or other reagents to destroy the complex form of the carrier element. Backwashing is the back extraction, with a fresh batch of the aqueous phase, of impurities carried over with batches of the extracting solvent. This backwashing step can remove many of the impurities from the combined organic solvent batches with little loss of the carrier element. Increased selectivity can also be achieved by changing the oxidation state of an impurity ion to one that does not form a suitable complex for extraction. A change in oxidation state can also be used as a stripping or back extraction method in which the carrier ion complex is destroyed at the altered oxidation state. The use of masking agents to form charged complexes with impurity radionuclides has already been mentioned. Its use as a back-extraction method is also possible. The use of salting-out agents can result in marked increases in sensitivity, and the presence of large concentrations of salts increases the distribution of many metal complexes toward the organic phase. The selection of a suitable salting-out agent can enhance the separation factor between the carrier element and the impurity radionuclides.

A large variety of extraction systems has been developed for radiochemical separations; for example, the book by Morrison and Freiser, listed in Section 6.4.2, describes more than 32 extraction systems that are useful for the separation of more than 65 elements. The NAS series on the Radiochemistry of the Elements includes solvent extraction procedures for almost all of the elements.

### 6.2.4   Distillation and Electrodeposition

Distillation and electrodeposition are two important methods of chemical separation used in radiochemistry. When pertinent, excellent separations are possible by these physical means. The major obstacle in their widespread use in radiochemistry is the applicability of these methods only to specific chemical elements or compounds; that is, distillation is practical for only those elements or their compounds that are volatile at the normal temperatures conveniently available in laboratories, and electrodeposition is practical primarily for those elements with low electromotive potentials. For these specific elements, however, radiochemical separations from the other elements are greatly enhanced.

### *Distillation*

The term distillation is most properly applied to the separation of a mixture of liquids by vaporization, but the term is also applied to the process in which a volatile compound is removed from a solution by evaporation. Thus distillation is a process for obtaining pure water. It is also useful for the removal of a specific radionuclide from an impure solution if the desired radionuclide can be converted into a volatile compound at a temperature lower than those for the impurities in the solution. In other cases the matrix or major interfering radionuclides can be removed by volatilization. Radionuclides that are gaseous at room temperature (e.g., the rare gases) can be volatilized during the dissolution of the irradiated sample in a closed system and collected by gas adsorption.

The decontamination factors in distillation separations can be modified by controlling several parameters; for example, chemical system, pressure, temperature, distillation rate, and condensing system. Distillation separations of volatile radionuclides are carried out by two general methods: the carrier vapor distillation train and the simple one-plate vacuum distillation.

For carrier vapor distillation ordinary distillation units may be used for solution samples. In radiochemical techniques all-glass systems are generally used to minimize the amount of adsorption of the carrier-free radionuclides. In some separations a stream of carrier gas, such as helium or nitrogen, may be used to facilitate removal of the volatile fraction. A train of vapor-collecting flasks may be introduced into the system as chemical traps for specific interfering radionuclides. The desired element is then either condensed as a liquid, chemically trapped by conversion into a nonvolatile form, or adsorbed as a gas on a solid adsorbent. Elements that have been separated by distillation methods from solution are listed in Table 6.4. They may be separated into groups of elements that are gaseous at room temperature, elements that are volatile at elevated temperature, and elements that have a

**Table 6.4** Elements for Which Radiochemical Distillation Separation Procedures Have Been Reported[a]

| Z | Element | Volatile Form | Z | Element | Volatile Form |
|---|---|---|---|---|---|
| 1 | Hydrogen | $HT, CH_3T$, etc. | 38 | Strontium | Sr on W filament |
| 2 | Helium | He | 43 | Technicium | $Tc_2O_7$ |
| 6 | Carbon | $CO_2, C_2H_2, CH_4,$ etc. | 44 | Ruthenium | $RuO_4$ |
| 7 | Nitrogen | $NH_3$ | 45 | Rhodium | Rh on W filament |
| 8 | Oxygen | $O_2, CO_2$ | 47 | Silver | Ag on Ta filament |
| 10 | Neon | Ne | 48 | Cadmium | Cd from Hg-amalgam |
| 14 | Silicon | $SiF_4$ | 49 | Indium | In |
| 15 | Phosphorous | P on W filament | 50 | Tin | Halides |
| 16 | Sulfur | $SO_2, H_2S$ | 51 | Antimony | Halides |
| 17 | Chlorine | $Cl_2$ | 52 | Tellurium | $TeBr_4$ |
| 18 | Argon | Ar | 53 | Iodine | $I_2$ |
| 22 | Titanium | $TiCl_4$ | 54 | Zenon | Zn |
| 23 | Vanadium | $VCl_4$ | 75 | Rhenium | $Re_2O_7$ |
| 24 | Chromium | $CrO_2Cl_2$ | 76 | Osmium | $OsO_4$ |
| 25 | Manganese | $HMnO_4$ | 79 | Gold | Au |
| 32 | Germanium | Halides | 80 | Mercury | $Hg, HgCl$ |
| 33 | Arsenic | Halides | 84 | Polonium | $Po_2S_3$, organic |
| 34 | Selenium | $SeBr_4$ | 86 | Radon | Rn |
| 35 | Bromine | $Br_2$ | 87 | Francium | Fr |
| 36 | Krypton | Kr | 92 | Uranium | $UF_6$ |

[a] Adapted from J. R. DeVoe, Application of Distillation Techniques to Radiochemical Separations, National Academy of Sciences—National Research Council Report NAS-NS-3108, August 1962.

conveniently volatile compound. Section 6.4 lists several references of specific radiochemical distillation methods for these elements.

The one-plate vacuum distillation is generally a closed system consisting of a distillation furnace and a vapor collector. A system for the vacuum distillation of metals is shown in Figure 6.3. The vapor collector is placed close to the furnace to reduce the dispersion of the vapor and therefore increase the yield of the distillation. The collector is often made of a thin, small-area plastic film held in a cold terminal. After the distillation the film serves as the mounting for the radiation measurement source. This technique is especially useful for the removal of a desired radionuclide in carrier-free form from a solid irradiated sample without the need to dissolve the sample. Furthermore, the deposited film can be transferred rapidly to the radiation detection system. This method has proved useful for the measurement of radioactivation products with half-lives as short as 1 sec.

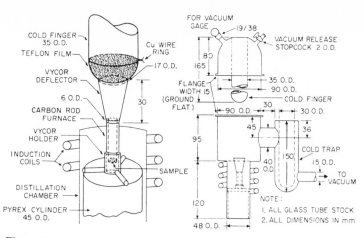

**Figure 6.3** Apparatus for vacuum distillation showing a furnace with the collector in place (*left*) and the complete apparatus showing the location of the furnace (*right*). (From J. R. DeVoe, Application of Distillation Techniques to Radiochemical Separations, National Academy of Sciences—National Research Council Report NAS-NS-3108, August 1962.)

The decontamination factor of a metal radioactivation product from a metal target can be estimated for the single-plate batch-type distillation from its relative volatility. The more volatile product 1 has a decontamination factor $D$ from the less-volatile activated matrix metal 2 given by

$$D = \frac{A_2^\circ}{A_2'} \tag{28}$$

where $A_2$ is the radioactivity of the matrix element present with the product element before, $A_2^\circ$, and after, $A_2'$, the separation. The yield $Y$ of the product is given by

$$Y = \frac{A_1'}{A_1^\circ} \tag{29}$$

The relative volatility $\alpha$, which is used to estimate the degree of separation, is defined as

$$\alpha = \frac{\bar{P}_1 X_2}{\bar{P}_2 X_1} \tag{30}$$

where $\bar{P}_1$ and $\bar{P}_2$ are the partial pressures of the two components and $X_1$ and $X_2$ are the respective mole fractions in the distillation pot. Assuming that Dalton's law of partial pressures applies,

$$\frac{\bar{P}_1}{\bar{P}_2} = \frac{Z_1}{Z_2} \tag{31}$$

where $Z$ is the respective mole fraction in the vapor. Thus

$$\alpha = \frac{Z_1 X_2}{Z_2 X_1} = \frac{M_1' M_2}{M_1 M_2'} \tag{32}$$

where $M$ = the respective number of moles in the distillation pot,

$M'$ = the respective number of moles in the vapor phase.

Since the specific activity $S$ of each component remains constant during the distillation,

$$\alpha = \frac{A_1' A_2}{A_1 A_2'} \tag{33}$$

If it is further assumed that Raoult's law is valid under these conditions, then

$$\bar{P}_1 = X_1 P_1 \quad \text{and} \quad \bar{P}_2 = X_2 P_2 \tag{34}$$

Thus

$$\alpha = \frac{P_1}{P_2} \tag{35}$$

where $P$ is the respective vapor pressure of each pure component.

Equation 35 implies that $\alpha$ is a constant. At the beginning of the distillation, when the yield is negligible, that is, when $A_2 \approx A_2^\circ$ and $A_1 \approx A_1^\circ$,

$$\alpha = \frac{A_1' A_2^\circ}{A_1^\circ A_2'} \tag{36}$$

As the distillation progresses, however, DeVoe* shows that the product $DY$ decreases according to the expression

$$\alpha' = DY = \frac{Y}{1 - (1 - Y)^{1/\alpha}} \tag{37}$$

Equation 37 further implies that $\alpha'$ should be independent of the interfering radionuclide concentration and that there is an optimum yield that depends on the degree of separation required.

**Electrodeposition**

The separation method of electrodeposition involves the deposition of a metal (or its compound) from solution on an electrode by the flow of an electric current. It is often the final separation of a metal, plated on a disk as the counting source and for yield determination, or an electrolytic separation to be followed by some other stoichiometric processes for final

* J. R. DeVoe, Application of Distillation Techniques to Radiochemical Separations, National Academy of Sciences—National Research Council Report NAS-NS-3108, 1962.

separation. Analytical electrodeposition has as its basis

  (a) the stoichiometric laws of electrolysis,
  (b) the thermodynamic relationships of electrode potentials,
  (c) the kinetics determined by the ion migration and diffusion in liquids.

### Electrolysis

Electrolysis is the migration of ions in an electrolyte solution due to an applied potential difference. If the potential difference between an electrode and the solution is sufficient to produce a current to flow (positively charged ions toward the cathode, negatively charged ions towards the anode), the chemical effects are governed by the law of electrolysis:

$$m = \frac{IEt}{F} \tag{38}$$

where $m$ = mass of the electrodeposited substance (g),

  $I$ = current (A),

  $E$ = equivalent weight (g/g-atom),

  $t$ = time of current flow (sec),

  $F$ = Faraday constant (96,500 C).

### Electrode Potential

If a metal electrode, immersed in a solution of its ions, is in reversible equilibrium with them, a steady potential with respect to the solution is established, given by the Nernst equation

$$E = E_0 + \frac{RT}{zF} \ln \{A\} \tag{39}$$

where $E$ = potential difference with respect to the solution (V),

  $E_0$ = standard electrode potential (V),

  $R$ = gas constant per mole,

  $T$ = absolute temperature (°K),

  $z$ = valence of the metal ions,

  $F$ = Faraday constant,

  $\{A\}$ = activity of the metal ions.

The standard electrode potential is defined as the potential difference in volts of the metal (in equilibrium with a solution of its ions at unit activity) with respect to a hydrogen electrode in reversible equilibrium with a solution of hydrogen ions at unit activity. Some standard electrode potentials of metals used in electrodeposition procedures are listed in Table 6.5.

**Table 6.5** Standard Potentials for Electrode Reactions at $25°$[a]
$$M^{n+} + ne = M$$

| Ion | $E_0$ (V) | Ion | $E_0$ (V) | Ion | $E_0$ (V) |
|-----|-----------|-----|-----------|-----|-----------|
| $Au^+$ | $+1.70$ | $Cd^{2+}$ | $-0.40$ | $Lu^{3+}$ | $-2.25$ |
| $Au^{3+}$ | $+1.50$ | $Fe^{2+}$ | $-0.44$ | $Am^{3+}$ | $-2.32$ |
| $Pt^{2+}$ | $+1.20$ | $Ga^{3+}$ | $-0.53$ | $Y^{3+}$ | $-2.37$ |
| $Pd^{2+}$ | $+0.99$ | $Cr^{3+}$ | $-0.71$ | $Mg^{2+}$ | $-2.37$ |
| $Hg^{2+}$ | $+0.86$ | $Zn^{2+}$ | $-0.76$ | $La^{3+}$ | $-2.52$ |
| $Ag^+$ | $+0.80$ | $Nb^{3+}$ | $-1.10$ | $Gd^{3+}$ | $-2.40$ |
| $Hg_2^{2+}$ | $+0.79$ | $V^{2+}$ | $-1.18$ | $Sm^{3+}$ | $-2.41$ |
| $Cu^+$ | $+0.52$ | $Mn^{2+}$ | $-1.18$ | $Nd^{3+}$ | $-2.44$ |
| $Cu^{2+}$ | $+0.34$ | $Zr^{4+}$ | $-1.53$ | $Ce^{3+}$ | $-2.48$ |
| $Bi^{3+}$ | $+0.20$ | $Ti^{2+}$ | $-1.63$ | $Na^+$ | $-2.71$ |
| $Sb^{3+}$ | $+0.10$ | $Al^{3+}$ | $-1.66$ | $Ca^{2+}$ | $-2.87$ |
| $Sn^{4+}$ | $+0.05$ | $Hf^{4+}$ | $-1.70$ | $Sr^{2+}$ | $-2.89$ |
| $2H^+$ | $\pm0.00$ | $U^{3+}$ | $-1.80$ | $Ba^{2+}$ | $-2.90$ |
| $Pb^{2+}$ | $-0.13$ | $Be^{3+}$ | $-1.85$ | $Ra^{2+}$ | $-2.92$ |
| $Sn^{2+}$ | $-0.14$ | $Np^{3+}$ | $-1.86$ | $Cs^+$ | $-2.93$ |
| $Mo^{3+}$ | $-0.20$ | $Th^{4+}$ | $-1.90$ | $Rb^+$ | $-2.93$ |
| $Ni^{2+}$ | $-0.25$ | $Pu^{3+}$ | $-2.07$ | $K^+$ | $-2.93$ |
| $Co^{2+}$ | $-0.28$ | $Sc^{3+}$ | $-2.08$ | $Li^+$ | $-3.05$ |

[a] From C. L. Wilson, D. W. Wilson, and C. R. Strouts, *Comprehensive Analytical Chemistry*, Vol. IIA (Elsevier, Amsterdam, 1964), p. 9.

Since the activity of metal ions is approximately proportional to their concentration, (38) relates the change in potential to the change in concentration at constant temperature:

$$\Delta E = k \, \Delta \ln C \tag{40}$$

where $k$ is a proportionality constant.

### Conductance and Diffusion in Liquids

The current-carrying capacity (conductance) of an electrolyte solution (at constant temperature) is given by Ohm's law:

$$I = \frac{E}{R} = CE \tag{41}$$

where $I$ = current (A),

$E$ = potential difference (V),

$R$ = resistance of the electrolyte ($\Omega$),

$C$ = conductance ($\Omega^{-1}$ = mho).

The conductivity or specific conductance $K$ of a given electrolyte of unit length and cross section and at a constant temperature is related to the ion mobility by the equation

$$\Lambda = KV \tag{42}$$

where $V$ is the volume containing 1 g equivalent of the electrolyte compound. As $V$ increases, $\Lambda$ approaches a limiting value $\Lambda_\infty$ corresponding to conditions in which all ions contribute maximum effort toward conductance. For a pure solution of a binary electrolyte

$$\Lambda_\infty = l_a + l_c \tag{43}$$

where $l_a$ and $l_c$ are the ionic mobilities ($cm^2/\Omega$) of the anion and cation, respectively. Since ionic mobilities increase with temperature, larger currents can be carried at higher temperatures.

The rate of electrodeposition is also influenced by diffusion, which results when the solution in contact with the electrode is depleted in concentration. This diffusion results in an activity (concentration) gradient near the electrode, described by Fick's law:

$$\frac{dN}{dt} = AD\frac{da}{dx} \tag{44}$$

where $dN/dt$ = rate of diffusion of the ions up to an area $A$ of the plane electrode,

$da/dx$ = activity gradient through the diffusion layer,

$D$ = diffusion coefficient.

In practice, vigorous stirring during electrodeposition decreases the thickness of the diffusion layer, thereby increasing the rate of diffusion. The rate of deposition can also be increased by adding another electrolyte that does not deposit. For uniform deposition as a radiation source and for stoichiometric measurement of the chemical yield the deposit must be in suitable microcrystalline form, free from occluded material and sufficiently adherent to withstand washing, drying, and weighing operations.

### Electrodeposition Separations

The separation of two cations by electrodeposition results from control of the applied potential to exceed the deposition potential of one of the cations but not the other. If $V$ is the overvoltage of the process (the potential to reduce the equilibrium value to the deposition value), the deposition potential is given by

$$E_D \simeq E_0 - V + \frac{RT}{zF}\ln C \tag{45}$$

or

$$E_D \simeq E_0 - V + 0.059 \log_{10} C \qquad (46)$$

for a univalent ion at 21°C. The deposition potential is thus a characteristic of the metal to be deposited and is dependent on the concentration of the metal in solution. The deposition potential is also a function of the current flowing through the solution; for example, Figure 6.4*a* shows the effect of increasing current on the deposition potential of two metals to be separated by electrodeposition of metal 1. For the range of currents shown in the figure only metal 1 will deposit if the applied voltage is equal to the value at the dashed line, for the deposition potential of metal 2 is not exceeded (negatively). However, as the deposition of metal 1 proceeds, each tenfold reduction in concentration of metal 1 reduces its deposition potential by 0.059 V, according to (46). Figure 6.4*b* shows that at some concentration of metal 1, for a given current, the deposition potential of metal 2 will be exceeded

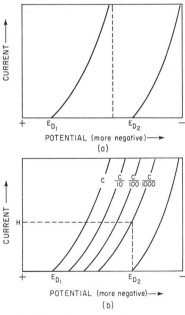

**Figure 6.4** (*a*) Current-voltage relationship during electrodeposition; (*b*) effect of concentration on electrodeposition curves. [From A. J. Lindsey, Electrode-position, Chapter 11, in C. L. Wilson and D. W. Wilson, Eds., *Comprehensive Analytical Chemistry*, Vol. 11 (Elsevier, Amsterdam, 1964).]

**Table 6.6**    Metals Grouped into Electrodeposition Classes[a]

---

A. Precious Metal Group
   Gold, platinum, mercury, silver

B. Copper Group
   Copper, bismuth, antimony, arsenic

C. Lead Group
   Lead, tin

D. Zinc Group
   Nickel, cobalt, cadmium, zinc

E. Mercury Cathode Group (amalgams)
   Alkali metals, rare earths; also includes Bi, Cd, Cr,
   Co, Cu, Ga, Ge, Au, In, Ir, Fe, Ag, Mo, Ni, Pd,
   Pt, Po, Re, Ru, Ag, Tc, Th, Sn, and Zn

F. Anodic Group (metals anodically deposited as oxides)
   Lead, cobalt, thallium, plutonium

---

[a] From A. J. Lindsey, Electrodeposition, Chapter II, in C. L. Wilson and D. W. Wilson, Eds., *Comprehensive Analytical Chemistry*, Vol. II (Elsevier, Amsterdam, 1964).

(negatively) and both metals would then deposit. For a given decontamination factor the pertinent decrease in concentration of the desired electrodeposited metal determines the maximum current $I$ to be used in the electrodeposition separation.

Many metals are conveniently separated by electrodeposition. Table 6.6 lists a number of them, grouped according to the potential series given in Table 6.5. Generally metals in a group are difficult to separate from one another solely by control of the applied potential. Complexing agents are often used to alter the relative deposition potentials.

## 6.3 RADIONUCLIDE ANALYSIS

The final part of a radiochemical analysis is the determination of the radioactivity content of the isolated carrier. This determination generally consists of three considerations:

1. The qualitative identification of the desired radionuclides.

2. The determination of the background and presence of any interfering radionuclides.

3. The quantitative calibration of the measured radiations.

Other factors involved in the quantitative measurement of a desired element include the yield of the chemical separation and the form in which the carrier element is prepared for the radioactivity measurement. These factors have already been discussed. In general the chemical and physical form in which the carrier is prepared for radioactivity measurement is dictated primarily by the type of radiation detector to be used and the type and energy of the radiations to be measured. The carrier element is usually prepared as a source that gives the maximum over-all counting efficiency consistent with the requirement to identify the radiation unambiguously. At very low activity levels the special low-background detection methods described in Chapter 5 may be required to obtain the necessary sensitivity.

### 6.3.1 Qualitative Identification

Radioactivation analysis, for determinations at maximum sensitivity, searches for the presence of particular elements. A typical objective for an activation analysis might be the following: Is there a measurable amount of element $x$ in sample $y$; for example, the often quoted analysis of "traces of arsenic in dead people's hair?" The attainment of the objective is undertaken in radioactivation analysis by choosing a suitable nuclear reaction for the element (in this example, fast or slow neutrons with arsenic) which gives a suitable activation product (i.e., $^{74}$As or $^{76}$As). Radiochemical analysis of the sample (the irradiated hair) is completed by determining whether the chemically separated carrier (arsenic) has any radiations that have the properties of the expected products ($^{74}$As or $^{76}$As). The properties of a radionuclide that determine its identity in radiochemical analysis include the following

1. Its method of production.
2. Its chemical behavior as an isotope of the carrier element.
3. Its half-life.
4. The types of radiation emitted.
5. The energies of these radiations.
6. The branching ratios of the radiations by type and energy.

The first property is especially useful in choosing a nuclear reaction that enhances sensitivity by greater yield, gives a more easily determined radionuclide, or reduces the interference from other elements. In the example given the choice of nuclear reactions might be $(n,\gamma)$ or $(\gamma,n)$, giving, respectively, the activation products 26.5-hr [76]As (decaying by emission of $\beta^-$ particles and gamma rays) and 18-d [74]As (decaying by emission of $\beta^+$- and $\beta^-$-particles and gamma radiations).

The property of isotopic chemical behavior is useful when the radionuclide radiations are difficult to measure. Repeated chemical separations that result in a constant specific activity give confidence of the atomic number of the radionuclide, but the possibility of quantitative coprecipitation of another radionuclide must always be evaluated. When the decontamination of the desired radionuclide is adequate, positive identification can usually be made by half-life or decay-scheme determinations alone.

A more stringent requirement for radioactivation analysis results from the objective: Here is a small sample; what's in it? The identification of all the elements in an unknown sample is a formidable task. The qualitative identification of radionuclides must, in this case, follow from a systematic separation of the chemical elements, individually or in chemically similar groups. Fortunately in most applications of radioactivation analysis the elements to be measured have been predetermined by the requirements for the analysis.

The qualitative identifications of radionuclides in instrumental activation analysis are more difficult because the chemical nature of the radioactive nuclides is not determined. The identification of the radionuclides must be made entirely from the radiations detected. The properties of half-life and the types and energies of the radiations are generally examined, although in the specific method of instrumental gamma-ray spectrometric analysis, qualitative identification of gamma-emitting radionuclides is made solely by measurement of the characteristic gamma-ray abundances and energies.

The radiochemical method of identification, even for samples not chemically separated, compares the measured half-lifes, types, and energies of radiation with "literature" values for these properties.

Half-life determinations are useful for the shorter lived radionuclides, where a significant reduction in radioactivity is measurable within a convenient period of counting. The half-lives for mixtures of radionuclides can be resolved from a total $\beta^-$-activity decay curve only when the number of nuclides is small (generally $< 4$) and if the half-lives are sufficiently different (by at least a factor of 2). Half-lives of mixtures of gamma-ray-emitting radionuclides can be more readily accomplished by measuring the half-lives of the gamma-ray energies spectroscopically separated with a multichannel analyzer. Tables of beta- and gamma-ray-emitting radionuclides ordered by half-life are available in many nuclear data sources (see Section 7.4.1).

The type of radiation emitted is an important property, since it determines the kinds of detectors required. Radionuclides such as [35]S or [45]Ca require beta-radiation detectors, since they emit no gamma radiation. Radionuclides such as $^{51}$Cr or $^{54}$Mn require x-ray detectors, since they decay solely by electron capture. Special counting methods, such as $\beta^-$-$\gamma$ or $\beta^+$-annihilation gamma radiation coincidence counting or magnetic-field separation of $\beta^+$ from $\beta^-$ radiations take advantage of the multiple types of radiation emitted by specific radionuclides. Although only a few activation-product radionuclide decay by alpha emission, for these cases the measurement of alpha-particle radiation becomes a useful means of identification.

Spectroscopic analysis of radiation is of practical value in radioactivation analysis primarily for gamma radiations. For those few nuclides that decay by alpha-emission, solid-state semiconductor detectors afford adequate resolution to measure the alpha-particle radiations by spectroscopic means alone, although in most cases the sample must be chemically treated to prepare the radionuclide as a thin source. The continuous spectrum property of beta radiation makes energy resolution difficult. The maximum beta-ray energy can usually be estimated only for single radionuclides or mixtures of only two or three beta emitters in which $E_{\beta^-}$ (max) is sufficiently different to give an absorption curve with separable slopes. The measurement of mono-energetic conversion electrons by semiconductor detectors is another useful means of nuclide identification.

Gamma-ray scintillation spectrometry is a primary method of radionuclide analysis. Its usefulness in determining the presence of two radionuclides with almost similar gamma-ray energies is limited only by the resolution of currently available scintillation detectors. As the efficiency of the semiconductor detectors is increased, their inherently greater resolution capability will increase their usefulness as gamma-ray spectrometers.

### 6.3.2 Quantitative Calibration

The measurement of radioactivity is the determination of the number of disintegrations of a given radionuclide occurring in unit time, which according to the radioactive decay law is proportional to the absolute number of radioactive atoms present.

$$D = -\frac{dN}{dt} = \lambda N \qquad (46)$$

The constant of proportionality $\lambda$ is also an experimentally determined value. Thus the measurement of the number of radioactive atoms present in a given source is determined by the measurements of the decay rate and the half-life. The measurement of decay rate, in turn, is determined by the measurement of a radiation-detection rate and an efficiency for detection $\epsilon$. In most cases of activation products from low-energy reactions the half-lives reported in the literature have been obtained by several methods of evaluating

$D$ and $N$ and are sufficiently reliable for most purposes in radioactivation analysis. Half-life values, however, are revised in the literature from time to time, and it is good practice to use the latest sources of nuclear data available.

The quantitative calibration of a radionuclide thus becomes the determination of the disintegration rate $D$ of the nuclide from a measured counting rate $A$ in a given detector with an over-all counting efficiency $\epsilon$, where

$$A = \epsilon D \qquad (47)$$

The determination of the disintegration rate of a given radionuclide may be made in two ways; on a relative basis or on an absolute basis.

The relative basis requires a standard source of the radionuclide, one in which the disintegration rate is known accurately. If the unknown sample ($u$) of the radionuclide is counted under conditions identical to those for the standard source ($s$), the ratio of the disintegration rates is equal to the ratio of counting rates, since the efficiency of counting for the two sources is identical:

$$\frac{\epsilon D_u}{\epsilon D_s} = \frac{D_u}{D_s} = \frac{A_u}{A_s} \qquad (48)$$

or

$$D_u = \left(\frac{A_u}{A_s}\right) D_s \qquad (49)$$

There are many ways to obtain standard sources of radionuclides. The simplest, of course, is to buy them, since many primary standard radionuclides are available commercially. However, primary standards cannot be obtained for many other radionuclides because of inconveniently short half-lives or extremely complex decay curves.

Standardization can be made in the laboratory by several means. Among the more popular methods are $4\pi$ absolute counting of beta-emitting radionuclides and coincidence counting of radionuclides that decay with emission of multiple radiations; for example, beta and gamma rays, two or more gamma rays, or gamma and x-rays. The $4\pi$ geometry counter, in which two identical proportional counters surround a "weightless" source mounted on an ultrathin film, achieves an over-all counting efficiency, $\epsilon > 99.5\%$, so that the disintegration rate is essentially equal to the counting rate.

In the coincidence counter two detectors are used, each sensitive to a particular radiation, for example, a proportional counter for beta rays, a scintillation counter for gamma-rays, or a thin scintillation counter for x-rays. A schematic diagram of a beta-gamma coincidence counter is shown in Figure 6.5. The counting rate of the beta counter is

$$A_\beta = \epsilon_\beta D \qquad (50)$$

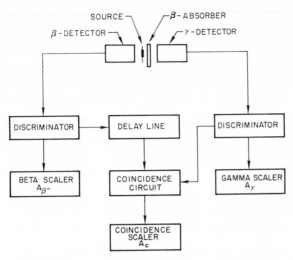

**Figure 6.5** Schematic diagram of a beta-gamma coincidence-counting system.

and the counting rate of the gamma counter is

$$A_\gamma = \epsilon_\gamma D \qquad (51)$$

Since a coincidence event counts only when beta and gamma counts occur simultaneously (during the coincidence time) in the two counters, the coincidence counting rate is

$$A_c = \epsilon_\beta \epsilon_\gamma D \qquad (52)$$

Rearrangement of (50) to (52) shows that

$$D = \frac{A_\beta A_\gamma}{A_c} \qquad (53)$$

A number of corrections is required in those counting systems. They include corrections for detector backgrounds, random coincidences, conversion electron decays, and dead-time losses. For some radionuclides the beta decay results in a metastable nuclide that does not emit a prompt gamma ray. Additional sensitivity can be achieved by adding a delay line between the beta detector and the coincidence circuit. The delay-line time can be adjusted to be equal to the delay time for the gamma emission from the particular metastable isomer.

The absolute method of counting determines the disintegration rate of the radionuclide by calculating the actual value of the over-all counting efficiency $\epsilon$ from evaluation of its many parameters. Although $\epsilon$ can be determined experimentally by counting known radioactivity standards, it can be estimated as a product of the following parameters:

$$\epsilon = (I)(G) f_W f_A f_C f_B f_H f_S \qquad (54)^*$$

where $(I) =$ the detection probability, the probability that a beta particle entering the counter gas will produce sufficient ionization to result in a count;

$(G) =$ the geometric relation between the source and the sensitive volume of the counter, given by the fraction of the solid angle of $4\pi$ srs which enter the counter from an isotropically emitting source;

$f_W =$ the factor for the absorption of the beta particles in the window of the counter and in the air between the source and window;

$f_A =$ the factor for the scattering effect of the air;

$f_C =$ the factor for the effect of any covering material over the sample—the covering of samples with thin protective films (e.g., cellophane, mylar) is standard practice in most laboratories to prevent contamination of the counter;

$f_B =$ the factor for the effect of backscattering of electrons moving away from the counter by the material supporting the source;

$f_H =$ the factor for the effect of the counter supports and shield in scattering electrons from outside the geometric solid angle into the counter;

$f_S =$ the factor for the effect of the finite mass of the source, self-absorption and self-scattering, which at each height in the source acts as an absorber and scatterer for beta particles originating in the mass below it.

For a given radionuclide $\epsilon$ must also be corrected for the radiation yield, the number of radiations of the type measured per disintegration, and the contribution from other radiations present in the decay scheme.

Several methods of end-window counting are employed to reduce the complexity and corresponding uncertainty of $\epsilon$ for absolute counting of samples. Many methods eliminate a great number of the parameters given in (54). An obvious way is to standardize a counter for each radionuclide by using a fixed set of conditions (e.g., sample size, location, and thickness) to

---

* A discussion of each of these parameters is given by E. P. Steinberg, Counting Methods for the Assay of Radioactive Samples, Chapter 5, in A. H. Snell, *Nuclear Instruments and Their Uses*, Vol. 1 (Wiley, New York, 1962).

count a known standard of the radionuclide. This is possible for a great number of radionuclides of interest in activation analysis. Other methods include preparation of essentially weightless samples on thin backing and counting in a position of known geometry. The parameters of (54) can be estimated for these conditions with high precision.

### 6.3.3  Scintillation Gamma-Ray Spectroscopy

The scintillation detector is generally associated with gamma-ray spectroscopy in which mixtures of gamma-ray emitting radionuclides can be resolved quantitatively by pulse-height analysis. The scintillation detector is also useful as a gamma-ray counting system, especially for the measurement of pure gamma-ray emitting radionuclides following radiochemical separations or in tracer measurements in which a single radionuclide is employed.

Gamma radiation measurements may be conveniently made in three ways:

1. Integral counting, in which all pulses whose magnitude is above some chosen threshold value are counted equally as they come from the photomultiplier tube.

2. Single-channel counting, in which only those pulse heights falling between two specified values are counted (e.g., the upper and lower values of a photopeak).

3. Multichannel counting, in which the pulses are sorted by height in step-function of value into a large number of electronic channels (generally ranging from 100 to more than 1000 channels), each counting only those pulses in the narrow pulse-height step.

Single and multichannel pulse-height spectrometers are used as an instrumental means for qualitative and quantitative gamma-ray energy analysis and as such represent a technique complementary to radiochemical analysis. Many activation analyses involving mixtures of several radioelements can be performed by gamma-ray spectroscopy without chemical separations. In many cases, even after chemical separations, gamma-ray spectroscopy is used to resolve the gamma-ray emitting radioisotopes of the isolated element.

### Integral Counting

The voltage applied to the dynodes of the photomultiplier tube affects the electron multiplication (or system gain) and thus the pulse height and counting rate of a scintillation detector in the same manner as in gas counters. The increases in counting rate with increases in applied voltage in a gas counter determines the "plateau curve." The same data for a scintillation counter is called an "integral spectrum," which is a characteristic of each gamma-ray emitting radionuclide. The integral spectrum for $^{85}$Sr ($E_\gamma =$ 0.514 MeV) is shown in Figure 6.6.

The pulse heights produced by the phototube are proportional to the amounts of the gamma-ray energy deposited in the crystal, and as the voltage is increased the pulses are amplified proportionally. The largest pulses, produced by the deposition of the total gamma-ray energy, are detected first. With further amplification the smaller pulses from Compton scatterings become large enough to be counted. At a sufficiently high voltage all pulses, including the noise pulses produced by random electrons in the photo-multiplier tube, are amplified enough to be counted. Figure 6.6 shows the long "plateau" from about 700 to 1100 V of the integral spectrum of $^{85}$Sr for a NaI(Tl) detector and the rapid rise of the noise above 1200 V.

Integral counting of individual gamma-ray emitting radionuclides is convenient in activation analysis because of the greater over-all counting efficiencies obtained, especially when used with the well-type scintillation counter. The counter used to produce the data for Figure 6.6 is operated with an applied voltage of 900 V. At this voltage the efficiencies for integral counting of several radionuclides are shown in Figure 6.7a. Over-all efficiencies for the 3 × 3 in. NaI(Tl) well-type counter range of about > 80%

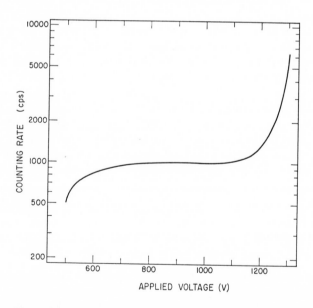

**Figure 6.6**   An integral spectrum for $^{85}$Sr ($E_\gamma = 0.514$ MeV) taken in a 3-in. NaI(Tl) well-type scintillation detector.

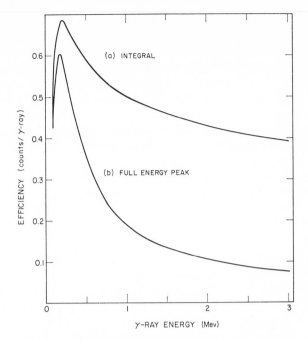

**Figure 6.7** Counting efficiencies for a 3 × 3-in. NaI(Tl) well-type scintillation detector: (*a*) Integral counting with baseline set at 0.05 MeV; (*b*) photopeak counting.

at 0.2 MeV to about 40% at > 2 MeV. Backgrounds for such counters are of the order of several hundred counts per minute.

### Single-Channel Counting

The single-channel pulse-height analyzer is a useful gamma-ray counter for chemically separated or tracer radionuclides. The two characteristics of the single-channel pulse-height analyzer are the following:

1. The "baseline"; the voltage which sets the minimum pulse height from the detector accepted for counting.

2. The "window"; the increment in voltage from the baseline voltage which sets the maximum pulse height accepted.

These characteristics are illustrated in Figure 6.8, a single-channel pulse height spectrum of [85]Sr. In this example the range of the base-line voltage was set to correspond to a range in gamma-ray energy of 0 to 1 MeV. The window was set at a constant width of 2% of the maximum baseline voltage.

Successive counts of the [85]Sr source, with increased baseline voltage of an equivalent of 20 keV for each count, produced the illustrated spectrum. The photopeak energy of 0.51 MeV for [85]Sr gamma rays was observed over the range of 0.41 to 0.61 MeV. The resolution of this detector system at 0.51 MeV is 13.7%. Photopeak counting of [85]Sr under these conditions can be conveniently made with high efficiency by setting the base line at 4.1 V and adjusting the window to a width of 2 V. The reduction in background from integral counting to photopeak counting is about a factor of ~ 10. The advantage of lower background is most pronounced when maximum sensitivity for the desired elements is required.

The total energy peak (photopeak) efficiency of a 3 × 3 in. NaI(Tl) well-type counter as a function of gamma-ray energy is shown in Figure 6.7b.

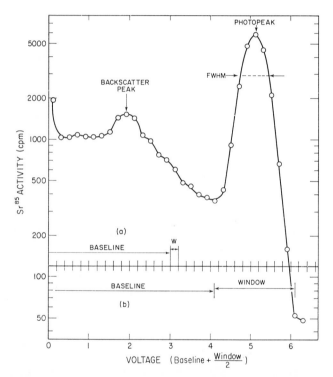

**Figure 6.8**  (a) Differential spectrum of [85]Sr in a 3-in. NaI(Tl) well-type scintillation detector made with a single-channel pulse-height analyzer window of 0.2 V, advancing the baseline 0.2 V on each count through a range of 10 V corresponding to a gamma-ray energy of 1 MeV; (b) the single-channel analyzer settings of baseline voltage of 4.1 V and window of 2 V for total photopeak counting of [85]Sr.

**Table 6.7**  Useful Standard Sources for Gamma-Intensity Calibrations[a]

| Nuclide | Half-Life | Gamma Energy (MeV) | Number of Gammas per Disintegration | Beta Energy (MeV) | Method of Determination of Disintegration Rate |
|---|---|---|---|---|---|
| $^{22}$Na | 2.60 y | 1.277 | 1.11[b] | 0.542 | $4\pi$ counting, |
|  |  | 0.511 | 2.00[b] |  | $\beta\text{-}\gamma$ coincidence |
| $^{24}$Na | 15.0 h | 1.38 | 1.00 | 1.39 | $4\pi$ counting, |
|  |  | 2.76 | 1.00 |  | $\beta\text{-}\gamma$ coincidence |
| $^{46}$Sc | 85 d | 0.89 | 0.995 | 0.36 | $4\pi$ counting, |
|  |  | 1.12 | 1.00 |  | $\beta\text{-}\gamma$ coincidence |
| $^{60}$Co | 5.27 y | 1.332 | 1.00 | 0.306 | $4\pi$ counting, |
|  |  | 1.172 | 1.00 |  | $\beta\text{-}\gamma$ coincidence |
| $^{95}$Nb | 35 d | 0.745 | 1.00 | 0.160 | $4\pi$, $\beta\text{-}\gamma$ coincidence |
| $^{131}$I | 8.14 d | 0.364 | 0.80 | 0.815 | $4\pi$ |
| $^{198}$Au | 2.69 d | 0.412 | 0.957 | 0.963 | $4\pi$ |
| $^{203}$Hg | 47 d | 0.279 | 0.81 | 0.208 | $4\pi$, $\beta\text{-}\gamma$ coincidence |
| $^{208}$Tl | 3.1 m[c] | 2.62 | 1.00 | 1.79 | $\alpha$-counting |
| $^{241}$Am | 500 y | 0.0596 | 0.357 | $\alpha$ | $\alpha$-counting |

[a] From E. P. Steinberg, Counting Methods for the Assay of Radioactive Samples, Chapter 5, in A. H. Snell, Ed. *Nuclear Instruments and Their Uses* (Wiley, New York, 1962).
[b] Number of gammas per positron.
[c] Supported by 10.6-h $^{212}$Pb.

The photopeak efficiency curve, which is a characteristic of each NaI(Tl) gamma-ray counter and the counting conditions, can be made from standard $\gamma$-ray emitting radionuclides, such as those listed in Table 6.7. The counting efficiency of nuclides for which standards are not available may be estimated by interpolation.

### Multichannel Analysis

The multichannel analyzer is used to determine spectroscopically the energies and intensities of the gamma rays emitted from a mixture of radio-nuclides. Each gamma-ray-emitting radionuclide produces a characteristic gamma-ray spectrum in a scintillation detector system. The measurement of a radionuclide consists of the identification of its characteristic spectrum and the quantitative determination of the spectrum intensity. Such measurements of mixtures of radionuclides are not simple, for the multiplicity of types of interaction of gamma rays with the detector material generally leads to complicated gamma-ray spectra.

CHARACTERISTIC SPECTRA. Several of the interaction processes lead to the escape of degraded photons after only a certain fraction of the gamma-ray energy has been deposited in the detector. Table 6.8 lists some of the processes that lead to the escape of the photon and the origin of nontotal energy peaks in the gamma ray spectrum. Figure 6.9 shows an example of the many peaks observed in the standard gamma-ray spectrum of $^{24}$Na which emits mono-energetic gamma rays of 2.75 and 1.37 MeV in cascade.

The most useful part of the gamma-ray spectrum is the photopeak, which represents the absorption of the total gamma-ray energy by the processes of photoelectric effect and multiple Compton scatterings. The relative weight of the photopeak with respect to the other processes is dependent not only on the gamma-ray energy and the scintillation material but also on the physical characteristics of the detection-system environment. Such properties as detector size and shape, sample geometry, density, and spacing with the crystal may be optimized for a given detection system. The shape of a spectrum may still be perturbed by effects resulting from the detector environment. Such effects include the loss of the iodine characteristic x-ray,

**Table 6.8**  Photon Escape and Origin of Peaks in Pulse-Height Spectra[a]

| Photon Escape Energy | Origin | Spectrum Peak Energy | Spectrum Peak Name |
|---|---|---|---|
| 0 | Total absorption | $E$ | Photopeak |
| 0.51 MeV | Pair production and escape of one annihilation photon | $E - 0.51$ | First pair escape peak |
| 1.02 MeV | Pair production and escape of both annihilation photons | $E - 1.02$ | Second pair escape peak |
| $E'' = E/(1 + 2\alpha)$ | Compton 180° scattering | $E_{ce} = E/(1 + \tfrac{1}{2}\alpha)$ | Compton edge |
| $E''$ to $E$ | Single Compton scattering | $E_{ce}$ to 0 | Single Compton distribution |
| 0 to $E$ | Multiple Compton scattering | $E$ to 0 | Multiple Compton distribution |
| $E_k = 29$ keV | Iodine K x-rays | $E - E_k$ | Iodine escape peak |
| $E$, with $E''$ re-enters | External Compton 180° scattering | $E'' = E/(1 + 2\alpha)$ | Backscatter peak |

[a] Adapted from J. B. Birks, *The Theory and Practice of Scintillation Counting* (Macmillan, New York, 1964), p. 478.

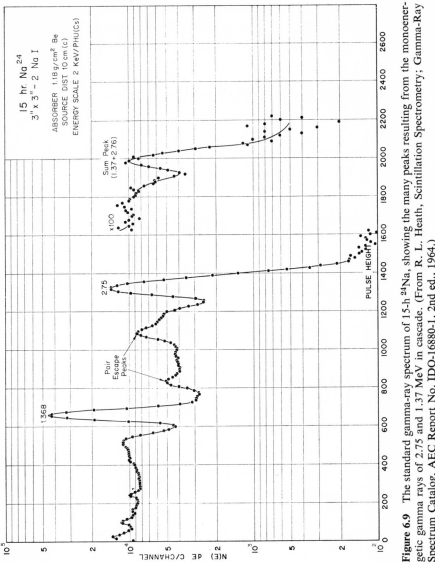

**Figure 6.9** The standard gamma-ray spectrum of 15-h $^{24}$Na, showing the many peaks resulting from the monoenergetic gamma rays of 2.75 and 1.37 MeV in cascade. (From R. L. Heath, Scintillation Spectrometry; Gamma-Ray Spectrum Catalog, AEC Report No. IDO-16880-1, 2nd ed., 1964.)

the scattering from the detector and shielding materials, the absorption of beta radiation, the annihilation of the positron following pair-production events, and coincidences of interactions.

*Iodine K X-Ray Escape Peak.* Figure 5.7 shows that gamma rays with energy below about 150 keV interact with NaI primarily by the photoelectric effect. Since (4.43) indicated that the probability of such interaction increases as $Z^5$, the ejection of $K$ electrons from the iodine atoms in NaI is the major photoelectric effect. The emission of the characteristic iodine 29 keV K x-ray follows. If the interaction occurs near the surface of the crystal, the x-ray

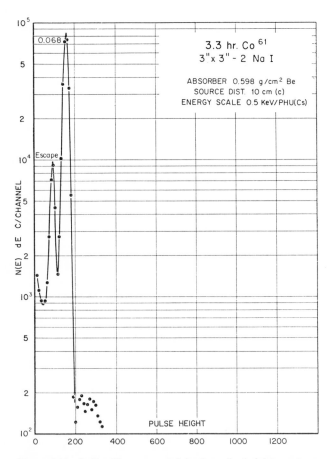

**Figure 6.10** Iodine K x-ray noted in the pulse-height spectrum of 99-m $^{61}$Co. (From R. L. Heath, Scintillation Spectrometry; Gamma-Ray Spectrum Catalog, AEC Report No: IDO-16880-1, 2nd Ed., 1964.)

may escape without further interaction in the crystal. Such events result in the absorption of the incident gamma ray, but the energy left in the detector is $(E_\gamma - 29)$ keV. Thus the "iodine escape peak" results in the spectrum of a gamma-ray-emitting radionuclides at an energy 29 keV less than the photo-peak energy. This effect is illustrated in the standard spectrum of 99-m $^{61}$Co $(E_\gamma = 0.068$ MeV) in Figure 6.10.

*Backscatter Peak.* Gamma rays emitted isotropically from a radionuclid, source are scattered by the materials around the source; for example, the detector shield, which generally constitutes the major mass of the systeme and internal materials other than the scintillation crystal act as scattering media for the gamma rays. Some of the gamma rays are scattered into the scintillation detector. Since the extent of Compton-scattered radiation entering the crystal is related to the atomic number of the shielding material and the source-detector geometry, the shield should be of high $Z$ and have internal dimensions as large as feasible. The effect of shield material and size on the backscatter peak in a spectrum of the $^{54}$Mn 0.835-MeV gamma ray is noted in Figure 6.11. The spectrum in the 6 × 6 in. Pb shield also shows the presence of 72 keV Pb K x-rays produced at the surface of the shield. Its intensity is substantially reduced in the larger shield, and it can also be reduced in small shields by lining them with one or more thin sheets of materials with decreasing atomic number.

*Beta-Ray Continuum.* The efficient absorption of beta radiations by NaI can introduce a beta spectrum component in a gamma-ray spectrum which can obscure the gamma-ray peaks. An example of this effect is shown in Figure 6.12, where curve 1 shows that the gamma-ray peaks of 3.6-h $^{92}$Y are almost completely obscured by the high energy (3.6 MeV) beta-ray con-tinuum. Curve 2 shows the considerable improvement obtained by inserting a beryllium absorber between the source and detector. Curve 3 shows a further improvement in the 1.4 MeV gamma-ray photopeak obtained by surrounding the sides of the detector with polystyrene. Beryllium is used as the absorber for the beta particles to minimize the production of Brems-strahlung. The polystyrene cap excludes the beta particles scattered from the air and surrounding materials into the sides of the detector. Minimum thickness of these absorbers is used to minimize the attenuation of the gamma-ray energies.

*Annihilation Radiation.* Section 4.3.3 described the process of annihilation of the positron following the pair-production process. The 0.511-MeV annihilation gamma rays appear in spectra of those radionuclides that decay by positron emission or emit gamma rays with energy above the threshold for pair production. The latter results from pair production in the walls of

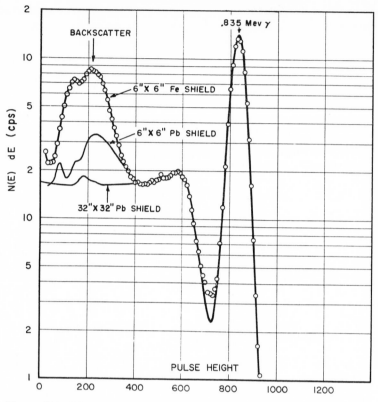

**Figure 6.11** The effect of detector-shield configuration on scattered component of pulse-height spectrum obtained with gamma rays of 0.835 MeV energy. (From R. L. Heath, Scintillation Spectrometry; Gamma-Ray Spectrum Catalog, AEC Report No. IDO-16880-1, 2nd ed., 1964.)

the detector shield or vicinity of the detector. The annihilation radiation peak at 0.511 MeV is shown in the spectrum of 3.08-h $^{45}$Ti in Figure 6.13.

*Summation Effects.* Some peaks observed in gamma-ray spectra are due to the summation of the energy deposited in a detector either by random or coincident gamma rays within the resolving time of the system or by true coincidences in the decay of a radionuclide.

A spectrum with random coincidences is shown in Figure 6.14. Typical resolving times are about 1 $\mu$sec for NaI(Tl) detectors. The count rate in the summation peak $N_s$ is estimated by

$$N_s = 2\tau N^2 \qquad (55)$$

where $N$ is the total count rate in the photopeak. Figure 6.14 also shows the effect of the source intensity on the random sum peak.

Many radionuclides emit two or more gamma rays in cascade following beta decay. If these gamma rays are detected simultaneously, their total energy will appear as a coincident sum peak in the spectrum. The count rate in the summation peak $N_c$ is estimated by

$$N_c = N_0 \epsilon_1 \epsilon_2 W \tag{56}$$

where $N_0$ = the count rate of the coincident pairs of gamma rays emitted by the source,

$\epsilon_1$ and $\epsilon_2$ = the respective detection efficiencies for two gamma rays,

$W$ = a correction factor for possible anisotropic angular dependence.

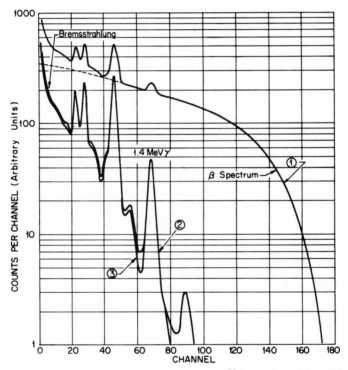

**Figure 6.12** Pulse-height spectrum of 3.6-h $^{92}$Y which decays with $E_\beta$ max = 3.6 MeV. Curve 1 shows the beta-spectrum obtained without absorbers. Curve 2 shows the effect of a 1.18 g/cm$^2$ beryllium absorber. Curve 3 shows the further effect of a 0.7 g/cm$^2$ polystyrene cap on the detector to absorb scattered beta particles. (From R. L. Heath, Scintillation Spectrometry: Gamma-Ray Spectrum Catalog, AEC Report No. IDO-16880-1, 2nd ed., 1964.)

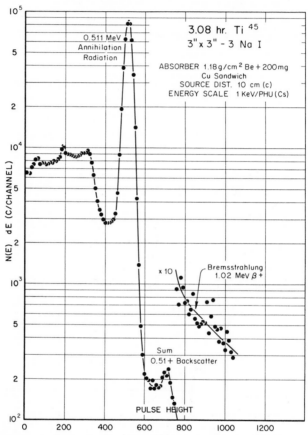

**Figure 6.13** Coincidence summing of annihilation radiation with 180°-scattered photons from the other annihilation quantum observed in the pulse-height spectrum of 3.08-h $^{45}$Ti. (From R. L. Heath, Scintillation Spectrometry; Gamma-Ray Spectrum Catalog, AEC Report No. IDO-16880-1, 2nd ed., 1964.)

Figure 6.15 shows a spectrum of $2 \times 10^4$-y $^{94}$Nb which decays with emission of 0.703 and 0.870-MeV gamma rays in cascade. The effect of source-detector geometry on the magnitude of the coincident sum peak is also shown.

Another case of coincident summing occurs for radionuclides decaying by positron emission. If one of the two annihilation photons enters the detector, the other, emitted in the opposite direction, can be scattered back into the detector. Figure 6.13 shows the sum peak for such events.

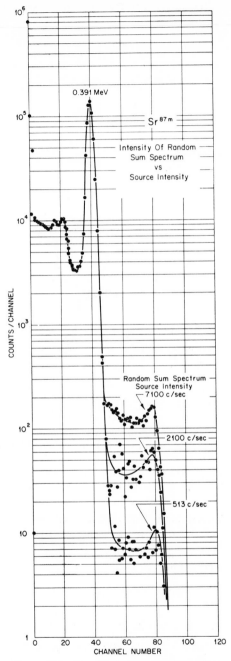

**Figure 6.14** The effect of the counting rate on the intensity of the random sum peak. (From R. L. Heath, Scintillation Spectrometry; Gamma-Ray Spectrum Catalog, AEC Report No. IDO-16880-1, 2nd ed., 1964.)

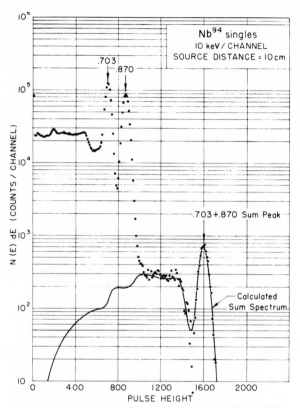

**Figure 6.15** The coincidence sum spectrum observed in the pulse-height spectrum of $2 \times 10^4$-y $^{94}$Nb and the effect of source-detector geometry. (From R. L. Heath, Scintillation Spectrometry: Gamma-Ray Spectrum Catalog, AEC Report No. IDO-16880-1, 2nd ed., 1964.)

QUANTITATIVE DETERMINATION OF GAMMA-RAY SPECTRA. The calibration of a NaI detection system for integral or single-channel counting was noted to be similar to that of a gas proportional counter. The over-all counting efficiency $\epsilon(E)$ may be defined as

$$\epsilon(E) = \frac{A}{D_\gamma} \tag{57}$$

where $A$ = counting rate for voltage from threshold to maximum,

$D_\gamma$ = rate of gamma-ray emission from the source.

The over-all counting efficiency for a constant source-detector system is dependent not only on the gamma-ray energy but also on the voltages applied.

Figure 5.8 showed efficiency curves for typical integral and single-channel gamma-ray detection systems. The constant voltage width (window) for single-channel counting is usually set to include all or most of the photopeak absorptions at all energies.

In the quantitative determination of gamma-ray spectra taken with multi-channel analyzers several efficiency parameters can be obtained.

For total gamma-radiation measurement in a well-defined source-detector geometry the detection efficiency $T(E)$ can be calculated. This efficiency is defined as the fraction of gamma rays emitted from the source which interacts with the detector and can be estimated for the known values of the absorption cross section of gamma rays $[\tau(E)]$ with NaI; for example, for a point source of radiation on the extended axis of a cylindrical detector the detector efficiency is given by

$$T(E) = \frac{1}{2}\left(\int_0^{\tan^{-1} r_0/(h_0+t_0)}\left\{1 - \exp\left[-\tau(E)\frac{t_0}{\cos\theta}\right]\right\}\sin\theta\,d\theta\right.$$

$$\left. + \int_{\tan^{-1} r_0/(h_0+t_0)}^{\tan^{-1} r_0/h_0}\left\{1 - \exp\left[-\tau(E)\left(\frac{r_0}{\sin\theta} - \frac{h_0}{\cos\theta}\right)\right]\right\}\sin\theta\,d\theta\right) \quad (58)$$

The geometry and terms of this equation are shown in Figure 6.16. The detector efficiency $T(E)$ is generally useful only for pure radionuclides because the spectrum of the gamma-ray is markedly affected by the scattered radiation entering the detector with degraded energy.

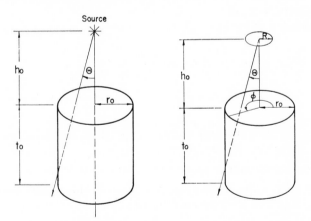

**Figure 6.16** The detector efficiency calculated for a point source and for a disk source of radiation and a cylindrical detector. (From R. L. Heath, Scintillation Spectrometry, Gamma-Ray Spectrum Catalog, AEC Report No. IDO-16880-1, 2nd ed., 1964.)

For analysis and measurement of spectra of mixed gamma radiations it is more convenient and precise to measure only the contribution to the spectra resulting from full energy absorption. The photopeak efficiency $\epsilon_p$ is defined as the probability that a gamma ray energy of $E$, emitted from the source, will appear in the photopeak of the observed pulse-height spectrum. Since total energy absorption includes multiple scattering processes as well as photoelectric effect absorption, the photopeak efficiency is difficult to calculate directly.

An indirect method for calculating the photopeak efficiency is given by

$$\epsilon_p = T(E)P \tag{59}$$

where $T(E)$ is the calculated over-all detection efficiency and $P$ is the fraction of the total number of events in the pulse-height spectrum which appears in the photopeak. The values of $P$ (the peak-to-total ratio) may be determined experimentally under conditions that minimize the contribution from scattered radiation. An example of the peak-to-total ratio calculation is shown in Figure 6.17, where $P$ is given by

$$P = \frac{N_p}{N_{\text{total}}} = \frac{\displaystyle\int_{E_1}^{E_2} N(E)\, dE}{\displaystyle\int_0^{E_2} N(E)\, dE} \tag{60}$$

The initial value of integration for $N_p$ is estimated by extrapolating the photopeak as a Gaussian curve.

In this notation of Heath[*] the emission rate of a single gamma ray $N_0$ is given by

$$N_0 = \frac{N_p}{T(E)PA} \tag{61}$$

where $N_p$ = the area under the photopeak (extrapolated as shown in Figure 6.17),

$T(E)$ = the total absolute detection efficiency for the source-detector geometry of the system,

$P$ = the appropriate value for the peak-to-total ratio,

$A$ = a correction factor for absorption of the gamma radiation by any beta absorber used in the system.

The calibration of a NaI detector system requires extensive control of the many factors that affect the pulse height and the counting rate as functions of the incident gamma radiation. A three-dimensional model of the response of

---

[*] R. L. Heath, Scintillation Spectrometry: Gamma-Ray Spectrum Catalog, AEC Report No. IDO-16880-1; 2nd ed., August 1964.

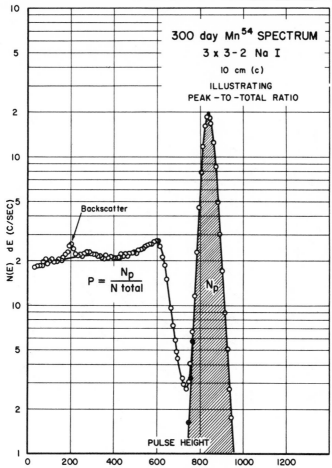

**Figure 6.17** The determination of $N_p$ in the peak-to-total ratio method for the evaluation of the photopeak efficiency for counting 300-day $^{54}Mn$. (From R. L. Heath, Scintillation Spectrometry: Gamma-Ray Spectrum Catalog, AEC Report No. IDO-16880-1, 2nd ed., 1964.)

a 3 × 3 in. NaI detector to monoenergetic gamma radiation is shown in Figure 6.18. Table 6.7 listed some of the radionuclides suitable for the calibration of NaI(Tl)-multichannel analysis gamma-ray scintillation spectrometers.

### 6.3.4 Semiconductor Conversion-Electron and Gamma-Ray Spectroscopy

The early uses of the relatively small depletion layer semiconductor detectors were for heavy charged particles such as fission fragments or alpha

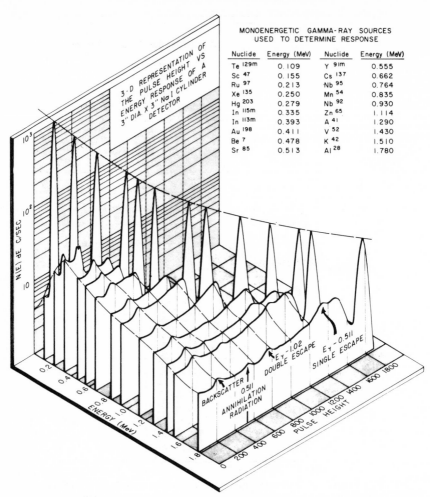

**Figure 6.18** A three-dimensional model of the response of a 3 × 3 in. NaI detector to monoenergetic gamma radiation. This model illustrates the interpolation scheme used to obtain the energy response of a given detector. (From R. L. Heath, Scintillation Spectrometry, Gamma-Ray Spectrum Catalog, AEC Report No. IDO-16880-1, 2nd ed., 1964.)

particles which have high specific ionizations. The development of the lithium drift detector, with sensitive-region widths of more than 1 cm, has made its general use for conversion-electron and gamma-ray spectroscopy feasible.

The continuous energy spectra of beta radiation make use of beta-ray spectroscopy as a means of identifying and measuring mixtures of beta-emitting nuclides by beta-ray energy resolution as difficult for semiconductors as it is for other detector types. However, monoenergetic conversion electrons (competing with gamma radiation as a de-excitation process) can not only be determined spectroscopically but the energy resolution ability of semi-conductors allows the conversion electrons from a given nuclide to be sorted by the small difference in the binding energy of the electrons in the several orbits. Figure 6.19 shows a conversion electron spectrum of $^{207}$Bi measured with a lithium-ion drifted detector at liquid nitrogen temperature. The $K$, $L$, and $M$ electron conversion lines have been resolved.

Energy resolutions of the order of 0.5 to 3% have been achieved for $K$ electrons, and the potential for conversion-electron spectroscopy as a measurement method in radioactivation is indeed great. Detector systems of lithium-ion drifted silicon and germanium semiconductors are readily available for electron spectroscopy. Present devices generally require cooling of the detector to below ∼100°K to reduce the leakage current and the use of vacuum chambers to prevent condensation on the detector.

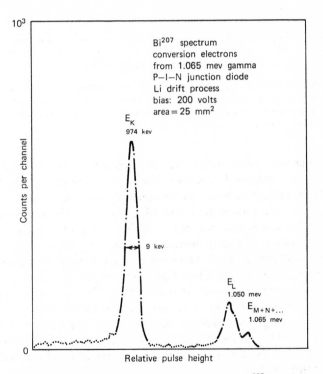

Figure 6.19 Conversion electron spectrum of $^{207}$Bi, taken with a Ge(Li) p-i-n radiation detector at liquid nitrogen temperature. [From J. L. Blankenship and C. J. Borkowski, *IRE Trans. Nucl. Sci.* **NS-9**, No. 3, 181 (1962).]

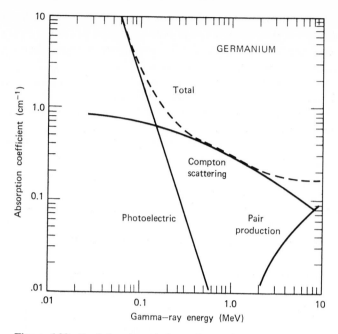

**Figure 6.20** Partial and total absorption coefficients for gamma rays in germanium. [From G. Dearnaley and D. C. Northrop, *Semiconductor Counters for Nuclear Radiations* (Wiley, New York, 1966), p. 16.]

For gamma-ray spectroscopy the semiconductor detector has several deficiencies. As a detector it operates simply as an ionization chamber. For the linear relationship between the incident photon energy and the pulse-height spectrum, the photon must lose all its energy in the depletion region. Silicon, with the low atomic number of 14, has a low absorption coefficient for gamma radiation. With thin detectors the ratio of total energy absorption to Compton scattering is small. The Li-drifted germanium n-i-p detector has improved matters considerably. Since the photoelectric absorption coefficient increases as $Z^5$, germanium, with atomic number 32, is about 40 times more effective than silicon. The n-i-p detectors also give greater sensitive volumes, but the small energy gap (0.66 eV at 300°K) of germanium necessitates operation at low temperature (usually $< 150°K$) to reduce the leakage current noise. The stability of the lithium ions in the crystal is also a problem at room temperatures.

Figure 6.20 shows that germanium detectors can be used for $\gamma$-ray spectroscopy in two energy ranges; below ~2 MeV, the photoelectric absorption defines the gamma-ray energy and above ~2 MeV absorption by pair production becomes important. Since the energies of most gamma-ray transitions from radioactivated nuclides are less than 2 MeV, the photoelectric effect (and total absorption by multiple Compton scatterings) contribute to the photopeak efficiency for measuring gamma radiation by pulse-height analysis.

A typical gamma-ray spectrum obtained with an 18-mm diameter Ge(Li) detector of 3.5-mm depletion depth is shown in Figure 6.21. The resolution of this detector is 3.6 keV for the 662 keV gamma-ray transition of $^{137}$Cs. This resolution of $\sim 0.5\%$ may be compared with the resolution of 7 to 10% obtained for NaI(Tl) crystals.

Figure 6.21 also shows the line width of a mercury pulser fed into the input of the preamplifier. The pulser is used as a calibration of the preamplifier. Since the gamma-ray pulse is subject to the same preamplifier effects as the pulse from the pulser, the excess in width of the gamma-ray pulse is due to the statistical fluctuations in the production and collection of the electron-hole pairs.

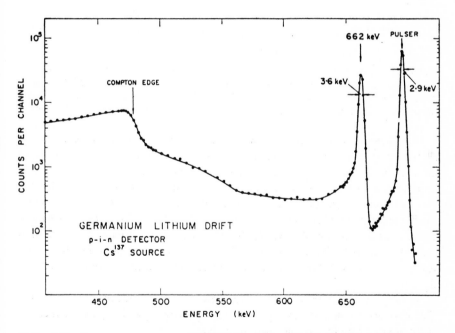

**Figure 6.21** Gamma-ray spectrum of $^{137}$Cs taken with a 2.5-cm diameter × 3.5-mm deep Ge(Li) semiconductor detector. The resolution of the detector system for the $^{137}$Cs photopeak is 0.54%, whereas the resolution for the pulser is 0.41%. [From G. T. Ewan and A. J. Tavendale, High-resolution Studies of γ-Ray Spectra Using Li-Drift Germanium γ-Ray Spectrometers, *Can. J. Phys.* **42**, 2286 (1964).]

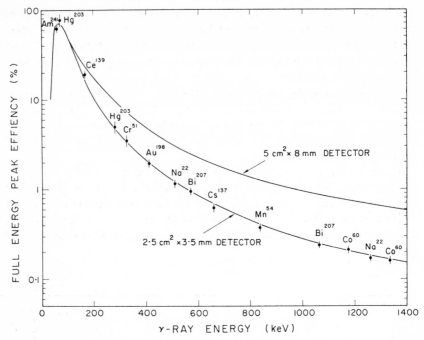

**Figure 6.22** Full-energy peak efficiencies for two sizes of Ge(Li) detectors. At 1 MeV the efficiency of about 1% for the larger detector is almost four times that of the smaller detector. [From G. T. Ewan and A. J. Tavendale, High-resolution Studies of $\gamma$-Ray Spectra Using Li-Drift Germanium $\gamma$-Ray Spectrometers, *Can. J. Phys.* **42**, 2286 (1964).]

In addition to resolution, however, the efficiency of the detector is an important characteristic of a gamma-ray spectrometer. Figure 6.22 shows an experimental full-energy peak efficiency curve for a 2.5 cm² × 3.5 mm depletion layer Ge(Li) detector. For comparison the corresponding efficiency curve is also shown for a 5 cm² × 8 mm detector, one of the more recently available larger Ge(Li) detectors. The larger detector is about four times more efficient at an energy of 1 MeV, the efficiency being greater than 1% and increasing to 10% for an energy of 0.3 MeV. The corresponding efficiencies for a 3 × 3 in. NaI(Tl) scintillation detector are several factors greater (see Section 5.3.2).

As larger and better semiconductor detectors are produced, it is expected that conversion-electron and gamma-ray spectroscopy will play an increasingly greater role in radioactivation analysis.

### 6.3.5 Proportional X-Ray Spectroscopy

The measurement of x-radiation in activation analysis has aroused some interest as an alternate to delayed neutron, positron, and prompt gamma

**Table 6.9**  X-ray-Emitting Neutron-Activation Products

Electron-Capture Products

| Element | Product | EC (%) | E (x-ray) (keV) |
|---------|---------|--------|-----------------|
| Argon | 35 d $^{37}$Ar | 100 | 3 |
| Chromium | 27.8 d $^{51}$Cr | 100 | 6 |
| Iron | 2.6 yr $^{55}$Fe | 100 | 7 |
| Copper | 12.8 hr $^{64}$Cu | 43 | 8 |
| Germanium | 11.4 d $^{71}$Ge | 100 | 10 |
| Selenium | 120 d $^{75}$Se | 100 | 11 |
| Bromine | 17.6 m $^{80}$Br | 5 | 11 |
| Palladium | 17 d $^{103}$Pd | 100 | 20 |
| Tin | 115 d $^{113}$Sn | 100 | 24 |
| Antimony | 2.8 d $^{122}$Sb | 3 | 25 |
| Europium | 9.3 h $^{152m}$Eu | 23 | 40 |
| Erbium | 10.3 h $^{165}$Er | 100 | 48 |
| Irridium | 72.4 d $^{192}$Ir | 4.4 | 63 |

Products with Internal Conversion Coefficient > 1

| Element | Product | Coefficient ($e_k/\gamma$) | E (x-ray) (keV) |
|---------|---------|----------------------------|-----------------|
| Cobalt | 10.5 m $^{60m}$Co | 41 | 6 |
| Germanium | 48 s $^{75m}$Ge | 1.4 | 10 |
| Selenium | 3.9 m $^{79}$Se | 7 | 11 |
|  | 57 m $^{81m}$Se | 9 | 11 |
| Bromine | 6.1 m $^{82m}$Br | 268 | 11 |
| Niobium | 6.3 m $^{94m}$Nb | 840 | 16 |
| Cadmium | 48.6 m $^{111m}$Cd | 1.5 | 22 |
| Tin | 14 d $^{117}$Sn | Large | 24 |
| Tellurium | 34 d $^{129m}$Te | Large | 26 |
| Xenon | 2.3 d $^{133m}$Xe | 4 | 29 |
| Cesium | 2.9 h $^{134m}$Cs | 2.6 | 30 |
| Dysprosium | 1.26 m $^{165m}$Dy | 3.4 | 45 |
| Erbium | 9.4 d $^{169}$Er | 106 | 48 |
| Thulium | 130 d $^{170}$Tm | 1.6 | 49 |
| Lutecium | 3.7 h $^{176m}$Lu | 1.2 | 53 |
| Platnium | 4.1 d $^{195m}$Pt | 9 | 65 |

radiations for those special cases in which the specific radiation can be detected in the presence of large amounts of other radiations.

X-rays were noted as resulting from the processes of orbital electron capture (see Section 1.4.2) or internal conversion (see Section 1.4.3). Both processes create vacancies in the inner electron shells (primarily the K shell). The transitions of outer electrons into these shells result in the emission of the characteristic x-radiations of the element.

X-ray spectroscopy has proved to be feasible in neutron activation analysis, especially when neutron capture leads primarily to radionuclides that decay by $\beta^-$ emission. However, only a few of the product radionuclides decay, some only partly, by electron capture. These radionuclides are listed in Table 6.9. Furthermore, radionuclides which have significant internal conversion coefficients generally are those only with high multipolarity isomeric transitions. Some of these radionuclides are also listed in Table 6.9.

The application of x-ray spectrometry for radioactivation analysis has been described by Shenberg, Gilat, and Finston,* who suggested such x-ray

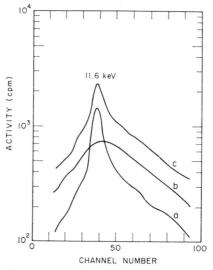

**Figure 6.23** Proportional x-ray spectra taken about 7 min after a 10-min irradiation of (a) 0.5 μg Br, (b) 100 μg Na, and (c) 100 μg Na + 0.5 μg Br. [From C. Shenberg, J. Gilat, and H. L. Finston, Use of X-ray Spectrometry in Activation Analysis: Determination of Bromine, *Anal. Chem.* **39,** 780–785 (1967).]

* S. Shenberg, J. Gilat, and H. L. Finston, Use of X-Ray Spectrometry in Activation Analysis: Determination Bromine, *Anal. Chem.* **39,** 780–785 (1967).

detectors as proportional counters, thin scintillation crystals, and semi-conductors for measurement of x-rays in the presence of large backgrounds of gamma radiations.

Bromine, for example, was shown to be suited to neutron activation analysis with x-ray spectroscopy; it forms the radionuclides 18-m $^{80}$Br and 4.5-h $^{80m}$Br which decays by isomeric transition to $^{80}$Br. Both nuclides emit x-rays; the excited state decays by a cascade of 48- and 36-keV gamma rays with a x-ray yield of about 1.2 bromine $K$ x-rays of 11.6 keV per disintegration, and the ground state decays approximately 5% by electron capture, following which an 11-keV selenium $K$ x-ray is emitted. In addition, the radionuclide 6-m $^{82m}$Br, also formed by neutron activation, decays by emission of a 46-keV gamma ray with a x-ray yield of about 0.7 bromine $K$ x-rays per disintegration.

Proportional x-ray spectrometry was applied to the analysis of bromine in matrices containing large amounts of interfering elements such as sodium. Figure 6.23 shows the proportional x-ray spectra of the 11.6 bromine $K$ x-ray from 0.5 $\mu$g bromine in the presence of 100 $\mu$g sodium. Accuracies of the order of 2 to 5% are attainable with Na/Br or K/Br ratios up to 500:1. This method is especially suited to nondestructive analysis of milligram quantities in organic matter.

## 6.4  BIBLIOGRAPHY

### 6.4.1  General References

ANALYTICAL CHEMISTRY
I. M. Kolthoff, P. J. Elving, and E. B. Sandell, *Treatise on Analytical Chemistry* (Interscience, New York, 1959–1968).
G. E. F. Lundell and J. I. Hoffman, *Outlines of Methods of Chemical Analysis* (Wiley, New York, 1958).
G. H. Morrison, Ed., *Trace Analysis, Physical Methods* (Wiley, New York, 1965).
C. L. Wilson and D. W. Wilson, *Comprehensive Analytical Chemistry* (Elsevier, Amsterdam, 1964).
J. H. Yoe and H. J. Koch, *Trace Analysis* (Wiley, New York, 1957).

RADIOCHEMISTRY
E. Bleuler and G. J. Goldsmith, *Experimental Nucleonics* (Rinehart, New York, 1957).
B. Brown, *Experimental Nucleonics* (Prentice-Hall, Englewood Cliffs, N.J., 1963).
G. D. Chase and J. L. Rabinowitz, *Principles of Radioisotope Methodology*, 2nd ed. (Burgess, Minneapolis, 1962), Chapters 8 and 9.
G. R. Choppin, *Experimental Nuclear Chemistry* (Prentice-Hall, Englewood Cliffs, N.J., 1961), Chapter 9.
R. A. Fairs and B. H. Parks, *Radioisotope Laboratory Techniques* (Pitman, London, 1958).

G. Friedlander, J. W. Kennedy, and J. M. Miller, *Nuclear and Radiochemistry*, 2nd ed. (Wiley, New York, 1964), Chapter 12.

N. R. Johnson, E. Eichler, and G. D. O'Kelley, *Nuclear Chemistry* (Wiley-Interscience, New York, 1963), Chapters 5 and 7.

R. T. Overman and H. M. Clark, *Radioisotope Techniques* (McGraw-Hill, New York, 1960).

## 6.4.2   Radiochemical Separations

GENERAL

S. Amiel, Rapid Radiochemical Separations, in L. Yaffe, Ed., *Nuclear Chemistry*, Vol. II (Academic, New York, 1968), Chapter 9.

H. L. Finston and J. Miskel, Radiochemical Separation Techniques, *Ann. Rev. of Nuclear Sci.* 5, 269–296 (1955).

F. Girardi, Radiochemical Separations for Activation Analysis, in J. R. DeVoe, Ed., *Modern Trends in Activation Analysis*, National Bureau of Standards Special Publication 312, 577–616, 1969.

Y. Kusaka and W. W. Meinke, Rapid Radiochemical Separations, National Academy of Sciences—National Research Council, NAS-NS 3104, 1961.

Subcommittee on Radiochemistry, National Academy of Sciences—National Research Council, A Series of Monographs on the Radiochemistry of the Elements, NAS-NS 3001–3058. Available from the Office of Technical Services, Department of Commerce, Washington, D.C.

T. T. Sugihara, Radiochemical Separations of Low-Level Radioactivity, *Analytical Chemistry* 3, 1–23 (Pergamon, Oxford, 1962).

ION EXCHANGE

K. A. Kraus and F. Nelson, Radiochemical Separations by Ion Exchange, *Ann. Rev. Nucl. Sci.* 7, 31–46 (1957).

R. Kunin, *Ion Exchange Resins*, 2nd ed. (Wiley, New York, 1958)

R. Kunin, *Elements of Ion Exchange* (Reinhold, New York, 1960).

F. C. Nachod and J. Schubert, *Ion Exchange Technology* (Academic, New York, 1956).

J. E. Salmon and D. K. Hale, *Ion Exchange: A Laboratory Manual* (Butterworth, London, 1959).

O. Samuelson, *Ion Exchangers in Analytical Chemistry* (Wiley, New York, 1953).

J. Schubert, Ion Exchange, *Ann. Rev. Phys. Chem.* 5, 413–448 (1954).

SOLVENT EXTRACTION

G. H. Morrison and H. Freiser, *Solvent Extraction in Analytical Chemistry* (Wiley, New York, 1957).

H. Frieser and G. H. Morrison, Solvent Extraction in Radiochemical Separations, *Ann. Rev. Nucl. Sci.* 9, 221–244 (1959).

OTHER

H. G. Cassidy, Fundamentals of Chromatography (Interscience, New York, 1957).

V. J. Coates, H. J. Noebels, and I. S. Fagerson, *Gas Chromatography* (Academic, New York, 1958).

W. D. Cooke, *Electroanalytical Methods in Trace Analysis* (Academic, New York, 1956), Vol. III.

D. H. Desty, *Gas Chromatography* (Academic, New York, 1958).

J. R. DeVoe, Application of Distillation Techniques to Radiochemical Separations, National Academy of Sciences—National Research Council, NAS-NS-3108, 1962.

A. I. M. Keulemans, *Gas Chromatography* (Reinhold, New York, 1957).

M. Lederer and E. Lederer, *Chromatography*, 2nd ed. (Elsevier, Amsterdam, 1957).

J. J. Lingave, *Electroanalytical Chemistry* (Interscience, New York, 1953).

A. J. Lindsey, Electrodeposition, Chapter II, in *Comprehensive Analytical Chemistry* (Elsevier, Amsterdam, 1964).

### 6.4.3   Radiometric Assay

R. L. Heath, Scintillation Spectrometry: Gamma-Ray Spectrum Catalog, 2nd ed., AEC Report No. IDO-16880, 2nd ed., August 1964.

G. D. O'Kelley, Ed., Application of Computers to Nuclear and Radiochemistry, National Academy of Sciences—National Research Council, NAS-NS-3107, 1962.

K. Siegbahn, Ed., *Alpha, Beta, and Gamma-Ray Spectroscopy* (North-Holland, Amsterdam, 1955), Vol. 1, Chapters V and VIIC, D.

E. P. Steinberg, Counting Methods for the Assay of Radioactive Samples, Chapter 5 in A. H. Snell, *Nuclear Instruments and Their Uses* (Wiley, New York, 1962).

## 6.5   PROBLEMS

1.  A carrier solution of cadmium containing 11.7 mg $Cd^{2+}$ was added to the solution of an irradiated sample. The final CdS, precipitated, filtered, and dried, weighed 8.8 mg. Determine the chemical yield (in percent) of the carrier element.

2.  Copper is an impurity in household aluminum. Two 100 mg of aluminum foil, one of which held 10.0 mg of added copper powder, were irradiated identically in a reactor. The annihilation gamma-rays were measured identically and corrected to a common time. The activity of the foil containing the added copper was 27,400 cpm; the activity of the other foil was 3800 cpm. Determine the concentration of copper (in percent) in the household aluminum.

3.  A separation procedure for two elements, chemically similar, was evaluated by a radiotracer method. To a solution containing 10 mg of each element was added a solution of $1.2 \times 10^4$ cpm of a long-lived radioisotope of each element. After the separation radiometric assay of the solution containing the desired element was 9430 cpm of its radioisotope and 1420 cpm of the other radioisotope. Calculate the separation factor for this procedure.

4.  A routine analysis for bromine by the $^{81}Br(n,\gamma)^{82}Br$ reaction concludes with the precipitation of AgBr deposited on a filter paper 3 cm² in area and counted in an end-window proportional counter. To correct the

counting rate for variability in absorption losses due to variability in chemical yield a calibration of the counting efficiency is made as a function of sample thickness. A series of samples is prepared with increasing amounts of $Br^-$ carrier to which is added $3.65 \times 10^4$ dpm of a standard carrier-free solution of $^{82}Br$; AgBr is precipitated from each sample, filtered, dried, weighed, and mounted for counting. An equal aliquot of the standard $^{82}Br$ is mounted in the same way. The integral gamma-ray activity of each sample is determined with a NaI scintillation counter before counting in the end-window proportional counter. The following data, corrected for background and decay, were obtained. Construct a calibration curve of counting efficiency as a function of sample thickness (in $mg/cm^2$) for $^{82}Br$ in AgBr for this proportional counter.

|  | AgBr | Net Counting Rate (cpm) | |
| --- | --- | --- | --- |
| Sample No. | Recovered (mg) | NaI Counter | Proportional Counter |
| 0 | ... | 7430 | ... |
| 1 | 2.03 | 6810 | 1860 |
| 2 | 4.35 | 6790 | 1955 |
| 3 | 6.54 | 6940 | 2015 |
| 4 | 8.56 | 6750 | 2035 |
| 5 | 13.55 | 7140 | 2040 |
| 6 | 17.65 | 6950 | 1890 |
| 7 | 22.2 | 7010 | 1710 |
| 8 | 25.7 | 6820 | 1525 |
| 9 | 36.8 | 7270 | 1190 |
| 10 | 45.7 | 7150 | 1010 |

5. The distribution coefficients for Sr and Cs in a certain ion exchange resin are 960 and 10, respectively. If a one liter solution containing 10 mg of each carrier element is shaken with 10 g of resin, what is the maximum $^{90}Sr/^{137}Cs$ activity ratio that will result in a solution phase of $^{137}Cs$ with less than 1 % $^{90}Sr$ contamination?

6. The data for the elution of a $^{85}Sr$ sample is given below. The interstitial volume of the ion-exchange column was 14.3 ml. Consecutive aliquots of 10-ml eluate were collected and counted under reproducible conditions.

(a) Plot the data as an elution curve and determine from the plot the values of $U^*$, $M^*$, $M^*/e$, and $U_a$.

(b) Calculate the distribution ratio $C$ and the number of theoretical plates $p$ for this column.

| Aliquot No. ($n$) | Volume Eluted $10(n - 0.5)$ | Activity of Aliquot (cpm) |
|---|---|---|
| 8 | 75 | 1840 |
| 9 | 85 | 2500 |
| 10 | 95 | 4550 |
| 11 | 105 | 8600 |
| 12 | 115 | 8400 |
| 13 | 125 | 3750 |
| 14 | 135 | 2320 |
| 15 | 145 | 1770 |
| 16 | 155 | 1540 |

7. The distribution ratio of iodine between water and carbon tetrachloride is 100. What volume of $CCl_4$ is required to remove the iodine carrier from a 100-ml aqueous solution with an extraction efficiency of 99%?

8. Determine the electrode potentials for the following half reactions and at the given concentrations:

$$Ag^+ + e^- = Ag \quad \text{for} \quad [Ag^+] = 10^{-2} \text{ M}$$
$$Cu^{2+} + 2e^- = Cu \quad \text{for} \quad [Cu^{2+}] = 10^{-8} \text{ M}$$
$$Np^{3+} + 3e^- = Np \quad \text{for} \quad [Np^{3+}] = 10^{-10} \text{ M}$$

9. The photopeak efficiency for a NaI scintillation detector for 47-d $^{203}$Hg was determined with an absolute measurement of a $^{203}$Hg source in a beta-gamma coincidence counter. The activities shown by the three counters were 854 cpm for betas, 647 cpm for gammas, and 533 for coincidences. The total activity in the photopeak measured in the NaI detector was 246 cpm. Calculate the photopeak efficiency for $^{203}$Hg in this system.

10. Examine the decay scheme of 25-m $^{128}$I [e.g., in Lederer, Hollander, and Perlman, *Table of Isotopes* (Wiley, New York, 1967), p. 272] and list the peaks expected in the spectrum of $^{128}$I in a NaI scintillation spectrometer. With the help of Figure 6.7 estimate the relative intensities of the pertinent peaks.

# Chapter 7

# Activation Analysis: Practices

In the preceding six chapters the several components of radioactivation analysis have been dissected and examined. The syntheses of these components, then, constitute the practices of activation analysis. A specific analysis is made by a series of suitable choices from among the many possibilities that have been examined. There is, of course, the fundamental question whether radioactivation analysis is an optimum or proper means for a particular chemical analysis or tracer project. This question is examined during the discussion of the uses of radioactivation analysis in Chapter 9. For a discussion of practices in activation analysis it is assumed that its choice as the analytical method has been made at least partly on the basis of the inherent sensitivity available with this method. The actual procedures to be used in a particular problem are chosen from a review of the many components of the

232

method. To evaluate these components a sensitivity calculation is generally made to fix or optimize the several parameters of the activation method within the available limits.

Parameters for the radioactivation are generally chosen to produce sufficient levels of radioactivity to meet the sensitivity requirements. The post-irradiation treatment and radiation measurement methods must be compatible with these levels. If high levels of radioactivity of the desired radionuclides are produced, the requirements for the radiation detection and measurement methods become more relaxed to achieve a given level of precision and accuracy. If the levels of radioactivity are low, sensitive methods of radiation measurements become more important. The choices of general activation analysis methods include radiochemical analysis, instrumental analysis, and several special techniques. The two major quantitative procedures are absolute and comparative activation analysis.

## 7.1 RADIOCHEMICAL ACTIVATION ANALYSIS

The general technique of radioactivation analysis may properly be called radiochemical activation analysis. It adds to the basic two-step concept of radioactivation analysis (i.e., the steps of radioactivation and radiation measurement) a third step of identifying the radiations by chemical element. This step reduces in most cases the uncertainty in chemical analysis resulting from evaluation only of radiation type, energy, and half-life. This is especially true in minor or trace element analysis in which only low levels of radionuclides are produced. The large number of radionuclides that decay with gamma-ray emissions in the energy range from about 0.5 to 1.5 MeV make the detection of specific radionuclides difficult. In many cases, however, the need for convenience, the need for speed, or the need to preserve a sample from chemical destruction makes instrumental activation analysis necessary or desirable. It may be concluded, that barring such needs, the identification of sought radionuclides by chemical element is a general aspect of radioactivation analysis.

### 7.1.1  The General Technique

The general technique of radioactivation analysis consists of executing a series of steps whose parameters are chosen to optimize the sensitivity and accuracy with minimum effort and cost. The major choices are:

1. the optimum nuclear reaction
2. a suitable irradiation facility
3. the preirradiation treatment required, if any

4. the conditions of irradiation
5. an adequate irradiation time
6. the postirradiation treatment
7. the optimum radiation measurement system
8. the desired precision and accuracy

Each of these choices is examined further.

### Choice of an Optimum Nuclear Reaction

The three independent variables in a nuclear reaction are the target nucleus, the irradiation particle, and the product nucleus. In practice these variables are not completely independent. In the general technique of activation analysis the target nucleus is specified as the element to be determined in the analysis. For elements with more than one major stable isotope, however, the target nucleus is still a variable; for example, if tin were the sought element, nuclear reactions could be examined for any of its 10 stable isotopes. The major considerations of "which isotope" include the isotopic abundance and the cross section for a particular reaction. In turn, the choice of irradiating particle may be determined by external parameters; for example, the availability of a particular local facility. It may also be dictated for a given target nucleus on the properties of the resultant product nucleus, such as its half-life and decay scheme. The activation product that exists long enough to be measured, and by radiations suitable for the measurement systems available, must be chosen.

The selection of the optimum nuclear reaction considers the physical, chemical, and nuclear properties not only of the isotopes of the sought elements and their activation products but also those of the matrix and its major elements. An evaluation of possible interfering reactions or competing activation products, as discussed in Chapter 8, should be made. Adequate sources of nuclear data, such as those listed in Section 7.4.2, should be available.

Thus the choice of the optimum nuclear reaction may involve a great deal of compromise in adjusting the three parameters. To review the considerations given in Chapter 3 for the sought element copper a comparison of the activations available for its two isotopes is made in Table 7.1 for four different nuclear reactions. The data illustrate the variety of radionuclides that can be produced with $\Delta Z = 0$ or $\pm 1$, with half-lives ranging from 5.1 min to 92 yr, and with large variations in types, abundances, and energies of the radiations.

### Choice of a Suitable Irradiation Facility

The major technical consideration in the choice of a suitable irradiation facility is dependent on the type of nuclear particle required for the chosen

**Table 7.1**  Data for Radioactivation of Copper Isotopes

| | $^{62}$Cu ($f = 0.691$) | | | | $^{65}$Cu ($f = 0.309$) | | | |
|---|---|---|---|---|---|---|---|---|
| | Cross Section ($b$) | Activation Product | Half-Life | Major Radiations ($E$, MeV) | Cross Section ($b$) | Activation Product | Half-Life | Major Radiations ($E$, MeV) |
| Slow neutrons (n,γ) | 4.5 | $^{64}$Cu | 12.8 h | 38% $\beta^-(0.57)$ 19% $\beta^+(0.66)$ 0.5% $\gamma(1.34)$ | 2.3 | $^{66}$Cu | 5.1 m | 91% $\beta^-(2.63)$ 9% $\gamma(1.04)$ |
| 14-MeV neutrons (n,p) | ? | $^{63}$Ni | 92 y | 100% $\beta^-(0.067)$ | 0.019 | $^{65}$Ni | 2.56 h | 58% $\beta^-(2.14)$ 25% $\gamma(1.48)$ |
| 12-MeV protons (p,n) | ~0.4 | $^{63}$Zn | 38.4 m | 80% $\beta^+(2.36)$ 8% $\gamma(0.67)$ | 0.8 | $^{65}$Zn | 245 d | 1.7% $\beta^+(0.32)$ 49% $\gamma(1.12)$ |
| 17-MeV photons (γ,n) | 0.85 | $^{62}$Cu | 9.8 m | 97% $\beta^+(0.66)$ | ? | $^{64}$Cu | 12.8 h | 38% $\beta^+(0.57)$ 19% $\beta^+(0.66)$ 0.5% (1.34) |

nuclear reaction. The types of irradiation particles have been classified as neutrons, charged particles, photons, and electrons. The corresponding practices are termed neutron activation, charged-particle activation, and photo (or electron) activation. A review of the sources of these irradiating particles was given in Chapter 3. The energy characteristics of the irradiating particles are a second consideration. A third, which is important in the selection of a particular facility for a given kind and energy of irradiating particle, is the flux or beam intensity required to achieve the desired sensitivity. Other important properties of a facility include the physical properties of the irradiation geometry and the sample, specific regulation and encapsulation requirements, the location of the facility, and the cost of the irradiation. The location of the facility with respect to the postirradiation measurements limits the useful range of half-lives that can be considered for the activation product.

For many reasons thermal neutrons are the most widely used of all the irradiating particles; for example:

1. The general availability of large fluxes of thermal neutrons in a variety of devices.

2. The large volumes of uniform neutron flux intensity.

3. The reduction of errors due to uncertainties of the energy spectrum.

4. The large cross sections generally prevalent for radiative capture of thermal neutrons.

5. The elimination of many radionuclides produced through endoergic reactions.

The most prolific source of thermal neutrons is the nuclear reactor. Service irradiations are readily available with neutron fluxes ranging from $10^{10}$ to $10^{14}$ n/cm$^2$-sec. Adequate sources of thermal neutrons can be obtained with suitable moderating conditions in cyclotrons, accelerators, and radioisotope sources. These devices, especially the specific low voltage (d,t) neutron generators, can also produce large quantities (to $\sim 10^{12}$ n/sec) of fast neutrons. Fast neutron activations are useful for such reactions as (n,p), (n,$\alpha$), and (n,2n), especially in those cases in which the (n,$\gamma$) reaction does not produce a suitable activation product. The disadvantages of fast neutron activations stem primarily from the small cross sections exhibited for these energies and the rapid loss in neutron energy with penetration into the sample.

Photon activation by the ($\gamma$,n) photoneutron reaction is similar to the fast-neutron (n,2n) activation reaction in producing the same product nucleus. Although photons can be readily produced with energies in excess of 14 MeV, the cross sections for photoneutron reactions are small, generally of the order of millibarns or fractions of millibarns.

Charged particles, although useful in many specific analyses, also present many problems in their use for activation. Cross sections for such reactions as (p,n), (p,d), (d,p), (d,n) vary widely with energy and are usually low compared with corresponding neutron energies. Additional problems in obtaining irradiations on charged particle accelerators include the mounting of the target in a suitable form with respect to beam losses and cooling requirements.

## Preirradiation Treatment

It was suggested in Section 6.1.1 that for radioactivation analysis, pre-irradiation treatment of samples should be reduced to the very minimum necessary. Yet several occasions were given for chemical treatment of samples before activation that were (a) too bulky for irradiation, (b) too inhomo-geneous for representative sampling, and (c) too radioactive after irradiation.

In general, the extent of preirradiation treatment of samples must consider the physical, chemical, and nuclear properties not only of the matrix con-stituents but also of the activation products. The physical properties may affect the bulk of the matrix, the minimum thickness of a solid attainable. Some irradiation facilities do not permit irradiation of powders, liquids, or volatile materials. The chemical considerations include hazards due to heat or radiation-decomposable materials or to explosive materials. Nuclear con-siderations include limitations on mass or thickness of sample for both neutron or charged-particle irradiations. The decrease with depth in intensity or energy of the irradiating particles may become of serious concern. Matrix major elements with high cross sections may make the activated sample hazardous for unshielded laboratory operations.

## Irradiation Conditions

The last of the problems involved in obtaining an irradiation is the preparation of the sample into a form suitable for irradiation and getting the sample container into (or on) and out of (or off) the irradiation facility. This aspect of radioactivation is not always so simple an operation as it may seem. Problems are especially prevalent in charged-particle accelerators in which irradiations with the primary beam are desired. The sample must become part of the vacuum system if no prior degradation of the beam is sought. In this case the target must be solid and rigid, be able to maintain a vacuum, and have a high heat conduction. Such properties are generally restricted to metals. Other materials require some form of encapsulation and methods or devices for maintaining the capsule in position during the irradiation. Many of the problems of accelerators are relieved with external-beam irradiations. The costs are generally a reduction in beam intensity and degrading of the particle energy.

Inhomogeneities in beam intensity are present in many accelerators, both for neutrons and charged particles. In many cases provisions are made to rotate the target during the irradiation to average out such inhomogeneities. This process is especially important when irradiations of several samples (e.g., samples with blanks and standards) are made simultaneously.

Irradiations in nuclear reactors result in much less stringent requirements. Neutron fluxes are constant over a much larger volume compared with beam irradiations. Temperature problems are also less stringent. Plastic and polyethylene containers are often used for irradiation capsules.

Precise timing of entry and removal of samples from an irradiation field is often a problem in activation analysis. The goals of rapid entry and removal transport, especially for short-lived activation products, are coupled to the goals of reproducible irradiation geometry. These goals are met in many cases by the introduction of a separate pneumatic transfer system in the irradiation facility. Such systems are referred to as rabbits; an appropriate description. A sophisticated pneumatic transfer system used in conjunction with a 14-MeV neutron generator is shown schematically in Figure 7.1. The system includes a dual-birotational sample target assembly and control to

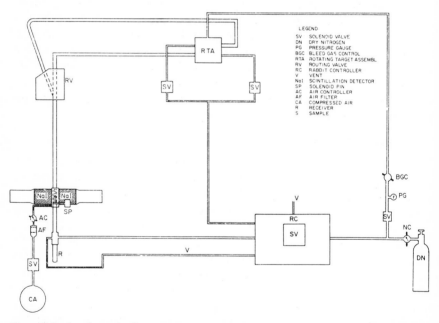

**Figure 7.1** A schematic diagram of the pneumatic transfer system at National Bureau of Standard's Radiochemical Analysis Section of the Institute for Materials Research. (From J. R. DeVoe, Ed., National Bureau of Standards Technical Note 404, U.S. Department of Commerce, 1966.)

allow simultaneous irradiation and sequential counting of an unknown and a standard sample between two $4 \times 3$ in. NaI(Tl) scintillation detectors connected to a multichannel pulse-height analyzer. Generally, for reactor irradiations, a simpler pneumatic transfer system is used to accelerate capsules containing combinations of samples, blanks, standards, and flux monitors into and out of the irradiation position.

## *Irradiation Time*

For a given sample in a given irradiation facility the maximum amount of radioactivity of a given radionuclide is the saturation activity, equal to the production rate when the length of the irradiation approaches infinity. Thus it is generally desirable to irradiate a sufficient length of time (five half-lives are equivalent to about 97% saturation) to approach the level of saturation activity. This level is especially useful when the trace elements sought are at the limits of sensitivity. Longer irradiations do not increase the useful levels of the desired product radionuclide, but they do increase the levels of radionuclides with longer half-life produced from the matrix constituents, the total radiation level of the sample, and the cost of irradiation.

Thus it is general practice to reduce the irradiation time to the minimum value consistent with the required sensitivity and accuracy. For short-lived product radionuclides irradiation time is generally not a significant factor; for example, for such products as 2.3-m $^{28}$Al, 3.77-m $^{52}$V, and 3.5-m $^{55}$Cr a 5-to-10-min irradiation produces a major fraction of the saturation activity.

For very long-lived product radionuclides irradiation time is essentially a linear function of the minimum amount of radioactivity desired. The value of $e^{-x}$ is approximated by the expansion

$$e^{-x} = 1 - x + \frac{x^2}{2!} - \frac{x^3}{3!} + \frac{x^4}{4!} - \cdots \tag{1}$$

For $x \ll 1$ terms in $x$ greater than the first power can be neglected. For $t \ll T_{1/2}$ the expansion of $e^{-\lambda t}$ reduces to

$$e^{-\lambda t} = 1 - \lambda t \tag{2}$$

and the saturation factor becomes

$$(1 - e^{-\lambda t}) = \lambda t \tag{3}$$

and

$$D^0 = R(\lambda t) \tag{4}$$

Thus the choice of the irradiation time is based on exceeding the minimum desired level of radioactivity balanced by the levels of interfering radionuclides and cost or other conditions of the irradiation.

### Postirradiation Treatment

Two major choices are made for postirradiation treatment of radioactivated samples:

1. Radiochemical versus instrumental measurement of the sought radionuclides.
2. Absolute versus comparative methods of quantitative radioactivation analysis.

These two choices are discussed in Section 7.1.3.

### The Optimum Radiation Measurement System

Radiation measurement systems that range from simple G.M. counters to elaborate high-resolution electron and gamma-ray multichannel spectrometers have been described. The optimum system for a particular radionuclide may be defined as the one that measures that radionuclide with the maximum identification and precision. Many radiochemical laboratories have but one radiation measurement system. For such laboratories it is by definition the optimum system. In laboratories that have detectors for beta and gamma radiation the choice of counter is generally made from considerations of the decay scheme of the radionuclide and whether or not to include radiochemical separations. Few laboratories that perform radioactivation analyses are equipped solely with high resolution gamma-ray energy spectrometers. Such laboratories would have difficulty in the analysis of radionuclides decaying by beta emission or in the analyses of gamma-ray energies from radionuclides in low abundance. For beta- or gamma-emitting radionuclides reduction of background generally results in increased sensitivity for low-level radionuclides.

### Precision and Accuracy

In quantitative analytical practices, of which radioactivation analysis is one, the general objective is the determination of some specific element or component. The quality of the determination is a function of the precision and accuracy obtained. These, in turn, can generally be improved to some given level if sufficient care and calibrations are made during the activation analysis. However, the choice of or need for precision and accuracy is also an important function of the requirements for the analysis. On a very general basis the need for precision and accuracy may be divided into four categories, according to the question to be answered:

1. Qualitative: is element $Z$ present in the samples?
2. Threshold: is element $Z$ present in amounts greater than some given amount?

3. Relative: is element $Z$ present in small (defined) or large (defined) amount?

4. Absolute: exactly how much of element $Z$ is present in the sample?

It is clear that the requirements for precision and accuracy vary greatly for these four categories of objectives. For the first, precision and accuracy are of little importance if the element can be positively identified above the minimum amount detectable. A discussion of minimum amount detectable is given in Chapter 8. Precision and accuracy become more important as the level present approaches the minimum amount detectable. Such analyses are frequently desired in forensic applications in which the presence of a particular element or material is factual evidence. These considerations apply with less latitude for threshold analysis, especially if the fixed level is significantly above the minimum detectable amount. More latitude is available for the relative analysis, but greater precision and accuracy are required if the defined amounts represent a range of lower to upper limits, as the level approaches either one. Quality control of impurities, in which the range in allowable impurity concentration may be large or small compared with the available precision of the analysis, is a typical example. The absolute determination attempts to measure the amount as precisely and accurately as convenient or possible. In many cases the attainable accuracy is much greater than the range of acceptable determinations, whereas in others every precaution that will decrease the total error is employed.

### 7.1.2  The Sensitivity Calculation

The sensitivity calculation is used to evaluate the parameters that influence the minimum amount of a given element that can be detected or measured for a given set of irradiation and analytical procedures. The calculation takes into account the properties of the element and the matrix, the properties of the irradiation facility and conditions, and the properties of the radiochemical treatment and radiation measurement systems to be used. Thus the sensitivity calculation serves as a "dry run" or a "paper run" of a particular radioactivation analysis in which the strategy of the determination is decided or the feasibility of the analysis is established. Basically, the calculation is a solution of the radioactivation equation under an assumed set of conditions. For maximum sensitivity the appropriate maximum or minimum values of the several parameters are evaluated.

In view of the importance of the sensitivity calculation in activation analysis, it is desirable to review each of the parameters in the activation equation. In general, the parameters are chosen to maximize the minimum amount of radioactivity detectable for the amount of irradiation obtainable.

**Table 7.2**    Parameters Affecting Maximum Sensitivity

---

**I. Radiation Measurement**
1. Detector background, $B$
2. Minimum resolvable counting rate above background, $A_m(t) \simeq 2B$
3. Energy resolution of the system, FWHM
4. Over-all counting efficiency for the radionuclide, $\epsilon = C_i/D$

**II. Postirradiation Processing**
1. Decay during elapsed time from end of irradiation to start of counting, $e^{\lambda t}$
2. Chemical yield of carrier, $Y$
3. Self-absorption in sample thickness, $F$

**III. Irradiation Conditions**
1. Cross section for chosen reaction, $\sigma$
2. Maximum flux or beam intensity, $\phi, \bar{J}$
3. Saturation factor, $(1 - e^{-\lambda T})$

**IV. Sample Conditions**
1. Isotopic abundance of chosen isotope, $f_i$
2. Maximum size of sample suitable for irradiation, $S$
3. Characteristics of the other constituents in sample

---

Table 7.2 lists many of the constant and experimental factors that determine the maximum sensitivity of a particular analysis. In choosing the values for the many parameters it is helpful to work backward, starting with the radiation measurement system and ending with the sample size required.

### Radiation Measurement

The minimum resolvable counting rate of an activated sample is influenced markedly by the counter background and efficiency and the levels of interfering radionuclides in the counting sample. With radiochemical separations the influence of interfering radionuclides can be reduced significantly, often to values approaching the intrinsic counter background. Of course, interferences from radioisotopes of the same element cannot be reduced by radiochemical methods.

For a radiation measurement whose background is $B$ cpm (this could be the value of a reagent blank run with the actual samples or the Compton scattering component of higher energy gamma rays from impurity radionuclides that contribute to a photopeak activity measurement), the minimum resolvable counting rate above background $A_m(t)$ is often assumed to be $2B$. Thus for a scintillation counter in which a blank shows a value of 200 cpm in the channels selected to measure the photopeak of the desired radionuclide the minimum net photopeak activity may be required to exceed 400 cpm.

For a low-background anticoincidence counter of 0.2 cpm background the minimum net activity could be set as low as 0.4 cpm. However, because of the long counting times required to obtain adequate statistics at such low counting rates, a minimum of 1 cpm is often used.

Associated with background in gamma-ray spectrometer systems is resolution of the system. In general the greater the resolution, the lower the background, since fewer channels are required to encompass the full-energy peak. The resolution of scintillation and semiconductor crystals has been given as FWHM, the full width of the full energy peak at half maximum. A more important parameter for maximum sensitivity is the over-all counting efficiency

$$\epsilon = \frac{C_i}{D} \qquad \text{(in } c/d) \qquad (5)$$

where $C_i$ is the number of counts of radiation $i$ and $D$ is the number of disintegrations of the nuclide in the same time interval. The over-all counting efficiency for a particular radionuclide depends both on the radiation-energy efficiency of the detector system and on the decay scheme of the radionuclide itself, which is independent of the measurement system. Thus a NaI scintillation detector may have an 20% efficiency for 1.04-MeV gamma rays, but the radionuclide 5.1-m $^{65}$Cu (see Table 7.1) decays only 9% with the emission of 1.04-MeV gamma rays. The over-all counting efficiency of this radionuclide in this counter would be

$$\epsilon = 0.20 \times 0.09 = 0.018 \qquad (c/d) \qquad (6)$$

Other counting conditions, such as sample thickness or counting geometry, can affect the over-all counting efficiency. Thus $\epsilon$ is usually determined from a calibration curve of $\epsilon$ versus $E_{radn}$ for a fixed set of counting conditions (see Figure 6.7).

### Postirradiation Processing

In radiochemical activation analysis the counting sample is prepared from the purified carrier and mounted in a form suitable for the detection system; for example, as a thin precipitate on filter paper or disks for end-window beta or flat-crystal gamma-ray counting or as a solution in a vial for well-crystal gamma-ray counting. The corresponding factors that affect maximum sensitivity are the elapsed time from end of irradiation to start of counting,

$$A^0 = A(t)e^{\lambda t} \qquad (7)$$

the chemical yield of the carrier,

$$Y = \frac{\text{amount of carrier recovered}}{\text{amount of carrier added}} \qquad (8)$$

and the loss of radiation in the sample by self-absorption,

$$F = \frac{\text{number of radiations emitted from sample}}{\text{number of radiations produced in sample}} \qquad (9)$$

Thus the disintegration rate of the radionuclide at the end of the irradiation is

$$D^0_{\min} = \frac{A_m(t)e^{\lambda t}}{\epsilon \cdot Y \cdot F \cdot 60} \qquad \text{(dps)} \qquad (10)$$

### Irradiation Conditions

The major parameters of the irradiation are the cross section for the chosen reaction, the maximum particle flux or beam intensity available, and the length of the irradiation. The first parameter is a fixed value; the second is fixed by the particular irradiation facility. Thus only the time of irradiation is generally a variable whose optimum value is evaluated. The basis for evaluation was reviewed in Section 7.1.1.

The minimum number of target nuclei measurable as $D^0_{\min}$ dps is then given as

$$n_{\min} = \frac{D^0_{\min}}{\sigma\phi(1 - e^{-\lambda T})} \qquad \text{(atoms)} \qquad (11)$$

### Sample Conditions

If $n_{\min}(i)$ is the minimum number of atoms of the isotope $i$ detectable by the foregoing parameters, the minimum weight (maximum sensitivity) of the element detectable is

$$W_{\min} = \frac{A \cdot n_{\min}(i)}{f \cdot \mathbf{N}} \qquad (12)$$

or, in summary, to show all the parameters that have been considered,

$$W_{\min} = \frac{A \cdot A_m(t) \cdot e^{\lambda t}}{f \cdot \mathbf{N} \cdot \epsilon \cdot Y \cdot F \cdot 60 \cdot \sigma \cdot \phi \cdot (1 - e^{-\lambda T})} \qquad (13)$$

where   $A$ = atomic number of the element,

$A_m(t)$ = minimum detectable counting rate at time of counting,

$e^{\lambda t}$ = decay from end of irradiation to start of counting,

$f$ = isotopic abundance,

$\mathbf{N}$ = Avogadro's number,

$\epsilon$ = over-all counting efficiency,

$Y$ = chemical yield,

$F$ = self-absorption losses,

$60$ = factor to convert dpm to dps,

$\sigma$ = reaction cross section,

$\phi$ = irradiation flux,

$(1 - e^{\lambda T})$ = saturation factor for the length of irradiation.

Generally it is the concentration of an element that is desired rather than the absolute amount in the sample, and therefore the amount of sample taken for the irradiation is a maximum for maximum sensitivity. If the sample weight is $S$ grams, the minimum concentration of the element measurable is

$$C_{min} = \frac{W_{min}}{S} \times UF \tag{14}$$

where $UF$ is the units factor; for example, if $C_{min}$ is desired in parts per million (ppm) or in mg/l, $UF = 10^{+6}$. If $C_{min}$ is desired in percent, $UF = 10^2$.

The characteristics of the other constituents in the sample may play an important part in the analysis such as creating a radiation hazard, reducing the available flux or beam intensity or energy (see Chapter 8), or requiring extensive radiochemical separations that reduce the attainable chemical yields. Such effects are estimated on an individual basis, since general rules are not sufficiently precise to evaluate the change in maximum sensitivity.

As an example of the sensitivity calculation, an activation analysis for the determination of copper in household aluminum will be evaluated. The following data or assumptions are compiled for $^{65}Cu(n,\gamma)^{66}Cu$ activation:

1. Counting system:
   NaI(Tl) scintillation detector, total $B$ (photopeak) = 100 cpm,
   $A_{min}(t) = 200$ cpm,
   over-all counting efficiency = 1.8%.
2. Postirradiation processing:
   elapsed time to counting = 8.3 min,
   chemical yield (rapid chemistry) = 48%,
   self-absorption in sample $\simeq 0$.
3. Irradiation conditions:
   cross section = 2.3 barn,
   flux available = $10^{12}$ n/cm²-sec,
   irradiation time = 10 min (to minimize contribution of 1.34-MeV gamma-rays from 12.8-h $^{64}Cu$).
4. Sample conditions:
   isotopic abundance of $^{65}Cu$ = 0.309,
   maximum sample size = 5 g,
   no major interfering constituents, except the aluminum 100% $^{27}Al(n,\gamma)$
   2.30-m $^{28}Al$ ($\sigma = 0.23$ b).

The calculation:

$$C_{min}(\%) = \frac{W_{min} \cdot 10^2}{S}$$

$$= \frac{63.54 \times 200 \times e^{+(0.693/5.1)8.3} \times 10^2}{5 \times 0.309 \times 6.023 \times 10^{23} \times 0.018 \times 0.48 \times 1.0 \times 2.3 \\ \times 10^{-24} \times 10^{12} \times (1 - e^{-(0.693/5.1)10})}$$

$$= 2.86 \times 10^{-4}\%$$

To estimate the total activity of the aluminum at the end of the irradiation

$$D^0 = n\sigma\phi(1 - e^{-\lambda t})$$

$$= \frac{5 \times 6.023 \times 10^{23}}{26.98} \times 2.3 \times 10^{-25} \times 10^{12} \times (1 - e^{-(0.693/2.30)10})$$

$$= 2.82 \times 10^{10} \text{ dps.}$$

### 7.1.3   The Postirradiation Treatment

The two choices for postirradiation treatment, namely,
(a) radiochemical versus instrumental radiation analysis and
(b) comparator versus absolute determination,
have already been described. The basic criterion may be considered the sensitivity with which the desired element is to be measured. The factors affecting sensitivity include the chemical, physical, and nuclear properties of the matrix elements as well as those of the desired element. If the radiations of the sought radionuclide can be measured quantitatively with the required precision and without the need to dissolve the sample and chemically separate the radionuclide from its interfering radionuclides, then, of course, the instrumental method of analysis would be much preferred. Where the sought radionuclide cannot be unambiguously identified among the many radionuclides produced, radiochemical separations are obviously required. It is the middle-range of this spectrum where difficulties in choice of method become evident. The sought radionuclide may be positively identified in a radiation-energy spectrometer, but its quantitative measurement must be made in the presence of large backgrounds of other radionuclides (e.g., large Compton-scattering counting rates of radionuclides with greater $\gamma$-ray energies in the photopeak channels of the sought radionuclide). In these cases the evaluation of the gain in precision of the measurement must be weighed in terms of the inconvenience or difficulties in completing a successful radiochemical separation within the counting time available before the radionuclide decays to unmeasurable levels. In many cases, especially for radionuclides with half-life about equal to the time of radiochemical processing,

the evaluation is difficult to make. The radiochemical experience of the analyst generally plays a strong role in the final decision in such cases.

The second of the two choices is generally much easier to make. The absolute assay technique determines the analysis by direct evaluation of the radioactivation equation, in which the radioactivity produced initially is given by

$$D^0 = n\sigma\phi(1 - e^{-\lambda t}) \tag{15}$$

For an elemental analysis the number of target nuclei in the sample is calculated from the measured values of the initial disintegration rate, the neutron flux or particle beam intensity, the irradiation time, and known values of the pertinent nuclear data, such as half-lives and reaction cross sections. Irradiation times are generally simple to measure, although problems may arise with slow insertions and removals into reactors, temporary stoppings of beams, etc. The disintegration rate should be measured in radiation-measurement equipment accurately calibrated for the particular radionuclides. The flux or beam intensity should be measured by suitable monitor irradiations, generally performed with the sample irradiation. In general, great care must be exercised to obtain accurate values of each of the parameters in the activation equation.

An evaluation of accuracies and precisions attainable by absolute gamma-ray counting and direct calculation of unknown element weights from the nuclear constants has been made by Girardi, Guzzi, and Pauly.*

The comparator method of radioactivation analysis is based on simultaneous analysis of a standard known sample for the desired elements in which the known and unknown samples are treated identically. Under such conditions the several common parameters of the activation are equal, even if they are not sufficiently precise.

$$\frac{D_u^0}{D_k^0} = \frac{n_u\sigma_u\phi_u(1 - e^{-\lambda t})}{n_k\sigma_k\phi_k(1 - e^{-\lambda t})} \tag{16}$$

and for equal irradiation times, irradiation flux conditions, and reaction cross section the number of target atoms of the unknown is given by

$$n_u = n_k \frac{D_u^0}{D_k^0} \tag{17}$$

In fact, the measured radionuclide of the two samples need not be corrected

* F. Girardi, G. Guzzi, and J. Pauly, Activation Analysis by Absolute Gamma-Ray Counting and Direct Calculation of Weights from Nuclear Constants, *Anal. Chem.* **36,** 1588–1594 (1964).

to the initial radioactivity but can be corrected to any convenient common time.

$$n_u = n_k \frac{D_u(t)}{D_k(t)} \tag{18}$$

The ease of the comparator method of activation analysis, which requires only relative measurements and eliminates the need for precise nuclear data, has resulted in its more wide-spread use compared with the absolute method. In this method, however, equal conditions of sample and comparator must be carefully maintained. The comparator samples must be prepared and encapsulated with the unknown samples so that they both experience identical irradiations under the same flux conditions. This is not always easy to achieve in facilities in which beam intensity or flux varies with time or position. The postirradiation treatment must yield samples whose counting rates are measured with the same or known efficiencies. Other precautions are listed in Chapter 8.

### 7.1.4  Substoichiometric Radiochemistry

A radiochemistry method for radioactivation analysis, which eliminates the requirement for chemical-yield determination, has been developed by Ruzicka and Stary.* Their method also increases the selectivity of the pertinent radiochemical separation procedures.

In the comparitor relationship of (18) the radiation measurement is corrected to its total chemical-yield value. The substoichiometric method is valid for only partial chemical yields, if the two are equal; that is, the weights of an element in the unknown sample and the comparitor standard will be proportional to the activities in the two counted samples

$$\frac{W_u}{W_k} = \frac{A_u(t)}{A_k(t)} \tag{19}$$

if the following two requirements are fulfilled:

1. The amount of carrier added to each of the two samples is equal.
2. The recovered fraction of the separated compound used for radiation measurement is equal.

Although the first condition is easily attained by decision, the second condition is much more difficult. The substoichiometric separation method

---

* J. Ruzicka and J. Stary, "A New Principle of Activation Analysis Separations—I: Theory of Substoichiometric Determinations," *Talanta* **10**, 287–293 (1963).

advanced by Ruzicka and Stary achieves the latter condition by an incomplete but constant fractional removal of the added carrier. This is accomplished by adding a smaller amount of a reagent than that which corresponds stoichiometrically to the amount of carrier added. For this method to be successful the reagent must react quantitatively with the carrier element to form a compound readily separable from the excess added carrier and from any interfering elements in the matrix. Separation processes have been described for (a) solvent extraction, (b) ion exchange, and (c) precipitation. Others, such as electrolysis, are possible. The method is especially useful for activation analysis because the addition of carriers (usually in milligram amounts) greatly simplifies the substoichiometric separation procedures.

### Solvent Extraction

For solvent extractions of the form

$$M + N(HA)_{org} \rightarrow (MA_N)_{org} + N(H) \tag{20}$$

with the conditions that more than 99.9% of the organic reagent HA has been used in forming the extractable complex $MA_N$, the amount of carrier added is of the order of $10^{-3}$ g/ml, and the ion molarities and organic reagent concentrations are of the order of $10^{-2}\ M$, the threshold pH of the determination is given by

$$pH \geq 5 - \frac{1}{N} \log K. \tag{21}$$

The pH values for the determination of several metal elements may be predicted from the values of $\log K$ given in Table 7.3. If a masking agent $H_nB$ is used to increase the selectivity by forming the complex $MB_s$, the threshold pH can be estimated from

$$pH \geq 5 - \frac{1}{N} \log K + \frac{1}{N} \log (1 + K_s[B]^s) \tag{22}$$

where

$$K_s = \frac{[MB_s]}{[M][B]_s} \tag{23}$$

### Ion Exchange

For ion-exchange separations the substoichiometric condition is achieved by the addition of a complexing agent with subsequent ion-exchange separation of the complexed form of the element. The complexing agent $H_nY$ must form a neutral or negatively charged complex MY with the stability condition, for systems in which the amount of complexing agent is about half the stoichiometric requirement, of

$$K_{MY} \geq 10^5 \sum_{n=0}^{n} \frac{[H]^n}{k_0 \cdots k_n} \tag{24}$$

**Table 7.3**    Log $K$-Values for Various Extraction Systems[a]

| Metal | Acetyl-acetone in Benzene | Benzoyl-acetone in Benzene | Dibenzoyl-methane in Benzene | Thenoyltri-fluoracetone in Benzene | 8-hydroxy-quinoline in Chloroform | Dithizone in Carbon Tetra-chloride |
|---|---|---|---|---|---|---|
| $Ag^I$ | | −7.8 | −8.6 | | −4.5[b] | 8.9 |
| $Al^{III}$ | −6.5 | −7.6 | −8.9 | −5.2 | −5.2 | |
| $Ba^{II}$ | | | | −14.4 | −20.9[c] | |
| $Be^{II}$ | −2.8 | −3.9 | −3.5 | −3.2 | −9.6 | |
| $Br^{III}$ | | | | −3.2 | −1.2 | 9.6 |
| $Ca^{II}$ | | −18.3 | −18.0 | −12.0 | −17.9[b] | |
| $Cd^{II}$ | | −14.1 | −14.0 | −11.4 | −5.3[c] | 2.1 |
| $Co^{II}$ | | −11.1 | −10.8 | −6.7 | −2.2[c] | 0.0 |
| $Cu^{II}$ | −3.9 | −4.2 | −3.8 | −1.3 | −1.8 | 9.6 |
| $Fe^{III}$ | −1.4 | −0.5 | −1.9 | 3.3 | 4.1 | |
| $Ga^{III}$ | −5.5 | −6.3 | −5.8 | | 3.7 | |
| $Hg^{II}$ | | | | | | 26.8 |
| $In^{III}$ | −7.2 | −9.3 | −7.6 | −4.3 | −0.9 | 4.8 |
| $La^{III}$ | | −20.5 | −19.5 | −10.5 | −16.4 | |
| $Mg^{II}$ | | −16.6 | −14.7 | | −15.1 | |
| $Mn^{II}$ | | −14.6 | −13.7 | | −9.3 | |
| $Ni^{II}$ | | −12.1 | −11.0 | | −2.2 | −0.6 |
| $Pb^{II}$ | −10.2 | −9.6 | −9.4 | −5.2 | −8.0 | 1.2 |
| $Pd^{II}$ | >2 | 1.2 | | | 15.0 | >26 |
| $Sc^{III}$ | −5.8 | −6.0 | −6.0 | −0.8 | −6.6[b] | |
| $Sr^{II}$ | | −20.0 | −20.9 | −14.1 | −19.7[c] | |
| $Th^{IV}$ | −12.2 | −7.7 | −6.4 | 0.8 | −7.2 | |
| $Tl^I$ | | | | −5.2 | | −3.3 |
| $U^{VI}$ | −5.2[b] | −4.7[b] | −4.1[b] | −2.0 | −1.6[b] | |
| $Zn^{II}$ | | −10.8 | −10.7 | | −2.4[c] | 2.7 |
| $Zr^{IV}$ | | | | 9.0 | | |

[a] From J. Ruzicka and J. Stary, A New Principle of Activation-Analysis Separations—I: Theory of Substoichiometric Determinations, *Talanta* **10**, 287–293 (1963).
[b] Complexes $MA_NHA$ are formed.
[c] Complexes $MA_NHA_2$ are formed.

where

$$k_n = \frac{[H][H_{n-1}Y]}{[H_nY]} \qquad (k_0 = 1) \qquad (25)$$

The selectivity of ion exchange substoichiometric separations is the same as for solvent extractions.

**Table 7.4**  Elements Suitable for Substoichiometric Determinations[a]

| Element | Radio-nuclide | Half-Life | Activation Cross Section, (barns) | Natural Abundance, (%) |
|---|---|---|---|---|
| Antimony | $^{122}Sb$ | 2.8 d | 6.8 | 57.25 |
| Arsenic | $^{76}As$ | 26.5 h | 5.4 | 100 |
| | $^{124}Sb$ | 60 d | 2.5 | 42.75 |
| Bismuth | $^{210m}Bi$ | 5.0 d | 0.019 | 100 |
| Cadmium | $^{115}Cd$ | 2.3 d | 1.1 | 28.86 |
| | $^{115m}Cd$ | 43 d | 0.14 | 28.86 |
| | $^{117}Cd$ | 2.9 h | 1.5 | 7.58 |
| Chromium | $^{51}Cr$ | 27.8 d | 15.9 | 4.31 |
| | $^{55}Cr$ | 3.6 m | 0.38 | 2.38 |
| Cobalt | $^{60m}Co$ | 10.5 m | 16 | 100 |
| | $^{60}Co$ | 5.27 y | 20 | 100 |
| Copper | $^{64}Cu$ | 12.8 h | 4.51 | 69.09 |
| | $^{66}Cu$ | 5.1 m | 1.8 | 30.91 |
| Gallium | $^{70}Ga$ | 21.1 m | 1.4 | 60.4 |
| | $^{72}Ga$ | 14.1 h | 5.0 | 39.6 |
| Gold | $^{198}Au$ | 2.7 d | 96 | 100 |
| Indium | $^{114m}In$ | 50 d | 56 | 4.28 |
| | $^{116m}In$ | 54 m | 155 | 95.72 |
| Iron | $^{55}Fe$ | 2.6 y | 2.8 | 5.82 |
| | $^{59}Fe$ | 45 d | 1.01 | 0.33 |
| Lead | | | | |
| Mercury | $^{197m}Hg$ | 24 h | 420 | 0.146 |
| | $^{197}Hg$ | 65 h | 880 | 0.146 |
| | $^{203}Hg$ | 47 d | 3.8 | 29.8 |
| Molybdenum | $^{99}Mo$ | 67 h | 0.51 | 23.78 |
| | $^{101}Mo$ | 15 m | 0.20 | 9.63 |
| Nickel | $^{65}Ni$ | 2.56 h | 1.52 | 1.08 |
| Palladium | $^{103}Pd$ | 17 d | 4.8 | 0.96 |
| | $^{109}Pd$ | 13.6 h | 10.4 | 26.71 |
| Platinum | $^{193m}Pt$ | 4.4 d | 90 | 0.78 |
| | $^{197}Pt$ | 18 h | 0.87 | 25.3 |
| Scandium | $^{46}Sc$ | 84 d | 12 | 100 |
| Silver | $^{108}Ag$ | 2.3 m | 45 | 51.35 |
| | $^{110m}Ag$ | 253 d | 3.2 | 48.65 |
| Thallium | $^{204}Tl$ | 3.9 y | 8 | 29.5 |
| | $^{206}Tl$ | 4.3 m | 0.10 | 70.5 |
| Thorium | $^{233}Th$ | 22.4 m | 7.33 | 100 |
| Vanadium | $^{52}V$ | 3.8 m | 4.5 | 99.76 |
| Zinc | $^{65}Zn$ | 245 d | 0.47 | 48.89 |
| | $^{69m}Zn$ | 13.9 h | 0.097 | 18.57 |
| | $^{69}Zn$ | 55 m | 1.0 | 18.57 |
| Zirconium | $^{95}Zr$ | 65 d | 0.076 | 17.40 |
| | $^{97}Zr$ | 17 h | 0.053 | 2.80 |

[a] From J. Ruzicka and J. Stary, A New Principle of Activation-Analysis Separations—I: Theory of Substoichiometric Determinations, *Talanta* **10**, 287–293 (1963).

*Precipitation*

For precipitation reactions the condition for substoichiometric separations with reagent HA is derived from the solubility product

$$S_{MA_N} = [M][A]^N = [M]\frac{K_{HA}^N [HA]^N}{[H]^N} \tag{26}$$

where

$$K_{HA} = \frac{[H][A]}{[HA]} \tag{27}$$

For activation analysis use, the corresponding threshold pH value is

$$pH > \frac{1}{N}(N\, pK_{HA} - pS_{MA_N} + 2 + 5N) \tag{28}$$

The selectivity may be further increased by masking interfering metals. In this case the pH of the determination is

$$pH > \frac{1}{N}[N\, pK_{HA} - pS_{MA_N} + 2 + 5N + \log(1 + K_s[B]^s)] \tag{29}$$

where $K_s$ is the stability constant of the water-soluble complex $MB_s$.

Substoichiometric separations are thus of use in activation analysis, since they not only eliminate the need for chemical-yield determination but they may also increase the selectivity of the separation procedure and result in faster or more sensitive analyses. Some of the radionuclides suitable for substoichiometric determination are listed in Table 7.4.

### 7.1.5   Automated Radiochemical Separations

In recent years increasing attention has been given to the possibility of automating chemical analyses to process large numbers of samples in the laboratory. Although it is obvious that automated radiochemical activation analysis would be a most welcome addition to the practices of activation analysis, few results have been reported on the application of automatic chemical separations. This is primarily because of the inherent difficulty in automating chemical operations that involve phase changes and chromatographic separations.

Automation does not necessarily imply improvement; it generally implies only convenience and economy, sometimes with a loss in quality. Activation analyses that cannot be done solely by instrumental methods cannot be automated for instrumental activation analysis. The need for analytical capability for very large numbers of samples has justified in many cases the development of automatic apparatus. This is of particular interest when a simple radiochemical procedure will separate an especially troublesome interference which prevents the measurement of a desired element. Automatic radiochemical activation analysis is being developed to cope with the general problem of the presence of sodium and chlorine in biological or seawater samples, in which their radionuclides may be present in concentrations greater than $10^4$ times those of the sought elements. In some cases removal of the $^{24}$Na and $^{38}$Cl radionuclides have allowed the desired elements to be measured by gamma-ray spectroscopy without further purification.

Automatic chemical separations of multiple samples can be achieved in two ways, as described by Girardi et al.*:

1. The sample is moved continuously along an "analytical line," which can be defined as the sequence of operations leading to the desired isolated compound in a form suitable for the proper detector at the end of the line.

2. The reagents are added to the sample, starting with a predisposed sequence and maximizing the handling of the sample in each reaction vessel.

In the first method the reagents are introduced and the waste products are extracted at the appropriate points on the line. Figure 7.2 shows the schematics for an automated system in which the samples move continuously along an analytical line and several possible phase separation methods. In this system the samples are separated from one another by pumping air or immiscible liquids between samples. The separation between the analytical line and the wastes can be achieved with phase separations. The time required to run $n$ samples is

$$T_n = T + nt \qquad (30)$$

where $T$ = time required to go through the analytical line,

$t$ = time separation between adjacent samples (i.e., the minimum time required to avoid the simultaneous detection of the two samples).

In the second method the transfer of each sample from one reaction vessel to the next is done only when required: for example, when a phase separation is needed. The detector monitors the end of the analytical line when the separation has been achieved. The operation on the sample is also shown schematically in Figure 7.2. The time required to run $n$ samples is again given by (30), where $t$ is now the longest residence time of the sample in a reaction vessel.

Although the two systems can be mixed when required to obtain the best results, preference is given to the second method, even though it is slower,

* F. Girardi, G. Guzzi, J. Pauly, and R. Pietra, The Use of an Automated System Including a Radiochemical Step in Activation Analysis, in *Proc. Modern Trends in Activation Analysis*, 337–343 (Texas A&M University, College Station, 1965).

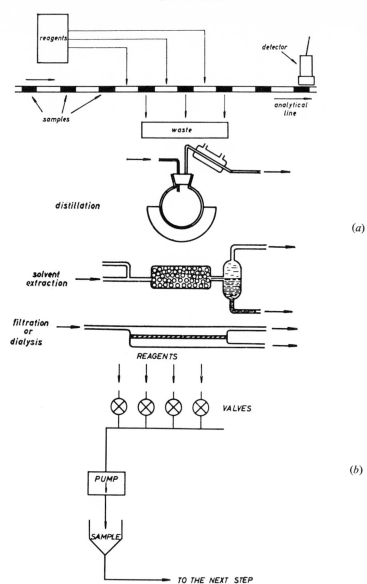

**Figure 7.2** Schematics of systems for automated radiochemical separations. The system in (*a*) is based on continuous movement of the samples along an analytical line with pertinent phase separation methods. The system in (*b*) is based on discontinuous movement of the samples between consecutive steps. [From F. Girardi, G. Guzzi, J. Pauly, and R. Pietra, The Use of an Automated System Including a Radiochemical Step in Activation Analysis, in *Modern Trends in Activation Analysis* (Texas A&M University, College Station, 1965), pp. 337–343.]

for the following reasons:

1. Solution of solid samples and adjustments for the required medium and molarity is simpler.
2. Chromatography separations can be done only by this method.
3. A wider variety of radiochemical separations is possible.
4. The apparatus is also useful for the study of separations by the first method.

Figure 7.3 shows a schematic drawing of a two-unit automatic system for radiochemical separations; the programming unit is connected to the operating unit by flexible cable. The programmer regulates the operating units, pumps, valves, etc., according to a determined time sequence. It also controls the flow rates of the reagents by a programming card which may be in the form of a printed circuit. A program of the scheme for dissolving biological materials is also shown in Figure 7.3. The operating unit contains the pumps and valves necessary to start, regulate, and stop the flow of the various reagents along the contained analytical line.

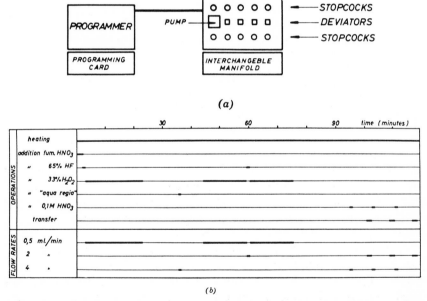

(a)

(b)

**Figure 7.3** (a) Schematic diagram of an automated system for radiochemical separations; (b) a programming card controls the time sequence and flow rates of the procedure, for example, for a dissolution scheme for biological materials. [From F. Girardi, G. Guzzi, J. Pauly, and R. Pietra, The Use of an Automated System Including a Radiochemical Step in Activation Analysis, in *Modern Trends in Activation Analysis* (Texas A&M University, College Station, 1965), pp. 337–343.]

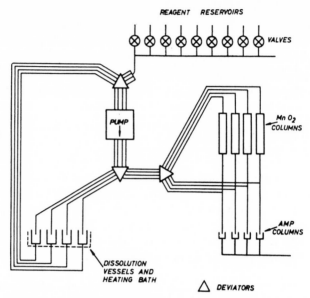

REAGENT RESERVOIRS

VALVES

PUMP

Mn O₂
COLUMNS

AMP
COLUMNS

DISSOLUTION
VESSELS AND
HEATING BATH

△ DEVIATORS

**Figure 7.4**  An analytical line for four simultaneous determinations of Cs in biological specimens, including dissolution of the samples, removal of $^{56}$Mn, and adsorption of $^{134m}$Cs on ammonium molybdophosphate (AMP). [From F. Girardi, G. Guzzi, J. Pauly, and R. Pietra, The Use of an Automated System Including Radiochemical Step in Activation Analysis, in *Modern Trends in Activation Analysis* (Texas A&M University, College Station, 1965), pp. 337–343.]

Figure 7.4 shows an analytical line for the determination of cesium in biological materials, which contain interfering amounts of sodium and manganese. Steps in the procedure are the dissolution of the sample, removal of $^{56}$Mn, and absorption of $^{134m}$Cs on ammonium molybdophosphate (AMP), in four independent parallel systems.

Radioactivity measurements for such systems are normally made with gamma-ray spectrometry and computer calculations. The radiochemical step reduces significantly the uncertainty in the amount of the desired elements when interfering elements are present in larger amounts. Figure 7.5 shows the sequence of improvement in resolution of tracer $^{137}$Cs with radiochemical separations from $^{56}$Mn and $^{24}$Na. The generalized system for automatic activation analysis, including chemical separations, is given in Figure 7.6, which shows the flow chart of the automatic system used to reduce the $^{56}$Mn and $^{24}$Na interferences in determining cesium in biological materials.

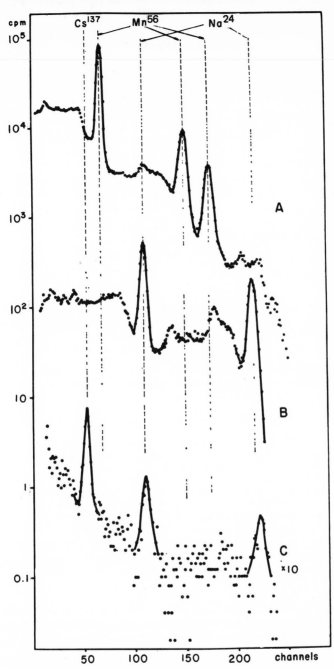

**Figure 7.5** Sequence of improvement of the resolution of trace amounts of $^{137}$Cs in $\gamma$-ray spectra: A, Before chemical separations; B, after removal of $^{56}$Mn on a $MnO_2$ column; and C, after removal of $^{24}$Na by holding $^{137}$Cs on an ammonium molybdophosphate column. [From F. Girardi, G. Guzzi, J. Pauly, and R. Pietra, The Use of an Automated System Including a Radiochemical Step in Activation Analysis, in *Modern Trends in Activation Analysis* (Texas A&M University, College Station, 1965), pp. 337–343.]

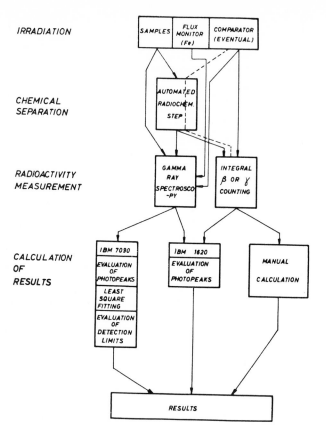

IRRADIATION

CHEMICAL
SEPARATION

RADIOACTIVITY
MEASUREMENT

CALCULATION
OF
RESULTS

**Figure 7.6** A flow sheet for an automated activation analysis system which includes radiochemical separations to concentrate the desired radionuclides or remove interfering radionuclides. [From F. Girardi, G. Guzzi, J. Pauly, and R. Pietra, The Use of an Automated System Including a Radiochemical Step in Activation Analysis, in *Modern Trends in Activation Analysis* (Texas A&M University, College Station, 1965), pp. 337–343.]

An example of the first method of automatic radiochemical separations, continuous measurement with samples separated by air bubbles, has been reported for the measurement of iodine in biological materials.* The automatic system, whose schematic arrangement is shown in Figure 7.7, consists

* D. Comar and C. LePoec, On the Use of an Automatic Chemical Treatment System in Activation Analysis of Biological Samples, in *Proceedings, Modern Trends in Activation Analysis* (Texas A&M University, College Station, 1965), pp. 351–356.

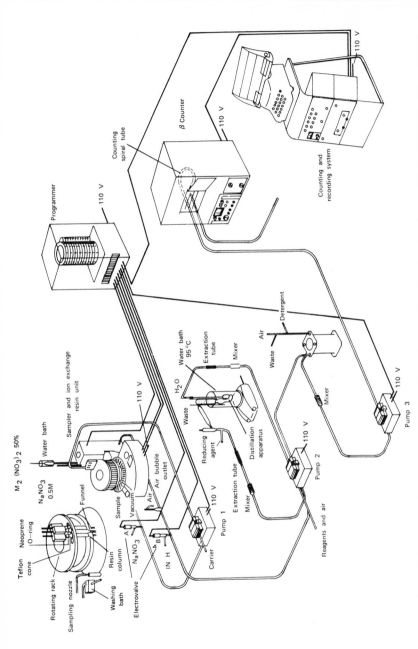

**Figure 7.7** Schematic components of an automatic chemical treatment system. [From D. Comar and C. LePoec, on the Use of an Automatic Chemical Treatment System in Activation Analysis of Biological Samples, in *Modern Trends in Activation Analysis* (Texas A&M University, College Station, 1965), pp. 351–356.]

of four stages of processing:

1. Taking of sample and dilution with inactive carrier.
2. Exchange and elution of the iodine on an ion-exchange column.
3. Purification by solvent extraction and distillation.
4. Radioactivity measurement.

Advantage is taken of the liberation of the radioiodine from organic compounds under neutron irradiation in a form that exchanges well on anionic resins. Some of the features of the automatic system, such as the sampler and ion-exchanger, the distillation apparatus, the counting system, and the measurement cycle of the programmer are shown in Figure 7.8.

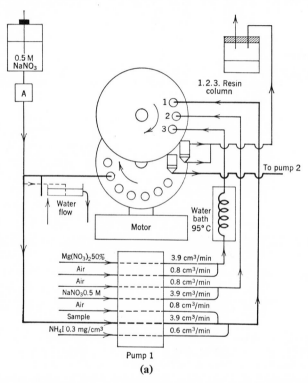

Figure 7.8   Sketches of some of the components of an automatic chemical treatment system for iodine in biological materials. (*a*) Samples and ion-exchange resin system; (*b*) distillation apparatus, (*c*) counting system; and (*d*) the measurement time cycle on the drum of the programmer. [From D. Comar and C. LePoec, on the Use of an Automatic Chemical Treatment System in Activation Analysis of Biological Samples, in *Modern Trends in Activation Analysis* (Texas A&M University, College Station, 1965), pp. 351–356.]

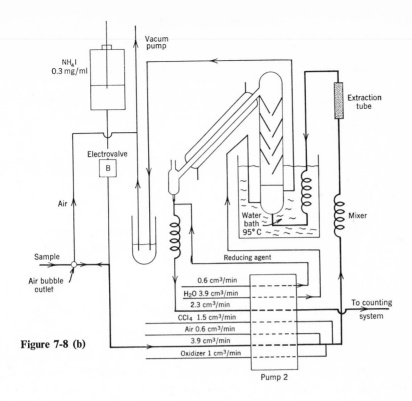

**Figure 7-8 (b)**

NH$_4$I
0.3 mg/ml

Vacuum pump

Electrovalve

B

Air

Sample

Air bubble outlet

Extraction tube

Water bath 95°C

Mixer

Reducing agent

0.6 cm$^3$/min
H$_2$O 3.9 cm$^3$/min
2.3 cm$^3$/min
CCl$_4$ 1.5 cm$^3$/min
Air 0.6 cm$^3$/min
3.9 cm$^3$/min
Oxidizer 1 cm$^3$/min

To counting system

Pump 2

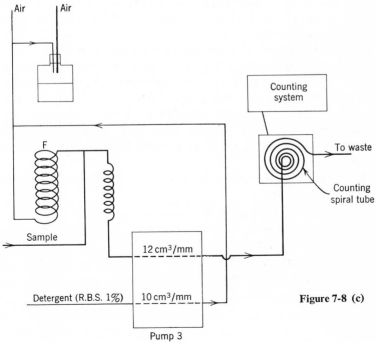

Air     Air

F

Sample

Counting system

To waste

Counting spiral tube

12 cm$^3$/mm

Detergent (R.B.S. 1%)     10 cm$^3$/mm

Pump 3

**Figure 7-8 (c)**

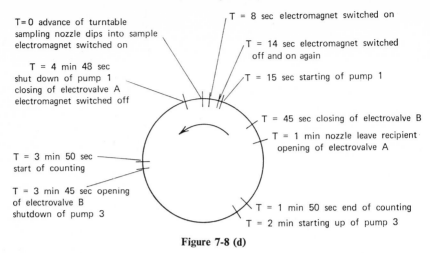

T = 0 advance of turntable
sampling nozzle dips into sample
electromagnet switched on

T = 4 min 48 sec
shut down of pump 1
closing of electrovalve A
electromagnet switched off

T = 3 min 50 sec
start of counting

T = 3 min 45 sec opening
of electrovalve B
shutdown of pump 3

T = 8 sec electromagnet switched on

T = 14 sec electromagnet switched
off and on again

T = 15 sec starting of pump 1

T = 45 sec closing of electrovalve B

T = 1 min nozzle leave recipient
opening of electrovalve A

T = 1 min 50 sec end of counting

T = 2 min starting up of pump 3

**Figure 7-8 (d)**

## 7.2 INSTRUMENTAL ACTIVATION ANALYSIS

Instrumental activation analysis may be defined as a chemical element analysis made by nuclear activation followed by the measurement of specific induced radioactivities without the use of radiochemical separations. The advantages of instrumental activation analysis are almost too obvious to warrant enumeration, primary ones being convenience, speed, economy, and nondestruction of the sample. In many cases the last consideration is of paramount importance.

The promise of being able to insert a sample into a "black box" at one end and receive a printed analysis of its chemical element composition at the other in less than one minute has indeed spurred considerable research and development in instrumental activation analysis. Although it may yet be some time before this promise is fulfilled, especially for analyses at maximum sensitivity, considerable progress has already been made toward the feasibility of instrumental activation analysis as a rapid accurate method of chemical analysis.

The successes are due primarily to the rapid strides made in the development of the following:

1. Steady-operating reactors and accelerators.

2. Automatic methods to transfer samples to irradiation sources and radiation detector systems.

3. Efficient scintillation and semiconductor radiation detectors coupled to high-resolution radiation-energy spectrometers.

4. High-speed computational facilities and methods to resolve quantitatively the complex decay curves and the complex spectra of mixtures of radiations into their component radionuclides.

Automatic activation analysis systems have been developed which can routinely program the entire analysis; that is, transfer a sample to an irradiation position for a predetermined irradiation time, transfer the sample after a predetermined decay period to a measurement position in a radiation detector system, compile a decay curve and/or radiation spectrum for a predetermined measurement time, and then compile the sample, irradiation, and radiation measurement data in a computer for print-out of the chemical analysis.

The principles of irradiation and radiation-energy spectroscopy have already been reviewed. The major development toward the success of instrumental activation analysis has been the resolution of mixtures of complex gamma-ray spectra without the benefit of prior separation by chemical element. The components of a spectrum of a pure radionuclide have also been reviewed. It was noted that the measurement of a full-energy peak (the photopeak) is limited by the presence of the Compton scattering continuum, and both were functions not only of the source but of the detector characteristics and source-detector geometry as well. In many methods only the photopeaks are used to identify and measure a given radionuclide; the Compton continuum is considered as an undesirable background in the measurement of photopeaks of lower energy. Other methods evaluate the Compton continuum (and other components of gamma-ray spectra) as part of the total response of a detector. Several methods to reduce the Compton contribution to the spectra have been developed. For gamma-ray energy above 1.02 MeV further complications develop from pair-production events and associated escape peaks, as noted in Table 6.8. Thus the resolution of gamma-ray spectra is indeed complicated.

Further complications appear when the number of important radionuclides contributing to the total spectrum is an unknown, when short-lived radionuclides are present, and when the total radioactivity is small. Methods to resolve mixed gamma-ray spectra, called *stripping*, have been developed which are based on analytical or graphical procedures, electronic means in multichannel analyzers, and numerical means by computer.

### 7.2.1 Spectrum Stripping

Spectrum stripping is the sequential determination of the amount of each radionuclide present in a gamma-ray spectrum of mixed radionuclides. The stripping is performed graphically or electronically by identifying the gamma-ray photopeak energies and successively subtracting the spectrum of each

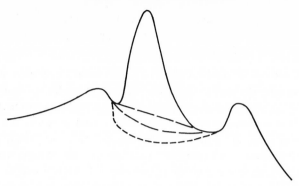

**Figure 7.9**   Simple subtractions of baseline from a full energy peak in a mixed $\gamma$-ray spectrum. Considerable errors may result, especially if the Compton continuum of a higher energy gamma ray represents a significant fraction of the total distribution in the pertinent full-energy channels.

radionuclide, starting with the highest gamma-ray energy photopeak. The amount of the spectrum subtracted is a measure of that radionuclide. Normalization of the spectrum subtracted is generally made by comparison to an appropriate standard spectrum of the pure radionuclide made under identical source-detector conditions.

Several methods are used for manual spectrum stripping. A simple one, used for spectra in which the photopeaks are clearly identified with little overlap, is the direct determination of the photopeak height (or area). For "background" subtraction a baseline extrapolated between the valleys adjacent to the peak is estimated. Figure 7.9 shows several possible ways to estimate the baseline, a straight line between the minima being the simplest, though not necessarily the best. Although the baseline may be drawn subjectively with considerable variation, when the photopeak is clearly defined, the fractional error in the photopeak height or area may be quite small. The error becomes more significant as the mixture of radionuclides increases and the lower energy peaks represent increasingly smaller fractions of the total counts in the pulse-height distribution.

A general method for determining gamma-ray intensities from pulse-height spectra has been the graphical stripping method. This method requires a catalog of calibrated pulse-height distribution graphs for each gamma-ray energy present in the spectrum. In graphical stripping the intensity of the highest energy gamma ray in the spectrum is determined by superimposing on the unknown spectrum the standard spectrum of that energy gamma ray. The normalization may be accomplished graphically with standard spectra plotted on transparent graph paper. In the absence of sum peaks, when the highest energy photopeak has its base on the abscissa of the pulse-height distribution curve, the intensity of the gamma ray can be measured directly by the peak height or area. A channel-by-channel subtraction is made from the total spectrum, and the gamma-ray of the next highest energy is determined. If the full-energy peak now has its base on the abscissa, its intensity can be measured directly by the peak height or area. The normalization is made with the superimposed standard spectrum, and again the remaining gamma-ray spectrum is obtained with channel-by-channel subtraction. The process is repeated until all identifiable photopeaks have been resolved.

An example of this method is shown in Figure 7.10, in which a spectrum obtained from a mixture of the radionuclides $^{113}$Sn, $^{137}$Cs, $^{54}$Mn, and $^{65}$Zn was resolved. The individual standard spectrum of each pure radionuclide is given in the upper part; the mixed spectrum in Box (a) of the lower part. Successive subtraction of the standard spectra show the resolution of the total spectrum by the graphical stripping method.

Another example of the graphical stripping method is shown in Figure 7.11. In this case the composite sample was synthesized with three radionuclides, each emitting a single gamma ray. To provide a moderately troublesome case the activity of the two lower energy radionuclides was made less than 10 % of the highest energy radionuclide. The data and stripping curves, produced from channel-by-channel subtraction of the standard spectra, given by Heath,* are shown in Figure 7.11. With such careful analysis, results with errors less than $\pm 3\%$ can be obtained.

The method is generally initiated after subtraction of the detector background spectrum. Other components, such as backscatter, Bremsstrahlung, annihilation, and sum peaks, which require some a priori information about the mixture of radionuclides in order to "clean" them from the total spectrum, are usually considered part of the total spectrum. Although summing effects of gamma-ray cascades may be calculated, their identification as a sum peak is often difficult. Once identified as due to two other $\gamma$-rays in the spectrum, the pulse-height distribution can be determined and subtracted after normalization of intensity at the sum peak.

The major problem, besides the tediousness of the graphical subtraction process, is the inherent loss in precision at the lower $\gamma$-ray energies because of the cumulative effect of the errors in the subtraction process. Quality control of the accuracy of the normalization at the lower energy peaks is somewhat subjective and is made by inspecting the residuals in the higher energy region and the "general appearance" of the partly stripped spectrum. Re-evaluation of the prior stripping measurements may be required when the residuals or the partly stripped spectrum become unsatisfactory.

* R. L. Heath, *Scintillation Spectrometry, Gamma-ray Spectrum Catalog*, AEC Report No. IDO-16408 (1958).

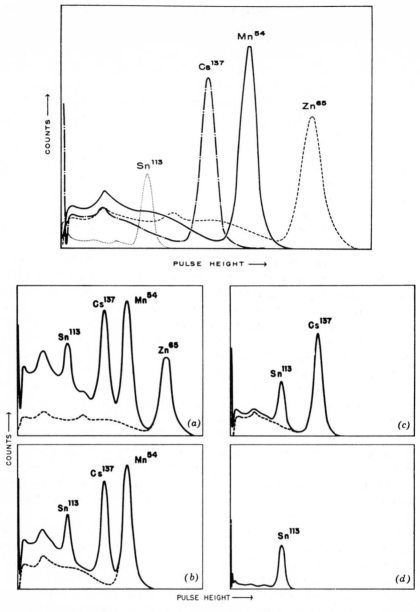

**Figure 7.10** The method of graphical stripping of a mixed $\gamma$-ray spectrum. The combined spectrum of the four radionuclides, whose pulse-height distributions are given individually in the upper figure, are graphically stripped in order of decreasing $\gamma$-ray energy in the lower figure. [From D. F. Covell, "Determination of Gamma-Ray Abundance Directly from the Total Absorption Peak," *Anal. Chem.* **31**, 1785 (1959).]

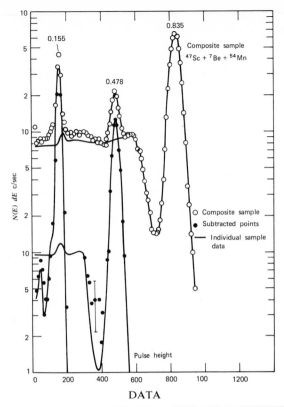

Figure 7.11 Graphical stripping of a composite gamma-ray spectrum in which the two lower energy sources were intentionally less than 10% of the higher energy source. (From R. L. Heath, *Scintillation Spectrometry Gamma-ray Spectrum Catalog*, AEC Report IDO-16408, 1958.)

| Nuclide | $E_\gamma$ (MeV) | Activity ($\gamma$ps) Single Spectrum | Activity ($\gamma$ps) Composite Spectrum | Percent Error |
|---------|------|----------------|-------------------|-------|
| $^{54}$Mn | 0.835 | 54,731 | — | 0.1 |
| $^{7}$Be | 0.478 | 4,256 | 4,393 | 3.2 |
| $^{47}$Sc | 0.155 | 3,633 | 3,567 | 1.8 |

The graphical stripping method has proved to be satisfactory to trained analysts, especially for the simpler spectra that show well-resolved full-energy peaks. The graphical method has also been considered flexible in that complicated spectra may be evaluated by diligent application of "fitting" by trial and error. Although subjectivity is always present in the analyses, for simple spectra the graphical stripping method can yield gamma-ray intensities with an accuracy of about ±3 to ±5%; for more complex spectra accuracies of the order of ±10% are more common.

One of the early methods for the analysis of gamma-ray spectra was described by Connally and Leboeuf,* who were able to resolve spectra of three or four radionuclides by integrating the full-energy peak area after sequentially correcting for the Compton background from the highest energy peaks, downward, in the spectrum. Another method of spectrum reduction involves the measurement of a calibrated fraction of the full-energy peak area.† The method is based on the validity of the assumption that the ratio of the area $N$ above a horizontally intersecting line of constant length $l$ is proportional to the total area of the peak $S$ and therefore varies directly with the gamma-ray intensity. Figure 7.12 shows the graphical definition of the fixed fraction $N$ of the total peak area $S$. This method has the advantages of greater speed compared with the graphical stripping method and does not require a catalog of standard spectra; it has the disadvantage of not correcting for the baseline level or shape.

A method for a more direct estimation of gamma-ray intensities in complex spectra is known as the "complement subtraction" method. This method uses

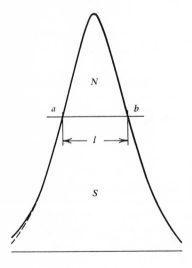

**Figure 7.12**  For well-resolved full-energy peaks, in which the Compton continuum is reasonably flat, the area $N$ defines a fixed fraction of the total peak area $S$. [From D. F. Covell, Determination of Gamma-Ray Abundance Directly from the Total Absorption Peak, *Anal. Chem.* **31**, 1785 (1959).]

---

* R. E. Connally and M. B. Leboeuf, Analysis of Radionuclide Mixtures, *Anal. Chem.*, **25** 1095–1100 (1953).
† L. D. McIsaac, USNRDL-TR-72 (1956).

the electronic capability of multichannel analyzers to assist in the measurement of the radionuclides from a mixed gamma-ray spectrum directly, without manipulation of the data or any mathematical procedures. In this method, as described by Lee,* standard spectra of known pure radionuclides are subtracted directly from the pulse-height data stored in the multichannel analyzer. The net activity of each radionuclide is determined by comparison of the counting times. The "complement" function of a multichannel analyzer subtracts the number of counts stored in each channel of the memory from the full capacity of the channel. The result, as seen on the analyzer's oscilloscope, is an inversion of the pulse-height spectrum. When a pure radionuclide standard is then counted, its pulse-height spectrum is added to the complemented memory. This results, in effect, in the subtraction of the standard spectrum, channel by channel, from the original spectrum. The end point for the matching of the known radionuclide spectrum to the mixed spectrum is achieved by matching a major photopeak on the oscilloscope screen. The disintegration rate as given by the height of the photopeak in the unknown spectrum is determined by the known disintegration rate of the standard and the ratio of counting times required to match the photopeak heights. The major problems in this method are the instrument dead-time differences between sample and standard counting rates, drift of the instrument between these counts, and availability of suitably calibrated standards, especially short-lived ones. This latter problem is mainly eliminated in the comparitor method of activation analysis. Timers to measure accurately the live-time counting times of the sample and standard are available.

These graphical stripping methods, and variations of them, have in common the limitations in accuracy resulting from the conversion of the basic digital data of multichannel analyzers into graphic form. They are subject to human bias and thus do not provide an analytical basis for determination of precision. Several methods have been developed to analyze gamma-ray spectra in its digital form; they have led to the use of high-speed electronic computers for gamma-ray spectral analysis.

An early development was a digital analysis program based on the three-dimensional surface, shown in Figure 6.18, which represents the response of a particular detector to monoenergetic gamma rays.† Analytical representation of the experimental pulse-height distributions of single gamma-ray spectra was made as a sum of several analytic functions. The components of the spectrum, listed in Table 6.8, are shown in Figure 7.13a. The photopeak

---

* W. Lee, Direct Estimation of Gamma-Ray Abundances in Radionuclide Mixtures, *Anal. Chem.* **31**, 800–806 (1959).
† R. L. Heath, Data Analysis Techniques for Gamma-Ray Scintillation Spectrometry, AEC Report IDO-16784 (1962).

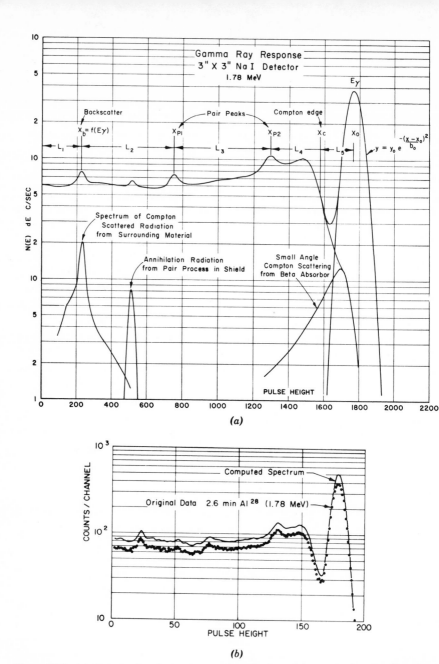

**Figure 7.13** (*a*) The details of structure and method used in computing the Compton distribution in a simple gamma-ray pulse-height spectrum; (*b*) comparison of computed response with observed spectrum for the 1.78-MeV gamma-ray of 2.6-m $^{28}$Al. (From R. L. Heath, Data Analysis Techniques for Gamma-Ray Scintillation Spectrometry, AEC Report IDO-16784, 1962.)

is represented by a Gaussian function of the form

$$\bar{y} = y_0 e^{-(x-x_0)^2/b_0}$$ (31)

where $\bar{y}$ = the calculated number of counts in channel $x$,

$x_0$ = the pulse height at the center of the distribution,

$y_0$ = the number of counts per channel at $x_0$,

$b_0$ = a parameter of the resolution, given by

$$b_0 = \frac{w_0^2}{\sqrt{2 \ln 2}}$$ (32)

where $w_0$ is the full width of the peak at $y_0/2$.

The other parts of the pulse-height distribution in Figure 7.13a are due primarily to the Compton process and pair production; their effects on gamma-ray spectra were described in Section 6.3.3. The analytical process divides the spectrum below the full-energy peak into five segments, $L_1, \ldots,$ $L_5$, bounded by 0, $X_b$, $Xp_1$, $Xp_2$, $X_c$, and $X_0$. For gamma-rays with $E_\gamma <$ 1.02 MeV, only three segments with bounds $X_b$, $X_c$, and $X_0$ are required to describe the Compton distribution adequately. Each segment is fitted to a series of the form

$$y = a + bx + \sum_{k=1}^{N_k} b_p \sin \frac{K \pi x}{L}$$ (33)

where the lower channel bound is used as the origin and the upper bound is $L$. A region of overlap is taken to ensure a smooth fit at the segment end points. The number of terms $k$ in the expansion is chosen to give a certain amount of smoothing to the data points.

The calculated count rate $N(E) \, dE$ in each channel as a function of gamma-ray energy represents a section through the three-dimensional surface perpendicular to the pulse-height axis. The points are fitted to a polynominal in gamma-ray energy. The number of terms in the polynomial is adjustable to give the best fit to the data points. This produces a set of polynomials in energy, one for each channel on the pulse-height scale. The process is repeated for each segment of the Compton distribution. A comparison of the calculated shape of a pulse-height spectrum for the radionuclide 2.6-m $^{28}$Al ($E_\gamma =$ 1.78 MeV) with the experimental spectrum in Figure 7.13b shows that the details of the experimental spectrum have been accurately reproduced by the shape-generation program. The agreements by individual channels are within the range of statistical variation of the experimental data. Corrections for coincidence sum effects and cascade gamma-rays have been added. The calculated sum spectrum of $^{94}$Nb was shown in Figure 6.15. These corrections and analyses of complex spectra usually require computer calculations.

Another method of analysis of gamma-ray spectra in digital form was described by Covell.* In this method the response of the pulse-height analyzer is represented graphically as rectangles whose areas are proportional to the numbers of counts in the channels. Figure 7.14 shows a full-energy peak plotted in histogram form. The response of the channel containing the greatest number of counts is defined as $a_0$. Succeeding channel responses, down the low amplitude side of the peak, are designated $a_1, a_2, a_3, \ldots, a_n$, down the high amplitude side of the peak, $b_1, b_2, b_3, \ldots, b_n$.

Summation of the responses from $b_n$ through $a_0$ to $a_n$ gives a value that represents the total counts contained in these channels, shown graphically as the area $P$ in Figure 7.14. If this area $P$ is divided by a line connecting the ordinate values of $a_n$ and $b_n$, the area above the line, given by $N$ in Figure 7.14, corresponds to the area $N$ in Figure 7.10. Thus, within the statistical variation of the counts in the pertinent channels, area $N$ is related to the total peak area and thus to the gamma-ray intensity.

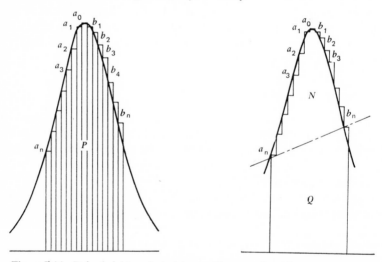

**Figure 7.14**   Pulse-height analysis data as a histogram in which the area $P$ represents the total counts contained in channels $a_n$ to $b_n$ and the intersect between $a_n$ and $b_n$ above which the area $N$ bears a constant relationship to the total area contained in the peak and therefore the gamma-ray intensity. [From D. F. Covell, Determination of Gamma-Ray Abundance Directly from the Total Absorption Peak, *Anal. Chem.* **31**, 1785, (1959).]

* D. F. Covell, Determination of Gamma-Ray Abundance Directly from the Total Absorption Peak, *Anal. Chem.* **31**, 1785–1790 (1959).

From Figure 7.14 it is noted that

$$P = a_0 + \sum_{i=1}^{n} a_i + \sum_{i=1}^{n} b_i \tag{34}$$

and

$$Q = P - N = \frac{(2n-1)(a_n + b_n)}{2} + (a_n + b_n) = (n + \tfrac{1}{2})(a_n + b_n) \tag{35}$$

By substitution

$$N = a_0 + \sum_{i=1}^{n} a_i + \sum_{i=1}^{n} b_i - (n + \tfrac{1}{2})(a_n + b_n) \tag{36}$$

Since (36) is an algebraic sum of $2n + 1$ independent terms, each of which may be considered to have a Poisson distribution, the variance of the number of counts may be estimated from

$$\text{var}\,(N) = a_0 + \sum_{i=1}^{n} a_i + \sum_{i=1}^{n} b_i + [(n - \tfrac{1}{2})^2 - 1](a_n + b_n) \tag{37}$$

which can be simplified to

$$\text{var}\,(N) = N + (n - \tfrac{1}{2})(n + \tfrac{1}{2})(a_n + b_n) \tag{38}$$

The standard deviation is thus

$$\sigma(N) = [N + (n - \tfrac{1}{2})(n + \tfrac{1}{2})(a_n + b_n)]^{\frac{1}{2}} \tag{39}$$

### 7.2.2 The Use of Computers

The development and availability of high-speed electronic computers has resulted in widespread uses of computers in nuclear physics, radiochemistry, and radioactivation analysis. Outstanding among the latter uses has been the development of more accurate, and obviously less tedious, methods for analysis of complex gamma-ray spectra. Other uses include analysis of multi-component radioactive decay curves, the introduction of decay constants as another parameter for analysis of time dependent gamma-ray spectra, optimization calculations, and the control of automatic activation analysis systems.

#### Computer Assistance in Spectrum Stripping

Computers can be used in gamma-ray spectrometry either as an aid in "analyst" resolution of a mixed spectrum or as a numerical calculation device for "computer" resolution. Modern large computers and improved "programs" can generally provide gamma-ray spectrum analysis as good as any by trained analysts, certainly in much less time and with much less chance of arithmetic errors. The investment in time and cost for a large computer system is still large, however, and is warranted only if large numbers of analyses are to be performed. Commercial services of proved reliability for smaller analytical needs are available.

An early use of computers to assist in smoothing gamma-ray spectra was described by Anders and Beamer.* A small digital computer was programmed to smooth, normalize to standard irradiation and sample weight conditions, and resolve time-dependent gamma-ray spectra, especially for short-lived radionuclides produced in short-irradiation-time neutron-activated samples. The use of standard spectra for stripping avoids the need for simultaneous irradiation of comparitor element standard samples. The samples are irradiated with flux monitors for a preset irradiation time and counted precisely at preset time intervals. Punched paper tape from the analyzer is used as input to the computer:

1. To combine data points to smooth the spectra and increase statistical significance.
2. To correct for detector dead time.
3. To convert data to activity values.
4. To subtract detector background.
5. To normalize spectra to preset sample size and neutron flux.
6. To output the corrected, normalized data as plots on semilog graph paper.

The spectra are then "stripped" by repeated subtraction of standard spectra until only statistical noise remains. Figure 7.15 shows the stripping sequence of the computer-adjusted data for the analysis of small amounts of manganese and vanadium in a $CaCO_3$ matrix. Standard spectra of 69 elements have been compiled by Anders (see Section 7.4.1).

A later use of computers to diminish statistical scatter in gamma-ray spectra by a smoothing technique has been described by Yule.† A computer program uses data convolution techniques to locate peaks in a spectrum, determine the areas, and estimate their energy as an aid in gamma-ray spectrum analysis. Convolution methods, applicable to polynomials, are based on the assumption that five data points define a peak that can be represented by a second or third degree polynomial; the second degree polynomials used in this method located peak tops with more accuracy.

The convolution method is illustrated in Figure 7.16, in which a mixed peak from $^{131}Ba$ and $^{135m}Ba$ show a distorted full-energy peak. The equation

* O. U. Anders and W. H. Beamer, Resolution of Time-Dependent Gamma Spectra with a Digital Computer and its Use in Activation Analysis, *Anal. Chem.* **33**, 226–230 (1961).
† H. P. Yule, Data Convolution and Peak Location, Peak Area, and Peak Energy Measurements in Scintillation Spectrometry, *Anal. Chem.* **38**, 103–105 (1966).

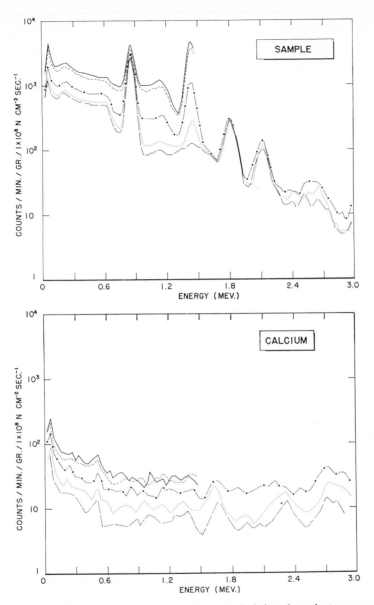

**Figure 7.15** Stripping of a computer-smoothed time dependent gamma-ray spectrum of a $CaCO_3$ sample. [From O. U. Anders and W. H. Beamer, Resolution of Time-Dependent Gamma Spectra with a Digital Computer and its Use in Activation Analysis, *Anal. Chem.* **33**, 226–230 (1961).]

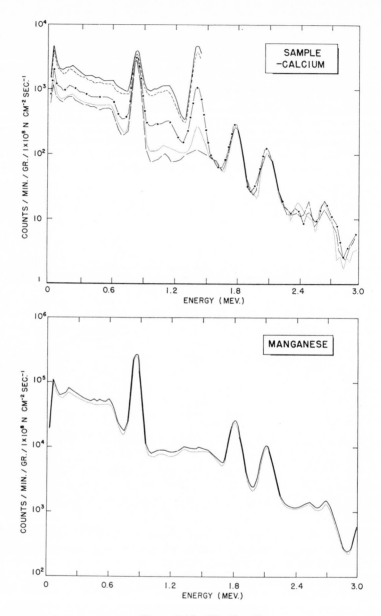

**Figure 7-15   (Continued)**

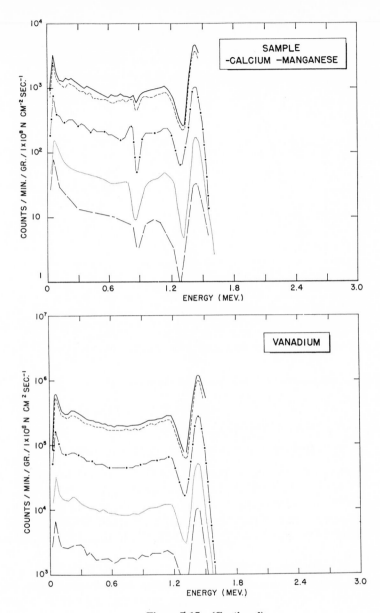

**Figure 7-15    (Continued)**

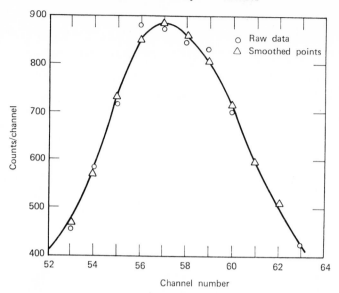

**Figure 7.16** Data points for a combined $^{131}$Ba and $^{135m}$Ba photopeak before and after convolution. [From H. P. Yule, Data Convolution and Peak Location, Peak Area, and Peak Energy Measurements in Scintillation Spectrometry, *Anal. Chem.* **38**, 103–105 (1966).]

used to produce the smoothed spectrum points was

$$35 \times D_i = -3 \times (C_{i-2} + C_{i+2}) + 12 \times (C_{i-1} + C_{i+1}) + 17 \times C_i \quad (40)$$

where $C_i$ = number of observed counts in channel $i$,

$D_i$ = number of counts in channel $i$ in the smoothed spectrum.

Different sets of coefficients produce the first or second derivative. Once the data have been convoluted, location of peaks is made by finding a sign change in the first derivative or a minimum in the second derivative. The method discards "peaks" that turn out to be Compton shoulders or statistically invalid, but it does list as "peaks" those due to backscatter or summing effects.

Computer programs have been written expressly to take advantage of the high resolution of solid-state detection systems; for example, Guzzi, Pauly, Girardi, and Dorpema* have reported a computer program to determine radionuclides and possible interferences in spectra from high resolution Ge-Li drifted detectors coupled to large memory multichannel analyzers.

---

* G. Guzzi, J. Pauly, F. Girardi, and B. Dorpema, Computer Program for Activation Analysis with Germanium Lithium Drifted Detectors, Euratom Report EUR 3469.e, 1967.

The program includes a scan of the spectrum data to determine peaks over five channels and evaluates them as photopeaks. For a peak with a maximum in channel $n$ the condition to be satisfied is

$$C_{n-2} < C_n - \sqrt{C_n} > C_{n+2} \tag{41}$$

The minimum on both sides of the maximum is chosen as channels $n - k$ and $n + k'$ for which

$$C_{n-k-1} \geq C_{n-k} - \sqrt{C_{n-k}} \tag{42}$$

$$C_{n+k'+1} \geq C_{n+k'} - \sqrt{C_{n+k'}} \tag{43}$$

For these photopeaks the area $S$ is calculated as

$$S = \sum_{n-k}^{n+k'} C - \frac{C_{n-k} + C_{n+k'}}{2} (k + k' + 1) \tag{44}$$

The standard deviation is

$$\sigma_s = \left( S + \frac{C_{n-k} + C_{n+k'}}{4} n^2 \right)^{1/2} \tag{45}$$

The peak is retained as a true photopeak if

$$\sigma_s < \frac{S}{2} \tag{46}$$

The energy corresponding to the peak is measured from the peak symmetry axis $A$, after subtraction of the "background"

$$A = \frac{\sum_{k_1}^{k_2} k(C_k)}{\sum_{k_1}^{k_2} C_k} \tag{47}$$

where the limits $k_1$ and $k_2$ are chosen so that the corrected channel contents $C_k$ are all larger than half the channel content of the maximum.

The program also tabulates the resolution of the detector as a function of energy and evaluates possible interferences.

### Gamma-Ray Spectral Analysis

By far the greatest contribution of the high-speed digital computer has been in the problem of gamma-ray spectral analysis. Many programs have been written which resolve complex gamma-ray spectra in one of several ways. One of the most widely used is based on the method of fitting a least-squares curve to a smoothed, continuous representation of the gamma-ray spectrum.

The method of least-squares curve fitting determines a function relating two independent variables for which sets of measured values have been

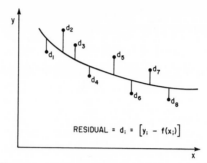

**Figure 7.17** Curve fitting by least-squares analysis. The curve represents the calculated $f(x_i)$ for which the summation of $(d_i)^2$ is a minimum.

obtained. An illustration of curve fitting in Figure 7.17 shows the function $y = f(x)$ drawn as the "best fit" through the eight experimental points. The residuals $d_i$ represent the difference between the value of $y$ observed and calculated from the function $f(x)$, where

$$d_i = y_i - f(x_i) \tag{48}$$

The best fit of a function to a set of points is defined as the one that produces a minimum in the square of the residuals. Thus the sum to be minimized is

$$S = \sum_{i=1}^{n} (d_i)^2 = \sum_{i=1}^{n} [y_i - f(x_i)]^2 \tag{49}$$

The method of least squares is based on the assumptions that the residuals are randomly distributed; that is, the variance of the measurements is a constant and the mean of the residuals is zero.

When no information about the precision of each value is available, the data are usually treated as having common variance. A pulse-height distribution may be smoothed to give a continuous curve $y = f(x)$, where $y$ is the number of counts per channel and $x$ is the channel number. The variance of the counts in each channel, however, is not a constant, but by the statistical nature of radioactive decay it is approximately equal to the number of counts in the channel (the count includes the magnitude of the background counts).

To restore the validity of the least-squares method each of the channel data values must be corrected by an appropriate weighting factor, which for the statistics of radioactive decay is inversely proportional to the variance of the number of counts. Then the sum to be minimized for the residuals of a pulse-height analysis curve is

$$S = \sum_{i=1}^{n} \frac{[y_i - f(x_i)]^2}{\text{var}\,(f(x_i))} \tag{50}$$

Although solutions for the spectral shape of $f(x)$ have been made by nonlinear least-squares fitting,* most programs use the linear form of the least-square formulas. In this case, for a multichannel analyzer spectrum of $n$ channels containing a mixed spectrum of $m$ radionuclides, there is a set of $n$ measurements $y_i$ of known variance. Each value is fitted by an expression that is a linear combination of $m$ known functions or quantities $a_{ij}$ with unknown coefficients $x_j$, the number of counts per radionuclide in the mixture. For $n \geq m$ there is a set of $n$ equations:

$$y_i = \sum_{j=1}^{m} a_{ij} x_j + d_i \tag{51}$$

where $d_i$ is the residual. Thus

$$S = \sum_{i=1}^{n} \frac{(\sum_{j=1}^{m} a_{ij} x_j - y_i)^2}{\text{var}\,(y_i)} \tag{52}$$

To minimize $S$ the function is differentiated with respect to each parameter $x_k$ and each derivative is set equal to zero. Thus

$$\frac{\partial S}{\partial x_k} = \sum_{i=1}^{n} \frac{2a_{ki}(\sum_{j=1}^{m} a_{ij} x_j - y_i)}{\text{var}(y_i)} = 0 \tag{53}$$

$$\sum_{i=1}^{n} \frac{\sum_{j=1}^{m} (a_{ki} a_{ij} x_j) - a_{ki} y_i}{\text{var}\,(y_i)} = 0 \tag{54}$$

By reversing the order of summation

$$\sum_{j=1}^{m} x_j \left[ \sum_{i=1}^{n} \frac{a_{ki} a_{ij}}{\text{var}\,(y_i)} \right] = \sum_{i=1}^{n} \frac{a_{ki} y_i}{\text{var}\,(y_i)} \tag{55}$$

for each value of $k$ from 1 to $m$. The solutions to these equations yield the values of $x_j$ (the calculated gamma-ray intensity of each radionuclide which minimizes the quantity $S$ for the observed pulse-height distribution).

The solutions to these equations are normally made with the use of matrix algebra. The set of equations given by (55) may be written out, for example,

* See, for example, R. O. Chester, R. W. Peelle, and F. G. Maienschein, Nonlinear Least-Squares Fitting Applied to Gamma-Ray Scintillation Detector Response Functions, in Applications of Computers to Nuclear and Radiochemistry, NAS-NS 3107, 1963, pp. 201–212.

for $n$ channels containing counts from $m = 4$ radionuclides, as

$$x_1 \sum a_{i1}a_{i1} + x_2 \sum a_{i1}a_{i2} + x_3 \sum a_{i1}a_{i3} + x_4 \sum a_{i1}a_{i4} = \sum a_{i1}y_i$$
$$x_1 \sum a_{i2}a_{i1} + x_2 \sum a_{i2}a_{i2} + x_3 \sum a_{i2}a_{i3} + x_4 \sum a_{i2}a_{i4} = \sum a_{i2}y_i$$
$$x_1 \sum a_{i3}a_{i1} + x_2 \sum a_{i3}a_{i2} + x_3 \sum a_{i3}a_{i3} + x_4 \sum a_{i3}a_{i4} = \sum a_{i3}y_i$$
$$x_1 \sum a_{i4}a_{i1} + x_2 \sum a_{i4}a_{i2} + x_3 \sum a_{i4}a_{i3} + x_4 \sum a_{i4}a_{i4} = \sum a_{i4}y_i$$

$$(56)$$

where summations are over the $n$ channels. This system of equations can be expressed in the form of matrices:

$$
\begin{bmatrix}
\sum a_{i1}a_{i1} & \sum a_{i1}a_{i2} & \sum a_{i1}a_{i3} & \sum a_{i1}a_{i4} \\
\sum a_{i2}a_{i1} & \sum a_{i2}a_{i2} & \sum a_{i2}a_{i3} & \sum a_{i2}a_{i4} \\
\sum a_{i3}a_{i1} & \sum a_{i3}a_{i2} & \sum a_{i3}a_{i3} & \sum a_{i3}a_{i4} \\
\sum a_{i4}a_{i1} & \sum a_{i4}a_{i2} & \sum a_{i4}a_{i3} & \sum a_{i4}a_{i4}
\end{bmatrix}
\begin{bmatrix} x_1 \\ x_2 \\ x_3 \\ x_4 \end{bmatrix}
=
\begin{bmatrix} y_1 \\ y_2 \\ y_3 \\ y_4 \end{bmatrix}
\qquad (57)
$$

in which the set of coefficients in the first brackets is represented by the matrix [A], the set of the variable $x_j$ represented by the one-column matrix (vector) (X), and the set of constant $y_i$ as the vector (Y). In matrix algebra (57) may be written

$$[A](X) = (Y) \qquad (58)$$

Multiplication of two matrices [A] and [B] can be made only if the number of columns in [A] is equal to the number of rows in [B]. The element in the $i$th row and $j$th column of the product matrix [P] is found by summing the products of the successive pairs of elements of the $i$th row of [A] and the $j$th column of [B]; for example, for $n$ columns in [A] and $n$ rows in [B]

$$P_{ij} = \sum_{k=1}^{n}(a_{ik})(b_{kj}) \qquad (59)$$

For the general case in which $m < n$, that is, the number of unknown radionuclides is less than the number of channels in the spectrum, the matrix [A] is not in a form suitable for multiplication. However, the matrix [A] can be transformed into a square matrix (one in which the number of rows and columns is equal) by multiplying it by its transpose matrix $[A^T]$. The transpose matrix is formed by interchanging the rows and columns of the matrix [A]. The multiplication of the (m × n) matrix [A] by the (n × m) matrix $[A^T]$ gives an (m × m) square matrix, which represents the square of the amplitude of a vector in $n$-space.

$$[A^T][A] = \sum_{i=1}^{n} A_i^2 \qquad (60)$$

Equation 58 then becomes

$$[A^T][A](X) = [A^T](Y) \qquad (61)$$

The operation of division is not defined for matrices, although for the special case of a square matrix [M] there exists an inverse matrix, [M$^{-1}$], whose product with M gives the unit matrix [I] in which all elements of the main diagonal are equal to unity and all other elements are zero:

$$[M][M^{-1}] = [M^{-1}][M] = [I] \tag{62}$$

An inverse matrix can be used in the solution of a system of linear simultaneous equations. The system given by (61) can be solved for (X) by multiplying both sides of the equation by the inverse of [A$^T$A]

$$(X) = [A^T A]^{-1}[A^T](Y) \tag{63}$$

Thus the elements of the product of the inverse matrix [A$^T$A]$^{-1}$ and the constant vector (Y) will be the unknown (X) values. The inversion of the matrix and calculation of the $X_j$ values is made by computer. Several numerical methods* are available for the solution of simultaneous linear equations by methods other than matrix inversion; for example, the Gauss-Jordan elimination method, the use of Cramer's rule, and the Gauss-Seidel iteration method.

The least-squares method applied to spectrum analysis also allows for an estimate of the quality of the over-all fit of the calculated $y_i$ to the observed $y_i$ by a $\chi^2$ test, where $\chi^2$ represents $(n - m)$ degrees of freedom. If the residuals between the calculated $y_i$ and the observed $y_i$ are given by $r_i$, then $\chi^2$ is defined as

$$\chi^2 = \sum_{i=1}^{n} \frac{r_i^2}{\text{var}(y_i)} \tag{64}$$

Since, in general, the number of radionuclides determined is much smaller than the number of channels of the analyzer, that is, $m \ll n$, the first $m$ terms of $\chi^2$ are small compared with the $(n - m)$ terms and a "goodness of fit" ratio $Q$ is defined as

$$Q = \frac{\chi^2}{n - m} \simeq 1 \tag{65}$$

Most computer programs include a calculation of $Q$ to check the goodness of fit of the ratio of observed to predicted errors. If $Q$ is much greater than one, several sources of error can be re-evaluated:

1. Errors in the program or calculation.
2. Insufficient counting statistics to obtain adequate var $(y_i)$.
3. Incorrect form or smoothing of the spectrum.

* See, for example, T. R. Dickson, *The Computer and Chemistry* (Freeman, San Francisco, 1968).

4. Insufficient standards for the radionuclides present.

5. Changes in gain or threshold values between sample and standards.

The latter effect is discussed in Section 8.3.2.

### Analysis of Multicomponent Decay Curves

The analysis of multicomponent decay curves has been programmed for computer solution in several forms. One of these, the least-squares method, is analogous to the resolving of gamma-ray spectra. When a mixture of radionuclides is counted over a period of time, the decay curve, after correction for detector background, is described by the equation

$$A(t) = \sum_{j=1}^{m} x_j^0 \, e^{-\lambda_j t} \tag{66}$$

where $A(t)$ = the total activity at time $t$,

$\quad x_j^0$ = the initial amount of radionuclide $j$,

$\quad \lambda_j$ = the decay constant of radionuclide $j$,

$\quad t$ = the elapsed time; $t = 0$ is normally the time at end of irradication or the time of first count.

The least-squares fitting is done with the set of equations

$$A_i = \sum_{j=1}^{m} x_j e^{-\lambda_j t_i} \tag{67}$$

weighted by the standard deviation $\sigma_i$ of $A_i$:

$$y_i = \frac{A_i}{\sigma_i} = \sum_{j=1}^{m} x_j \frac{e^{-\lambda_j t_i}}{\sigma_i} \tag{68}$$

The vector $(\mathbf{X})$ is calculated by (63) and the "goodness of fit" of the vector by (65).

A method described by Nervik* avoids the figure loss associated with the inversion of the matrix $[\mathbf{A}^T][\mathbf{A}]$ by the processes of orthogonalization and diagonalization. The first process is equivalent to multiplying the matrix from the right by an upper triangular matrix $[\mathbf{U}]$ with unit diagonal elements to give a new matrix $[\mathbf{B}]$, where

$$[\mathbf{A}][\mathbf{U}] = [\mathbf{B}] \tag{69}$$

$$[\mathbf{B}^T] = [\mathbf{U}^T][\mathbf{A}^T] \tag{70}$$

* W. E. Nervik, Brunhilde—A Code for Analyzing Multicomponent Radioactive Decay Curves, in *Applications of Computers to Nuclear and Radiochemistry*, NAS-NS 3107, 1963, pp. 9–25.

This converts (64) to

$$[\mathbf{B}^T][\mathbf{B}][\mathbf{U}]^{-1}(\mathbf{X}) = [\mathbf{U}^T][\mathbf{A}^T](\mathbf{Y}) \tag{71}$$

The product $[\mathbf{B}^T][\mathbf{B}]$ is a diagonal matrix $[\mathbf{D}]$, since the off-diagonal terms from the cross products of the mutually orthogonal matrices $[\mathbf{B}^T]$ and $[\mathbf{B}]$ are all zero. Thus

$$(\mathbf{X}) = [\mathbf{U}][\mathbf{D}]^{-1}[\mathbf{U}^T][\mathbf{A}^T](\mathbf{Y}) \tag{72}$$

The reduction in rounding errors results from the substitution of each element in the diagonalized matrix by its reciprocal. From (63) and (72) it follows that

$$[\mathbf{U}][\mathbf{D}]^{-1}[\mathbf{U}^T] = [\mathbf{A}^T\mathbf{A}]^{-1} \tag{73}$$

which is called the "error matrix" and can be used to calculate the deviation of each point from the calculated curve.

A modification to this method of least-squares analysis, in which a provision for determining the half-lives of the radionuclides is added, is described by Cumming.* A problem arises because the decay constants in (68) are not in linear form. The solution involves an expansion of the exponential form in terms of small changes $\delta x_j$ and $\delta \lambda_j$ from a set of initial guesses $x_j{}^0$ and $\lambda_j{}^0$. Then from

$$(x_j{}^0 + \delta x_j)e^{-(\lambda_j{}^0 + \delta \lambda_j)t_i} \approx (x_j{}^0 + \delta x_j)e^{-\lambda_j{}^0 t_i} - x_j{}^0\, \delta \lambda_j t_i e^{-\lambda_j t_i} \tag{74}$$

a solution for the $\delta \lambda$ terms is possible. An iterative procedure is used until a preset degree of convergence is attained.

An alternate method to least-squares analysis is the method of Fourier transformation, as described by the Gardners.† The method assumes that the exponential series of (67) can be represented by a Laplace integral equation

$$A(t) = \sum_{j=1}^{m} x_j{}^0 e^{-\lambda_j t_i} = \int_0^\infty g(\lambda)e^{-\lambda t}\, d\lambda \tag{75}$$

where $g(\lambda)$ is a sum of delta functions, each of which would theoretically be zero except at specific values of $\lambda_j$ of radionuclides which contribute to $A(t)$. The statistical nature of $A(t)$ results, however, in a frequency distribution of $\lambda_j$ as a function of $g(\lambda)$. Thus a plot of $g(\lambda)$ versus $\lambda$ will indicate a radionuclide when a significant peak occurs. The $\lambda_j$ at $g(\lambda)_{max}$ is the decay constant and the height of the peak is proportional to the coefficient $x_j{}^0$. The function $g(\lambda)/\lambda$ is obtained by Fourier transformation of the variables $\lambda = e^{-y}$ to

* J. B. Cumming, CLSQ; the Brookhaven Decay Curve Analysis Program, in *Applications of Computers to Nuclear and Radiochemistry*, NAS-NS 3107, 1963, pp. 25–33.
† D. G. Gardner and J. C. Gardner, Analysis of Multicomponent Decay Curves by Use of Fourier Transforms, in *Applications of Computers to Nuclear and Radiochemistry*, NAS-NS 3107, 1963, pp. 33–40.

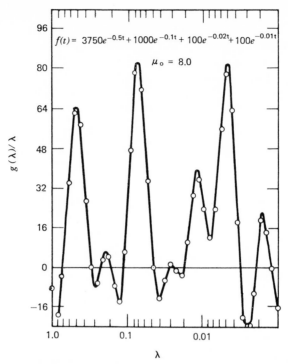

**Figure 7.18** Resolution of a four-component decay curve by a Fourier transform analysis method. (From D. G. Gardner and J. C. Gardner, Analysis of Multicomponent Decay Curves by Use of Fourier Transforms, in *Applications of Computers to Nuclear and Radiochemistry*, NAS-NS 3107, 1963, pp. 33–40.)

yield a function $g(e^{-y})$, which is evaluated by numerical solution of the Fourier transform equations. Since

$$g(e^{-y}) \, dy = \frac{g(\lambda)}{\lambda} \, d\lambda \tag{76}$$

The value $g(\lambda)/\lambda$ is plotted against $y = \ln \lambda$. A resolution of a four-component decay curve by this method is shown in Figure 7.18. The decay constants are readily determined by the program as well as the large error peak noted in the figure at $\lambda \simeq 0.0035$.

### Optimization Procedures

Another use of computers in instrumental activation analysis is for optimizing irradiation conditions and postirradiation decay times.

Computational methods have been devised to optimize the length of irradiation and decay before counting to increase selectivity and resolution. A program developed by Isenhour and Morrison* determines the optimum times of irradiation and decay of any element in a complex mixture, taking into account the nuclear reactions with both sought and interfering elements.

From the general equation for radioactivation and subsequent decay, the activity ratio of the radionuclide $j$ in a mixture of $N$ radionuclides is

$$R = \frac{A_j}{\sum_{i=1}^{N} A_i} = \frac{I_j n_j \sigma_j \phi_j (1 - e^{-\lambda_j t}) e^{-\lambda_j T}}{\sum_{i=1}^{N} I_i n_i \sigma_i \phi_i (1 - e^{-\lambda_i t}) e^{-\lambda_i T}} \tag{77}$$

where $A_i$ = net activity of radionuclide $i$,

$\quad I_i$ = fraction of events produced in the energy range to be measured,

$\quad n_i$ = number of target nuclei,

$\quad \sigma_i$ = reaction cross section,

$\quad \phi_i$ = flux of irradiating particles giving reaction for $\sigma_i$,

$\quad \lambda_i$ = decay constant,

$\quad t$ = irradiation time,

$\quad T$ = decay time.

The maximum selectivity (based on instantaneous counting rates) for radionuclide $j$ occurs when $R_j$ is a maximum. The maximum value of $R_j$ is determined from the solution of the two partial derivative equations:

$$\left(\frac{\partial R_j}{\partial t}\right)_T = 0 \quad , \quad \left(\frac{\partial R_j}{\partial T}\right)_t = 0 \tag{78}$$

which lead to

$$\sum_{i=1}^{n} K_i (\lambda_i - \lambda_j)(1 - e^{-\lambda_i t}) e^{-\lambda_i T} = 0 \tag{79}$$

$$\sum_{i=1}^{n} K_i [\lambda_j e^{-\lambda_j t}(1 - e^{-\lambda_i t}) - \lambda_i e^{-\lambda_i t}(1 - e^{-\lambda_i t})] e^{-\lambda_i T} = 0 \tag{80}$$

where

$$K_i = I_i N_i \sigma_i \phi_i$$

The solution of these two simultaneous equations for $t$ and $T$, which give the optimum values for the irradiation and decay times, respectively, may be obtained by a method of successive approximations, e.g., the Newton-Raphson method. If $x_0$ is an approximate value of $x$ for $f(x) = 0$, a better

* T. L. Isenhour and G. H. Morrison, A Computer Program to Optimize Times of Irradiation and Decay in Activation Analysis, *Anal. Chem.* **36**, 1089–1092 (1964).

value is given by the Newton-Raphson method as

$$x = x_0 - \frac{f(x_0)}{f'(x_0)} \tag{81}$$

This new value of $x$ is substituted for $x_0$ and $x_1$ is computed. This is repeated until the value $x_n$, where $\Delta x$ is insignificantly small. The Isenhour and Morrison method computes values of $R_j$ by maximizing each variable $t$ and $T$ separately until the computation produces no significant change (less than the third significant figure) of $R_j$. To initiate the start of the computer process with reasonable values of $R_j$, a matrix of $R_j$ values is computed over the range of irradiation and decay times and the values giving the largest $R_j$ are selected to start the computation. A manual describing the entire program with detailed instructions for its use has been prepared.*

### 7.2.3  Automated Activation Analysis

The concept of automation for routine activation analysis stems from the possibility that both the irradiation and radiation measurement processes can be automated. The impetus for automation results from the potential of activation analysis for rapid, accurate, low cost analysis of large numbers of similar samples in many scientific, biological, and industrial problems. The simplest approach to automated activation analysis is the adaptation of instrumental activation analysis. The addition of radiochemical methods would, of course, offer the many advantages of chemical element identification. The possibility of automated radiochemical separations has been reviewed in Section 7.1.5.

An ideal automatic system would send samples at predetermined time intervals into an irradiation facility, such as a reactor, a fast neutron generator, or an accelerator, under reproducible irradiation conditions. At the end of the irradiation the samples would be sent, after an appropriate delay period, into a radiation-spectrometer system that would measure the appropriate radiations and print out the elemental analysis. A computer would be used to program the flow of samples to the irradiation and measurement systems in optimum sequence and compute the resulting data. Programs are underway to achieve the required "reproducible irradiation conditions" and the necessary degree of resolution of complex gamma-ray spectra.

A model automated activation analysis system is the Mark II system in operation at the Activation Analysis Research Laboratory at Texas A&M

---

* Cornell Material Science Center Report No. 202.

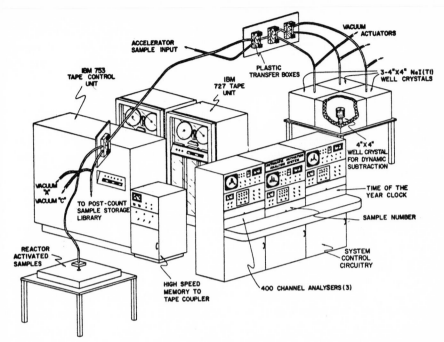

**Figure 7.19** The Texas A&M University Mark II automated activation analysis system. [From R. E. Wainerdi et al., The Design and Use of an Advanced, Integrated, Automatic System and Computer Program for Nuclear Activation Analysis, in *Radiochemical Methods of Analysis*, Vol. II (International Atomic Energy Agency, Vienna, 1935), pp. 149–159.]

University.* This system, sketched in Figure 7.19, is coupled to a large computer system to reduce the output data to qualitative and quantitative analyses. The system can be used to measure the induced radioactivity from individual samples irradiated by charged particles in a cyclotron or from simultaneous activation of many samples irradiated by neutrons in a reactor. The major components of the system are shown in functional sequence in Figure 7.20. Samples from either irradiation facility are transferred, in order, to each of three matched NaI(Tl) gamma-ray detector systems. Each detector is coupled to a multichannel analyzer. The data for each sample, which consist of the time of day at the start and end of the counting period, as well as the data stored in the memory of the pulse-height analyzer, are recorded onto magnetic tape for computer reduction of the accumulated data.

* R. E. Wainerdi, L. E. Fite, D. Gibbons, W. W. Wilkens, P. Jimenez, and D. Drew, "The Design and Use of an Advanced, Integrated, Automatic System and Computer Program for Nuclear Activation Analysis," in *Radiochemical Methods of Analysis*, Vol. II (International Atomic Energy Agency, Vienna, 1965), pp. 149–159.

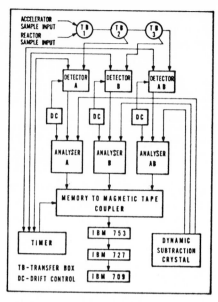

**Figure 7.20** A block diagram of the components of the Texas A&M University Mark II automated activation analysis system. [From R. E. Wainerdi et al., The Design and Use of an Advanced, Integrated, Automatic System and Computer Program for Nuclear Activation Analysis, in *Radiochemical Methods of Analysis*, Vol. II (International Atomic Energy Agency, Vienna, 1965), pp. 149–159.]

An early computer program developed for the reduction of these data was the AA-6 code. In this program the computational analysis can be made by the peak area ratio method, a least-squares method, or a locally designed quadratic programming method. The AA-6 program included an option for numerical methods which allowed for positive constraints when the least-squares method is selected. It was noted that the unrestricted least-squares method often resulted in computation of negative masses for elements in the samples. The AA-6 program constrained each of the matrix coefficients for the library elements to be positive or zero during computation of "best fit." The quadratic function is in the form of (61), for which the difference $Q$ is minimized:

$$Q = [A^T][A](X) - [A^T](Y) \tag{82}$$

Another advantage of this automatic system is flexibility of the program to prepare libraries. In general, 3 to 10 standards of each element, with varying weights, can be activated and counted. The program then corrects all the standards of each element to common weight and time parameters and uses the average spectrum as the standard. This feature is of interest as a quality-control check over large groups of similar samples. The deviations of each of the spectra of the standard element from the average spectrum can be used as a comparison of the computed spectrum of a sample to its actual spectrum.

A more recent general-purpose activation analysis computer program, "Hevesy," has been developed for the reduction of data from the automatic activation analysis systems. The major components of the Hevesy program, described by Yule,* are shown in Figure 7.21. After computing the yields of the full-energy gamma-ray photopeak of the standard for each of the three detectors, the computer program determines the concentration levels in ppm for each of the elements of interest in the unknown samples.

An estimate has been made of the cost per sample for processing large numbers of samples in such an automatic facility. The estimate is based on determinations of activation costs, computer costs, and a time use charge based on the cost of the total system and the time for each analysis. The time for analysis includes such items as transfer of samples to and from the counting library ($\sim$4 sec), data readout ($<\frac{1}{16}$ sec), and computation time ($\sim$4 sec). The number of samples per second from which $\gamma$-ray spectral data can be collected is

$$\frac{N}{T} = \frac{3}{C + 4} \tag{83}$$

where $C$ is the counting time per sample in seconds. It is noted that for $C =$ 60 sec, the system could handle the data of 4040 samples/24-hr day.

The concept of automatic instrumental activation analysis lends itself to the concept of central-facility irradiation and radiation measurement services for activation analysis. Such services might logically be offered by large nuclear facility operators; for example, at least one commercial organization† offers a variety of irradiation services, that is, nuclear reactor, d-t neutron generator, and an electron linear accelerator. Samples forwarded to the facility are irradiated according to prearranged instructions and counted on suitable equipment to obtain gamma-ray spectra. The data are transmitted by data phone services to the client for digital reproduction and subsequent computer analysis by a method chosen by the client. Such activation analysis

* H. P. Yule, "Hevesy," A Computer Program for Analysis of Activation Analysis Gamma-Ray Spectra, in *Modern Trends in Activation Analysis* (National Bureau of Standards, Gaithersbuurg, Md., 1968) pp. 1108–1110.
† Gulf General Atomic Corp., Activation Analysis Center, La Jolla, Calif.

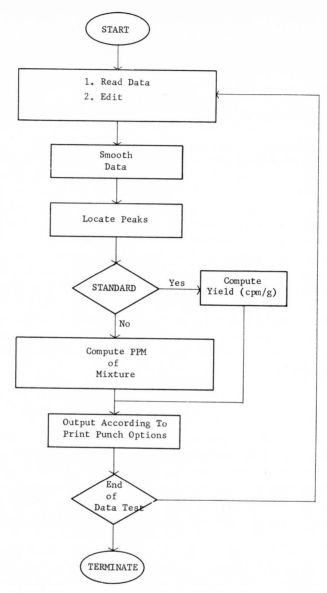

**Figure 7.21** The general purpose activation analysis computer program, "Hevesy." [From H. P. Yule, "Hevesy," A Computer Program for Analysis of Activation Analysis Gamma-Ray Spectra, in *Modern Trends in Activation Analysis* (National Bureau of Standards, Gaithersburg, Md., 1969) pp. 1108–1110.]

services are especially efficient when a user requires only occasional analyses, unusual irradiation conditions, or when the desired product radionuclides have half lives too short to allow for transportation delays.

## 7.3  SPECIAL PRACTICES

Although most activation analysis practices are based on radiochemical or gamma-ray spectroscopy determinations of the radionuclides produced by thermal neutron irradiation, many special practices are used to broaden the general scope of radioactivation analysis.

Some were developed to compensate for limitations inherent in routine neutron activation analysis. Examples are reduction in production rate of unwanted radionuclides or supression of Compton effect contributions to $\gamma$-ray spectra. These practices are examined in Chapter 8.

Others were developed as specific uses of activation analysis. Examples are on-line activation analysis, surface analysis, isotopic analysis, and stable-tracer activation analysis. These applications are examined in Chapter 9.

Still other special practices were developed as methods of radioactivation analysis complementary to neutron activation analysis. These practices may utilize other stages of nuclear reactions or nuclide transformations with charged particles and photons.

Effects from the several stages of nuclear reactions may be used as means of chemical element analysis.

1. Transfer of irradiation energy; elastic or inelastic scattering.
2. Creation of a compound nucleus; absorption effects.
3. De-excitation of the compound nucleus; release of prompt radiation.
4. Decay of the compound nucleus; radioactivity.

Elastic scattering methods have not been extensively developed for activation analysis. Chemical analysis by atomic scattering or absorption of irradiation beams has been made with x-ray defraction, Mossbauer effect, and atomic absorption analytical methods. Inelastic scattering reactions, such as the $(\gamma,\gamma')$ and $(n,n')$ to produce metastable isotopes, are useful as an activation analysis method. The nuclear absorption of irradiation particles may be used as an activation analysis practice in the "die-away of pulses" technique. The rate of capture of pulsed-beam particles (over periods of microseconds) is a function of the absorption cross section and thereby a measure of chemical composition.

A greater effort has developed in applying the measurement of the prompt radiations released by compound nuclei as a means of radioactivation analysis.

The development of pulsing capability of many reactors and accelerators has lead to the development of pulsed activation analysis. This capability is especially useful for measurement of activation products with half-life smaller than about 1 min.

Special irradiation techniques have been developed to increase the use of charged particle accelerators for activation analysis purposes. The use of fast neutron, $^3$He, $^3$H, and e$^-$ (photon) irradiations has increased notably.

### 7.3.1  Prompt-Radiation Activation Analysis

The measurement of the prompt radiation particle or photon released from the compound nucleus following absorption of an irradiating particle is an alternate method of activation analysis, complementary to radioactivation analysis. This method has also proven successful for the measurement of delayed neutrons from fissionable elements and of the prompt photoneutrons from $(\gamma,n)$ reactions.

The measurement of the prompt gamma ray following neutron capture in the $(n,\gamma)$ reaction has been extensively developed. In this method the gamma-ray intensity is dependent only upon the radiative capture cross section and not upon the half-life of the product nucleus. Thus prompt-radiation activation analysis differs from radioactivation analysis primarily in that the ground-state properties of the product nucleus is of no consequence. A product nucleus may have any of these difficulties:

1. Too short a half-life: for example, $^{11}$B$(n,\gamma)$ 0.22-sec $^{12}$B.
2. Too long a half-life: for example, $^9$Be$(n,\gamma)$ 2.7 × 10$^6$-y $^{10}$Be.
3. A stable composition: for example, $^1$H$(n,\gamma)^2$H; or $^{12}$C$(n,\gamma)^{13}$C.
4. A target nuclide of very small isotopic abundance:
   for example, $^2$H, $f = 0.00015$.
5. Too few or no gamma radiations:
   for example, $^{32}$P, $^{35}$S, $^{45}$Ca, $^{55}$Cr, $^{89}$Sr.

Elements that are especially amendable to prompt gamma-ray activation analysis are listed in Table 7.5. These include elements with isotopes that have either isotopic abundance greater than 50% or a radiative neutron capture cross section greater than 100 b and produce stable product nuclides. Also listed are the isotopes that produce radionuclides with half-lives either less than 1 min or greater than 100 yr. A significant number of elements meets these criteria.

The prompt gamma-ray activation analysis method stems from the radiative capture process which results in the decay of the compound nucleus by the emission of characteristic gamma radiation, either as a single photon with kinetic energy equal to the excitation energy less the recoil energy or, more

**Table 7.5**  Elements with Isotopes Suitable for Prompt
Gamma Radiation Activation Analysis[a]

| Nuclide | Isotopic Abundance (%) | (n,$\gamma$) Cross Section (b) | Product Half-Life |
|---|---|---|---|
| $^1$H | 99.985 | 0.33 | S |
| $^7$Li | 92.58 | 0.036 | 0.85 s |
| $^9$Be | 100 | 0.009 | $2.7 \times 10^6$ y |
| $^{11}$B | 80.22 | 0.005 | 0.02 s |
| $^{12}$C | 98.89 | 0.0034 | S |
| $^{14}$N | 99.63 | 0.08 | S |
| $^{17}$O | 99.759 | 0.0002 | S |
| $^{19}$F | 100 | 0.01 | 11 s |
| $^{20}$Ne | 90.92 | ? | S |
| $^{24}$Mg | 78.70 | 0.03 | S |
| $^{28}$Si | 92.21 | 0.08 | S |
| $^{32}$S | 95.0 | ? | S |
| $^{35}$Cl | 75.53 | 44 | $3 \times 10^5$ y |
| $^{39}$K | 93.10 | 2.2 | $1.3 \times 10^9$ y |
| $^{40}$Ca | 96.97 | 0.2 | $7.7 \times 10^4$ y |
| $^{48}$Ti | 73.94 | 8.0 | S |
| $^{52}$Cr | 83.76 | 0.8 | S |
| $^{56}$Fe | 91.66 | 2.5 | S |
| $^{58}$Ni | 67.88 | 4.4 | $8 \times 10^4$ y |
| $^{90}$Zr | 51.46 | 0.1 | S |
| $^{113}$Cd | 12.26 | 20,000 | S |
| $^{143}$Nd | 12.17 | 330 | $2.4 \times 10^{15}$ y |
| $^{149}$Sm | 13.83 | 41,500 | S |
| $^{152}$Sm | 26.72 | 210 | 47 h |
| $^{151}$Eu | 47.82 | 2,800 | 9.3 h |
|  |  | 5,900 | 12.4 y |
| $^{153}$Eu | 52.18 | 320 | 16 y |
| $^{155}$Gd | 14.73 | 58,000 | S |
| $^{157}$Gd | 15.68 | 240,000 | S |
| $^{161}$Dy | 18.88 | 600 | S |
| $^{162}$Dy | 25.53 | 150 | S |
| $^{163}$Dy | 24.97 | 125 | S |
| $^{164}$Dy | 28.18 | 2,000 | 75.4 s |
|  |  | 700 | 2.35 h |
| $^{167}$Er | 22.94 | 700 | S |
| $^{169}$Tm | 100 | 125 | 125 d |
| $^{199}$Hg | 16.84 | 2,000 | S |

[a] Based on either isotopic abundance $f < 0.5$ or (n,$\gamma$) product with half-life $T_{1/2} < 1$ m or $> 10^2$ y or with (n,$\gamma$) cross section $\sigma > 100$ b. Data from GE Chart of the Nuclides, 8th ed., March 1965.

likely, by a cascade of two or more photons with the same total energy. By (1-7) the excitation energy $E^*$ is

$$E^* = \Delta M^* c^2 = [M(Z^{A+1})^* - M(Z^{A+1})]c^2 \qquad (84)$$

where $\Delta M^*$ is the difference in mass between the compound nucleus and the ground-state product nucleus. The mass of the compound nucleus is

$$M(Z^{A+1})^* = \frac{E_n}{c^2} + M(n) + M(Z^A) \qquad (85)$$

in which $E_n$ is the kinetic energy of the captured neutron. Since

$$Q = [M(Z^A) + M(n) - M(Z^{A+1})]c^2 \qquad (86)$$

the excitation energy is

$$E^* = E_n + Q \qquad (87)$$

The rate of production of compound nuclei from a target element irradiated with neutrons is given by (2-34):

$$\frac{dN^*}{dt} = n_i \sigma_i \phi \qquad (88)$$

where $n_i$ = number of target nuclei of isotope $i$ in the element,

$\quad \sigma_i$ = corresponding $n$-capture cross section,

$\quad \phi$ = neutron flux.

Since the number of nuclei activated is generally infinitesimally small compared with the number present in the sample, $n_i$ is a constant during the irradiation and the rate of compound nucleus formation is constant,

$$R = n_i \sigma_i \phi \qquad (89)$$

If $\bar{g}$ prompt gamma rays are emitted per compound nucleus de-excitation, the rate of photon emission is

$$\frac{dN_\gamma}{dt} = \bar{g} R \qquad (90)$$

The experimental arrangements for prompt gamma-ray activation analysis consist of a neutron source and a gamma-ray spectrometer, shielded from the neutron source. Nuclear reactors may be used as the neutron source in two ways:

1. Withdrawal through a beam port in the reactor of a collimated fast neutron beam incident on the sample.

2. Placing the sample in the beam port with a bismuth shield between the sample and the core to screen the direct gamma radiation in the core from reaching the detector.

A (d,t) neutron generator is also useful as a neutron source, since the generator can be pulsed. A delay time may then be incorporated into the detection system to match the emission time of de-excitation with the off-time generator.

The prompt gamma-ray spectra can be measured with single or multiple NaI(Tl) scintillation spectrometer systems. The single crystal spectrometer

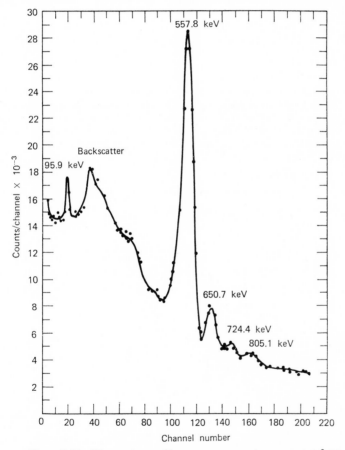

**Figure 7.22** The spectrum of low-energy prompt gamma rays from cadmium. [From R. C. Greenwood and J. Reed, Scintillation Spectrometer Measurements of Capture Gamma Rays from Natural Elements, in *Modern Trends in Activation Analysis* (Texas A&M University, College Station, 1961), pp. 166–171.]

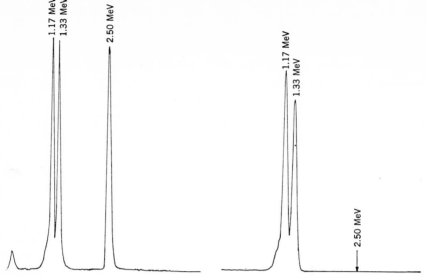

**Figure 7.23** (a) The spectrum of $^{60}$Co made with a sum-coincidence spectrometer; (b) the removal of the sum peak by the additional coincidence requirement of a fast/slow sum-coincidence spectrometer. [From W. G. Lussie and J. L. Brownlee, the Measurement and Utilization of Neutron-Capture Gamma Radiation, in *Modern Trends in Activation Analysis* (Texas A&M University, College Station, 1965), pp. 194–199.]

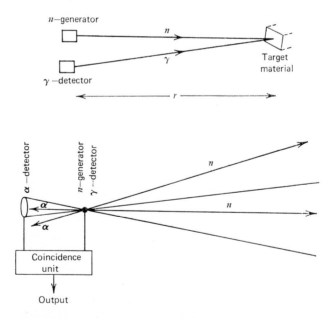

**Figure 7.24** A neutron generator, prompt gamma-ray detector system using a $(\alpha, \gamma)$ coincidence unit to improve the sensitivity of the system. [From D. Taylor, *Neutron Irradiation and Activation Analysis* (George Newnes, London, 1964).]

has been used successfully to record standard capture gamma-ray spectra for many elements; for example, the spectrum for low-energy gamma rays from cadmium is illustrated in Figure 7.22. Because of the presence of gamma rays in the spectrum from extraneous sources, such as the surrounding sample holders, shields, and fast neutron inelastic-scatter gamma rays, the net spectrum is obtained from a combination of four spectrum measurements under different irradiation conditions, with and without the sample in place.

Several methods may be useful in increasing the sensitivity for prompt $\gamma$-ray measurement. One of these is a sum-coincidence spectrometer in which pulses from two detectors are added in a summing unit. These pulses are analyzed to seek those summed pulses corresponding to the total excitation energy from the gamma-ray cascade. An additional slow coincidence circuit increases sensitivity even further by reducing random fast coincidences. Fast and fast/slow sum-coincidence spectra of $^{60}$Co are shown in Figure 7.23. The additional coincidence requirement of the fast/slow system is responsible for the absence of the sum-coincidence peak seen in the fast-system spectrum.

Another method, useful for (d,t) neutron generators, is an $(\alpha,\gamma)$ coincidence spectrometer. Since the alpha particle produced in the $^3$H(d,n)$^4$He reaction is emitted in the opposite direction to the neutron, a small alpha-particle detector placed behind the neutron generator defines the number of neutrons emitted in a similar small solid angle. A schematic arrangement of this system is shown in Figure 7.24. The sensitivity is improved by a reduction in the background of accepted gamma-ray pulses by a factor of about 100. Coincidences of gamma rays with alpha particles indicate gamma rays from neutron interactions within the solid angle between the generator and the sample.

### 7.3.2 Pulsed-Neutron Activation Analysis

The pulsing feature of some nuclear reactors has been used to increase sensitivity for activation analysis, especially for elements that produce short-lived radionuclides. Such reactors, for example the TRIGA shown in Figure 3.7, which have self-regulating power operation, can be safely and reproducibly pulsed at frequencies of about once every 6 min. The self-regulation results from the negative temperature gradient of uranium-zirconium hydride fuel elements used in these reactors. The excursion pulse produces high-peak neutron fluxes; for example, a reactor that operates at steady power of 250 kW with a total neutron flux of about $8 \times 10^{12}$ n/cm$^2$-sec can be pulsed to a peak output of 900 MW to yield a peak neutron flux of about $2.8 \times 10^{16}$ n/cm$^2$-sec. Peak fluxes of $10^{17}$ n/cm$^2$-sec can be attained.

The advantages of using such high-intensity neutron pulses for radioactivation analysis have been reported by Yule and Guinn.* It was noted in (2) that for irradiation times short compared with an activation product half-life ($\lambda t \lesssim 0.05$) the saturation factor $(1 - e^{-\lambda t})$ can be approximated by $\lambda t$. For a short burst of neutrons the activation equation may be reduced to

$$D_0' = N\sigma(nvt)'\lambda \qquad (91)$$

where $D_0'$ = the activity induced by the neutron pulse,

$(nvt)'$ = the integrated neutron flux.

The shape of a 900 MW reactor neutron pulse, shown in Figure 7.25, is approximately Gaussian, with a full width of about 15 msec at half maximum. For a Gaussian-shaped pulse the integrated neutron flux is given by

$$(nvt)' = 1.064 \times \phi_{\max} \times \text{FWHM} \qquad (92)$$

where $\phi$ equals the pulse peak neutron flux.

The integrated pulse energy for a 900-MW pulse is thus about $900 \times 0.015 \simeq 13.5$ MW-sec, and the integrated neutron flux is $1.064 \times 2.8 \times 10^{16} \times 0.015 \simeq 4.5 \times 10^{14}$ n/cm². 

At steady-state operation the saturation activity of the same radionuclide is given by

$$D^\infty = N\sigma(nv) \qquad (93)$$

The ratio of activities of pulsed activation to saturation activation at constant power is

$$\frac{D_0'}{D^\infty} = \frac{(nvt)'\lambda}{(nv)} \qquad (94)$$

For a 900-MW pulse the enhancement ratio, which compares the activation from a single pulse to saturation at steady operation at 250 kW, is

$$\frac{D_0'}{D^\infty} = \frac{4.5 \times 10^{14}}{8 \times 10^{12}} \times \frac{0.693}{T_{1/2}(\text{sec})} = \frac{40}{T_{1/2}(\text{sec})} \qquad (95)$$

In actual operation of a pulsing reactor, appreciable tailing of the pulse occurs before the control rod is reinserted. Empirically, a 900-MW peak pulse has an integrated energy of about 19.5 MW-sec instead of 13.5 MW-sec. The enhancement ratio therefore becomes

$$\frac{D_0'}{D^\infty} = \frac{54}{T_{1/2}(\text{sec})} \qquad (96)$$

---

* H. P. Yule and V. P. Guinn, Enhancement of Neutron Activation Analysis Sensitivities by Use of Reactor Pulses, in *Radiochemical Methods of Analysis* (International Atomic Energy Agency, Vienna, 1965), Vol. II, pp. 111–121.

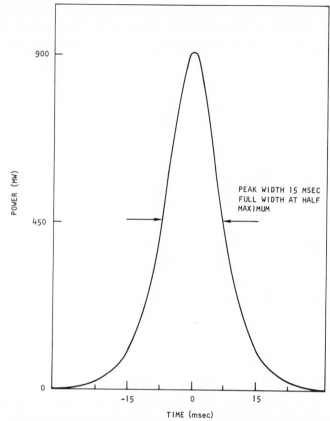

**Figure 7.25** The shape of a reactor power pulse reaching a peak power of 900 MW. [From H. P. Yule and V. P. Guinn, Enhancement of Neutron Activation Analysis Sensitivities by Use of Reactor Pulses: Experimental Results for 13 Elements, in *Proc.*, *Radiochemical Methods of Analysis* (International Atomic Energy Agency, Vienna, 1965), pp. 111–112.]

This ratio implies that the production of a radionuclide with a half-life of 1 sec in a single reactor pulse would be 54 times greater than the production from saturation irradiation (about 10 sec) at maximum steady operation power. Radionuclides with half-lives greater than 54 sec would have enhancement ratios less than one.

In systems in which the irradiated sample can be transferred and counted rapidly, the method of pulsed neutron activation is obviously of benefit for those elements giving product nuclides with a half-life smaller than about 1 min. Table 7.6 lists 14 elements, with product half-lives <60 sec, whose

**Table 7.6**   Experimental Results for Fourteen Elements in a Series of 19.5 ± 0.5 MW-sec Reaction Pulses[a]

| Element | Reaction | Product | Half-Life (sec) | Gamma-Ray Energy (MeV) | Net Photopeak (cpm/g at $t_0$) | Calculated Detection Limit $(\mu g)$[b] | Ratio, Pulse Activity to Steady Saturation Activity |
|---|---|---|---|---|---|---|---|
| O | n,p | $^{16}$N | 7.35 | 6.1 | $2.1 \times 10^7$ | 48. | 10. |
| F | n,$\gamma$ | $^{20}$F | 11. | 1.63 | $8.3 \times 10^9$ | 0.12 | 5.2 |
| F | n,p | $^{19}$O | 29. | 0.20 | $4.5 \times 10^8$ | 2.2 | 2.8 |
| Na | n,p | $^{23}$Ne | 38. | 0.44 | $1.4 \times 10^8$ | 7.0 | 2.2 |
| Na | n,$\alpha$ | $^{20}$F | 11. | 1.63 | $1.1 \times 10^8$ | 9.0 | 5.0 |
| Mg | n,p | $^{25}$Na | 60. | 0.38 | $4.8 \times 10^6$ | 210. | 2.7 |
| S | n,p | $^{34}$P | 12.4 | 2.1 | $1.0 \times 10^6$ | 1000. | 7.7 |
| Sc | n,$\gamma$ | $^{46m}$Sc | 20. | 0.140 | $2.6 \times 10^{12}$ | 0.00035 | 3.6 |
| Cr | n,p | $^{52}$V | 225. | 1.44 | $5.0 \times 10^6$ | 20. | 0.25 |
| As | n,p | $^{75m}$Ge | 49. | 0.139 | $5.2 \times 10^6$ | 190. | 1.3 |
| Se | c | $^{77m}$Se | 17. | 0.160 | $5.0 \times 10^{11}$ | 0.0020 | 6.4 |
| Y | n,n' | $^{89m}$Y | 16.1 | 0.915 | $9.5 \times 10^9$ | 0.10 | 1.9 |
| Tb | n,2n | $^{158m}$Tb | 11. | 0.111 | $8.9 \times 10^5$ | 1100. | 8.8 |
| W | c | $^{183m}$W | 5.3 | 0.105 | $1.2 \times 10^{10}$ | 0.086 | 7.5 |
| Pt | n,$\gamma$ | $^{199m}$Pt | 14. | 0.39 | $4.5 \times 10^8$ | 2.2 | 5.1 |
| Au | n,n' | $^{197m}$Au | 7.2 | 0.279 | $1.1 \times 10^{11}$ | 0.0089 | 9.2 |

[a] From H. P. Yule and V. P. Guinn, Enhancement of Neutron Activation Analysis Sensitivities by Use of Reactor Pulses, in *Radiochemical Methods of Analysis* (International Atomic Energy Agency, Vienna, 1965), Vol II, pp. 111–122.
[b] Detection limit defined as 1000 photopeak cpm at $t_0$ for $T_{1/2} < 1$ min, 100 photopeak cpm at $t_0$ $T_{1/2}$ of 1 to 60 min.
[c] For these two elements the observed activity is formed concurrently, to different degrees, by (n,$\gamma$), (n,n'), and (n,2n) reactions.

enhancement ratios have been determined. The enhancements for these elements range from 30 to 1000%. Figure 7.26 shows these measured ratios as a function of half-life compared with the ratio given by (96). A least-squares fit shows that the data correspond best to an enhancement ratio of $70/T_{1/2}$ (sec) instead of the expected $54/T_{1/2}$ (sec).

### 7.3.3   Nonthermal-Neutron Activation Analysis

Until recently activation analysis has been dominantly performed by irradiation of samples with thermal neutrons in nuclear reactors. Activations with thermal neutrons have been particularly popular for several reasons:

1. Relatively large cross sections for thermal neutron absorption.
2. Specificity of radionuclides produced by thermal neutron capture.
3. Production of easily measured $\beta^-$-decaying radionuclides.
4. Availability of large thermal neutron fluxes.
5. Ability to irradiate relatively large samples without excessive degradation of neutron energy or losses in flux.

6. Convenience of pneumatic transfer systems to insert and remove samples rapidly from the irradiation position.

All of these advantages of neutron irradiations become problems to varying degrees with high-energy neutron, charged particle and photon irradiations. By correspondence to the above list some of these problems are the following:

1. Relatively lower cross sections; whereas thermal neutron cross sections are generally in the range of 0.1 to 10 b, high-energy particle cross sections are in the range of 1 to 100 mb.

2. Nonspecificity of high-energy particle reactions; at energies above about 6–10 MeV, many endoergic nuclear reactions become possible; for example, with protons (p,n), (p,d), (p,α), (p,2n).

3. Neutron-deficient radionuclides are produced; although this may be exactly the reason a particular activation analysis reaction is chosen, in general, these radionuclides, some of which decay solely by electron capture, are more difficult to measure quantitatively.

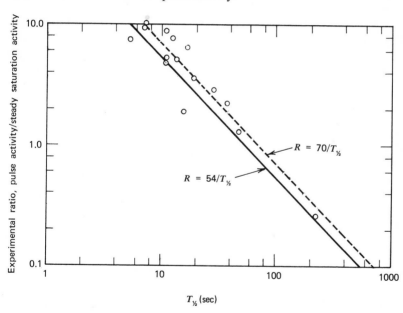

**Figure 7.26** Experimental values of ratio of pulse-produced activity to steady-state saturation activity; the ratios show an average value of $70/T_{1/2}$. [From H. P. Yule and V. P. Guinn, Enhancement of Neutron Activation Analysis Sensitivities by Use of Reactor Pulses: Experimental Results for 13 Elements. In *Proc. Radiochemical Methods of Analysis* (International Atomic Energy Agency, Vienna, 1965), pp. 111–112.]

4. Lack of high-intensity beams without concommitant target heating problems.

5. Degradation of particle energy with depth of penetration, limiting the sample thickness that can be irradiated.

6. Need for vacuum chambers for internal-beam irradiations and source mounts for external irradiations, which also limit the speed with which samples can be inserted and removed.

In spite of all these problems and others associated with high-energy particle irradiations, increasing use is being made of nonthermal energy neutron irradiations in activation analysis. Some of the reasons for this growth include the following:

1. Some elements cannot be determined conveniently by thermal neutron activation analysis. A list of some of them is given in Table 7.7. Among the reasons:

(a) too small a cross section for $(n,\gamma)$ reactions;

(b) too short a half-life of the product radionuclide;

(c) too long a half-life of the product radionuclide.

2. Engineering improvements in high-energy particle accelerator irradiation conditions:

(a) rotating target assemblies that can hold several samples on a cooled mounting such that identical irradiations can be made of standards, blanks, and samples;

(b) higher output or longer lived target assemblies that produce the irradiating beams of particles.

Table 7.7  Elements Difficult for Thermal Neutron
Activation Analysis

| Element | Difficulties |
|---------|--------------|
| H | $f < 0.01$; $\sigma < 1$ mb; $T_{1/2} > 1$ yr |
| He | $T_{1/2} \ll 1$ min |
| Be | $T_{1/2} \gg 1$ yr |
| B | $T_{1/2} \ll 1$ min |
| C | $\sigma < 1$ mb; $T_{1/2} \gg 1$ yr |
| N | $f < 0.01$; $\sigma < 1$ mb; $T_{1/2} < 1$ min |
| O | $f < 0.01$; $\sigma < 1$ mb; $T_{1/2} < 1$ min |
| F | $T_{1/2} < 1$ min |
| Ne | $T_{1/2} < 1$ min |
| Tc | $T_{1/2} < 1$ min |
| Pb | $T_{1/2} \gg 1$ yr; $T_{1/2} \ll 1$ min; $\sigma < 1$ mb |

3. Development of irradiation facilities expressly designed or better suited for activation irradiations:
   (a) the (d,t) 14-MeV neutron generator;
   (b) the small $^3$He cyclotron;
   (c) the increased availability of $^3$H as irradiation beams of particles:
   (d) the electron linear accelerator.

The two problems shared in common by all high-energy particle irradiations are the dependence of cross section on irradiating particle energy and the degradation of irradiating particle energy with depth of penetration. A discussion of thin and thick target cross section was given in Section 2.2.1, excitation functions in Section 2.2.6, and the stopping of charged particles in matter in Section 4.3. A method to reduce the complexity of charged-particle activation analysis of thick samples to almost the simplicity of thermal-neutron activation analysis was described by Ricci and Hahn.* They suggest the use of an average cross section $\bar{\sigma}$ to reflect the change in cross section with depth of penetration into a sample. The production rate of a radionuclide by a beam of particles striking a thin target (defined in Section 2.2.1 as one in which the attenuation of the incident particles is negligible) was given as

$$R_i = I_0 n \sigma_i x \qquad (2.6)$$

The production rate of all reaction products (defined in terms of a total cross section in Section 2.2) was given as

$$\Delta I = I_0(1 - e^{-n\sigma_T x}) \qquad (2.10)$$

which is valid when the thickness of sample $x$ is greater than the range of the beam particles.

Ricci and Hahn consider the activation system in which the energy loss in the sample is appreciable; that is, the variation of $\sigma_x$ with $x$ is significant. In a target of thickness greater than the total range $R$ of the charged particles $(x > R)$, the disintegration rate of the produced radionuclide is given by

$$D = I_0 n \int_0^R \sigma_x \, dx \qquad (97)$$

The integral cross section of (97) can be converted to an expression in terms of energy degradation with distance as

$$\int_0^R \sigma_x \, dx = \int_{E_0}^0 \sigma_E \left(\frac{dx}{dE}\right) dE \qquad (98)$$

---

* E. Ricci and R. L. Hahn, Theory and Experiment in Rapid, Sensitive Helium-3 Activation Analysis, *Anal. Chem.* **37**, 742–748 (1965), and Sensitivities for Activation Analysis of 15 Light Elements with 18-MeV Helium-3 Particles, *Anal. Chem.* **39**, 794–797 (1967).

where $\sigma_E =$ the excitation function cross section,

$dx/dE =$ the negative reciprocal of the stopping power of the sample material for the charged particles.

As a simplication for activation analysis purposes, since generally both $\sigma_E$ and $dx/dE$ are unknown and difficult to measure for specific matrices over specific energy intervals, Ricci and Hahn proposed a cross section parameter related to the integral cross section and independent of the matrix material. They define an average cross section $\bar{\sigma}$ as

$$\bar{\sigma} = \frac{\int_0^R \sigma_x \, dx}{\int_0^R dx} = \frac{\int_{E_0}^0 \sigma_E \left(\frac{dx}{dE}\right) dE}{\int_{E_0}^0 \left(\frac{dx}{dE}\right) dE} \tag{99}$$

They show, for charged particles of interest in activation analysis, that for a stopping power

$$-\frac{dE}{dx} = \frac{k}{E} \ln \frac{E}{I(Z)} \tag{100}$$

where $k = f(z,Z,N) =$ a constant for a given irradiating particle and a given amount of a given material,

$I(Z) =$ effective ionization potential for the matrix atoms, also constant for a given material.

Therefore, since

$$k \ln \frac{E}{I(Z)} \simeq \text{constant}, \tag{101}$$

(99) reduces without excessive loss in accuracy to

$$\bar{\sigma} \simeq \frac{\int_{E_0}^0 \sigma_E E \, dE}{\int_{E_0}^0 E \, dE} \tag{102}$$

They suggest that since $\bar{\sigma}$ is, to a good approximation, independent of the chemical composition of the matrix and a constant value for a given nuclear reaction and a given irradiation particle energy, it is a satisfactory parameter for use in activation analysis. An illustration of the variation of average cross section $\bar{\sigma}$ with atomic number of matrix $Z$ for the reactions $^{16}O(^3He,p)^{18}F$ and $^{16}O(^3He,n)^{18}Ne \rightarrow {}^{18}F$ is given in Figure 7.27. It is noted that over the range of elements from $Z = 4$ to $Z = 95$, $\bar{\sigma}$ changed by $8\%$, and from $Z = 4$

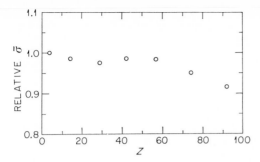

**Figure 7.27** The variation of the average cross section $\bar{\sigma}$ with matrix atomic number $Z$ for the reactions $^{16}O(^3He,p)^{18}F$, and $^{16}O(^3He,n)^{18}Ne \rightarrow ^{18}F$. [From E. Ricci and R. L. Hahn, "Theory and Experiment in Rapid, Sensitive Helium-3 Activation Analysis, *Anal. Chem.* **37**, 742–748 (1965).]

to $Z = 57$ $\bar{\sigma}$ changed by only 3%. Thus with the definition of $\bar{\sigma}$ in (99) the activation equation (97) is revised to

$$D = I_0 n\bar{\sigma}R \tag{103}$$

Some examples of charged-particle ranges in aluminum are shown in Figure 7.28.

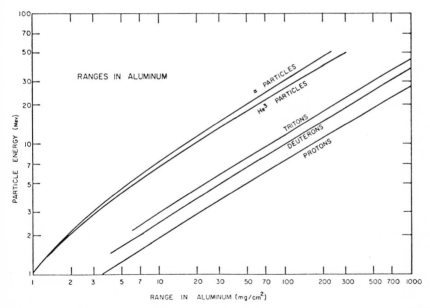

**Figure 7.28** Range of irradiation particles in aluminum. [From R. S. Tilbury, *Activation Analysis with Charged Particles*, NAS-NS 3110, undated.]

Equation 103 is especially suited to the comparitor method of activation analysis. Since $\bar{\sigma}$ is independent of matrix material the ratio of induced activities is given by

$$\frac{D_u}{D_s} = \frac{n_u I_u R_u}{n_s I_s R_s} \tag{104}$$

Thus for activations in which the standards are not in the same matrix form as the sample the beam intensities for the two irradiations can be normalized by Faraday cup measurement and the ranges can be obtained from range-energy relationships. For sufficiently similar matrices of standards and samples (104) reduces directly to the comparitor relation given by (18).

### Fast-Neutron Activation Analysis

Neutron generators are being extensively used in activation analysis. The features and limitations of neutron generators were reviewed in Section 3.1.2. It was noted that (d,t) 14-MeV neutron generators were useful when access to nuclear reactors was limited or when short-lived or non-(n,$\gamma$)-products were desired. It was also noted that the major limitation to the neutron generator as a convenient facility for activation analysis was inability of producing tritium targets with high neutron yields over long periods of time. An inherent loss of tritium occurs by the exchange of tritium with deuterons from the beam and the subsequent diffusion of the liberated tritium out of the target.

A method for reloading targets has been described by Hollister[*] which permits a generator to operate at higher neutron outputs for longer periods of time. The method reverses the process of tritium depletion by changing the ion beam from deuterium to tritium and rejuvenating the target by replacing the accumulated deuterium. A curve of neutron output as a function of time for a 4-Ci/in.[2] target is compared in Figure 7.29 with the curve for target reloading in which the output is maintained within 25% of maximum. Another type of generator source is the "sealed tube" which accelerates a mixture of deuterium and tritium. Tubes are available with constant neutron output of $10^{10}$ to $10^{12}$ n/sec for about 100 h. An example of a sealed-off neutron generator is shown in Figure 7.30. Higher voltage accelerators, such as Van de Graaff accelerators and cyclotrons are able to maintain steady-state neutron output yields.

Fast neutron generators are used extensively in activation analysis for convenient production of short-lived radionuclides in which the complementary functions of rapid transit to and from irradiation position and

* H. Hollister, Target Reloading in Neutron Generators, *Nucleonics* **22**, No. 6, 68–69 (1964).

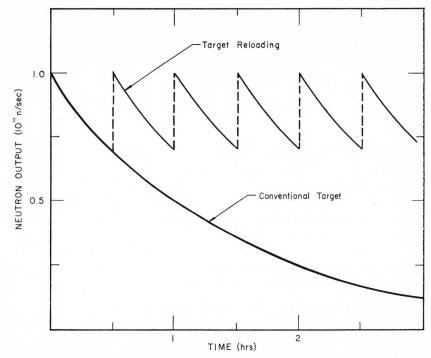

**Figure 7.29** Comparison of neutron output as a function of time for target reloading and conventional targets. [From H. Hollister, Target Reloading in Neutron Generators, *Nucleonics* **22**, No. 6, 68–69 (1964).]

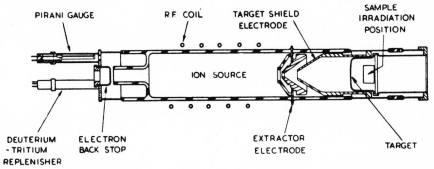

**Figure 7.30** A sealed-off neutron tube. The tube provides an output of $10^{10}$ n/sec over a life of about 100 hours. [From J. E. Bounden, P. D. Lomer, and J. D. Wood, High-Output Neutron Tubes, in *Modern Trends in Activation Analysis* (Texas A&M University, College Station, 1965), pp. 182–185.]

automated radiation spectroscopy make the analyses rapid, reproducible, and relatively cheap. The generators are also used extensively for analysis of elements of atomic number below 10, since these elements are not easily determined by activation with thermal neutrons.

Several compilations of sensitivities and cross sections for 14-MeV neutron activation analysis have been prepared, some of which are listed in Section 7.4.2.

### Charged-Particle Activation Analysis

The advantages and disadvantages of charged-particle activation analysis have been reviewed in Section 7.3. The characteristics of several types of particle and accelerator were reviewed in Chapter 3. The longer availability of proton, deuteron, and alpha-particle accelerators has resulted in a considerable usage of these particles in activation analysis practices. Recent availability of accelerators that can produce useful beams of $^3$He and $^3$H particles has increased the capabilities of charged-particle activation analyses. The properties of these two irradiation particles are examined further.

$^3$He ACTIVATION. The advantages of using $^3$He ions as an activation method in activation analysis were noted by Markowitz and Mahoney.* They showed that $^3$He reactions were especially useful for activation analysis of oxygen and other light elements that are difficult to determine by neutron activation. The specific advantage of $^3$He irradiations results from the low binding energy of 7.7 MeV for the $^3$He nucleus. This low binding energy results in many reactions with $^3$He being exoergic. Thus, for light nuclei with low Coulomb barriers, nuclear reactions can take place with relatively high cross sections at low $^3$He energies; for example, a comparison of $^3$He and $^4$He reactions with oxygen shows

$$^{16}O(^3He,p)^{18}F, \qquad Q = +2.0 \text{ MeV},$$
$$^{16}O(^4He,p)^{19}F, \qquad Q = -8.1 \text{ MeV}.$$

Advantages of $^3$He reactions over deuteron reactions include the lack of $(n,\gamma)$ reactions produced by the high neutron background surrounding deuteron accelerators; these $(n,\gamma)$ reactions give the same product as $(d,p)$ reactions. Another advantage of $^3$He reactions is the production of $\beta^+$-emitting radionuclides, which allow radiation measurement of the annihilation radiation. Comparison with proton irradiations shows that comparable proton reactions are generally endoergic with smaller cross sections.

A further enhancement of sensitivity for $^3$He activation analysis results from the fact that many of the reaction products are so neutron deficient that

---

* S. S. Markowitz and J. D. Mahoney, Activation Analysis for Oxygen and Other Elements by Helium-3-induced Nuclear Reactions, *Anal. Chem.* **34**, 329–335 (1962).

they decay with short half-life to a longer lived radionuclide; for example, the observed excitation function for the production of $^{18}F$ from $^3He$ irradiation of $^{16}O$ is the sum of two nuclear reactions

direct:     $^{16}O(^3He,p)$ 110-m $^{18}F$

secondary:     $^{16}O(^3He,n)$ 1.6-s $^{18}Ne \xrightarrow{\beta^+} {}^{18}F$

The excitation function for these reactions is shown in Figure 7.31. The maximum cross section of 0.4 b occurs at a $^3He$ energy of 7.5 MeV. The excitation functions determined for $^{12}C$ are also shown in Figure 7.31. The maximum cross section for the $^{12}C(^3He,\alpha)^{11}C$ reaction is 0.32 b at $E_{3He} = 10$ MeV. The high cross sections at $E_{3He} < 6$ MeV, without interference from the $^{12}C(^3He,d)^{13}N$ reaction, suggests that carbon, nitrogen, and oxygen can be determined, nondestructively and simultaneously, in the same sample. The three components would be produced by the reactions

$$
\begin{aligned}
^{12}C(^3He,\alpha)^{11}C \qquad & Q = +1.9 \text{ MeV,} \\
^{14}N(^3He,\alpha)^{13}N \qquad & Q = +10 \text{ MeV,} \\
^{18}O(^3He,p \text{ or } n)^{18}F \qquad & Q = +2.0 \text{ MeV.}
\end{aligned}
$$

A list of more than 150 nuclear reactions of elements from lithium, $Z = 3$, to calcium, $Z = 20$, with $^3He$ particles of energy less than 8 MeV, is provided by Markowitz and Mahoney. The availability of commercial "small" $^3He$ cyclotrons should result in increased use of $^3He$ activation analysis.

$^3H$ ACTIVATION.    The advantages of $^3H$ as an irradiating particle are similar to those of $^3He$, and therefore interest in $^3H$ irradiations is increasing as available facilities for $^3H$ irradiations increase. The $^3H$ is produced in nuclear reactors by the high cross section (940 b) reaction

$$^6Li(n,\alpha)^3H$$

which produces tritons with energy of 2.74 MeV. The tritons are used as irradiating particles on materials placed in contact with the lithium. This method has been shown to be useful for determination of oxygen by the reaction

$$^{16}O(t,n)^{18}F$$

This use of charged-particle irradiations in a nuclear reactor can combine the advantage of the former for light-element determination with the low-cost irradiations obtained in the latter. The 2.74-MeV energy of the triton particles limits their usefulness to elements of atomic number up to 11, since the Coulomb barrier for the element $_{12}Mg$ is greater than 2.74 MeV. This problem may be alleviated by the development of accelerators that can produce usable beams of tritons of higher energy with the safety requirement of preventing the radioactive tritium from escaping the irradiation facility.

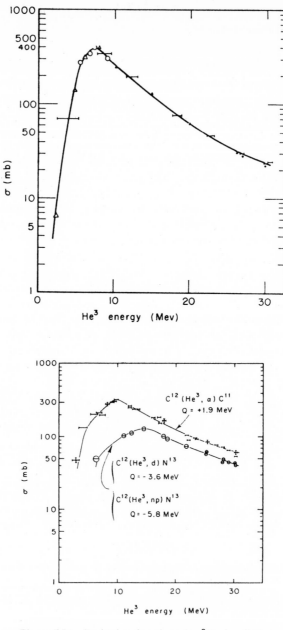

**Figure 7.31** Excitation functions for $^3$He irradiations with $^{16}$O (*upper*) and $^{12}$C (*lower*). [From S. S. Markowitz and J. D. Mahony, Activation Analysis for Oxygen and Other Elements by Means of $^3$He-induced Nuclear Reactions, *Anal. Chem.* **34**, 329–335 (1962).]

## Photon Activation Analysis

In general, photon activation analysis has been used as a supplement to the more widely used neutron activation analysis, primarily because of the low cross sections of photonuclear reactions. However, for those cases in which photonuclear reactions are advantageous as an analytical method, photon activation analysis is suitable. The increasing availability of electron linear accelerators should increase the use of photon activation analysis even further.

Photons, produced as monoenergetic radiation or as Bremsstrahlung by electron irradiation of high $Z$ metals, have the advantage of neutrons in extensive penetration of large samples, in producing metastable isomers by inelastic scattering, and can be used in lieu of neutrons when neutron-activation products are not suitable.

Two types of photon activation analysis result from the general threshold energy of about 6 to 12 MeV for the $(\gamma,n)$ reaction. With accelerators that produce photons with energy less than about 6 MeV, photon activation analysis can be acheived by the inelastic scattering reaction $(\gamma,\gamma')$ and measurement of the de-excitation of metastable isomers. Higher energy machines allow photon activation analysis by photonuclear reactions which generally produce neutron-deficient radionuclides that decay by position emission or orbital-electron capture.

INELASTIC SCATTERING. Many of the heavier elements in the period table consist of one or more isotopes that have metastable isomeric states with a measurable half-life. A list of stable nuclides of isotopic abundance greater than 1%, with metastable states of half-life greater than 1 sec, is given in Table 7.8. These elements are amenable to low-energy photon activation analysis; sensitivities reported for 16 of these elements are listed in Table 7.9. The values for the minimum concentration determinable were calculated for an irradiation of a 30-ml solution sample irradiated for 1 hr or to saturation for short-lived products, counting up to 1 hr with a detector with a background counting rate of 300 cpm, and with a concentration standard deviation of ±10%.

PHOTONEUTRON REACTIONS. Two special cases exist for low-energy photo-activation analysis with prompt-neutron radiation measurement. These are the photoneutron reactions of beryllium and deuterium, which have the low neutron binding energy of 1.67 and 2.23 MeV, respectively.

Other elements require photon energies in the range of about 10 to 25 MeV, corresponding to the energy range in which most elements have their peak cross section for the $(\gamma,n)$ reaction. At higher photon energies reactions such as $(\gamma,2n)$, $(\gamma,p)$, $(\gamma,d)$ may produce interfering radionuclides. At best, peak

## Table 7.8    Isomeric States of Stable Nuclides[a]

| Isomer | Half-Life | $E(IT)$ (MeV) | Conversion Ratio $(e/\gamma)$ |
|---|---|---|---|
| 79mSe | 17.5 s | 0.161 | 0.8 |
| 77mBr | 4.8 s | 0.21 | — |
| 87mSr | 2.83 h | 0.388 | 0.28 |
| 89mY | 16 s | 0.91 | 0.01 |
| 93mNb | 13.6 y | 0.030 | ? |
| 103mRh | 57 m | 0.040 | 40 |
| 109mAg | 40 s | 0.088 | 10.3 |
| 111mCd | 48.6 m | 0.150 + 0.246 | 1.5 + 0.054 |
| 113mIn | 100 m | 0.393 | 0.43 |
| 115mIn | 4.50 h | 0.355 (95%) + $(\beta^-)$ | 0.8 |
| 117mSn | 14.0 d | 0.159 + 0.162 | v. large + 0.10 |
| 119mSn | 250 d | 0.024 + 0.065 | 5.2 + ? |
| 125mTe | 58 d | 0.035 + 0.110 | 11 + 160 |
| 129mXe | 8.0 d | 0.040 + 0.197 | 7 + 11 |
| 131mXe | 11.8 d | 0.164 | 29 |
| 135mBa | 28.7 h | 0.268 | 3.8 |
| 137mBa | 2.55 m | 0.662 | 0.11 |
| 167mEr | 2.3 s | 0.208 | 0.50 |
| 176mLu | 3.7 h | $(\beta^- + \gamma)$ | — |
| 178mHf | 4.3 s | 0.089 + 0.427 + 0.326 + 0.214 + 0.093 | ? |
| 179mHf | 18.6 s | 0.161 + 0.217 | 19 + 0.05 |
| 180mHf | 5.5 h | 0.058 + 0.093 + 0.215 + 0.333 + 0.444 + 0.501 | 0.54 + 1.1 + 0.12 + 0.04 + 0.02 + 0.04 |
| 183mW | 5.3 s | 0.046 + 0.052 + 0.105 + 0.160 | ? |
| 190mOs | 9.9 m | 0.038 + 0.187 + 0.319 + 0.500 + 0.614 | >200 + 0.2 + 0.04 0.02 + 0.01 |
| 191mIr | 4.9 s | 0.418 + 0.129 | large + 1.9 |
| 193mIr | 12 d | 0.080 | ? |
| 195mPt | 4.1 d | 0.031 + 0.099 + 0.129 | >7 + 9 + large |
| 197mAu | 7.2 s | 0.130 + 0.279 | <1.2 + 0.29 |
| 199mHg | 43 m | 0.159 + 0.375 | 0.25 + >1.1 |
| 204mPb | 66.9 m | 0.289 + 0.375 + 0.622 0.899 + 0.912 | 0.37 + 0.04 + ? 0.008 + 0.054 |

[a] For isomeric states with $T_{1/2} > 1$ sec and $f > 1\%$. [Adapted from C. L. Lederer, J. M. Hollander, and I. Perlman, *Table of Isotopes* (Wiley, New York, 1967).]

**Table 7.9** Sensitivity of Photoactivation Analysis[a]

| Element | $A_0$ (cpm)[b] | Minimum Weight Determinable (mg) | Minimum Concentration Determinable (ppm) |
|---------|------|------|---------|
| Se | 460 | 3.3 | 110 |
| Sr | 32 | 3.2 | 110 |
| Y | 460 | 77 | 2,600 |
| Rh | 50 | 62 | 2,100 |
| Ag | 300 | 3.0 | 100 |
| Cd | 64 | 1.0 | 33 |
| In | 32 | 0.21 | 7 |
| Sn | 32 | 3,200 | 110,000 |
| Ba | 160 | 200 | 6,700 |
| Er | 3,000 | 10 | 330 |
| Lu | 32 | 6.4 | 210 |
| Hf | 440 | 0.11 | 3.7 |
| Ir | 1,300 | 1.3 | 43 |
| Pt | 32 | 64 | 2,100 |
| Au | 1,000 | 2.5 | 83 |
| Hg | 74 | 37 | 1,200 |

[a] Adapted from H. R. Lukens, J. W. Otvos, and C. D. Wagner, Formation of Metastable Isomers by Photoactivation with the Van de Graaff Accelerator, *Intern. J. Appl. Radiation Isotopes*, **11**, 30–37 1961).

[b] The required activity at irradiation end for a 1-hr irradiation with Bremsstrahlung from a 3-MeV, 1 mA electron beam, counted with a detector of 300 cpm background, and for $\pm 10\%$ standard deviation.

cross sections for $(\gamma,n)$ reactions range from a few to a few hundred milli-barns. Some representative values from Koch's *Activation Analysis Handbook* are given in Table 7.10.

**Table 7.10** Nuclear Data for $(\gamma,n)$ Activation[a]

| Element | Product | $T_{1/2}$ | $E_{th}$ (MeV) | Peak Cross Section $E_\gamma$ (MeV) | $\sigma$ (mb) |
|---------|---------|-----------|----------------|-------------------|---------------|
| H | Prompt n | ... | | | |
| Be | Prompt n | ... | 1.63 | 9 | 1.6 |
| C | $^{11}$C | 20.5 m | 18.7 | 22.5 | 10 |
| N | $^{13}$N | 10.0 m | 10.7 | 22.5 | 15 |
| O | $^{15}$O | 124 s | 15.5 | 24.2 | 11.4 |
| F | $^{18}$F | 111 m | 10.4 | 12 | 2.5 |
| Ne | $^{19}$Ne | 18 s | 16.9 | 21.5 | 7.7 |
| Na | $^{22}$Na | 2.58 y | 12.05 | 18.5 | $\approx 13$ |
| Mg | $^{23}$Mg | 12 s | $\approx 17$ | 19.5 | 8.4 |
| Al | $^{26m}$Al | 6.5 s | 13.4 | 19.5 | 22 |
| Si | $^{27}$Si | 4.2 s | 16.9 | 20.9 | 21 |
| P | $^{30}$P | 2.6 m | 12.33 | 19 | 16.4 |
| S | $^{31}$S | 2.6 s | $\approx 11.5$ | 20.1 | 19 |
| Cl | $^{34m}$Cl | 32.4 m | 9.95 | 17.6 | 4.4 |
| Ar | $^{39}$Ar | 260 y | $< 10$ | 20 | 38 |
| K | $^{38}$K | 7.7 m | 13.2 | 17.5 | 5.4 |
| Ca | $^{39}$Ca | 0.9 s | 15.8 | 19.3 | 15 |
| Cr | $^{49}$Cr | 42 m | 13.4 | 19 | 52 |
| Mn | $^{54}$Mn | 314 d | 10.0 | 19 | 100 |
| Fe | $^{53}$Fe | 9 m | 13.8 | 18.3 | 67 |
| Co | $^{58m}$Co | 9 h | 10.25 | 16.9 | 130 |
| Ni | $^{57}$Ni | 36 h | 11.7 | 18.5 | 57 |
| Cu | $^{62}$Cu | 9.9 m | 11 | 17 | 850 |
| Zn | $^{63}$Zn | 38 m | 11.6 | 18.5 | 123 |
| As | $^{74}$As | 18 d | 10.1 | 17 | 91 |
| Br | $^{78}$Br | 6.5 m | 10 | 16 | $\approx 100$ |
| | $^{80}$Br | 18 m | 11 | 18 | 88 |
| | $^{80m}$Br | 4.5 h | 11 | 18 | 42 |
| Rb | $^{86}$Rb | 18.7 d | 9.3 | 17.5 | 230 |
| Sr | $^{85}$Sr | 64 d | 11.5 | 15.9 | 160 |
| Y | $^{88}$Y | 108 d | 11.8 | 16.3 | 190 |
| Zr | $^{89}$Zr | 79 h | 11.8 | 17.8 | 120 |
| | $^{89m}$Zr | 4.4 m | 12.3 | 17.8 | 150 |
| Nb | $^{92}$Nb | 10.1 d | 8.86 | 17 | 230 |
| Mo | $^{91}$Mo | 15.6 m | 13.1 | 18.7 | 120 |
| | $^{91m}$Mo | 65 s | 13.2 | 18.7 | 24 |
| Rh | $^{102}$Rh | 206 d | 9.4 | 16.5 | 210 |
| Ag | $^{108}$Ag | 2.4 m | 9.05 | 16.5 | 320 |

**Table 7.10** (Continued)

| Element | Product | $T_{1/2}$ | $E_{th}$ (MeV) | Peak Cross Section $E_\gamma$ (MeV) | $\sigma$ (mb) |
|---------|---------|-----------|----------------|-------------------------------------|---------------|
| In | $^{114}$In | 72 s | 9.05 | 15 | 70 |
|  | $^{114m}$In | 50 d | 9.05 | 15 | 350 |
| Sb | $^{122}$Sb | 2.8 d | 9.3 | 15 | 362 |
| I | $^{127}$I | 13.2 d | 9.1 | 15.2 | 450 |
| Lu | $^{174}$Lu | 165 d | 7.77 | 16 | 225 |
| Ta | $^{180m}$Ta | 8.1 h | 7.6 | 14 | $\approx 600$ |
| Au | $^{196}$Au | 6.1 d | 8.0 | 14.5 | $\approx 600$ |
| Bi | $^{208}$Bi | $7.5 \times 10^5$ y | 7.4 | 13.5 | $\approx 450$ |
| Th | $^{231}$Th | 25.6 h | $\approx 6$ | 14.2 | 990 |
| U | $^{232}$U | 73.6 y | 7.0 | 14 | 1670 |
|  | $^{237}$U | 6.75 d | $\approx 6.0$ | 15.2 | 1290 |
| Pu | $^{238}$Pu | 89 y | 6 | 13.6 | 1580 |

[a] From R. C. Koch, *Activation Analysis Handbook* (Academic, New York, 1960).

## 7.4 BIBLIOGRAPHY

The literature of radioactivation has become large. Several books that detail many professional aspects of the field have been published. A greater number of review articles covering bibliographies of advances in either the techniques or the applications in specific areas have appeared over the years. Only a few are listed in the text. The immenseness of the radioactivation analysis literature has been made strikingly noticible by the publication of a bibliography by the National Bureau of Standards. This bibliography is recommended as a starting point for the search of information in any aspect of radioactivation analysis. Some of the more general sources of information are given in this section.

### 7.4.1 General Works on Activation Analysis

PROFESSIONAL BOOKS

H. J. Bowen and D. Gibbons, *Radioactivation Analysis* (Oxford, London, 1963).

R. C. Koch, *Activation Analysis Handbook* (Academic, New York, 1960).

J. M. A. Lenihan and S. J. Thompson, Eds., *Advances in Activation Analysis*, Vol. 1 (Academic, New York, 1969).

W. S. Lyon, Jr., Ed., *Guide to Activation Analysis* (Van Nostrand, Princeton, N.J., 1964).

M. Rakovič, (translated by D. Cohen) *Activation Analysis* (CRC Press, Cleveland, 1970).

D. Taylor, *Neutron Irradiation and Activation Analysis* (Van Nostrand, Princeton, N.J., 1964).

SYMPOSIA PROCEEDINGS

M. W. Brown, Ed., *Modern Trends in Activation Analysis* (A&M College of Texas, 1961), 178 pp. *Proc., 1961 International Conference*, College Station, December 15–16, 1961.

J. R. DeVoe, Ed., *Modern Trends in Activation Analysis*, Special Publication 312, 2 vols (National Bureau of Standards, Gaithersburg, 1969).

J. P. Guinn, Ed., *Modern Trends in Activation Analysis* (Texas A&M University, 1965), 390 pp. *Proc. 1965 International Conference*, College Station, April 19–22, 1965.

J. M. A. Lenihan and S. J. Thomson, Eds., *Activation Analysis: Principles and Applications* (Academic, London, 1965), 211 pp. *Proc., NATO Advanced Study Institute*, Glasgow, Scotland, August 1964.

B. R. Payne, Ed., *Radioactivation Analysis Symposium* (Butterworth, London, 1960), 141 pp. *Proc., Radioactivation Analysis Symposium*, Vienna, June 1–3, 1959.

*Nuclear Activation Techniques in the Life Sciences* (International Atomic Energy Agency, Vienna, 1967) 709 pp. *Proc., Symposium on Nuclear Activation Techniques in the Life Sciences*, Amsterdam, May 8–12, 1967.

Radiochemical Methods of Analysis (International Atomic Energy Agency, Vienna, 1965) 2 vols., *Proc., Symposium on Radiochemical Methods of Analysis*, Salzburg, October 19–23, 1964.

BIBLIOGRAPHIES

W. Bock-Werthmann and W. Schulze, *Aktivierungsanalyse*, Atomic Energy Document (Gmelin-Institute), AED-C-14-1 (plus supplements −2, −3), 1961 (1962–1963).

D. Gibbons et al., *Radioactivation Analysis, A Bibliography*, U.K. Atomic Energy Authority Report A.E.R.E.-1/R 2208 (plus supplement 1), 1957 (1960).

G. L. Lutz, R. J. Boreni, R. S. Maddock, and W. W. Meinke, *Activation Analysis: A Bibliography*, National Bureau of Standards, Technical Note 467, 2 parts (plus revision), 1968 (1969).

H. D. Raleigh, *Activation Analysis: A Literature Search*, USAEC Report TID-3575, 1963.

REVIEW ARTICLES

R. F. Coleman, Activation Analysis, A Review, *Analyst* **92**, 1–19, (1967).

W. R. Corliss, Neutron Activation Analysis (USAEC booklet in a series on Understanding the Atom, 1963) 27 pp. Single copies may be obtained free by writing USAEC, P.O. Box 62, Oak Ridge, Tennessee 37831.

F. Girardi, Some Recent Developments in Radioactivation Analysis, *Talanta* **12**, 1017–1041 (1965).

D. Mapper, Radioactivation Analysis, Chapter 9 in A. A. Smales and L. R. Wager, *Methods in Geochemistry* (Interscience, New York, 1960), pp. 297–357.

H. D. Raleigh, Activation Analysis—Principles and Applications, *Atomics* **16**, (No. 4) 15–33 (1963).

G. W. Reed, Activation Analysis Applied to Geochemical Problems in P. H. Abelson, *Researches in Geochemistry* (Wiley, New York, 1959).

E. V. Sayre, Methods and Applications of Activation Analysis in *Ann. Rev. Nucl. Sci.* **13** (Annual Reviews, Stanford, 1963), pp. 145–162.

A. A. Smales, Neutron Activation Analysis, Chapter 19 in J. H. Yoe and H. J. Koch, *Trace Analysis* (Wiley, New York, 1957), pp. 518–546.

John W. Winchester, Radioactivation Analysis in Inorganic Geochemistry, in F. A. Cotton, *Progress in Inorganic Chemistry*, Vol. 2 (Interscience, New York, 1960), pp. 1–32.

## 7.4.2 Sources of Data

NUCLEAR DATA

D. J. Hughes and J. A. Harvey, Neutron Cross Sections, BNL-325 (1958).

R. C. Koch, *Activation Analysis Handbook*, (Academic, New York, 1960).

*Nuclear Data Tables*, National Academy of Sciences, National Research Council, 1960.

*Chart of the Nuclides*, 10th ed., 1969. Single copies available from General Electric Co., Schenectady, New York 12305.

C. M. Lederer, J. M. Hollander, and I. Perlman, *Table of Isotopes*, (Wiley, New York, 1967).

K. Way et al., Nuclear Data Group, Nuclear Data Sheets, available from National Academy of Science—National Research Council, Washington, D.C., 1961 on.

STANDARD $\gamma$-RAY SPECTRA

O. U. Anders, Gamma Ray Spectra of Neutron Activated Elements, Dow Chemical Co., Midland, Mich., 1961.

C. E. Crouthamel, *Applied Gamma Ray Spectrometry*, (Pergamon, New York, 1960).

R. Heath, Scintillation Spectrometry, Gamma-Ray Spectrum Catalog, 2nd ed., AEC Report No. IDO-16880, 1964.

STANDARD SENSITIVITIES

A. S. Gillespie and W. W. Hill, Sensitivities for Activation Analysis with 14-MeV Neutrons, *Nucleonics* **19** No. 11, 170–173 (1961).

E. Ricci and R. L. Hahn, Sensitivities for Activation Analysis of 15 Light Elements with 18-MeV Helium Particles, *Anal. Chem.* **39**, 794–797 (1967).

## 7.4.3 Specific Practices

SUBSTOICHIOMETRIC RADIOCHEMISTRY

J. Ruzicka and J. Stary, *Substoichiometry in Radiochemical Analysis*, (Pergamon, Oxford, 1968).

APPLICATION OF COMPUTERS TO ACTIVATION ANALYSIS

T. R. Dickson, *The Computer and Chemistry* (Freeman, San Francisco, 1968).

D. Gibbons, Computer Methods in Activation Analysis, in Lenihan and Thompson, Eds., *Activation Analysis* (Academic, London, 1965), Chapter 15.

G. Guzzi, J. Pauly, F. Girardi, and B. Dorpema, Computer Program for Activation Analysis with Ge(Li)-Drifted Detectors, Euratom Report EUR 3469.e, 1967.

D. L. Morrison, Computers Applied to Nuclear Chemistry, in L. Yaffe, Ed., *Nuclear Chemistry*, Vol II (Academic, New York, 1968), Chapter 11.

G. D. O'Kelley, Ed., Applications of Computers to Nuclear and Radiochemistry, NAS-NS 3107, 1963. Available from CFSTI, Springfield, Va.

P. C. Stevenson, Processing of Counting Data, NAS-NS 3109, 1966. Available from CFSTI, Springfield, Va.

AUTOMATIC ACTIVATION ANALYSIS

D. Comar and C. LePoec, On the Use of an Automatic Chemical Treatment System in Activation Analysis of Biological Samples, in *Modern Trends in Activation Analysis* (Texas A&M University, College Station, 1965), pp. 351–356.

F. Girardi, G. Guzzi, J. Pauly, and R. Pietra, The Use of an Automated System including a Radiochemical Step in Activation Analysis, in *Modern Trends in Activation Analysis* · (Texas A&M University, College Station, 1965), pp 337–343.

F. Girardi, M. Merlini, J. Pauly, and R. Pietra, Progress Towards Automated Radiochemical Separations, in *Radiochemical Methods of Analysis*, Vol. II (International Atomic Energy Agency, Vienna, 1965), pp. 3–14.

R. L. Heath, Gamma-Ray Spectrometry and Automated Data Systems for Activation Analysis, in J. R. DeVoe, Ed., *Modern Trends in Activation Analysis* (National Bureau of Standards Special Publication No. 312, Vol 2, 1969), pp. 959–1031.

L. T. Skegg, Ed., *Automation in Analytical Chemistry* (Mediad, New York, 1966+).

PROMPT RADIATION ACTIVATION ANALYSIS

R. C. Greenwood and J. Reed, Scintillation Spectrometer Measurements of Capture Gamma Rays from Natural Sources, in *Modern Trends in Activation Analysis*, (Texas A&M University, College Station, 1961).

W G Lussie and J. L. Brownlee, Measurement and Utilization of Neutron-Capture Gamma Radiation, in *Modern Trends in Activation Analysis*, (Texas A&M University College Station, 1965).

D. Taylor, *Neutron Irradiation and Activation Analysis* (Van Nostrand, Princeton, N.J., 1964), Chapter 3.

CHARGED PARTICLE AND PHOTON ACTIVATION ANALYSIS

R. S. Tilbury, Activation Analysis with Charged Particles, NAS-NS 3110, undated.

C. A. Baker, Gamma-activation Analysis: A Review, *Analyst* **92**, 601–610 (1967).

S. S. Markowitz and J. D. Mahony, Activation Analysis for Oxygen and Other Elements by Means of $He^3$-induced Nuclear Reactions, *Anal. Chem.* **34**, 329–335 (1962).

## 7.5  PROBLEMS

1.  Determine the maximum sensitivity for measuring the manganese content in household aluminum foil under the irradiation conditions given in Section 7.1.2. Assume measurement of 2.58-h $^{56}$Mn with same photopeak background, an efficiency of 15%, counted 4 hr after irradiation, and a chemical yield of 90%.

2. With the data in Table 7.1,
   (a) choose the optimum nuclear reaction for copper analysis if there were available locally only
       (1) a thermal neutron reactor and a gamma-ray scintillation detector,
       (2) a 14-MeV neutron generator and a low-background beta counter,
       (3) a high intensity 4-MeV proton accelerator and any radiation detection system,
       (4) a 30-MeV electron accelerator and a gamma-ray scintillation detector;
   (b) determine the initial gamma-ray radioactivity (in $\gamma$ps) of the two product radionuclides in a copper disk of 150 mg/cm² irradiated by the entire beam of 10 $\mu$A of 12-MeV protons for 1 hr.

3. Hydrogen is essentially 100% $^1$H. The cross section for radiative neutron capture is 0.33 b and the 2.225-MeV prompt gamma-ray can be determined with an efficiency of 10% in a coincidence spectrometer. During a 500-sec irradiation of a 2-g sample with a neutron flux of $10^{10}$ n/cm²-sec a total of 800 net counts was obtained in an analyzer channel corresponding to the 2.225-MeV gamma ray. Determine the percent of $^1$H as %$H_2O$ in the sample. How much 12.26-y $^3$H would be produced in the sample during the irradiation? Deuterium has a neutron capture cross section of 0.0057 b and an isotopic abundance of 0.015%.

4. One gram of a tissue was irradiated in a neutron flux of $10^{13}$ thermal neutrons/cm²-sec and $2 \times 10^{12}$ fast neutrons/cm² sec for a period of 24 hr. The cross section for thermal neutron radiative capture by $^{75}$As ($f = 1.00$) is 4.3 b. The cross section for the (n,2n) inelastic scattering reaction of $^{75}$As with fast neutrons is 0.5 b. The decay schemes for $^{74}$As and $^{76}$As are shown in Figure 4.2.
   (a) If a 1-mg comparitor sample of pure arsenic was irradiated identically to the tissue sample, calculate the initial activity of the two arsenic radionuclides (in dps) in the comparitor sample.
   (b) If you could distinguish, in a gamma-ray spectrometer the gamma-ray energies 0.56, 0.60, and 0.64 MeV, how many $\gamma$ps of each transition would be emitted by the comparitor sample 24 hr after the end of the irradiation?
   (c) If the above gamma-rays were not distinguishable, how else could you measure the activity of each arsenic radionuclide
       (a) with only a gamma-ray spectrometer?
       (b) with only a beta counter?
       (c) what other techniques could be employed?
   (d) If 26.4 hr after the end of the irradiation the 0.56 MeV gamma ray

measurement of the tissue sample was $4.72 \times 10^3$ $\gamma$ps, what was the arsenic concentration (in ppm) in the tissue?

5. Neutron activation analysis was used to determine the concentration of vanadium in a standard dry kale sample. A 100-g aliquot was dissolved by digestion with $HClO_4$ and adjusted to a volume of 50 ml. A comparator solution of 2 $\mu$g V/ml was also prepared. A 5-ml aliquot of each solution was irradiated identically for 5 min in a thermal neutron flux of $5 \times 10^{10}$ n/cm²-sec. The nuclear reaction 99.76% $^{51}V$ (n,$\gamma$) 3.77-m $^{52}V$ has a cross section of 4.9 b for thermal neutrons. Following the irradiation, the vanadium was chemically separated from the kale solution by solvent extraction with a chemical yield of 50.2% and adjusted to a counting volume of 5 ml. The two samples were counted identically by photopeak counting of the 1.433-MeV gamma ray which is coincident with the $\beta^-$-decay of $^{52}V$. The decay curves, corrected for counter background and long-lived impurities, showed initial activities $A^0 = 9650$ cpm for the kale sample and $A^0 = 38,500$ cpm for the comparator solution. From these data determine the concentration of vanadium (in ppm) in the dry kale standard.

Chapter 8

# Activation Analysis: Limitations

Activation analysis, like all other analytical methods, cannot meet the multitude of requirements of analytical chemistry. Included among the inconveniences of activation analysis are the following:

1. The need to obtain irradiations at expensive, sometimes remote, nuclear facilities.

2. The need to work with radioactive materials and the concommitant need to observe radiological health regulations and precautions.

3. The need at times for exacting radiochemical separations.

4. The need at times to unscramble complicated radiation measurement data.

5. The inability to determine the chemical form of a desired element.

These inconveniences, alone, preclude the widespread use of radioactivation analysis as a general analytical chemistry method. The question "why use it at all?" is deferred to the introduction to Chapter 9, in which the advantages and disadvantages of activation analysis as a method for trace element determination are reviewed. That there is a vast activation analysis effort throughout the world, in spite of the inconveniences listed above, may be considered sufficient evidence that the method is quite useful within its realm of analytical chemistry and radioactivation tracing methods.

Nuclear irradiation facilities do exist at many accessible locations such as universities, national laboratories, and industrial companies. Service irradiations are readily available at most of them. With proper planning problems

**Table 8.1**   Sources of Errors in Radioactivation Analysis[a]

| Source | Estimated Error (%) |
|---|---|
| 1. Chemical | |
|   a. Sample weight | ±1 |
|   b. Standard weight | ±2 |
|   c. Yield determination | ±2 |
| 2. Irradiation | |
|   a. Self shielding (correction $<50\%$) | ±4 |
|   b. Flux depression | ±2 |
|   c. Thermal enhancement | ±2 |
|   d. Absolute value of thermal flux | ±5 |
|   e. Value of cadmium ratio | ±2 |
|   f. Irradiation time ($<1$ min) | ±3 |
|   g. Nonhomogeneity of neutron flux | ±1 |
| 3. Nuclear constants | |
|   a. Half life | ±2–10 |
|   b. Decay scheme | ±2–50 |
|   c. Cross section | ±5–30 |
| 4. Nuclear reactions | |
|   a. Competing | Variable |
|   b. Interfering | Variable |
| 5. Counting | |
|   a. Detector calibration | ±3 |
|   b. Counting rate ($<10^3$ cps) | ±4 |
|   c. Geometrical factors | ±1 |

[a] Adapted from J. P. Call, J. R. Weiner, and G. G. Rocco, The Accuracy of Radioactivation Analysis, in *Modern Trends in Activation Analysis* (Texas A&M University, College Station, 1965), pp. 253–258.

of time, distance, and decay, radiation levels can be reduced to values for which radiological health problems are negligible. Radiochemical procedures for activation analysis are becoming simpler, and the techniques for unscrambling complex radiation spectral data by computer methods have become more than adequate in many cases. These, then, are not the general limitations to the widespread use of radioactivation analysis.

As a chemical analysis method, activation analysis shares in common with all other analytical methods the limitations due to errors in chemical and physical manipulations and data processing. Many of the sources of these errors have already been described in the discussion of their respective topics. Chemical errors, as in sampling techniques, weighing of samples and standards, and chemical-yield determination, physical errors, as in irradiating sample and standard in different locations or counting them in different geometry, and manual errors, as in recording a reading or value improperly or in "slipping" a decimal point, are examples of common errors that can limit the usefulness of an analysis. They are also the ones that can, with careful laboratory practices, be eliminated or reduced to acceptable levels.

Not in common with other analytical chemistry methods are the sources of error inherent in the nuclear aspects of this analytical method. These sources are prevalent in both the production of the desired radionuclides and the measurement of their radioactivity. Table 8.1 lists some of the general sources of error in radioactivation analysis. Some of the more important problems in the nuclear aspects of radioactivation analysis are examined in this chapter. When possible, a review of the methods developed to reduce the effect of these limitations is included.

## 8.1 CHEMICAL LIMITATIONS

By far the most serious limitation of activation analysis as an analytical chemistry method is its inability to distinguish between chemical forms of a given element. Except for the analysis of trace elements, the organic chemist and the biochemist find little use for the method of activation analysis. The elements carbon, hydrogen, and oxygen are contained in hundreds of thousands of organic compounds. Extensive efforts have gone into the search for analytical methods to distinguish between these compounds. Many of them contain the same composition of the three elements and differ solely by their molecular arrangement. Physical methods are generally used to distinguish between such isomers.

Sometimes the determination of a particular organic chemical is desired when it is present in small quantity in an inorganic matrix; for example, the presence of labile viruses or pesticide residues in water may be of concern at concentrations less than $10^{-8}$ g/l. Although carbon constitutes about 50% of the chemical composition of a virus, a single virus may be of interest in some studies. A colony of viruses may weigh less than $10^{-9}$ g. Chlorine or bromine is a major constituent of many of the type A pesticides that have safe concentration recommendations of the order of 50 ng/l or less. If no other organic matter is known to be in a particular sample of virus or if no other chlorine compounds are known to be in a particular sample of water for pesticide residue analysis, the determination of these materials can be made with great precision by activation analysis methods. It is obvious that such conditions are met only infrequently, and even then great care must be maintained to avoid contamination of the sample before irradiation.

Elements that may exist in different oxidation states in inorganic compounds or in aqueous solution are subject to the same limitations in activation analysis if knowledge of the chemical form of the element is important; for example, manganese is an important trace element. Under suitable conditions it could exist in a sample in one or more of the following oxidation states:

| $+2$ | $Mn^{2+}$ | manganous |
| $+3$ | $Mn^{3+}$ | manganic |
| $+4$ | $MnO_2$ | manganese dioxide |
| $+6$ | $MnO_4^{2-}$ | manganate |
| $+7$ | $MnO_4^{-}$ | permanganate |

Another important trace element is iodine, which may exist in water in several oxidation states:

| $-1$ | $I^-$ | iodide |
| $0$ | $I_2$ | iodine |
| $+1$ | $IO^-$ | hypoiodite |
| $+5$ | $IO_3^-$ | iodate |
| $+7$ | $IO_4^-$ | periodate |

One precaution that has already been noted for iodine, and is required for other elements that may exist in several oxidation states, is the change in oxidation state that may result during the activation irradiation.

The "hot-atom" chemistry of many elements during irradiations has been extensively studied. The hot-atom refers to the atom with sufficient recoil energy following the decay of the compound nucleus to rupture its chemical bonds and reassociate with other neighboring atoms upon its relaxation. The Szilard-Chalmers process, in which a chemical method is used to separate the reassociated radioactive element from the target form, is often used to prepare high specific-activity radionuclides by $(n,\gamma)$ reaction irradiations; for example, from acid or neutral solutions of permanganate ions the $^{56}Mn$

**Table 8.2** Elements Which Can Undergo
the Szilard-Chalmers Process

| P | As | RE |
|---|----|----|
| Cl | Se | W |
| V | Br | Re |
| Cr | Mo | Os |
| Mn | Rh | Ir |
| Fe | Pd | Pt |
| Co | In | Au |
| Ni | Sn | Pb |
| Cu | Sb | Bi |
| Zn | Te | U |
| Ga | I | |

activity from a $(n,\gamma)$ irradiation is removable as $MnO_2$. Elements which have had $(n,\gamma)$ products enriched by the Szilard-Chalmers process are listed in Table 8.2.

Obviously for these and possibly other elements which may be in unknown chemical forms in samples care must be exercised in chemical separation procedures to avoid loss of all or part of the radioactive product isotope from the element in the sample.

## 8.2 NUCLEAR LIMITATIONS

Nuclear limitations occur during the activation part of an analysis in two forms: (a) by interfering nuclear reactions and (b) by changes in irradiation conditions. The former changes the proportionality between the amount of the sought element and the amount of the measured radionuclide. Interfering nuclear reactions can increase or decrease the amount of the desired radionuclide with respect to the sought element. Changes in irradiation conditions also change the proportionality between the sought element and the measured radionuclide. In this case the rate of production of the desired radionuclide is altered by changes in energy or flux of the irradiating particles. Both problems exist especially in the absolute method of activation analysis. Attention is normally given to these problems when the absolute method of activation analysis is used. However, although the comparitor method of activation analysis is generally more reliable than the absolute method, systematic errors may occur frequently in the former method because of nonrecognition that the samples and standards may undergo different irradiation conditions and that elements in the samples, not present in the pure standards, may contribute to the total amount of the measured radionuclide by interfering nuclear reactions. One method developed for reducing the extent of interfering nuclear reactions with matrix elements is the variable-energy activation analysis method.

### 8.2.1   Interfering Nuclear Reactions

Interfering nuclear reactions are defined as nuclear reactions that alter the linear relationship between a sought element and a desired radionuclide. The effects of interfering reactions are dependent on the elements present in the matrix and their nuclear properties. Three types of interfering nuclear reaction may be encountered in activation analysis:

1. Primary reactions, the most significant of the three, are reactions in which the desired radionuclide is produced from an element in the matrix other than the sought element.

2. Secondary reactions, significant in nonreactor irradiations, are reactions in which the resulting particles from primary reactions produce the desired radionuclides by nuclear reaction with other elements in the matrix.

3. Second-order reactions—two types: (a) enhancement of the desired product by decay of a short-lived product of a neighboring element to the sought element; (b) reduction of the desired product by activation if the primary product has a high cross section for nuclear reactions with the irradiating particles.

An example of all three of these interfering reactions occurs in the activation analysis of organic materials in which the determination of the trace elements chlorine, sulfur, phosphorous, and silicon is often required. Besides the obvious interest of these elements for biochemical studies, a specific example is of interest in the field of nuclear engineering.

Because of the general inertness of organic matter (C,H,O) to neutron activation, organic liquids are useful as nuclear reactor coolants and moderators; for example, terphenyls are organic liquids that possess excellent thermal properties for use in organic-cooled reactors. However, silicon, phosphorous, sulfur, and chlorine are usual contaminants of terphenyl and lead to undesirable radioactivity in the coolant on continuous circulation through the reactor. Therefore these elements are usually determined as a quality control process for terphenyl produced for coolants in nuclear reactors.

Irradiation of the four trace-elements impurities with thermal neutrons leads to the following useful (n,$\gamma$) reactions products:

$$^{37}_{17}Cl(n,\gamma)37.3\text{-m }^{38}Cl$$
$$^{34}_{16}S(n,\gamma)87.9\text{-d }^{35}S$$
$$^{31}_{15}P(n,\gamma)14.3\text{-d }^{32}P$$
$$^{30}_{14}Si(n,\gamma)2.62\text{-h }^{31}Si$$

In nuclear reactor irradiations the following interfering reactions can also occur:

1. Primary:

$$^{35}Cl(n,p)^{35}S$$
$$^{35}Cl(n,\alpha)^{32}P$$
$$^{32}S(n,p)^{32}P$$
$$^{34}S(n,\alpha)^{31}Si$$
$$^{31}P(n,p)^{31}Si$$

2. Secondary:

$$^{36}S(\gamma,n)^{35}S$$
$$^{36}S(\alpha,d)^{38}Cl$$
$$^{30}Si(\alpha,d)^{32}P$$

3. Second order:
   (a) an increase in product by

$$^{30}Si(n,\gamma)^{31}Si \xrightarrow{\beta^-} {}^{31}P(n,\gamma)^{32}P$$
$$^{36}S(n,\gamma)^{37}S \xrightarrow{\beta^-} {}^{37}Cl(n,\gamma)^{38}Cl$$

   (b) a decrease in product by

$$^{31}Si(n,\gamma)^{32}Si$$
$$^{32}P(n,\gamma)^{33}P$$
$$^{35}S(n,\gamma)^{36}S$$
$$^{38}Cl(n,\gamma)^{39}Cl$$

The nuclear data for the more important of these reactions are given in Figure 8.1. In actual practice the interfering reactions of importance are

$$^{35}Cl(n,p)^{35}S$$
$$^{32}S(n,p)^{32}P$$
$$^{35}Cl(n,\alpha)^{32}P$$

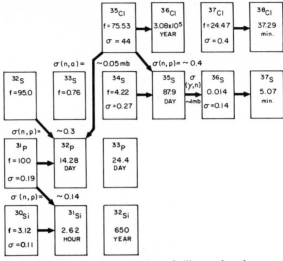

**Figure 8.1** Nuclear data for the activation of silicon, phosphorous, sulfur, and chlorine. The stable isotopes list the isotopic abundance in percent and the neutron capture cross section in barns. The radioactive isotopes list the half-life.

One solution to the activation analysis of terphenyl for these four elements is given by the following comparitor activation analysis procedure:

1. From a short irradiation determine the amount of silicon from $^{31}$Si and the amount of chlorine from $^{38}$Cl.

2. From a long irradiation of the same sample measure the amounts of total $^{35}$S and $^{32}$P.

3. Calculate from the amount of chlorine the amount of $^{35}$S produced by the $^{35}$Cl(n,p)$^{35}$S reaction.

4. Subtract this amount from the total measured $^{35}$S.

5. Calculate the amount of sulfur from the net $^{35}$S.

6. Calculate from the amount of chlorine the amount of $^{32}$P produced by the $^{35}$Cl(n,α)$^{32}$P reaction.

7. Calculate from the amount of sulfur the amount of $^{32}$P produced by the $^{32}$S(n,p)$^{32}$P reaction.

8. Subtract both contributions from the total measured $^{32}$P.

9. Calculate the amount of phosphorous from the net $^{32}$P.

The types of potential primary and second-order interfering nuclear reactions possible in activation analysis, summarizing the nuclear data given by Koch* as of 1960, is listed in Table 8.3. The major potential primary

* R. C. Koch, *Activation Analysis Handbook* (Academic, New York, 1960).

interfering reactions are the (n,p) and (n,$\alpha$) reactions. The fissionable elements Th, U, and Pu may contribute to the production of medium-weight radio-nuclides which are also fission products.

For specific cases an estimate of the extent of primary reaction interferences can be made if the pertinent nuclear data and the approximate amount of the interfering element are known. In practice, when primary interferences are suspected, comparitor samples of the interfering elements are usually included in the irradiation package.

The calculation of second-order interferences has been described by Ricci and Dyer* for the reactions

$$Z^A(n,\gamma)Z^{A+1}\begin{cases} \xrightarrow{\beta-} (Z+1)^{A+1}(n,\gamma)(Z+1)^{A+2} \\ \xrightarrow[\text{EC}]{\beta+} (Z-1)^{A+1}(n,\gamma)(Z-1)^{A+2} \end{cases} \tag{1}$$

They give as an example the analysis for zinc and nickel in a copper matrix, the second-order interferences to measurement of $^{65}$Zn and $^{65}$Ni, respectively:

$$^{63}Cu(n,\gamma)^{64}Cu\begin{cases} \xrightarrow{38\% \, \beta-} {}^{64}Zn(n,\gamma)^{65}Zn \\ \xrightarrow[43\% \text{ EC}]{19\% \, \beta+} {}^{64}Ni(n,\gamma)^{65}Ni \end{cases} \tag{2}$$

It is noted that $^{65}$Ni is also produced by the primary interfering reaction with fast neutrons

$$^{65}Cu(n,p)^{65}Ni \tag{3}$$

In such cases second-order interference can be neglected in the presence of first-order interference.

A list of 42 cases of potential second-order enhancement of desired radio-nuclide production found by Ricci and Dyer is reproduced in Table 8.4. Estimates of the interference in 23 of these cases, for which the nuclear data were available, were made from the following set of equations:

$$\frac{dN_1}{dt} = -N_1\phi_{\text{th}}\sigma_1 \tag{4}$$

$$\frac{dN_2}{dt} = -N_1\phi_{\text{th}}\sigma_1 - \lambda_2 N_2 - N_2\phi_{\text{th}}\sigma_2 \tag{5}$$

$$\frac{dN_3}{dt} = F_2\lambda_2 N_2 - \phi_{\text{th}}\sigma_3, \tag{6}$$

$$\frac{dN_4}{dt} = N_3\phi_{\text{th}}\sigma_3 - \lambda_4 N_4 - N_4\phi_{\text{th}}\sigma_4 \tag{7}$$

* E. Ricci and F. F. Dyer, Second-Order Interference in Activation Analysis, *Nucleonics* **22**, No. 6, 45–50 (1964).

**Table 8.3  Nuclear Data for Potential Interfering Nuclear Reactions**[a]

| Element | (n,γ) Product | (n,p) Nuclide | f(%) | σ(mb) | (n,α) Nuclide | f(%) | σ(mb) | Second Order Nuclide | f(%) | σ(n,γ) | T½ | (n,f) |
|---|---|---|---|---|---|---|---|---|---|---|---|---|
| Be | 10Be | 10B | 18.8 | <200 | 13C | 1.11 | 0.4 | | | | | |
| C | 14C | 14N | 99.6 | 1750 | 17O | 0.04 | | | | | | |
| F | 20F | 20Ne | 90.9 | | 23Na | 100 | | | | | | |
| Ne | 23Ne | 23Na | 100 | 33.9 | 26Mg | 11.3 | | | | | | |
| Na | 24Na | 24Mg | 78.6 | 190 | 27Al | 100 | 116 | | | | | |
| Mg | 27Mg | 27Al | 100 | 79 | 30Si | 3.05 | 46 | 26Mg | 11.3 | 0.06 | 9.45 m | |
| Al | 28Al | 28Si | 92.3 | 220 | 31P | 100 | 146 | | | | | |
| Si | 31Si | 31P | 100 | 86 | 34S | 4.22 | 138 | 30Si | 3.05 | 0.4 | 2.62 h | |
| P | 32P | 32S | 95.0 | 310 | 35Cl | 75.4 | 190 | | | | | |
| S | 35S | 35Cl | 75.4 | 190 | 38Ar | 0.06 | | | | | | |
| | 37S | 37Cl | 24.6 | 5 | 40Ar | 99.6 | | 36S | 0.017 | 0.14 | 5.04 m | |
| Cl | 38Cl | 38Ar | 0.06 | | 41K | 6.91 | 31 | | | | | |
| | 37Ar | | | | 40Ca | 97.0 | | | | | | |
| Ar | 41Ar | 41K | 6.91 | 81 | 44Ca | 2.06 | | 40Ar | 99.6 | 0.53 | 110 m | |
| K | 42K | 42Ca | 0.64 | | 45Sc | 100 | 0.0055 | | | | | |
| Ca | 445Ca | 45Sc | 100 | | 48Ti | 73.5 | | | | | | |
| Sc | 56Sc | 46Ti | 8.0 | 4.1 | | | | | | | | |
| Ti | 51Ti | 51V | 99.8 | 27 | 54Cr | 2.38 | 53 | 50Ti | 5.34 | 0.14 | 5.8 m | |
| V | 52V | 52Cr | 83.8 | 78 | 55Mn | 100 | 0.37 | | | | | |
| Cr | 51Cr | | | | 54Fe | 5.84 | <1.5 | | | | | |
| | 5Cr | 55Mn- | 100 | | 58Fe | 0.31 | | 54Cr | 2.38 | 0.38 | 3.5 m | |
| Mn | 56Mn | 56Fe | 91.7 | 0.44 | 59Co | 100 | 35 | | | | | |
| Fe | 59Fe | 59Co | 100 | 22 | 62Ni | 3.66 | 5.7 | | | | | |

| Element | Nuclide | (n,p) | % | σ | (n,α) | % | σ | (n,γ) | % | σ | t½ | |
|---|---|---|---|---|---|---|---|---|---|---|---|---|
| Co | $^{60,60m}$Co | $^{60}$Ni | 26.2 | 5 | $^{63}$Cu | 69.1 | <20 μb | $^{63}$Cu | 69.1 | 4.3 | 12.8 h | |
| Ni | $^{63}$Ni | $^{63}$Cu | 69.1 | 3.1 | $^{66}$Zn | 27.8 | 7.6 | $^{62}$Ni | 3.66 | 15 | 125 y | |
| Ni | $^{65}$Ni | $^{65}$Cu | 30.9 | 19 | $^{68}$Zn | 18.6 | 105 | $^{64}$Ni | 1.16 | 1.6 | 2.56 h | |
| Cu | $^{64}$Cu | $^{64}$Zn | 48.9 | <10 μb | | | | $^{63}$Cu | 69.1 | 4.3 | 12.8 h | |
| Cu | $^{66}$Cu | $^{66}$Zn | 27.8 | 80 | $^{69}$Ga | 60.2 | | | | | | |
| Zn | $^{65}$Zn | | | | | | | $^{64}$Zn | 48.9 | 0.44 | 245 d | |
| Zn | $^{69,69m}$Zn | $^{69}$Ga | 60.2 | 24 | $^{72}$Ge | 27.4 | 12.3 | $^{68}$Zn | 18.6 | 0.1 | 13.8 h | * |
| Ga | $^{70}$Ga | $^{70}$Ge | 20.6 | 130 | | | | $^{69}$Ga | 60.2 | 1.4 | 21 m | * |
| Ga | $^{72}$Ga | $^{72}$Ge | 27.4 | 65 | $^{75}$As | 100 | | $^{70}$Zn | 0.62 | 0.085 | 2.2 m | * |
| Ge | $^{71}$Ge | | | | $^{74}$Se | 0.87 | 38 | $^{70}$Ge | 20.6 | 3.9 | 114 d | * |
| Ge | $^{75}$Ge | $^{75}$As | 100 | 11.8 | $^{78}$Se | 23.5 | | $^{74}$Ge | 36.7 | 0.21 | 82 m | * |
| Ge | $^{77}$Ge | | | | $^{80}$Se | 49.8 | 10 | | | | | |
| As | $^{76}$As | $^{76}$Se | 9.02 | | $^{79}$Br | 50.5 | | | | | | |
| Se | $^{75}$Se | | | | $^{78}$Kr | 0.35 | | | | | | |
| Se | $^{81,81m}$Se | $^{81}$Br | 49.5 | | $^{84}$Kr | 56.9 | | $^{80}$Se | 49.8 | 0.5 | 18.2 m | ** |
| Se | $^{83}$Se | | | | $^{86}$Kr | 17.4 | | | | | | |
| Br | $^{80,80m}$Br | $^{80}$Kr | 2.27 | | | | | | | | | |
| Br | $^{82}$Br | $^{82}$Kr | 11.6 | | $^{85}$Rb | 72.1 | 64 | | | | | |
| Kr | $^{85,85m}$Kr | $^{85}$Rb | 72.1 | | $^{88}$Sr | 82.6 | | $^{84}$Kr | 56.9 | 0.1 | 4.36 h | ** |
| Kr | $^{87}$Kr | $^{87}$Rb | 27.9 | | | | | $^{86}$Kr | 17.4 | 0.06 | 78 m | ** |
| Rb | $^{86}$Rb | $^{86}$Sr | 9.86 | | $^{89}$Y | 100 | 70 | | | | | |
| Rb | $^{88}$Rb | $^{88}$Sr | 82.6 | 17.7 | | | | $^{87}$Rb | 27.9 | 0.12 | 17.8 m | ** |
| Sr | $^{87m}$Sr | | | | $^{90}$Zr | 51.5 | 194 | | | | | |
| Sr | $^{89}$Sr | $^{89}$Y | 100 | | $^{92}$Zr | 17.1 | | | | | | |
| Y | $^{90}$Y | $^{90}$Zr | 51.5 | 247 | $^{93}$Nb | 100 | 9 | | | | | |

$^{89}$Sr(n,γ)$^{90}$Sr

*(Continued overleaf)*

**Table 8.3** (Continued)

| Element | (n,γ) Product | (n,p) Nuclide | f(%) | σ(mb) | (n,α) Nuclide | f(%) | σ(mb) | Second Order Nuclide | f(%) | σ(n,γ) | $T_{1/2}$ | (n,f) |
|---|---|---|---|---|---|---|---|---|---|---|---|---|
| Zr | $^{95}$Zr | | | | $^{98}$Mo | 23.8 | | | | | | * |
| | $^{97}$Zr | | | | $^{100}$Mo | 9.62 | | | | | | * |
| Nb | $^{94m}$Nb | $^{94}$Mo | 9.12 | | | | | | | | | |
| Mo | $^{93m}$Mo | | | | $^{96}$Ru | 5.7 | | | | | | |
| | $^{99}$Mo | | | | $^{102}$Ru | 31.3 | | $^{98}$Mo | 23.8 | 0.45 | 66 h | * |
| | $^{101}$Mo | | | | $^{104}$Ru | 18.3 | | | | | | * |
| Tc | $^{99m}$Tc | $^{99}$Ru | 12.8 | | | | | | | | | |
| | $^{100}$Tc | $^{100}$Ru | 12.7 | | $^{103}$Rh | 100 | 63 | | | | | |
| Ru | $^{103}$Ru | $^{103}$Rh | 100 | | $^{106}$Pd | 27.1 | | | | | | * |
| | $^{105}$Ru | | | | $^{108}$Pd | 26.7 | 2.3 | | | | | * |
| Rh | $^{104,104m}$Rh | $^{104}$Pd | 9.3 | 130 | $^{107}$Ag | 51.4 | | | | | | |
| | $^{103}$Pd | | | | $^{106}$Cd | 1.22 | | | | | | |
| Pd | $^{109}$Pd | $^{109}$Ag | 48.7 | 11.5 | $^{112}$Cd | 24.1 | 1.35 | $^{108}$Pd | 26.7 | 10 | 13.5 h | * |
| | $^{111}$Pd | | | | $^{114}$Cd | 28.9 | 0.13 | | | | | * |
| Ag | $^{108}$Ag | $^{108}$Cd | 0.87 | | | | | | | | | |
| | $^{110m}$Ag | $^{110}$Cd | 12.4 | | $^{113}$In | 4.2 | 2.7 | $^{109}$Ag | 48.7 | 3.2 | 270 d | |
| Cd | $^{111m}$Cd | | | | $^{114}$Sn | 0.65 | | | | | | |
| | $^{115,115m}$Cd | $^{115}$In | 95.8 | 15.5 | $^{118}$Sn | 24.0 | | | | | | * |
| | $^{117}$Cd | | | | $^{120}$Sn | 33.0 | | | | | | * |
| In | $^{114,114m}$In | $^{114}$Sn | 0.65 | | | | | | | | | |
| | $^{116m}$In | $^{116}$Sn | 14.2 | | | | | | | | | |
| Sn | $^{121}$Sn | $^{121}$Sb | 57.2 | | $^{124}$Te | 4.61 | | | | | | * |

| | | | | | | | | | | | | | |
|---|---|---|---|---|---|---|---|---|---|---|---|---|---|
| Sn | $^{123,123m}$Sn | $^{123}$Sb | 42.8 | | $^{126}$Te | 18.7 | | $^{122}$Sn | 4.71 | 0.16 | | | * |
| | $^{125,125m}$Sn | | | | $^{128}$Te | 31.8 | | | | | | | * |
| Sb | $^{122,122m}$Sb | $^{122}$Te | 2.46 | | | | | | | | | | |
| | $^{124}$Sb | $^{124}$Te | 4.61 | | | | | | | | | 40 m | |
| Te | $^{127,127m}$Te | $^{127}$I | 100 | 11.7 | $^{127}$I | 100 | 18.4 | | | | | | * |
| | $^{129,129m}$Te | | | | | | | | | | | | * |
| | $^{131,131m}$Te | | | | | | | | | | | | * |
| I | $^{128}$I | $^{128}$Xe | 1.92 | | | | | | | | | | |
| Xe | $^{133,133m}$Xe | $^{133}$Cs | 100 | | $^{130}$Xe | 4.08 | | | | | $^{135}\text{Xe}(n,\gamma)^{136}\text{Xe}$ | | * |
| | $^{135}$Xe | | | | $^{132}$Xe | 26.9 | | | | | | | * |
| | $^{137}$Xe | | | | $^{134}$Xe | 10.4 | | | | | | | * |
| Cs | $^{134,134m}$Cs | $^{134}$Ba | 2.42 | | | | | | | | | | |
| Ba | $^{133,133m}$Ba | $^{139}$La | 99.9 | 4 | $^{136}$Ba | 7.81 | | $^{138}$Ba | 71.7 | 0.5 | | | * |
| | $^{139}$Ba | | | | $^{138}$Ba | 71.7 | | | | | | | |
| La | $^{140}$La | $^{140}$Ce | 99.5 | 12 | | | | $^{139}$La | 99.9 | 8.2 | | 84 m | * |
| Ce | $^{141}$Ce | $^{141}$Pr | 100 | | $^{136}$Ce | 0.19 | | | | | | 40 h | * |
| | $^{143}$Ce | | | | $^{142}$Ce | 11.1 | | | | | | | * |
| Pr | $^{142}$Pr | $^{142}$Nd | 27.1 | 13.5 | | | 2.6 | | | | | | |
| Nd | $^{147}$Nd | | | | $^{144}$Nd | 23.9 | 8.9 | | | | | | * |
| | $^{149}$Nd | | | | $^{146}$Nd | 17.2 | | | | | | | * |
| | $^{151}$Nd | | | | | | | | | | | | |
| Sm | $^{153}$Sm | $^{153}$Eu | 52.2 | | $^{150}$Sm | 7.47 | 3.2 | | | | | | * |
| | $^{155}$Sm | | | | $^{152}$Sm | 26.6 | | | | | | | * |
| | | | | | $^{154}$Sm | 22.5 | | | | | | | |
| Eu | $^{152,152m}$Eu | $^{152}$Gd | 0.20 | | | | | | | | $^{152}\text{Eu}(n,\gamma)^{153}\text{Eu}$ | | |
| Gd | $^{159}$Gd | $^{159}$Tb | 100 | | $^{156}$Gd | 20.5 | 3.6 | | | | | | * |
| | $^{161}$Gd | | | | $^{158}$Gd | 24.9 | | | | | | | * |
| | | | | | $^{162}$Dy | 25.5 | | | | | | | |
| | | | | | $^{164}$Dy | 28.2 | | | | | | | |

*(Continued overleaf)*

**Table 8.3** (Continued)

| Element | (n,γ) Product | (n,p) Nuclide | (n,p) f(%) | (n,p) σ(mb) | (n,α) Nuclide | (n,α) f(%) | (n,α) σ(mb) | Second Order Nuclide | Second Order f(%) | σ(n,γ) | T½ | (n,f) |
|---|---|---|---|---|---|---|---|---|---|---|---|---|
| Tb | $^{160}$Tb | $^{160}$Dy | 2.30 | | | | | | | | | |
| Dy | $^{165,165m}$Dy | $^{165}$Ho | 100 | | | | | | | | | |
| Ho | $^{166}$Ho | $^{166}$Er | 33.4 | | $^{168}$Er | 27.1 | | $^{164}$Dy | 28.2 | 2100 | 139 m | |
| Er | $^{169}$Er | $^{169}$Tm | 100 | | $^{169}$Tm | 100 | | | | | | |
| | $^{171}$Er | | | | $^{172}$Yb | 21.8 | | | | | | |
| | | | | | $^{174}$Yb | 31.8 | | | | | | |
| Tm | $^{170}$Tm | $^{170}$Yb | 3.03 | | | | | $^{168}$Er | 27.1 | 2   $^{170}$Tm(n,γ)$^{171}$Tm | 7.4 d | |
| Yb | $^{175}$Yb | $^{175}$Lu | 97.4 | 3.4 | $^{178}$Hf | 27.1 | 2.0 | | | | | |
| | $^{177}$Yb | | | | $^{180}$Hf | 35.4 | | | | | | |
| Lu | $^{176m}$Lu | $^{176}$Hf | 5.15 | | $^{180}$Ta | 0.012 | | $^{176}$Yb | 12.7 | 5.5 | 1.9 h | |
| | $^{177}$Lu | $^{177}$Hf | 18.4 | | $^{182}$W | 26.4 | | | | | | |
| | | | | | $^{184}$W | 30.6 | | | | | | |
| Hf | $^{179m}$Hf | $^{181}$Ta | 100 | | $^{185}$Re | 37.1 | | $^{180}$Hf | 35.4 | 10   $^{182}$Ta(n,γ)$^{183}$Ta | 44.6 d | |
| | $^{181,181m}$Hf | | | | | | | | | | | |
| Ta | $^{182,182m}$Ta | $^{182}$W | 26.4 | | | | | | | $^{187}$W(n,γ)$^{188}$W | | |
| W | $^{185,185m}$W | $^{185}$Re | 37.1 | 3.9 | $^{188}$Os | 13.3 | | $^{186}$W | 28.4 | 34 | 24 h | |
| | $^{187}$W | $^{187}$Re | 62.9 | | $^{190}$Os | 26.4 | 0.57 | | | | | |
| Re | $^{186}$Re | $^{186}$Os | 1.59 | | | | | | | | | |
| Os | $^{188}$Os | $^{188}$Os | 13.3 | 2.7 | $^{191}$Ir | 38.5 | | | | $^{193}$Os(n,γ)$^{194}$Os | | |
| | $^{191}$Os | $^{191}$Ir | 38.5 | | $^{194}$Pt | 32.8 | 2.4 | $^{190}$Os | 26.4 | 8 | 16 d | |
| | $^{193}$Os | $^{193}$Ir | 61.5 | | $^{196}$Pt | 25.4 | | $^{192}$Os | 41.0 | 1.6 | 31 h | |
| Ir | $^{192,192m}$Ir | $^{192}$Pt | 0.78 | | $^{197}$Au | 100 | 0.43 | | | | | |
| | $^{194}$Ir | $^{194}$Pt | 32.8 | 3.9 | | | | | | | | |

| | Isotope | f | σ | Isotope | f | σ | Isotope | f | σ | T½ |
|---|---|---|---|---|---|---|---|---|---|---|
| Pt | 193mPt | | | 196Hg | 0.15 | | | | | |
| | 195mPt | | | 198Hg | 10.0 | | 193Ir | 61.5 | 130 | 19 h |
| | 197Pt | 100 | 2.42 | 200Hg | 23.1 | | 196Pt | 25.4 | 0.8 | 18 h |
| Au | 197Au | 100 | | | | | | | | |
| | 198Hg | 10.0 | | 198Au(n,γ)199Au | | | | | | |
| Hg | 203Hg | 29.5 | | 203Tl | | | | | | |
| | 205Hg | 70.5 | | 205Tl | 4.9 | | | | | |
| Tl | 204Tl | 1.48 | | 204Pb | | | | | | |
| | 206Tl | 23.6 | | 206Pb | | | 202Hg | 29.8 | 3.8 | 47 d |
| | | | | 206Pb | 23.6 | | | | | |
| | | | | 208Pb | 52.3 | 1.58 | | | | |
| Pb | 209Pb | 100 | 1.33 | 209Bi | 100 | 0.77 | | | | |
| Bi | 210mBi | | | Natural 210mBi | | | | | | |
| Th | 233Th | 100 | | 233Th(n,γ)234Th | | | | | | |
| U | 239U | ≤100 | 3.0 | 235U | ~0 | | | | | |
| | 239Pu | | | | | | | | | |
| Pu | 240Pu | | | 240Pu(n,γ)241Pu | | | | | | |

[a] Data from R. C. Koch, *Activation Analysis Handbook* (Academic, New York, 1960). f is the isotopic abundance, in percent, σ(mb) is the cross section, in millibarns, for 14 or 14.5 MeV neutrons, σ(n,γ) is the activation cross section, in barns, T½ is the half-life of the product nucleus that decays to the sought (n,γ) product, and (n,f) signifies possible production as a fission product from heavy element fission.

Table 8.4    Activation Analysis Affected by Second-Order Interference[a]

| Analysis | Measured Nuclide | | First (n,$\gamma$) Product | | Mode of Decay |
|---|---|---|---|---|---|
| P in Si | 14.3 d | $^{32}$P | 2.62 h | $^{31}$Si | $\beta-$ |
| Cl in S | 37.3 m | $^{38}$Cl | 5.1 m | $^{37}$S | $\beta-$ |
| Sc in Ca | 84 d | $^{46}$Sc | 165 d | $^{45}$Ca | $\beta-$ |
| Mn in Cr | 2.58 h | $^{56}$Mn | 3.5 m | $^{55}$Cr | $\beta-$ |
| Co in Fe | 5.27 y | $^{60}$Co | 45 d | $^{59}$Fe | $\beta-$ |
| Cu in Ni | 12.9 h | $^{64}$Cu | 92 y | $^{63}$Ni | $\beta-$ |
| Zn in Cu | 245 d | $^{65}$Zn | 12.9 h | $^{64}$Cu | 38 % $\beta-$ |
| Ga in Zn | 21 m | $^{70}$Ga | 55 m | $^{69}$Zn | $\beta-$ |
| Ga in Zn | 14.1 h | $^{72}$Ga | 2.5 m | $^{61}$Zn | $\beta-$ |
| As in Ge | 26.5 h | $^{76}$As | $^{75m}$Ge → 82 m $^{75}$Ge | | $\beta-$ |
| Ga in Ge | 14.1 h | $^{72}$Ga | $^{71m}$Ge → 11 d $^{71}$Ge | | EC |
| As in Se | 26.5 h | $^{76}$As | 120 d | $^{75}$Se | EC |
| Y in Sr | 64.2 h | $^{90}$Y | 50.4 d | $^{89}$Sr | $\beta-$ |
| Ag in Pd | 249 d | $^{110m}$Ag | 13.6 h | $^{109}$Pd | $\beta-$ |
| Cd in Ag | 49 m | $^{111m}$Cd | 24 s | $^{110}$Ag | $\beta-$ |
| Pd in Ag | 13.6 h | $^{109}$Pd | 2.4 m | $^{108}$Ag | $\beta+$, EC |
| Cd in In | 2.3 d | $^{115}$Cd | 72 s | $^{114}$In | $\beta+$, EC |
| Sb in Sn | 2.8 d | $^{122}$Sb | 25 h | $^{121}$Sn | $\beta-$ |
| Sb in Sn | 60 d | $^{124}$Sb | 40 m | $^{123}$Sn | $\beta-$ |
| Sb in Te | 2.8 d | $^{122}$Sb | 17 d | $^{121}$Te | EC |
| La in Ba | 40.2 h | $^{140}$La | 83 m | $^{139}$Ba | $\beta-$ |
| Ce in La | 32.5 d | $^{141}$Ce | 40.2 h | $^{140}$La | $\beta-$ |
| Eu in Sm | 16 y | $^{154}$Eu | 46.7 h | $^{153}$Sm | $\beta-$ |
| Gd in Eu | 200 d | $^{153}$Gd | 13 y | $^{152}$Eu | $\beta-$ |
| Sm in Eu | 46.7 h | $^{153}$Sm | 13 y | $^{152}$Eu | $\beta+$, EC |
| Tb in Gd | 73 d | $^{160}$Tb | 18 h | $^{159}$Gd | $\beta-$ |
| Eu in Gd | 16 y | $^{154}$Eu | 200 d | $^{153}$Gd | EC |
| Ho in Dy | 27.2 h | $^{166}$Ho | 2.3 h | $^{165}$Dy | $\beta-$ |
| Tb in Dy | 73 d | $^{160}$Tb | 144 d | $^{159}$Dy | EC |
| Tm in Yb | 127 d | $^{170}$Tm | 32 d | $^{169}$Yb | EC |
| Ta in Hf | 115 d | $^{182}$Ta | 43 d | $^{181}$Hf | $\beta-$ |
| Lu in Hf | 3.7 h | $^{176m}$Lu | 70 d | $^{175}$Hf | EC |
| Re in W | 90 h | $^{186}$Re | $^{185m}$W → 74 d $^{185}$W | | $\beta-$ |
| Re in W | 17 h | $^{188}$Re | 24 h | $^{187}$W | $\beta-$ |
| W in Re | 24 h | $^{187}$W | 90 h | $^{186}$Re | 3.83 %EC |
| Ir in Os | 74 d | $^{192}$Ir | 15 d | $^{191}$Os | $\beta-$ |
| Pt in Ir | 4.4 d | $^{193m}$Pt | 74 d | $^{192}$Ir | $\beta-$ |
| Pt in Ir | 4.1 d | $^{195m}$Pt | 19 h | $^{194}$Ir | $\beta-$ |
| Os in Ir | 32 h | $^{193}$Os | 74 d | $^{192}$Ir | EC |
| Ir in Pt | 74 d | $^{192}$Ir | 3.0 d | $^{191}$Pt | EC |
| Tl in Hg | 3.9 y | $^{204}$Tl | 47 d | $^{203}$Hg | $\beta-$ |
| Au in Hg | 64.8 h | $^{198}$Au | 65 h | $^{197}$Hg | EC |

[a] From E. Ricci and F. F. Dyer, "Second-Order Interference in Activation Analysis," Nucleonics 22, No. 6, 45–50 (1964).

where the subscripts 1, 2, 3, and 4 refer, respectively, to the nuclides $Z^A$, $Z^{A+1}$, $(Z \pm 1)^{A+1}$, $(Z \pm 1)^{A+2}$ in (1) and

$N_i$ = the number of atoms of the respective subscripts,

$\lambda_i$ = the decay constant,

$\sigma_i$ = the effective reactor neutron-capture cross section,

$\varphi_{th}$ = the thermal-neutron flux,

$F_2$ = the fraction of disintegrations of nuclide 2 which produces nuclide 3.

The solution of these equations gives the number of atoms of the interfering nuclide $N_4$ in the chain after an irradiation time $t$ as

$$N_4 = N_1^0 \phi_{th}^2 \sigma_1 \sigma_3 \lambda_2 F_2 \sum_{i=1}^{4} C_i e^{-\Lambda_i t} \qquad (8)$$

where $N_i^0$ = initial number of atoms of nuclide $Z^A$.

The constants $C_i$ are given by

$$C_i = \prod_{j=1}^{4} \frac{1}{\Lambda_j - \Lambda_i}, \qquad j \neq i \qquad (9)$$

and the parameter $\Lambda$ is considered as a modified decay constant and/or removal rate,

$$\Lambda = \lambda + \phi_{th}\sigma \qquad (10)$$

Actual values of the interference also depend on the neutron-energy spectrum in a reactor irradiation facility. Ricci and Dyer included both resonance and thermal-neutron reactions in the calculations. The effective neutron-capture cross sections were defined as

$$R_e = \phi_{th}\sigma_e \qquad (e = 1,3) \qquad (11)$$

$$\sigma_e = g\sigma_{0,e} + \frac{\phi_r}{\phi_{th}} (0.45\sigma_{0,e} + I_{0,e}) \qquad (12)$$

where $R$ = rate/atom for the neutron-capture reaction $e$,

$\sigma_{0,e}$ = 2200 m/sec cross section,

$I_{0,e}$ = resonance integral (from $\sim$0.5 eV) for the reaction,

$\phi_{th}$ = thermal neutron flux,

$\phi_r$ = resonance neutron flux,

$g$ = a factor that accounts for non-1/v-dependence of the thermal-neutron capture cross section (in most cases $g \simeq 1$).

The data from these calculations are given in eight graphs in their paper. An example of the analysis of antimony in tin illustrates the extent of such second-order interference. For the irradiation conditions $\phi_{th} = 2.0 \times 10^{14}$ n/cm²-sec, $\phi_{th}/\phi_r = 15$, and $t = 2.0 \times 10^5$ sec, analysis may be made for either or both 2.8-d $^{122}$Sb and 60-d $^{124}$Sb, but because of second-order inter-ference each nuclide yields a different result. For a 1-g sample which shows apparent results of 2.0 ppm for $^{122}$Sb and 1.4 ppm for $^{124}$Sb, interferences from graphs are determined as 1.0 $\mu$g for $^{122}$Sb and 0.38 $\mu$g for $^{124}$Sb. The correct concentration of antimony in the tin sample is 1.0 ppm.

### 8.2.2  Irradiation Conditions

Activation irradiations in nuclear reactors are generally made in the volumes of highest neutron flux; these are the volumes that have the flux distribution described in Section 3.1.4. The typical reactor neutron spectrum has thermal neutrons with a Maxwellian distribution of energies and reson-ance neutrons following a 1/v distribution down to an energy of about 0.4 eV. The insertion of a condensed material into the volume of the neutron flux creates a perturbation in the thermal and resonance neutron fluxes. Høgdahl* has described this *flux perturbation* effect as the total of two partial effects. In the vicinity of the neutron-absorbing material a flux gradient exists in which the neutron flux decreases in the direction toward the sample. This *flux depression* effect reflects the inability of the absorbed neutrons to scatter back into the moderator and maintain a constant neutron flux. The neutron flux also decreases with penetration into the sample as neutrons are absorbed from the flux. This is known as *self-absorption*. The over-all effect is also generally called *neutron self-shielding*.

The results of neutron self-shielding during a reactor irradiation of samples, comparators, and blanks can often lead to loss of identical irradiation conditions required for the comparator activation analysis method. In addition to the self-shielding effects of solids on thermal neutron and resonance-neutron fluxes, an enhancement of the thermal neutron flux, as described by Reynolds and Mullins,† is possible by moderation with aqueous solutions.

### Thermal-Neutron Self-Shielding

An analysis of neutron self-shielding has been made by Zweifel,‡ who defines the self-shielding factor $f$ as

$$f = \frac{\bar{\phi}_s}{\phi_\infty} \tag{13}$$

---

* O. T. Høgdahl, Neutron Absorption in Pile Neutron Activation Analysis, University of Michigan Phoenix Project Report No. MMPP-226-1, 1962.
† S. A. Reynolds and W. T. Mullins, Neutron Flux Perturbation in Activation Analysis, *Inter. J. Appl. Radiation Isotopes* **14**, 421–425 (1963).
‡ P. F. Zweifel, Neutron Self-Shielding, *Nucleonics* **18**, No. 11, 174–175 (1960).

where $\bar{\phi}_s$ = average flux inside the sample,

$\phi_\infty$ = constant flux which would be there if the sample were absent.

The rate of neutron capture in small, absorbing samples is treated by considering the neutron balance outside the sample and determining the absorption rate from the net neutron flow into the sample. For a uniform isotropic neutron source in an infinite medium surrounding the absorber, an effective absorption cross section, $\Sigma_{a,\text{eff}}$, is defined as

$$\Sigma_{a,\text{eff}} = f\Sigma_a \tag{14}$$

such that the absorption rate is given by

$$R_a = \phi_\infty \Sigma_{a,\text{eff}} \tag{15}$$

For solid materials the self-shielding factor $f$ may be obtained graphically from curves given by Zweifel as a function of a shape parameter, $\tau$, where

$$f = \frac{1}{2\tau} \qquad \text{for} \qquad \tau \geq 2 \tag{16}$$

The shape parameter $\tau$ is defined as

$$\tau = 2\frac{V}{S}n\bar{\sigma} \tag{17}$$

where $V$ = volume of sample,

$S$ = surface area of the sample,

$n\bar{\sigma}$ = average macroscopic thermal-neutron cross section of the element.

For small values of $\tau(\tau < 2)$ the sample shape is significant, and approximations of $f$ for slabs, spheres, and cylinders given as

$$f_{\text{slab}} \simeq 1 - \frac{\tau}{2}\left(0.9228 + \ln\frac{1}{\tau}\right) \qquad \tau = n\bar{\sigma}t \tag{18}$$

$$f_{\text{sphere}} \simeq 1 - \tfrac{9}{8}\tau \qquad \tau = \tfrac{2}{3}n\bar{\sigma}r \tag{19}$$

$$f_{\text{cyl}} \simeq 1 - \tfrac{4}{3}\tau \qquad \tau = n\bar{\sigma}r \tag{20}$$

where $n$ = atom density (atoms/cm³),

$\bar{\sigma}$ = average thermal neutron cross section (cm²),

$t$ = thickness (cm),

$r$ = radius (cm).

In practice (18) can be used for foil samples, (19), for irregular-shaped samples, and (20), for wires. If the value of $\tau$ for a given sample is greater than

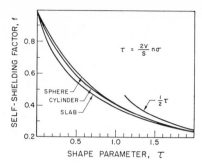

**Figure 8.2** Neutron self-shielding factors for infinite slabs, cylinders, and spheres. [From P. F. Zweifel, Neutron Self-Shielding, *Nucleonics* **18**, No. 11, 174–175 (1960).]

about 0.05, the self-shielding factor can be read from the appropriate curve in Figure 8.2.

For the more typical case of a finite cylinder, that is, a bottle or capsule, as commonly used in activation analysis irradiations, Gilat and Gurfinkel* suggested an interpolation between the shapes of the infinite cylinder and the infinite slab. For a cylinder of radius $r$ and length $h$ the self-shielding factor is

$$f(\text{finite cyl}) = \frac{r \times f(\text{slab}) + h \times f(\text{infinite cyl})}{r + h} \tag{21}$$

### Resonance-Neutron Self-Shielding

Attempts to calculate the extent of resonance-neutron self-shielding have not been so successful as those for thermal neutrons. To calculate the epithermal-neutron self-shielding factor $f_r$ Høgdahl showed that both the infinitely dilute resonance integral $I$ and the effective resonance integral $I^{\text{eff}}$ must be known for the nuclide of interest. He defines them as

$$I = \sum I_i + 0.44\sigma_0 \tag{22}$$

$$I^{\text{eff}} = \sum I_i^{\text{eff}} + f''0.44\sigma_0 \tag{23}$$

where the summations are over the respective dilute and effective resonances,

$0.44\sigma_0 = $ the cross section for the $1/v$ tail of the spectrum,

$f'' = $ the absorption factor for neutrons with energies outside the resonance peaks.

* J. Gilat and Y. Gurfinkel, Self-Shielding in Activation Analysis, *Nucleonics* **21**, No. 8, 143–144 (1963).

A review of the calculation of the dilute resonance integral has been given by Macklin and Pomerance.* A tabulation of their values for several nuclides is listed in Table 8.5. The effective resonance integrals for major-constituent nuclides can be calculated with reasonable accuracy, for example, as given by Chernick and Vernon.† In neutron activation analysis, however, the effective resonance integrals are often desired for minor-constituent nuclides and the needed $I^{\text{eff}}$ will be a function of the chemical composition of the sample. In addition to this problem, it is also important to know which of two approximations is necessary: the "narrow resonance" case in which scattering collisions within the absorber reduce the neutron energy below the resonance energy or the "narrow resonance, infinitely heavy absorber" case in which the collisions do not change the neutron energy. Further refinements can include the Doppler effect of the absorber atoms which broaden the resonances and increase the absorption.

Høgdahl concludes that although the resonance self-shielding factor cannot be calculated rigorously the effect can be made insignificant by irradiating in a reactor position where the cadmium ratio CR is $\geq 50$. As a test to decide whether a calculation of $f_r$ is necessary, he derives the condition that

$$f - f_{\text{th}} \leq \frac{2}{\text{CR}} \tag{24}$$

where $f_{\text{th}}$ = the self-shielding factor for thermal neutrons,

$f$ = the total self-shielding factor.

Because of these complexities in estimating the extent of self-shielding of resonance neutrons for purposes of minor-constituent activation analysis, Reynolds and Mullins described an empirical method that compares the absorption with the known self-shielding factors in cobalt. They define an "effective thickness," $t_{\text{eff}}$, for any absorber of resonance-energy neutrons, such that

$$t_{\text{eff}} = \frac{t I_x n_x}{I_{\text{Co}} n_{\text{Co}}} \tag{25}$$

where $I_x$ = resonance integral for elements $x$ (b),

$t$ = thickness of foil, radius of wire, or $\frac{2}{3}$ radius of sphere (cm),

$I_{\text{Co}}$ = resonance integral of cobalt, (75 b),

$n_x$ = atom density of element $x$ (atoms/cm$^3$),

$n_{\text{Co}}$ = atom density of cobalt.

* R. L. Macklin and H. S. Pomerance, Resonance Capture Integrals, *Proc. U.N. Intern. Conf. Peaceful Uses of Atomic Energy*, **5**, 96–101, 1956.
† J. Chernick and R. Vernon, Some Refinements in the Calculation of Resonance Integrals, *Nucl. Sci. Engr.* **4**, 649–672 (1958).

**Table 8.5** Resonance Capture Integrals[a]
(barns)

| Nuclide or Natural Element | Calculated | | Measured | |
|---|---|---|---|---|
| | Resonance | $1/v$ | Activation | Absorption |
| $_3$Li | | 31 | | 28 |
| $_{11}^{23}$Na | $0.12^{th}$ | 0.25 | ~0.24 | 0.27 |
| $_{12}$Mg | | 0.027 | | 0.9 |
| $_{13}^{27}$Al | | 0.095 | ~0.16 | 0.18 |
| $_{14}$Si | | 0.06 | | 0.5 |
| $_{15}^{31}$P | | 0.084 | ~0.092 | <2 |
| $_{16}$S | | 0.22 | | 0.6 |
| $_{17}$Cl | | 14.3 | | 12 |
| $_{19}$K | | 0.89 | | 1.1 |
| $_{20}$Ca | | 2.0 | | 2 |
| $_{21}^{45}$Sc | | 10.6 | ~10.7 | |
| $_{22}$Ti | | 2.5 | | 3.0 |
| $_{23}^{51}$V | | 2.2 | ~2.2 | |
| $_{24}$Cr | | 1.3 | | 1.9 |
| $_{25}^{55}$Mn | $12^{th}$ | 5.9 | ~11.8 | 10.8 |
| $_{26}$Fe | | 1.1 | | 2.1 |
| $_{27}^{59}$Co | $90^{th}$ | 16.3 | 49.3 | 48 |
| $_{28}$Ni | | 2.0 | | 4 |
| $_{29}$Cu | | 1.6 | 3.7 | 4 |
| $_{29}^{63}$Cu | $0.8^s$ | 1.9 | 4.4 | |
| $_{29}^{65}$Cu | $0.8^s$ | 0.92 | 2.2 | |
| $_{30}$Zn | | 0.48 | | 2 |
| $_{31}^{69}$Ga | $20^s$ | 0.62 | 9.2 | |
| $_{31}^{71}$Ga | | 1.58 | 15 | |
| $_{33}^{75}$As | $170^{th}$ | 1.8 | 36.8 | 33 |
| $_{35}^{79}$Br | $62^s$ | 4.8 | 147 | |
| $_{38}$Sr | | 0.53 | | 16 |
| $_{39}^{89}$Y | | 0.62 | 0.91 | |
| $_{40}$Zr | | | | 3 |
| $_{41}^{93}$Nb | | 0.47 | 3.87 | 8.3 |
| $_{42}$Mo | | 1.1 | | 13 |
| $_{45}^{103}$Rh | $1080^{th}$ | 66 | 656 | 575 |
| $_{46}$Pd | | 3.6 | | 23 |
| $_{47}$Ag | | 27 | | >650 |
| $_{47}^{107}$Ag | $82^s$ | 13 | 74 | |
| $_{47}^{109}$Ag | $1206^{th}$ | 37 | 1160 | |
| $_{49}^{113}$In | | 26 | 1050 | |
| $_{49}^{115}$In[b] | $1752^{th}$ | 87 | 2640 | |
| $_{50}$Sn | | 0.26 | | 4.3 |
| $_{51}^{121}$Sb | | 3.0 | 162 | |
| $_{51}^{123}$Sb | | 1.1 | ~138 | |

**Table 8.5** (Continued)

| Nuclide or Natural Element | Calculated | | Measured | |
|---|---|---|---|---|
| | Resonance | $1/v$ | Activation | Absorption |
| $_{52}$Te | | 2.0 | | 36 |
| $_{53}^{127}$I | 150$^s$ | 2.9 | 140 | >90 |
| $_{56}$Ba | | 0.53 | | 7.5 |
| $_{57}$La | | 4.0 | | 11 |
| $_{59}^{141}$Pr | | 4.9 | 11.3 | |
| $_{62}^{152}$Sm$^b$ | | 61 | >1750 | |
| $_{63}^{153}$Eu$^b$ | | 1400$^c$ | 950 | |
| $_{72}$Hf | 2778$^{th}$ | 51 | ~1750$^d$ | 1300 |
| $_{72}^{180}$Hf | | 4.4 | 21.8 | |
| $_{73}^{181}$Ta | 555$^s$ | 9.4 | 590 | |
| $_{74}^{186}$W | 180$^{th}$ | 15 | 355 | |
| $_{75}^{185}$Re | 1630$^{th}$ | 44 | 1160 | |
| $_{75}^{187}$Re | 310$^s$ | 33 | 305 | |
| $_{77}^{191}$Ir$^b$ | 7680$^{th}$ | 420 | 3500 | |
| $_{77}^{193}$Ir$^b$ | 976$^{th}$ | 57 | 1370 | |
| $_{78}$Pt | | 3.7 | | 69 |
| $_{79}^{197}$Au | 1390$^{th}$ | 45 | 1558$^e$ | |
| $_{80}$Hg | | $c$ | | 31 |
| $_{81}^{203}$Tl | | 4.8 | 129 | |
| $_{81}^{205}$Tl | | 0.04 | 0.5 | |
| $_{82}$Pb | | 0.08 | | 0.1 |
| $_{83}$Bi | | 0.014 | | 0.5 |
| $_{90}^{232}$Th | 83$^s$ | 3.2 | 69.8 | |
| $_{92}^{233}$U$^f$ | 650$^g$ | 234 | | |
| $_{92}^{235}$U$^f$ | >260$^s$ | 255$^c$ | 271 | |
| $_{92}^{238}$U | 290$^s$ | 1.2 | 282 | |
| $_{94}^{239}$Pu$^f$ | 250$^g$ | $c$ | | |

[a] From R. L. Macklin and H. S. Pomerance, "Resonance Capture Integrals," *Proc., U.N. Intl. Conf. Peaceful Uses of Atomic Energy.* **5**, 96–101 (1956).

[b] A near thermal resonance leads to considerable dependence on the details of the cadmium absorber that was used.

[c] The capture cross section is not $1/v$ near thermal energies.

[d] Capture gamma radiation used to determine the cross section; some self-protection in the sample.

[e] Adopted as the standard, using $\sigma_{th} = 98 \times 10^{-24}$ cm$^2$.

[f] Fission integral.

[g] Graphical estimate. For $^{239}$Pu a cutoff at 2.0 eV was used and no separate $1/v$ contribution.

[s] Estimated from average level parameters.

[th] Estimated from parameters of the first large resonance and the thermal cross section.

By fitting empirical data on self-shielding of epicadium neutrons (i.e., $E_n > \sim 0.4$ eV) in cobalt from data of Eastwood and Werner,[*] they obtain a "resonance-neutron self-shielding" factor

$$f_r = -0.29 \log t_{\text{eff}} = -0.29 \log \frac{tI_x n_x}{I_{Co} n_{Co}} \tag{26}$$

Resonance integrals for many elements have been tabulated.[†] Reynolds and Mullins note that (26) is inaccurate for values of $f_r$ near unity. They obtain better results by taking the effective cross section for the most important resonance as $\frac{1}{10}$ the peak value and using (18), (19), or (20).

For irradiation samples in reactors in which both thermal-neutron and resonance-neutron absorptions are important a total self-shielding factor $f'$ can be determined as

$$f' = \frac{fR_{\text{th}}}{R_t} + \frac{f_r R_r}{R_t} \tag{27}$$

where $R_{\text{th}}$ = thermal neutron reaction rate = $\phi_{\text{th}} \sigma$,

$R_r$ = resonance neutron reaction rate = $\phi_r I$,

$R_t$ = total reaction rate = $R_{\text{th}} + R_r$.

### Elimination of Absorption Effects

As an alternate to calculating the effects of neutron absorption in reactor-neutron activations, the irradiations can be designed to eliminate or reduce the problems of neutron absorption. Høgdahl summarizes several methods. Many are available; the choice depends on such parameters as chemical composition, physical form, irradiation facility available, and degree of accuracy desired. The limitations of the following methods due to these parameters are reviewed by Høgdahl:

1. The method of dilution: the samples and standards are diluted with materials that have low absorption cross sections for reactor neutrons. For irradiation positions in which the cadmium ratio is $<50$ both thermal- and resonance-neutron absorptions must be considered.

2. The internal standard, method I: a standard is prepared such that the effects of neutron absorption in the standards and samples are approximately equal. This can be done by adding a known amount of the desired

---

[*] T. A. Eastwood and R. D. Werner, Resonance and Thermal Neutron Self-Shielding in Cobalt Foils and Wires, *Nucl. Sci. Engr.* **13**, 385–390 (1962).

[†] For example, A. E. McCarthy et al., ANL Reactor Physics Constants Center Newsletter No. 1, 1961.

element to a matrix that duplicates the effects of chemical composition, weight, and geometric form. The method is generally useful only if the element can be mixed homogeneously with the sample.

3. The internal standard, method II: an improvement to method I has been described by Leliaert, Hoste, and Eeckhaut,[*] in which a second element is added to the two samples as an internal standard. The method is useful if the cadmium ratio of the desired element is large ($>50$) or if it has no large resonance peaks. The method is also useful even when the two samples cannot be irradiated simultaneously, since the internal standard normalizes the two irradiations as flux monitors.

4. The method of extrapolation: the determined weight of a desired element in samples and in diluted solutions is plotted as a function of sample weight. An extrapolation to zero weight gives the determined weight corrected for absorption effects. Although this method requires considerable caution and a homogeneous neutron flux over the volume of samples and standards, the extrapolation process smoothes out small variations in the flux.

## Enhancement of Thermal Neutron Activation

Many neutron activations are carried out in solution form for convenience, rapid postirradiation chemistry, or to avoid nonidentical irradiation conditions of samples and comparitors. Reynolds and Mullins have examined the effect on the neutron flux by moderation of epithermal neutrons within

**Table 8.6**  Enhancement of Thermal Activation[a]

| Material | Container Volume (ml) | Ratio, with and without Water |
|---|---|---|
| Mn (in Al)[b] | 1.5 | 1.064 |
| Au (in Al)[b] | 1.5 | 1.071 |
| Cu | 1.5 | 1.087 |
| Ag | 1.5 | 1.063 |
| Ni | 1.5 | 1.062 |
| Mn (in Fe)[b] | 30 | 1.21 |
| Ni | 30 | 1.20 |

[a] From S. A. Reynolds and W. T. Mullins, Neutron Flux Perturbation in Activation Analysis. *Intern. J. Appl. Radiation Isotopes* **14**, 421–425 (1963).
[b] Dilute alloy.

[*] G. Leliaert, J. Hoste, and Z. Eeckhaut, Activation Analysis of Vanadium in High Alloy Steels Using Manganese as Internal Standard, *Anal. Chem. Acta.* **19**, 100–107 (1958).

aqueous samples. They evaluated the effects of thermal neutron "enhancement" by irradiating wires of various materials with and without water in two commonly used types of container, a 1.5-ml vial and a 30-ml bottle. Their data for the enhancement of the thermal neutron activation are given in Table 8.6. From their calculations, which involved geometrical approximations, they concluded that the thermal neutron enhancement for elements in solution was about 5% for the 1.5-ml volume and about 12% for the 30-ml volume.

Table 8.7  Comparison of Neutron Flux Perturbation Factors for Solid and Solution Samples[a]

| Material | Weight (mg) or $t$ (cm)[b] | Solution Volume (ml) | Active Nuclide | $f_{th}$ | $f_r$ | $f'$ Calc. | $f'$ Obs. |
|---|---|---|---|---|---|---|---|
| $Na_2CO_3$ | 50 | 1.5 | $^{24}Na$ | 1.00 | 1.00 | 0.95 | 0.96 |
| $Sc_2O_3$ | 10 | 1.5 | $^{46}Sc$ | 0.95 | 0.73 | 0.90 | 0.85 |
| Cr | 40 | 1.5 | $^{51}Cr$ | 0.98 | 1.00 | 0.93 | 0.86 |
| Mn | 75 | 30 | $^{56}Mn$ | 0.93 | 0.58 | 0.82 | 0.82 |
| Fe wire | 0.022 | 1.5 | $^{59}Fe$ | 1.00 | 1.00 | 0.95 | 0.93 |
| Co wire | 0.051 | 30 | $^{60}Co$ | 0.91 | 0.47 | 0.79 | 0.80 |
| Co–Al[c] | 115 | 30 | $^{60}Co$ | 1.00 | 0.89 | 0.89 | 0.93 |
| Ni wire | 0.081 | 30 | $^{65}Ni$ | 0.98 | 1.00 | 0.87 | 0.86 |
| Cu | 25 | 1.5 | $^{64}Cu$ | 0.98 | 0.71 | 0.92 | 0.90 |
| Zn | 70 | 1.5 | $^{65}Zn$ | 0.99 | 1.00 | 0.94 | 0.98 |
| $As_2O_5$ | 35 | 30 | $^{76}As$ | 0.99 | 0.73 | 0.83 | 0.85 |
| Se | 50 | 1.5 | $^{75}Sc$ | 0.98 | 0.82 | 0.86 | 0.85 |
| $NH_4Br$ | 175 | 1.5 | $^{82}Br$ | 0.99 | 0.62 | 0.79 | 0.83 |
| Zr foil | 0.013 | 1.5 | $^{95}Zr$ | 1.00 | 1.00 | 0.95 | 0.97 |
| Ag foil | 0.013 | 1.5 | $^{110}Ag$ | 0.91 | 0.33 | 0.66 | 0.57 |
| In foil | 0.013 | 1.5 | $^{114}In$ | 0.86 | 0.50 | 0.77 | 0.81 |
| In foil | 0.013 | 30 | $^{115m}In$ | 0.86 | 0.30 | 0.50 | 0.53 |
| $(NH_4)_2Ce(NO_3)_6$ | 75 | 1.5 | $^{143}Ce$ | 0.99 | 1.00 | 0.94 | 0.94 |
| Ta foil | 0.004 | 1.5 | $^{182}Ta$ | 0.99 | 0.50 | 0.69 | 0.69 |
| W foil | 0.013 | 1.5 | $^{187}W$ | 0.94 | 0.53 | 0.77 | 0.78 |
| Re foil | 0.015 | 1.5 | $^{188}Re$ | 0.88 | 0.45 | 0.78 | 0.79 |
| $Na_2IrCl_6$ | 40 | 1.5 | $^{194}Ir$ | 0.90 | 0.53 | 0.78 | 0.87 |
| Au foil | 0.003 | 1.5 | $^{198}Au$ | 0.96 | 0.43 | 0.74 | 0.73 |

[a] From S. A. Reynolds and W. T. Mullins, "Neutron Flux Perturbation in Activation Analysis," *Intern. J. Appl. Radiation Isotopes* **14**, 421–425 (1963).
[b] Weight of powder or pieces, or thickness (diameter) of foil or wire.
[c] Dilute alloy.

The over-all effects of neutron flux perturbation for samples in solution were determined for 21 elements in 23 materials. To compare induced activities in solution with those in solids, they set $f'$ in (27) equal initially to 0.95 for the 1.5-ml volume and 0.85 for the 30-ml volume to take into account the enhancement. The total $f'$ was then computed from (27). Their data, given in Table 8.7, indicate both the magnitude of the errors that can arise in activation analysis by neglecting self-shielding and thermal-neutron enhancement and the extent with which these potential errors can be avoided by the computations established for cases in which experimental data are unavailable.

### 8.2.3 Variable-Energy Activation Analysis

The technique of variable-energy activation analysis could logically have been included in the description of activation analysis practices in Chapter 7. Its major contribution, however, is its ability to increase sensitivity of activation analysis through reduction of interferences. It is therefore included here as a means of affecting the nuclear limitations of activation analysis.

Variable-energy activation analysis is the technique for activation of samples with irradiating particles of selected energies so that variations in reaction threshold energy, coulomb barriers, and excitation functions can be used to increase the selectivity of activation. An increase in selectivity implies an improvement in the ratio of the activity produced from the desired element to those produced by interfering elements in the matrix.

Although the method is generally applicable to any type of irradiating particle, its use in neutron activation analysis has been extensively developed. Steele,* for example, reviewed the use of neutrons produced by several reactions in a variable-energy cyclotron and a neutron-generator accelerator. Interfering reactions have been noted to produce the same radionuclide from elements other than the sought element. The example has already been given in which the nuclear reactions

$$^{31}P(n,\gamma)^{32}P$$

$$^{32}S(n,p)^{32}P$$

$$^{35}Cl(n,\alpha)^{32}P$$

all lead to the same radioelement. Steele gives as a practical example of the value of variable-energy neutron activations the determination of chromium in the presence of manganese. Irradiation of these two elements with 14-MeV

---

* E. L. Steele, Variable-Energy Neutron Activation Analysis, in *Modern Trends in Activation Analysis* (Texas A&M University, College Station, 1965), pp. 102–106.

neutrons from a (d,t) neutron generator result in the production of 3.76-m $^{52}$V by the reactions

$$^{52}Cr(n,p)^{52}V$$

$$^{55}Mn(n,\alpha)^{52}V$$

In a manner similar to the method for determining phosphorous in a matrix containing large amounts of sulfur and chlorine, manganese and chromium can be determined when together in a sample by first making a separate determination of manganese from the $^{55}Mn(n,\gamma)^{56}Mn$ reaction with thermal neutrons and subtracting the equivalent $^{52}V$ contribution from the total $^{52}V$ formed during the simultaneous 14-MeV neutron irradiation. The thermal neutrons require a thermalizing tank (e.g., for water or paraffin) around the sample irradiation position. The respective contributions to the $^{52}V$ photopeak from $^{52}Cr$ and $^{55}Mn$ irradiation with 14-MeV neutrons are shown in Figure 8.3*a*. The manganese measurement must also be examined for possible interference from the $^{56}Fe(n,p)^{56}Mn$ if iron is present in the sample. Although the method is tedious, the technique does not provide a useful determination of chromium when the manganese concentration is much larger.

The variable-energy method makes use of the nuclear data that although the threshold energy for the $^{55}Mn(n,\alpha)^{52}V$ reaction is only 0.6 MeV, compared with 3.1 MeV for the $^{52}Cr(n,p)^{52}V$ reaction, the coulomb barrier for the former reaction is 8.6 MeV compared with 4.3 MeV for the latter. Since integrated cross sections for charged-particle emission nuclear reactions are generally small up to the irradiating energy equal to the coulomb barrier, an enhancement of the $^{52}V$ photopeak from the $^{52}Cr$ reaction relative to the $^{55}Mn$ reaction would be expected. Figure 8.3*b* shows the gamma-ray spectrum of the sample irradiated with a beam of 7-MeV deuterons on a $D_2O$ target producing fast neutrons of 10-MeV maximum energy. The enhancement of the fraction of the $^{52}V$ photopeak due to the $^{52}Cr(n,p)^{52}V$ reactions is clearly seen. Thus chromium can be determined in a sample containing significant amounts of manganese by activation with neutrons of maximum energy less than about 10 MeV.

Other examples cited by Steele include the following:

1. Fluorine in the presence of oxygen for F/O > 10:

$$^{19}F(n,\alpha)^{16}N, \qquad E_{th} = 1.6 \text{ MeV},$$

$$^{16}O(n,p)^{16}N, \qquad E_{th} = 10.2 \text{ MeV}.$$

2. Iron in the presence of cobalt:

$$^{56}Fe(n,p)^{56}Mn, \qquad E_{th} = 2.9 \text{ MeV}, \qquad V_c = 4.65 \text{ MeV},$$

$$^{59}Co(n,\alpha)^{56}Mn, \qquad E_{th} = 0.4 \text{ MeV}, \qquad V_c = 8.35 \text{ MeV}.$$

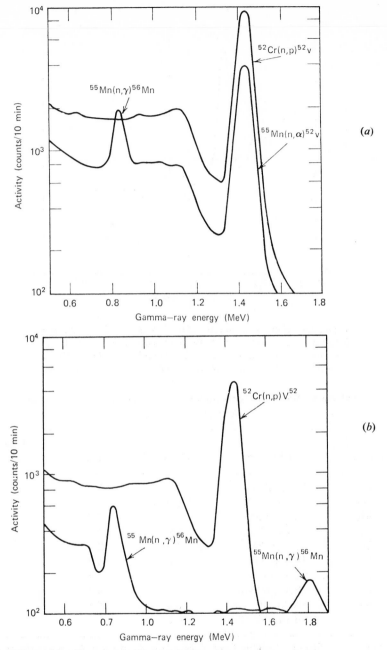

**Figure 8.3** Gamma-ray spectra of chromium-manganese samples irradiated with (*a*) 14-Mev neutrons, (*b*) 10-Mev(max) neutrons. [From E. L. Steel, Variable-Energy Neutron Activation Analysis, in *Modern Trends in Activation Analysis* (Texas A&M University, College Station, 1965).]

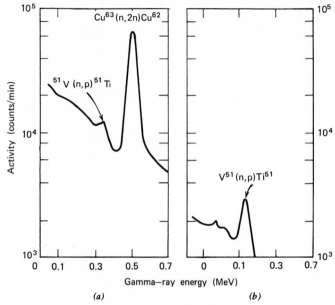

**Figure 8.4** Gamma-ray spectra of 0.8 % V–99.2 % Cu irradiated with (a) 14-MeV neutrons, (b) 10-MeV(max) neutrons. [From E. L. Steele, Variable-Energy Neutron Activation Analysis, in *Modern Trends in Activation Analysis* (Texas A&M University, College Station, 1965).]

**Table 8.8**   Some Elements Suitable for Variable-Energy Neutron Activation Analysis[a]

| Element | Reaction | $E_{th}$ (MeV)[a] | $V_c$ (MeV)[b] | $\sigma$ (14 MeV) (mb)[c] |
|---------|----------|---------|---------|---------|
| F  | $^{19}F(n,p)^{19}O$       | 4.0  | 2.0 | 135 |
| Na | $^{23}Na(n,p)^{23}Ne$     | 3.6  | 2.3 | 34  |
| Mg | $^{24}Mg(n,p)^{24}Na$     | 4.7  | 2.5 | 190 |
| Al | $^{27}Al(n,p)^{27}Mg$     | 1.8  | 2.7 | 79  |
| Si | $^{28}Si(n,p)^{28}Al$     | 3.9  | 2.9 | 220 |
| V  | $^{51}V(n,p)^{51}Ti$      | 1.6  | 4.2 | 27  |
| Cr | $^{52}Cr(n,p)^{52}V$      | 3.1  | 4.4 | 78  |
| Fe | $^{56}Fe(n,p)^{56}Mn$     | 2.9  | 4.7 | 110 |
| As | $^{75}As(n,p)^{75\,m}Ge$  | 0.4  | 5.5 | 12  |
| Br | $^{79}Br(n,n')^{79\,m}Br$ | 0.62 | 0   | ... |

[a] From E. L. Steele, Variable-Energy Neutron Activation Analysis, in *Modern Trends in Activation Analysis* (Texas A&M University, College Station, 1965).
[b] Calculated from $V_c = 0.9Z/(1 + A^{1/3})$ (for the product nuclide).
[c] From R. C. Koch, *Activation Analysis Handbook* (Academic, New York, 1960).

3. Phosphorous and vanadium in the presence of copper: Reduction of the $^{62}Cu$ background from $^{63}Cu(n,2n)^{62}Cu$ reactions so that the photopeak of 2.27-m $^{28}Al$ from the $^{31}P(n,\alpha)^{27}Al$ reaction and of 5.8-m $^{51}Ti$ from the $^{51}V(n,p)^{51}Ti$ reaction could be measured with sufficient precision. A comparison of the gamma-ray spectra obtained following irradiation of samples with 14-MeV neutrons from a (d,t) generator and 10-MeV neutrons from a 7-MeV (d,d) generator are shown in Figure 8.4. The elimination of the $^{62}Cu$ annihilation gamma-ray photopeak is notable. Some elements for which variable-energy neutron irradiations are useful are listed in Table 8.8. Steele concluded that for neutron production in the 4-to-12-MeV range, deuterons between 1 and 10 MeV would be the most useful accelerator particles, even though there is a tendency for thermal-neutron products to appear. If thermal-neutron interferences are present, they may be eliminated by fast-neutron production with $^3He$ accelerator particles.

## 8.3 GAMMA-RAY SPECTROSCOPY LIMITATIONS

The potential for gamma-ray spectroscopy in developing radioactivation analysis into an instrumental method of chemical analysis warrants the extensive efforts to decrease the limitations of the resolution of complex gamma-ray spectra. The two major limitations that affect the sensitivity with which desired elements can be identified and measured by the stripping and computer-resolving methods described in Section 7.2 are the extent of the Compton-scattering "background" and the stability of the gamma-ray spectrometer electronic equipment.

Thus two parallel efforts are underway to decrease these limitations. Several means are employed to reduce or suppress the Compton-scattering components of a gamma-ray spectrum, thereby enhancing the relative magnitude of the full-energy gamma-ray peaks. These include compensation techniques and coincidence and anticoincidence circuits. The primary limitation of electronic gamma-ray spectrometers is the drift of the gain and threshold settings in the spectrometer. These drifts are especially troublesome in computer resolution of complex gamma-ray spectra.

The quality control of gamma-ray spectrometers is an important aspect of instrumental activation analysis. Criteria and methods for quality control of spectrometers have been reported. Methods to control drift in spectrometers are part of several computer programs and units can be built into the spectrometer to limit drift.

### 8.3.1 Compton Suppression

With the development of gamma-ray spectroscopy as a means of identification and measurement of gamma-ray-emitting radionuclides came attempts to increase the resolution and sensitivity of photopeak analysis. The ratio of full-energy gamma-ray peaks to the Compton-scattering continuum was steadily improved as larger scintillation crystals were manufactured. However, with increasing size also came increasing background and decreasing resolution.

Several methods have been devised to suppress the contribution of Compton scattering to $\gamma$-ray spectra. An early method was the use of two crystals in coincidence;* a schematic diagram of the method is shown in Figure 8.5. The pulse in detector $D$ at a fixed angle $\theta$ from the incident photon into detector $X$ is used for a coincident pulse-height count of the recoil-electron energy in the scintillator. Since the energy of backscattered photons is about $\frac{1}{2}$ mc² for a large range of values of $h\nu \geq$ mc², the dependence of pulse height on $\theta$ is small for angles between 135 and 180°, allowing the use of a large solid angle for detector $D$. A ring counter, shown by $DD'$, may be used to increase the detection efficiency. Although the resolution was good, the low efficiencies due to the small solid angle made this method unsuited to activation analysis.

Another method was the use of an anticoincidence system in which pulses not equal to the full energy of the incident photons were discarded. This is accomplished by surrounding the primary detector with a cluster of scintillation crystals† or a tank of liquid scintillator‡ with high efficiency for

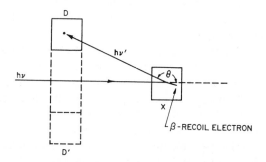

**Figure 8.5** Schematic for a coincidence method to obtain Compton-suppressed gamma-ray spectra. [From R. Hofstadter, and J. A. McIntyre, Measurement of Gamma-Ray Energies with Two Crystals in Coincidence, *Phys. Rev.* **78**, 619–620 (1950).]

* R. Hofstadter and J. A. McIntyre, Measurement of Gamma-Ray Energies with Two Crystals in Coincidence, *Phys. Rev.* **78**, 619–620 (1950).

† R. D. Albert, An Anticoincidence Gamma-Ray Scintillation Spectrometer, *Rev. Sci. Instr.* **24**, 1096–1101 (1953).

‡ P. R. Bell, Scintillation Spectrometer with Improved Response, *Science* **120**, 625–626 (1954).

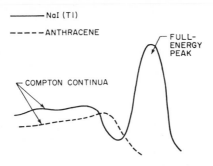

**Figure 8.6** Comparison of responses to monoenergetic gamma rays by NaI(Tl) and anthracene crystals.

interactions with the scattered radiations. The anticoincidence circuit blocks pulses from the primary detector if a simultaneous pulse is detected in the shield detectors. Thus only those pulses that correspond to full-energy absorption in the primary detector reach the pulse-height analyzer. These systems also require extensive collimation of the $\gamma$-ray source, with corresponding losses in counting efficiencies, thereby making them unsuited for activation analysis of low-activity samples.

A third method for a two-detector gamma-ray spectrometer to improve the photopeak sensitivity is called Compton subtraction.* The second crystal in this method is an anthracene crystal, which responds primarily to the Compton process. The relative gamma-ray spectrum for $E_\gamma < 2$ MeV in sodium iodide and anthracene is sketched in Figure 8.6. In the Compton-subtraction spectrometer the Compton continua from the two crystals are equalized in counting rate and in energy scales such that subtraction of the pulses in a count-rate circuit records the full-energy peaks from the NaI scintillator without the corresponding Compton continuum. Because of the better geometry for detection afforded by this method with increased sensitivity and resolution, the method has been suitable for activation analysis. An example of the Compton reduction possible with this technique is illustrated in Figure 8.7, which show a gamma-ray spectrum of activated impurities in commercial-grade aluminum. Improvements in this method have been made; for example, the substitution of a large plastic detector for anthracene to improve the detection efficiency and the use of a lead grating to achieve better compensation of the Compton effect over a broader energy range.†

* D. H. Peirson, A Two-Crystal Gamma-Ray Scintillation Spectrometer, *Nature* **173,** 990–991 (1954).
† D. DeSoete and J. Hoste, A Compton Compensated Gammaspectrometer, *Radiochimica Acta* **4,** 35–38 (1965).

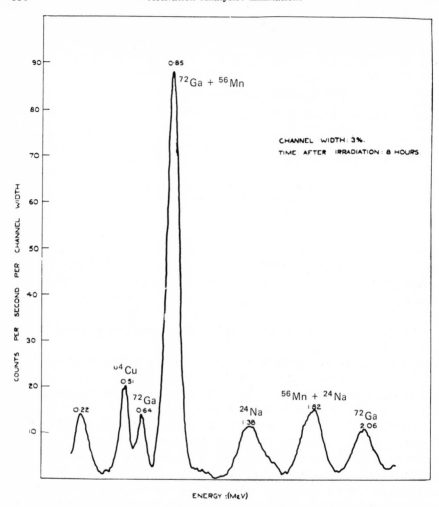

**Figure 8.7** Gamma-ray spectrum of an irradiated aluminum sample using a Compton-subtraction gamma-ray spectrometer. [From D. H. Peirson, Radiochemical Analysis by Gamma-Ray Spectrometry, *Atomics* **7**, 316–322 (1956).]

Other improvements in Compton suppression has been attained in the special techniques of $\gamma$-$\gamma$ coincidence spectrometry and sum-coincidence spectrometry, described in Section 7.3.1. One of the more advanced, and also more expensive, detector systems utilizes both coincidence and anticoincidence methods for multidimensional gamma-ray spectroscopy.* The detector

---

\* R. W. Perkins and D. E. Robertson, Selective and Sensitive Analysis of Activation Products by Multidimensional Gamma-Ray Spectroscopy, in *Modern Trends in Activation Analysis* (Texas A&M University, College Station, 1965), pp. 48–57.

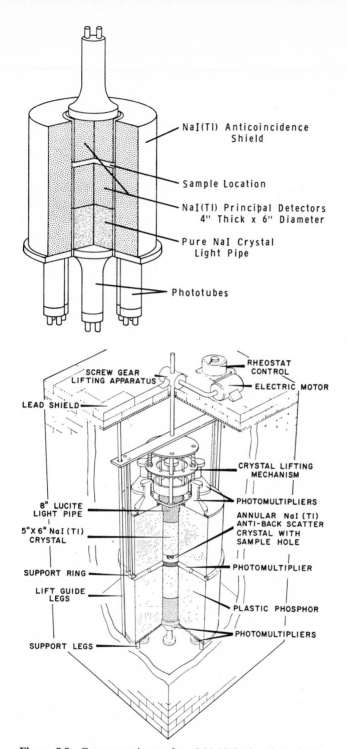

**Figure 8.8** Cut-away views of a 2-NaI(Tl)-phosphor shield detector system for multidimensional gamma-ray spectrometry. [From R. W. Perkins and D. E. Robertson, in *Modern Trends in Activation Analysis* (Texas A&M University, College Station, 1965), pp. 48–57].

system, shown in Figure 8.8, utilizes two NaI(Tl) detectors, 5 in. thick and 6 in. in diameter, in coicindence circuit, contained in a plastic phosphor, 15 in. thick and 30 in. in diameter, which serves as an anticoincidence shield for Compton and background suppression. The multidimensional spectrometer is a 4096-channel analyzer in a 54 × 64 channel grouping. The sample is sandwiched between the two principle detectors. An event in one detector is recorded in the appropriate multichannel grouping. An event when two photons are emitted simultaneously, each interacting with a separate detector, is stored in an energy-energy plane at a point uniquely characteristic of the two energies. If the total energy is not absorbed in the two detectors, interaction with the plastic phosphor cancels the total event. The application of this system for an activation analysis of biological material in which seven trace elements were resolved in a single energy-energy plane count is illustrated in Figure 8.9.

The development of the superior-resolution semiconductor detector for instrumental activation analysis, with their much smaller ratios of full-energy absorption to Compton-scattering events, makes the search for improved

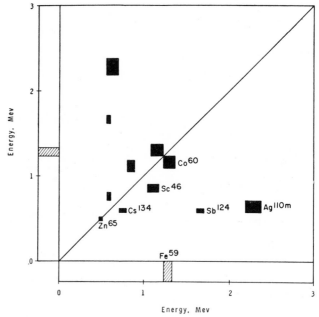

**Figure 8.9** Energy areas used in multidimensional gamma-ray spectrometry to determine seven elements in biological samples. [From R. W. Perkins and D. E. Robertson, in *Modern Trends in Activation Analysis* (Texas A&M University, College Station, 1965), 48–57.]

Compton suppression techniques even more potentially rewarding. Besides the obvious improvements in ratio obtained with larger coaxial Ge(Li) detectors, the coincidence and anticoincidence methods used for scintillation spectrometry should be directly applicable to semiconductor detectors.

### 8.3.2 Gain and Threshold Shift

The hopes for rapid, automatic, instrumental activation analysis depend not only on the ability of modern electronic equipment to identify, resolve, and measure characteristic radiation with sufficient sensitivity and precision but also on the ability of this equipment to remain in calibration over extended periods of time. Among the general limitations of the method of least-squares analysis of complex gamma-ray spectra by computer methods are the problems caused by the preparation of poor standards, the presence of unsuspected radionuclides, and the absence of some of those included in the reference library. Poor library standards can be avoided by proper choice of materials free from interfering impurities and of ensuring satisfactory irradiation and measurement conditions relative to samples. The stripping limitations can be reduced by continuous improvement in identification and resolution procedures of the computer programs and reference libraries.

A major limitation, which is an inherent characteristic of the electronic equipment, is the changes in calibration that result from shifts (drift) in the gain and threshold of the spectrometers. The calibration of a gamma-ray spectrometer relates the pulse height (the photon energy absorbed by the crystal) to a channel number in the spectrometer. A perfectly stable analyzer would have a constant value of energy increment per channel (gain) and a constant intercept, the channel for $E_\gamma = 0$ (threshold or baseline).

Modern, sophisticated gamma-ray spectrometers lack perfect stability; they are subject to drifts in threshold and in channel number for a given $E_\gamma$. High stability instruments can show drifts in threshold of $\pm 0.5$ channel and in gain of $\pm 0.5\%$ $\Delta E$/channel over a period of a week. Efforts to control the limitations due to drift have taken two forms: (a) instrumental methods to improve stability by positive control of the spectrum position of a reference peak and (b) corrections for the gain and threshold shifts as part of the data analysis program.

### Control of Drift

A review of the theory of stabilization for spectrometers has been made by Dudley and Scarpatetti.* They derive the basic features of stabilization which

* R. A. Dudley and R. Scarpatetti, Stabilization of a Gamma Scintillation Spectrometer Against Zero and Gain Shifts, *Nucl. Instr. Methods* **25**, 297–313 (1964).

include not only the magnitude and residual errors of an automatic stabilization process but also show the influence of the process on the resolution of the spectrometer. Automatic stabilization systems can be constructed to position photopeaks in gamma-ray scintillation spectra with modern spectrometers in peak channel numbers within stabilities of less than 0.1 % over indefinite periods of time and without causing significant alterations to the gamma-ray spectra. Positive control is made by setting the position of a reference peak in the spectrometer. Several methods have been used as reference peaks:

1. Specific photopeaks in specific spectra.
2. Light pulses introduced into the scintillator from an external lamp.
3. Light pulses introduced by internal or external alpha-ray sources.
4. Electronic pulses.

Dudley and Scarpatetti show that an ideal reference for $\gamma$-ray spectrometry is a gamma ray uniquely distinguished from all other gamma rays being detected. They note the following as advantages:

1. The reference cannot drift.
2. It is not altered by changes in sample gamma-ray spectra.

They also note that a second criterion for positive control of drift is the point in the spectrometer system from which the correction signal is taken. An ideal point for the origin of the correction signal is where pulses are already in digital form. At such a point the signal would be drift-free. Several reported methods of stabilization achieve this second criterion.

The principal components of a stabilization system that meets the two criteria for automatic correction of gain and threshold drifts in gamma-ray scintillation spectrometer are shown in Figure 8.10. A reference gamma source is incorporated within a small plastic scintillator. The source, $^{228}$Th, offers several pairs of $\beta$, $\gamma$ radiations for coincidence use by choice of radiations from the descendant radionuclides; for example, the 0.239-MeV gamma ray of $^{212}$Pb and the 2.614-MeV gamma ray of $^{208}$Tl are used for control of threshold and gain, respectively. When one of them interacts within the NaI(Tl) crystal, it produces a reference pulse in the spectrometer. However, the coincident beta pulse from the plastic scintillator opens the coincidence gate, allows the stabilization circuitry to operate if the gamma pulse meets the required amplitude criterion, and prevents the reference pulse from being stored in the memory. The stabilization circuit monitors the address scaler in the spectrometer and adjusts the threshold (zero) and gain to hold the reference $\gamma$-ray peaks in the preassigned channels.

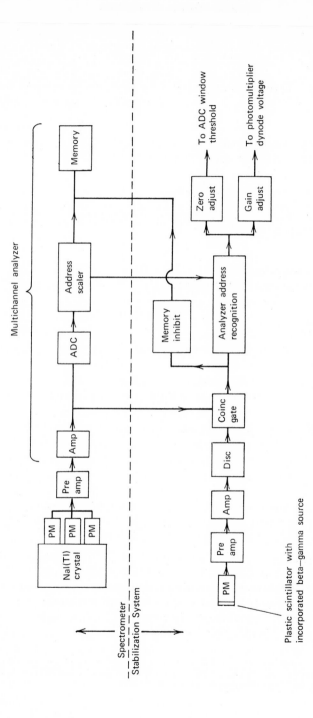

**Figure 8.10** A schematic diagram of a threshold and gain stabilized gamma-ray spectrometer. [From R. A. Dudley and R. Scarpetetti, Stabilization of a Gamma Scintillation Spectrometer Against Zero and Gain Drifts, *Nucl. Instr. and Methods* **25**, 297–313 (1964).]

## Corrections for Drift

The basis for correcting for gain and threshold shifts was described by Schonfeld, Kibbey, and Davis.* Figure 8.11 shows a schematic representation of the $E_\gamma$ versus channel number relation with greatly exaggerated shifts in threshold and gain. With the superscript $s$ referring to standard samples, the figure shows

the energy for channel 1 is $E_1{}^s$,
the energy for the last channel $n$ is $E_n{}^s$,
the energy for any channel $i$ is $E_i{}^s$,
the threshold (zero) channel is $\eta^s$.

Later, when an unknown sample is counted, the gain and threshold shifts produce the changes

channel 1 corresponds to $E_1$,
channel $n$ corresponds to $E_n$,
the threshold has shifted to channel $\eta$,

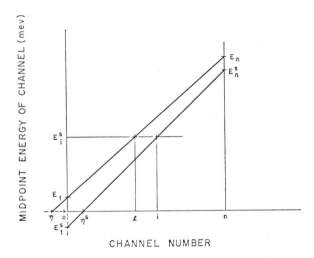

**Figure 8.11** Influence of threshold and gain shifts on the relation between gamma-ray energy and channel number. [From E. Schonfeld, A. Kibbey, and W. Davis, Determination of Nuclide Concentrations in Solutions containing Low Levels of Radioactivity by Least-Squares Resolution of the Gamma-Ray Spectra, ORNL-3744 (1965).]

---

* E. Schonfeld, A. Kibbey, and W. Davis, Determination of Nuclide Concentrations in Solutions Containing Low Levels of Radioactivity by Least-Squares Resolution of the Gamma-Ray Spectra, ORNL-3744 (1965).

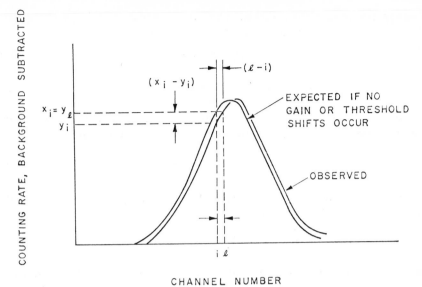

**Figure 8.12** Shift in location of gamma-ray spectrum due to shifts in threshold and gain. [From E. Schonfeld, A. Kibbey, and W. Davis, Determination of Nuclide Concentrations in Solutions containing Low Levels of Radioactivity by Least-Squares Resolution of the Gamma-Ray Spectra, ORNL-3744 (1965).]

the channel corresponding to energy $E_i^s$ has shifted from $i$ to $l$, the gain has shifted from $F = 1.00$ to

$$F = \frac{E_n^s - E_i^s}{E_n - E_1} \tag{28}$$

To obtain an estimate of the effects of gain and threshold shifts on counting rates in specified channels the relations in (28) can be expressed by the linear equations for the sample and standards, respectively, as

$$E_l = E_1 + \left(\frac{E_n - E_1}{n - 1}\right)(l - 1) \tag{29}$$

$$E_i^s = E_1^s + \left(\frac{E_n^s - E_1^s}{n - 1}\right)(i - 1) \tag{30}$$

These two equations can be combined to express the shift in the number of channels at channel $i$ in terms of the gain $F$ and the two threshold channels $\eta$ and $\eta^s$:

$$(l - i) = (F - 1)(i - \eta^s) + (\eta - \eta^s) \tag{31}$$

The change in location of a $\gamma$-ray peak due to gain and threshold changes is shown in Figure 8.12. For a net counting rate $y_i$, observed in channel $i$ (with background subtracted), the counting rate $x_i$, which would have occurred, had there been no gain and threshold shift, is given by

$$x_i = y_i + (x_i - y_i) = y_i + (y_l - y_i) \tag{32}$$

$$x_i = y_i + \left(\frac{y_l - y_i}{l - i}\right)(l - i) \tag{33}$$

Schonfeld, Kibbey, and Davis incorporate a numerical correction in their least-squares analysis program with the use of the approximation

$$\frac{y_l - y_i}{l - i} \sim \frac{y_{i+1} - y_{i-1}}{2} \tag{34}$$

By substituting for $(l - i)$ in (31) and using the approximation in (34), we obtain from (33):

$$x_i = y_i + \left(\frac{y_{i+i} - y_{i-1}}{2}\right)[(F - 1)(i - \eta^s) + (\eta - \eta^s)] \tag{35}$$

The unknown parameters $F$, $\eta$, and $\eta^s$ are incorporated into the least-squares analysis program [see (7.49)], except that $\eta^s$ can be evaluated, by fitting the data on standards through (30) for the linearity relation as

$$\eta^s = 1 - \frac{(n - 1)E_1^s}{E_n^s - E_1^s} \tag{36}$$

Another method of correcting for gain shifts has been reported by Helmer et al.* They incorporate a gain-shift program which consists of successive polynomial least-squares fits to sets of three channels to provide interpolated values around the original points. The original pulse-height scale is adjusted to place the photopeaks at the proper position on the standard pulse-height versus gamma-ray energy scale but with the constraint that the total pulse-height distribution area be conserved. This is accomplished by a "gain-shift" ratio $r$ in which the channel number is shifted by

$$i = rl \tag{37}$$

Then

$$x_i = \frac{1}{r} y_l \tag{38}$$

where $1/r$ is chosen for the area under the curve such that

$$\int x_i di = \int y_l dl \tag{39}$$

* R. G. Helmer, R. L. Heath, D. D. Metcalf, and G. A. Cazier, A Linear Least-Squares Fitting Program for the Analysis of Gamma-Ray Spectra Including a Gain-Shift Routine, AEC Report IDO-17015 (1964).

To correct also for shift in threshold (37) is altered to

$$i = r(l - a)$$ (40)

where $a$ is the desired shift in the zero channel.

### 8.3.3 Quality Control of Spectrometers

The limitations of gamma-ray spectroscopy for the identification, resolution, and measurement of gamma-ray emitting radionuclides have been described. Several methods to reduce the extent of these limitations have also been described. A gamma-ray spectrometer used as a high-precision analytical chemistry instrument requires long-term stability as well as adequate calibration. A considerable amount of painstaking laboratory practice is needed to calibrate a gamma-ray spectrometer suitably for high-precision measurement of radionuclides. Because of the long-term instabilities inherent in scintillation and semiconductor detectors and their associated electronic equipment, constant attention to the adequacy of the calibration is required. Activation analysis practices tend toward increased reliance on computer data processing, especially for analyses that require high precision on a routine basis. For such objectives some form of quality control is needed to ensure reproducibility of analyses within some desired range. Quality control of spectrometer performance may be made by periodic recalibration of the system or by continuous automatic methods.

An analysis of the performance specifications for multichannel analyzers has been made by Heath.* Besides the requirements for gain and threshold stability, already reviewed, he listed specifications of spectrometers for precision data reduction for the following properties:

1. Memory capacity of the analyzer (a sufficient numbers of channels for both spectra and data reduction).

2. Integral linearity of the analog-to-digital converter (the maximum deviation in any channel from a linear relation of channel position and input pulse amplitude).

3. Differential linearity of the channels (the deviation in channel width).

4. Noise modulation (the noise level of the pulse amplifier in percent of full output).

5. Accuracy of the analog-to-digital converter (the ability to store pulses of one size in one channel).

6. Accuracy of the live timer.

A review of performance testing of a multichannel analyzer system and of

* R. L. Heath, Digital Analysis of Pulse-Height Spectra, *Nucleonics* **20**, No. 5, 67–69 (1962).

methods for calibration and routine checking of calibration was made by Crouch and Heath.* These procedures are especially useful when several analyzers supply data to a common computer program. It is convenient if each of the analyzer systems can be standardized to a reproducible performance level; however, the operating characteristics of each system must be known precisely. The performance parameters tested are generally the ones listed above. The tests are made with a stable, linear pulse generator.

Calibration of the pulse-height analyzer takes into account the non-linearity of the detector crystals. For this purpose an absolute pulse-height scale, based on electrical calibration, is used. Performance and calibration are checked daily. The photomultiplier high voltage is adjusted to position a standard gamma-ray radionuclide at its proper channel position. Live timer, linearity, and zero shift are checked with another radionuclide standard. After this standard source has been analyzed its spectrum is superimposed over that radionuclide standard read into the analyzer from an external memory. A comparison of heights and positions of photopeaks points out any changes in performance parameters.

Covell† describes several proposed methods for automatic correction of spectrometer drifts. These are based on feedback loops in the instrument or normalization techniques for the resultant data, as reviewed in Section 8.3.2. Covell pointed out that these methods do not provide for early recognition of erratic performance but tend to mask such performance. His method of stabilization is augmented by regular critical examination of selected instrument performance factors as a quality-control procedure.

Quality control is noted as a technique for maintaining the quality of a product at a prescribed level and not the establishment of the quality. In the process of correcting system deficiencies, however, the level of quality may be raised. In establishing a quality-control procedure, the characteristics of the output products to be controlled must first be determined. "Normal" performance is determined from the central tendency (arithmetic mean) of the values for the selected characteristics and the dispersion (standard deviation) of these values. If the capabilities of the system, as reflected in these properties of the output product, are acceptable, periodic sampling can be used to compare the values for individual output units with the system norm. If the sampling is done so that the sample values generally represent the output product, an analysis of these values will provide a periodic estimate, within the limits due to sampling error, of the system performance.

* D. F. Crouch and R. L. Heath, Evaluation and Calibration of Pulse-Height Analyzer Systems for Computer Data Processing, paper 2–5, in Applications of Computers to Nuclear and Radiochemistry, NAS-NS 3107 (1962).
† D. F. Covell, Quality Control for the Gamma-Ray Scintillation Spectrometer, *Nucl. Instr. Methods* **22,** 101–108 (1963).

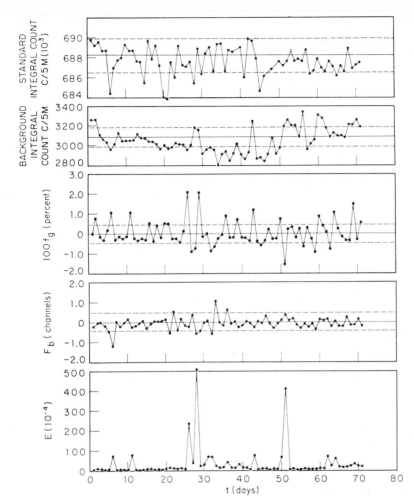

**Figure 8.13** A quality control chart for a NaI(Tl) scintillation spectrometer system over a six-month period. [From D. F. Covell, A Computer-Coupled Quality Control Procedure for Gamma-Ray Scintillation Spectrometry, *Nucl. Instru. Methods* **47**, 125–132 (1967).]

Covell* has developed a computer-coupled quality-control procedure for gamma-ray scintillation spectrometry based on a response characteristic which is a composite of the transfer characteristics of the detector, the photomultiplier, and the pulse-height analyzer. The specific parameters examined

---

* D. F. Covell, A Computer-Coupled Quality Control Procedure for Gamma-Ray Scintillation Spectrometry, *Nucl. Instr. Methods* **47**, 125–132 (1967).

are the following:

1. Integral counts of a standard source.
2. Integral background counts.
3. Computed shift in gain, $f_g$.
4. Computed shift in baseline, $F_b$.
5. Minimum value of instrumental aberrations, $E$.

The instrumental aberrations are determined by comparing a current standard spectrum of a pure radionuclide with a reference standard spectrum of that radionuclide. The value $E$ is defined as

$$E = \sum_i e_i = \frac{\sum (A_i - a_i^*)^2}{A_i} \tag{41}$$

where $e_i$ = the weighted square of the difference between the corrected current standard and the reference standard data of channel $i$,

$A_i$ = the channel $i$ datum of the reference standard spectrum,

$a_i^*$ = the channel $i$ datum of the current standard spectrum after correction for shifts.

The analyses of the sampling data for a particular instrument can be evaluated by plotting the individual sample values on a chart that also shows the system norm and dispersion. From such a control chart trends or excursions deviating from the normal value or from the established standard deviation can be readily detected. An example of a quality-control chart for a NaI(Tl) detector, 256-channel pulse-height analyzer over a six-month period is shown in Figure 8.13. The normal value is shown by the solid line and the one standard deviation values are shown by the dashed line.

## 8.4  SENSITIVITY  LIMITATIONS

The sensitivity calculation, described in Section 7.1.2, is generally used to estimate the minimum amount of an element that can be determined in a sample under a set of assumed or real conditions. In many compilations of sensitivities for various activation-analysis practices reported in the literature the calculations were made with a set of conditions that may not be realizable in the laboratory. In addition, sensitivities are often calculated on the basis of some minimum "detectable" level for a given radiation-measurement system. The definition of this detectable level may vary considerably; it is influenced by such factors as the half-life of the radionuclides, the presence of interfering radionuclides, and the experience of the analyst. These factors

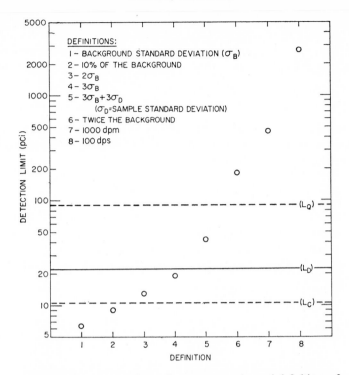

**Figure 8.14** A comparison of some commonly used definitions of "detection limit" which shows a range of nearly three orders of magnitude. [From L. A. Currie, Limits for Qualitive Detection and Quantitative Determination; Application to Radiochemistry, *Anal. Chem.* **40**, 586–593 (1968).]

may limit the actual sensitivity achievable to some value considerably less than a standard estimated sensitivity value.

An examination of the definition of a "detection limit" has been made by Currie.* He notes the use of such approximately equivalent terms as lower limit of detection, detection sensitivity, sensitivity, minimum detectable activity, and limit of guarantee for purity. More important, however, is the confusion that may exist in terms that do not refer to *detection* limits but rather to *determination* limits, the latter with reference to some given relative standard deviation. A variety of criteria for estimating detection limits is reported in the literature. A comparison of some commonly used definitions of "detection limit" is shown in Figure 8.14 for a hypothetical measurement

* L. A. Currie, Limits for Qualitative Detection and Quantitative Determination; Application to Radiochemistry, *Anal. Chem.* **40**, 586–593 (1968).

of a long-lived $\gamma$-emitting radionuclide counted for 10 min with an efficiency of 10% in a detector having a background of 20 cpm.

The detection limits, plotted in increasing order, encompass nearly three orders of magnitude. For an analysis of the lower limits of any measurement system Currie defines three specific quality levels:

1. $L_C$, the "decision limit," the net signal level (instrument response) above which an *observed* signal may be reliably recognized as *detected*.

2. $L_D$, the "detection limit," the *true* net signal level which may be expected, a priori, to lead to detection.

3. $L_Q$, the "determination limit," the level at which the measurement precision will be satisfactory for quantitative determination.

The magnitudes of these three levels for the same hypothetical measurement are also shown in Figure 8.14. These three limits may be derived from the principles of qualitative and quantitative analysis limits.

For limits in qualitative analysis there are two fundamental aspects to the detection problem:

1. For a given *net* observed signal $S$ has a *real* signal been detected; that is, is the true mean $\mu_S > 0$? This aspect involves an a posteriori decision based on the observation $S$ and a definite criterion for detection. The net observed signal is generally obtained as the difference of a pair of measurements of the sample plus background $(S + B)$ and the background $B$ alone, such that $S = (S + B) - B$.

2. For a completely specified measurement system the minimum true signal $\mu_S$ is estimated such that it may be expected to yield a sufficiently large net observed signal $S$ that it will be detected. This value is thus an a priori estimate of the detection capability of the detection system.

The a posteriori binary decision, known as hypothesis testing, is subject to two kinds of error:

$\alpha$. Error of the first kind—deciding that the radionuclide is present when it is not.

$\beta$. Error of the second kind—failing to decide that it is present when it is.

The acceptable value for $\alpha$, together with the standard deviation $\sigma_0$ of the net signal when $\mu_S = 0$, establishes the critical level $L_C$ on which decisions may be based. To be *detected* an observed signal $S$ must exceed $L_C$.

The a priori detection limit $L_D$ may be established from $L_C$, the acceptable level, $\beta$, for the error of the second kind, and the standard deviation $\sigma_D$, which characterizes the probability distribution of the net signal when its true value $\mu_S$ is equal to $L_D$.

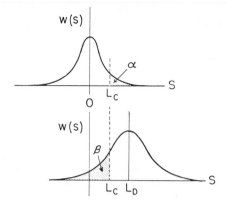

**Figure 8.15** The relationships between $L_C$, $L_D$, and the probability distributions for $\mu_s = 0$ and $\mu_s = L_D$. [From L. A. Currie, Limits for Qualitative Detection and Quantitative Determination; Application to Radiochemistry, *Anal. Chem.* **40**, 586–593 (1968).]

The relationships between $L_C$, $L_D$, and the probability distributions for $\mu_S = 0$ and $\mu_S = L_D$ are given in Figure 8.15. In this figure $\mu_B$ represents the true mean of the blank distribution and $\mu_{S+B}$ represents the true mean of the observed (sample + blank) distribution. The critical level is given as

$$L_C = k_\alpha \sigma_0 \tag{42}$$

and the detection limit as

$$L_D = L_C + k_\beta \sigma_D \tag{43}$$

where $k_\alpha$ and $k_\beta$ are abscissas of the standardized normal distribution corresponding to probability levels $1 - \alpha$ and $1 - \beta$, respectively. The probability distribution, when $\mu_S = 0$, intersects $L_C$ such that the fraction, $1 - \alpha$, corresponds to the (correct) decision, "not detected"; $L_D$ is defined so that the probability distribution, when $\mu_S = L_D$, intersects $L_C$ such that the fraction $1 - \beta$ corresponds to the (correct) decision, "detected."

For quantitative analysis neither a binary decision, based on $L_C$, nor an upper limit or a wide-confidence interval is satisfactory. A result that is satisfactorily close to the true value is required. Thus to establish the determination limit $L_Q$, where $\mu_S = L_Q$, the standard deviation $\sigma_Q$ must be small compared with $\mu_S$; for example, the relative standard deviation $\sigma_Q/\mu_S$ may be 10%. The determination limit may be given as

$$L_Q = k_Q \sigma_Q \tag{44}$$

where $1/k_Q$ is the requisite relative standard deviation.

**Activation Analysis: Limitations**

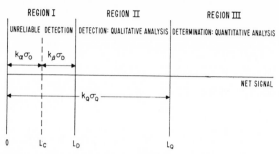

**Figure 8.16** The three principal analytical regions. [From L. A. Currie, Limits for Qualitative Detection and Quantitative Determination: Application to Radiochemistry, *Anal. Chem.* **40**, 586–593 (1968).]

The three principal analytical regions are summarized in Figure 8.16. In summary, the levels $L_C$, $L_D$, and $L_Q$ are determined by the error-structure of the measurement system, the risks $\alpha$ and $\beta$, and the maximum acceptable relative standard deviation for quantitative analysis. For the set of conditions $\alpha = \beta = 5\%$, $\sigma = \sigma_0 =$ constant, and $k_Q = 10$, Currie derived a set of "working" expressions for the three levels, summarized in Table 8.9 for equivalent observations of sample and blank (paired observations) and for a long history of observations of the blank which makes $\sigma^2 = \sigma_S^2$ only ("well-known" blank).

In the application of these principles to the measurement of radionuclides, Currie expresses (42), (43) and (44) in terms given by the Poisson distribution (see Section 5.5.2). For a large number of counts, with a "well-known" blank $\bar{B}$ from $n$ observations, the variance is given by

$$\sigma^2 = \sigma_{S+B}^2 + \sigma_B^2 = (\mu_S + \mu_B) + \frac{\mu_B}{n} \tag{45}$$

Since $\sigma_0^2$ is the variance when $\mu_S = 0$ and $\sigma_D^2$ is the variance when $\mu_S = L_D$,

**Table 8.9** "Working" Expressions for $L_C$, $L_D$, $L_Q{}^a$

|  | $L_C$ | $L_D$ | $L_Q$ |
|---|---|---|---|
| Paired Observations | 2.33 $\sigma_B$ | 4.65 $\sigma_B$ | 14.1 $\sigma_B$ |
| "Well-known" Blank | 1.64 $\sigma_B$ | 3.29 $\sigma_B$ | 10 $\sigma_B$ |

$^a$ Assumptions: $\alpha = \beta = 0.05$; $k_Q = 10$; $\sigma = \sigma_0 =$ constant. [From L. A. Currie, Limits for Qualitative Detection and Quantitative Determination, *Anal. Chem.* **40**, 586–593 (1968).]

Currie derives

$$L_C = k_\alpha \sigma_0 = k_\alpha (\mu_B + \sigma_B^2)^{1/2} \tag{46}$$

and

$$L_D = L_C + k_\beta \sigma_D = L_C + k_\beta (L_D + \sigma_0^2)^{1/2} \tag{47}$$

Solution of (46) and (47) leads to

$$L_D = L_C + \frac{k_\beta^2}{2} \left[ 1 + \left( 1 + \frac{4L_C}{k_\beta^2} + \frac{4L_C^2}{k_\alpha^2 k_\beta^2} \right)^{1/2} \right] \tag{48}$$

Thus estimates of $\mu_B$ and $\sigma_0$ allow calculation of $L_C$ and $L_D$ for selected values of $\alpha$ and $\beta$. If $k_\alpha = k_\beta = k$, (48) simplifies to

$$L_D = k^2 + 2L_C \tag{49}$$

The determination limit $L_Q$ is given by

$$L_Q = k_Q \sigma_Q = k_Q (L_Q + \sigma_0^2)^{1/2} \tag{50}$$

which reduces to

$$L_Q = \frac{k_Q^2}{2} \left[ 1 + \left( 1 + \frac{4\sigma_0^2}{k_Q^2} \right)^{1/2} \right] \tag{51}$$

Convenient "working" expressing for these limits are summarized in Table 8.10 for several blank conditions. Note that this blank may include more

**Table 8.10**  "Working" Expressions for Radiation Measurement[a]

|  | $L_C$ | $L_D$ | $L_Q$ |
|---|---|---|---|
| **Paired Observations** $\sigma_B^2 = \mu_B$ (counts) | $2.33\sqrt{\mu_B}$ | $2.71 + 4.65\sqrt{\mu_B}$ | $50\left[ 1 + \left( 1 + \frac{\mu_B}{12.5} \right)^{1/2} \right]$ |
| **"Well-known" Blank** $\sigma_B^2 = 0$ (counts) | $1.64\sqrt{\mu_B}$ | $2.71 + 3.29\sqrt{\mu_B}$ | $50\left[ 1 + \left( 1 + \frac{\mu_B}{25} \right)^{1/2} \right]$ |
| **Zero Blank** $\mu_B = 0$ (counts) | $0$ | $2.71$ | $100$ |
| Asymptotic Ratio[b,c] $S/\sigma_B$ | $1.64$ | $3.29$ | $10$ |

[a] From L. A. Currie, Limits for Qualitative Detection and Quantitative Determination, *Anal. Chem.* **40**, 586–593 (1968).

[b] For "well-known" blank case; multiply by $\sqrt{2}$ for paired observations.

[c] Correct to within 10% if $\mu_B \geq 0$, 67, 2500 counts, respectively, for each of the three columns. For paired observations $\mu_B \geq 0$, 34, 1250 counts, respectively.

than the detector background. If interfering radionuclides contribute to the blank level, the expressions in Table 8.10 take the form

$$\mu_B = \mu_b + \mu_I \tag{52}$$

$$\sigma_B{}^2 = \sigma_b{}^2 + \sigma_I{}^2 \tag{53}$$

where $b$ refers to the detector background and $I$, to the interference.

The detection limit for radiation measurement may be expressed as a minimum detectable mass $M_D$ by the equation

$$L_D = KM_D \tag{54}$$

where $K$ represents an over-all calibration factor relating the detector response to the mass present. In the case of nuclear activation

$$K = P(\sigma)S(\lambda,\tau)T(\lambda,t,\Delta t)\epsilon(x) \tag{55}$$

where $P$ = production rate,

$\sigma$ = reaction cross section,

$S$ = saturation factor,

$\lambda$ = product decay constant,

$\tau$ = irradiation time,

$T$ = time function = $(e^{-\lambda t}/\lambda)(1 - e^{-\lambda \Delta t}) \sim e^{-\lambda t} \cdot \Delta t$,

$t$ = delay time before counting,

$\Delta t$ = counting time,

$\epsilon$ = over-all counting efficiency,

$x$ = a variable detector parameter.

A similar expression is written for the mean number of counts from an interfering radionuclide

$$\mu_I = m_I[P(\sigma_I)S(\lambda_I,\tau)T(\lambda_I,t,\Delta t)\epsilon_I(x)] \tag{56}$$

and the mean number of background counts by

$$\mu_b = b(x)\,\Delta t \tag{57}$$

Thus the general expression for the sensitivity of an activation analysis (the minimum detectable mass), which has no restrictions on product half-life or on interfering radionuclides, is

$$m_D = \frac{k^2 + 2k[\mu_b + \sigma_b^{-2} + \mu_I + \sigma_I{}^2]^{1/2}}{P(\sigma)S(\lambda,\tau)T(\lambda,t,\Delta t)\epsilon(x)} \tag{58}$$

Alternate methods for analyzing the limit of detection of the half-life of a radionuclide, and therefore of the limit of identification for short-lived radionuclides, in activation analysis are given by Sterlinski* and for minimizing activation-analysis counting errors by application of binomial distribution statistics, by Jurs and Isenhour.†

## 8.5 BIBLIOGRAPHY

### 8.5.1 General References

J. P. Cali, Ed., *Trace Analysis of Semiconductor Materials* (Pergamon, New York, 1964).

L. A. Currie, Limits for Qualitative Detection and Quantitative Determination; Application to Radiochemistry, *Anal. Chem.* **40**, 586–593 (1968).

P. C. Jurs and T. L. Isenhour, Binomial Distribution Statistics Applied to Minimizing Activation Analysis Counting Errors, *Anal. Chem.* **39**, 1388–1394 (1967).

R. C. Plumb and J. E. Lewis, How to Minimize Errors in Activation Analysis, *Nucleonics* **13**, No. 8, 42–46 (1955).

A. A. Smales, The Scope of Radioactivation Analysis, *Atomics* **4**, No. 3, 55–63 (1953).

G. W. Smith, D. A. Becker, G. J. Lutz, L. A. Currie, and J. R. DeVoe, Determination of Trace Elements in Standard Reference Materials by Neutron Activation Analysis, *Analytica Chemica Acta* **38**, 333–340, (1967).

S. Sterlinski, The Limit of Identification for Short-Lived Radioisotopes in Activation Analysis, *Nuclear Instr. Methods* **47**, 329–341 (1967).

### 8.5.2 Nuclear Limitations

T. A. Eastwood and R. D. Werner, "Resonance and Thermal Neutron Self-Shielding in Cobalt Foils and Wires," *Nucl. Sci. Engr.* **13**, 385–390 (1962).

O. T. Høgdahl, Neutron Absorption in Pile Neutron Activation Analysis, paper SM 55/3 in *Radiochemical Methods of Analysis*, Vol. I (International Atomic Energy Agency, Vienna, 1965); also in University of Michigan Phoenix Memorial Lab Report MMPP-226-1 (1962).

G. Leliaert, J. Hoste, and Z. Eeckhaut, Activation Analysis of Vanadium in High Alloy Steels Using Manganese as Internal Standard, *Anal. Chem. Acta.* **19**, 100–107 (1958).

A. E. McCarthy et al., ANL Reactor Physics Constants Center Newsletter No. 1 (1961).

S. A. Reynolds and W. T. Mullins, Neutron Flux Perturbation in Activation Analysis, *Intern. J. Appl. Radiation Isotopes* **14**, 421–425 (1963).

* S. Sterlinski, The Limit of Identification for Short-Lived Radioisotopes in Activation Analysis, *Nuclear Instr. Methods* **47**, 329–341 (1967).

† P. C. Jurs and T. L. Isenhour, Binomial Distribution Statistics Applied to Minimizing Activation Analysis Counting Errors, *Anal. Chem.* **39**, 1388–1394 (1967).

E. L. Steele, Variable-Energy Neutron Activation Analysis, in *Modern Trends in Activation Analysis* (Texas A&M University, College Station, 1965) pp. 102–106.

P. F. Zweifel, Neutron Self-Shielding, *Nucleonics* **18**, No. 11, 174–175 (1960).

### 8.5.3 Spectroscopy Limitations

G. D. Albert, An Anticoincidence Gamma-Ray Scintillation Spectrometer, *Rev. Sci. Instr.* **24**, 1096–1101 (1953).

P. R. Bell, Scintillation Spectrometer with Improved Response, *Science* **120**, 625–626 (1954).

D. F. Covell, A Computer-Coupled Quality Control Procedure for Gamma-Ray Scintillation Spectrometry, *Nucl. Instr. Methods* **47**, 125–132 (1967).

D. F. Covell, Quality Control for the Gamma-Ray Scintillation Spectrometer, *Nucl. Instr. Methods* **22**, 101–108 (1963).

D. F. Crouch and R. L. Heath, Evaluation and Calibration of Pulse-Height Analyzer Systems for Computer Data Processing, paper 2–5, in Applications of Computers to Nuclear and Radiochemistry, NAS-NS 3107 (1962).

D. DeSoete and J. Hoste, A Compton Compensated Gammaspectrometer, *Radiochimica Acta* **4**, 35–38 (1965).

R. A. Dudley and R. Scarpatetti, Stabilization of a Gamma Scintillation Spectrometer Against Zero and Gain Shifts, *Nucl. Instr. Methods* **25**, 297–313 (1964).

R. L. Heath, Digital Analysis of Pulse-Height Spectra, *Nucleonics* **20**, No. 5, 67–69 (1962).

R. G. Helmer, R. L. Heath, D. D. Metcalf, and G. A. Cazier, A Linear Least-Squares Fitting Program for the Analysis of Gamma-Ray Spectra Including a Gain-Shift Routine, IDO-17015 (1964).

R. Hofstadter and J. A. McIntyre, Measurement of Gamma-Ray Energies with Two Crystals in Coincidence, *Phys. Rev.* **78**, 619–620 (1950).

D. H. Peirson, A Two-Crystal Gamma-Ray Scintillation Spectrometer, *Nature* **173**, 990–991 (1954).

R. W. Perkins and D. E. Robertson, Selective and Sensitive Analysis of Activation Products by Multidimensional Gamma-Ray Spectroscopy, in *Modern Trends in Activation Analysis* (Texas A&M University, College Station, 1965) pp. 48–57.

E. Schonfeld, A. Kibbey, and W. Davis, Determination of Nuclide Concentrations in Solutions Containing Low Levels of Radioactivity by Least-Squares Resolution of the Gamma-Ray Spectra, ORNL-3744 (1965).

### 8.6 PROBLEMS

1. Two 1-g samples of terphenyl are available for neutron activation analysis for chlorine, sulfur, and phosphorous. One 1-g sample is irradiated in a high flux ($10^{14}$ thermal $n/cm^2$-sec) reactor for a very short period of time (1 min). The second 1-g sample is irradiated in a lower flux ($10^{12}$ thermal $n/cm^2$-sec and $10^{12}$ fast $n/cm^2$-sec) for 10 days. The short irradiation is used to determine the chlorine content by the $^{37}Cl$ $(n,\gamma)$ $^{38}Cl$ reaction.

The longer irradiation is used to produce sufficient 87-d $^{35}$S and 14.2-d $^{32}$P for radioactivity measurement. The initial activity of $^{38}$Cl was $4.50 \times 10^5$ dps, $^{35}$S was $2.76 \times 10^4$ dps, and $^{32}$P was $1.63 \times 10^6$ dps. From these data and the appropriate activation data

(a) determine the concentration (in ppm by weight) of chlorine in the terphenyl;
(b) show that the fraction of observed $^{32}$P produced from the chlorine is negligible;
(c) determine the concentration (in ppm by weight) of sulfur and phosphorous in the terphenyl sample (ignore any second-order reactions);
(d) determine the maximum Cl/S and S/P weight ratios that will allow comparative analyses with pure sulfur and phosphorus comparators if the radioactivity interference is to be less than 2 %.

2. Lithium ions are diffused into germanium semiconductor detectors to increase the usable depletion depths for gamma-ray spectroscopy. Proton activation analysis may be used to determine the lithium content of the semiconductor as a function of depth. A thin wafer of germanium was sliced from a detector and the following data were obtained for the lithium analysis by proton activation analysis:

Sample size: 10 mg/cm$^2$.

Irradiation: p energy = 2.24 MeV; average beam intensity = $10^{12}$ p/sec (within the area of the sample); irradiation time = 1 hr.

Observed activity: 37 dps, measured 4 hr after the irradiation, and corrected for background, counter efficiency and chemical yield.

Nuclear data: lithium: MW = 6.94; $^7$Li: $f = 92.58\%$; cross section for the reaction $^7$Li (p,n) $^7$Be,

$$\int_{E_{th}=1.9}^{2.24} \sigma(E) \, dE = 0.51 \text{ b}$$

Possible interference: beryllium: MW 9.01; $^9$Be: $f = 100\%$; cross section for the reaction, $^9$Be (p,t) $^7$Be,

$$\int_{E_{th}}^{18.5} \sigma(E) \, dE = 12 \text{ mb}$$

(a) Determine the concentration of lithium (in ppm) in the germanium semiconductor wafer.
(b) Show from the appropriate masses (see Table 1.3) and the data above that the interference from any beryllium in the wafer is negligible for this irradiation.

3. A sample of steel is to be analyzed for vanadium by using manganese as an internal standard. Two approximately equal steel samples, sample 1 weighing $a$ mg and sample 2 weighing $b$ mg, are irradiated simultaneously in a suitable neutron flux. Before the irradiation $c$ mg of vanadium is mixed homogeneously with sample 2, $c$ being large compared with the amount of vanadium in the samples. After irradiation the measured activities of $^{56}$Mn at common time $t$, following decay of the $^{52}$V, was $A(t)_1(\text{Mn})$ and $A(t)_2(\text{Mn})$ in samples 1 and 2, respectively. An activation correction factor $\alpha$ can be defined as

$$\frac{A(t)_1(\text{Mn})}{A(t)_2(\text{Mn})} = \frac{a}{b} \times \alpha$$

(a) For the activities corrected to the end of the irradiation $t_0$, where

$A(t_0)_1 = $ total activity of sample 1,

$A(t_0)_2 = $ total activity of sample 2,

$A(t_0)_1(\text{Mn}) = $ initial activity of $^{56}$Mn in sample 1,

$A(t_0)_2(\text{Mn}) = $ initial activity of $^{56}$Mn in sample 2,

$\quad A(t_0)_1(\text{V}) = $ net initial activity of $^{52}$V in sample 1

$\qquad\qquad = A(t_0)_1 - A(t_0)_1(\text{Mn})$,

$\quad A(t_0)_2(\text{V}) = $ net initial activity of $^{52}$V in sample 2

$\qquad\qquad = A(t_0)_2 - A(t_0)_2(\text{Mn})$,

show that the concentration of $^{52}$V in sample 1, $S(t_0)_1(\text{V})$, is given by

$$S(t_0)_1(\text{V}) = \frac{1}{c}\left[\alpha(A(t_0)_2(\text{V})) - \frac{b}{a}(A(t_0)_1(\text{V}))\right]$$

(b) Give an equation expressing the vanadium content, in percent, in terms of $A(t_0)_1(\text{V})$, $a$, and $S(t_0)_1(\text{V})$.

4. If $\alpha = E_\gamma/mc^2$ for an incident $\gamma$-ray in a two-crystal coincidence spectrometer and $A = E_\beta/mc^2$ for the recoil electron following a Compton scattering event in which the scattered photon leaves at an angle $\phi$, show from equations (4–44) and (4–46) that the energy of the original photon can be determined from the energy of the recoil electron by

$$\alpha = \frac{A}{2}\left[1 + \left(1 + \frac{2}{A\sin^2\phi/2}\right)^{1/2}\right]$$

5. Compare the minimum detection mass, $m_D$ in grams, of potassium produced by thermal neutrons or by linac bremsstrahlung for the cases when the potassium is interference free and is in the presence of 1 gram of sodium.

The following data are available:

(a)

| Reaction | Target Nuclide | Product | Flux |
|---|---|---|---|
| $(n,\gamma)$ | $^{41}$K$(f = 0.069)$ | 12.4-h $^{42}$K | $10^{13}$ n/cm²-sec |
| $(n,\gamma)$ | $^{23}$Na$(f = 1.00)$ | 15-h $^{24}$Na | $10^{13}$ n/cm²-sec |
| $(\gamma,n)$ | $^{39}$K$(f = 0.931)$ | 7.7-m $^{38}$K | $10^{14}/E$ (quanta/cm²-min-MeV) |
| $(\gamma,n)$ | $^{23}$Na$(f = 1.00)$ | 2.6-y $^{22}$Na | $10^{14}/E$ (quanta/cm²-min-MeV) |

(b) Irradiation time and counting time—$10^3$ min or two half-lives, whichever is less.

(c) Delay time—negligible.

(d) Nuclear cross sections—use literature values.

(e) Interference correction factor—$f_I = 1\%$.

(f) Detection-paired observations, NaI detector.

| Reaction | Background, $B$ | Overall Efficiency, $\epsilon$ | |
|---|---|---|---|
| | | K | Na |
| $(n,\gamma)$ | 12 cpm | 0.014 | 0.0097 |
| $(\gamma,n)$ | 20 cpm | 0.32 | 0.29 |

# Chapter 9

# Activation Analysis: Applications

Activation analysis may readily be classified as a branch of analytical chemistry. Under such a classification the applications of activation analysis can be summarized simply as the detection and determination of the chemical elements. This summary is both too broad and too restrictive. It is too broad in that there are several chemical elements that are not readily detected and measured by activation analysis. It is too restrictive in that the method is useful for purposes other than the analysis of chemical elements.

Several comparative evaluations have been made of the many methods available for analysis of the chemical elements. Almost all show that activation analysis, for one reason or another, does not compete well with other methods

**380**

available for analyses of elements which constitute a significant fraction of the sample composition (e.g., $>1\%$). These evaluations also show that activation analysis becomes of increasing interest as the desired elements approach trace concentrations (e.g., $<10^{-4}$ g/g). One may conclude that the major attribute of activation analysis is sensitivity. There are other advantages. There are also disadvantages. For a given analysis the choice of activation analysis is usually made after comparison of this method with other available trace element methods suitable for the element, the matrix, and other analytical conditions.

Activation analysis is also useful as a "radioactive tracer" method in which no radioactivity need be added to the system being traced. The radiotracer method has many applications in scientific, biological, and engineering studies. The advantage of being able to avoid adding radioactive materials to a particular system, yet obtain the benefits of radiotracer applications, is indeed great. In this method the tracing is done with stable-nuclide tracers which are determined by activation analysis of samples taken from the system.

Many special applications have been developed for activation anaylsis. One is *comparative analysis*, a method of great interest in classification of materials by element composition. Use of this method is made, for example, in forensic science, to "fingerprint" classes of materials such as paints, poisons, and plants. Another special application is *surface analysis*, in which the characteristics of a surface, ranging from an oxide film on a metal to the chemical composition of the surface of the moon, are examined. The nuclear nature of the activation process allows methods for *isotopic analysis*. The production of short-lived activation products permits the development of *on-line activation analysis* as an industrial-quality and process-control method, especially useful if coupled with an automatic regulation system.

## 9.1 TRACE ELEMENT ANALYSIS

Many analytical techniques are available for the analysis of trace elements in all kinds of sample. An element may be said to be present in trace quantity when its concentration in a sample is between $10^{-4}$ and $10^{-14}$ g/g, a range of about 10 orders of magnitude. Samples may vary in weight from $<1$ $\mu$g to $>1$ kg, from sources ranging from cosmic to deep earth, and from materials freshly produced to petrified. Samples for trace element analysis must be protected from such problems as losses by decomposition or adsorption, contamination by extraneous materials and determination in the presence of excessive backgrounds and interferences.

These characteristics make the choice of a particular analytical method for a particular analysis a function of an evaluation of the advantages, disadvantages, and limitations of all trace-analysis methods available to the analyst. To assist in such an evaluation a summary of analytical methods for trace elements is given. The criteria for choosing activation analysis are reviewed. The advantages and disadvantages of activation analysis as a trace-element analytical method may be evaluated from its comparison with the other methods.

### 9.1.1 Survey of Analytical Methods for Trace Elements

The decision to use radioactivation analysis for a specific analysis in a given matrix is usually made by analysts who are familiar with the advantages and limitations of the method. It is generally made after comparison of this method with other trace-element analytical methods, and therefore it is advantageous to understand the principles, advantages, and limitations of the other trace-element analytical methods.

Many ways to characterize analytical chemical methods appear in the literature; for example, one pertinent grouping has been given by Pickering,* in which the principles are divided into the following:

1. Phase separations: for example, precipitation, distillation, and gas evolution.

2. Chemical equilibria: for example, acid-base titration, oxidation-reduction, and complex-formation.

3. Electrical transformations: for example, electrodeposition, polarography, and conductivity measurements.

4. Physical properties: for example, refractive index, thermal conductivity, weight losses, and optical rotation.

5. Energy transitions: for example, emission and absorption of electromagnetic radiation by atomic and molecular species.

For trace-element analysis it is the last category of analytical methods that generally has the prerequisite sensitivity for accurate determinations. In fact, we can consider radioactivation analysis to be a nuclear variant of the energy transitions group.

The variants of the energy transitions group, for atomic emission or absorption of electromagnetic radiation, can be classified in several ways:

1. Emission versus absorption.
2. Wavelength of the radiations.
3. Methods of intensity measurement.

* W. F. Pickering, *Fundamental Principles of Chemical Analysis* (Elsevier, Amsterdam, 1966).

Emission spectroscopy is based on exciting atoms of the desired element to an excited state by an external energy source and measuring the characteristic radiation emitted by the excited atoms. The general process consists of three parts:

1. The excitation by some source of energy. The energy source may be electrical discharge, or x-radiation.
2. The dispersion of the emitted radiations in its component wavelengths. The dispersion may be achieved by prisms or diffraction gratings.
3. Measurement of the intensity of the radiation at one or more characteristic wavelengths. The method may be photographic or electronic. The instruments are photometers, spectrophotometers, or spectrographs.

Absorption spectroscopy is based on the removal of a characteristic radiation from the external energy source. Radiations generally used for spectroscopic analysis range from x-rays to microwaves. The properties of these radiations are summarized in Table 9.1.

Emission spectroscopic methods generally used for trace-element analysis include the following:

1. Flame photometry.
2. Electric discharge (visible) emission spectroscopy.
3. X-ray fluorescence.

Absorption spectroscopic methods generally used for trace-element analysis include the following:

1. Colorimetry (spectrophotometry).
2. Fluorimetry.
3. Atomic absorption spectroscopy.

The basic features of each of these methods are examined briefly.

**Table 9.1** Radiations Used for Spectrochemical Analysis

| Type | Frequency (Hz) | Energy (eV) | Energy (kcal/mole) |
| --- | --- | --- | --- |
| X-rays | $10^{18}$ | $4 \times 10^3$ | $10^5$ |
| Vacuum ultraviolet | $10^{17}$ | 400 | $10^4$ |
| Ultraviolet | $1.5 \times 10^{15}$ | 6 | 140 |
| Visible | $7.5 \times 10^{14}$ | 3 | 70 |
| Near infrared | $10^{14}$ | 0.4 | 10 |
| Microwave | $3 \times 10^{10}$ | $1.2 \times 10^{-4}$ | $10^{-3}$ |

### Emission Spectroscopy Methods

FLAME PHOTOMETRY.    Flame photometry is one of the simplest of the emission spectroscopic methods. The thermal energy of a flame is sufficient to excite the least-bound electrons in the atoms of many elements. The sample in solution is atomized by the flame (e.g., acetylene-oxygen) which produces a vapor in which some of the atoms are excited. The ratio of atoms in the excited state $N_e$ to the number in the ground state $N_g$, is given by

$$\frac{N_e}{N_g} = \frac{P_e}{P_g} e^{-E^*/kT} \tag{1}$$

where $P_e$ and $P_g$ are the statistical weights of the two states,
    $E^*$ = excitation energy,
    $k$ = Boltzmann constant,
    $T$ = the absolute temperature.

The emitted light is dispersed by a prism or a grating and is measured with a photometer.

The advantages of flame photometry rest primarily in its simplicity and low cost. Its major disadvantage is the limited number of elements detectable, which can be measured only one at a time. Table 9.2 lists the flame-spectra data for the elements measurable by flame photometry. In practice, however, the method is used primarily for the alkali metals (Li, Na, K, Rb, Cs) and also for the alkaline earth elements (Mg, Ca, Sr, Ba). For the other elements in Table 9.2 trace concentrations are generally measured by other emission methods, such as arc-spark, or by atomic absorption spectroscopy. A comparison of the sensitivities for sodium and zinc indicates the marked differences in the method based on electronic structure; for example, at $2000°K$ the ratio $N_e/N_g$ is about $10^{-5}$ for sodium and about $10^{-14}$ for zinc.

ARC-SPARK SPECTROSCOPY.    Electric discharges as the source for emission spectroscopy is one of the oldest and still widely used methods for trace-element chemical analysis. The electric discharge may be obtained in the form of a spark or a dc arc. The arc method can use either a cathode or an anode sample-holding electrode.

The spark source is generally more reliable for trace-element analysis; the spark produced from an ac transformer with output of 10 to 100 kV produces spectra with many lines in the relatively high energy region of the spectrograph. Provisions for controlling the time of the spark aids in quantitative reproducibility. The dc arc source, which produces temperatures up to $5000°C$, results in greater excitation (sensitivity) than the spark source. This greater sensitivity makes it suitable for trace-element identification (qualitative analysis). The difficulty in stabilizing the arc makes it less reproducible than the spark source.

**Table 9.2**  Sensitivities by Flame Spectrometry[a]

| Element | Minimum Detected (ppm) | Wavelength, (mμ) | Intensity |
|---|---|---|---|
| Barium | 1 | 745[b] | 100 |
| | | 830[b] | 50 |
| Boron | 5 | 548[b] | 20 |
| | | 521[b] | 15 |
| Cadmium | 500 | 326.1 | 0.2 |
| Calcium | 0.3 | 624[b] | 300 |
| | | 554[b] | 200 |
| Cesium | 0.1 | 852.1 | 1,000 |
| Chromium | 3 | 359.3 | 30 |
| | | 645[b] | 30 |
| Cobalt | 5 | 350.2 | 20 |
| | | 352.7 | 20 |
| Copper | 1 | 324.8 | 100 |
| | | 327.4 | 90 |
| Gallium | 1 | 417.2 | 100 |
| Gold | 50 | 267.6 | 2 |
| Indium | 1 | 451.1 | 100 |
| Iron | 10 | 373.6 | 10 |
| Lead | 300 | 405.8 | 0.3 |
| Lithium | 0.05 | 670.8 | 2,000 |
| Magnesium | 10 | 285.2 | 10 |
| | | 370.8 | 10 |
| Manganese | 1 | 403.4 | 100 |
| | | 561[b] | 70 |
| Mercury | 50 | 253.6 | 2 |
| Nickel | 3 | 352.4 | 30 |
| Palladium | 50 | 340.5 | 2 |
| | | 363.5 | 2 |
| Potassium | 0.05 | 766.5 ⎫ 769.9 ⎭ | 2,000 |
| Rubidium | 0.1 | 780.0 | 1,000 |
| Ruthenium | 30 | 372.7 | 3 |
| | | 378.6 | 3 |
| Silver | 2 | 338.3 | 50 |
| Sodium | 0.01 | 589.0 ⎫ 589.6 ⎭ | 10,000 |

Also determinable are the rare earths, selenium and tin.
[a] P. T. Gilbert, Jr., R. C. Hawes, and A. O. Beckman, *Anal. Chem.*, **22,** 772 (1950). [From S. Siggia, *Survey of Analytical Chemistry* (McGraw-Hill, New York, 1968), p. 19.]
[b] An oxide band with its maximum value at this wavelength.

One major advantage of arc-spark emission spectroscopy is its usefulness for most trace elements in any type of sample. Quantitative analysis can be made by focusing the radiations emitted from the sample on a thin slit and dispersing them by prism or diffraction grating. Photographic recording of the separated lines is a common means of analysis. The intensity of the selected trace-element lines is determined by optical density measurement with a microphotometer and comparison by calibration to standard samples. Relative accuracies of $\pm 5\%$ are achieved by photographic recording. For quantitative analysis direct-reading photoelectric instruments which can read selected line intensities with relative accuracies of $\pm 2\%$ are used. These instruments can also be programmed to determine any number of desired elements in given samples. The emission data can be printed as analytical data within a few minutes of arcing, although in many cases the fluctuations in arc intensity may require that measurements be integrated over a long period of time.

Table 9.3 lists the approximate limit of sensitivity by dc-arc spectro-chemical analysis. Actual results will vary widely by sample matrix, interferences, and prior concentration or separation steps. The major limitation to the method is the need for ensuring comparibility between samples and standards. For accurate analysis samples and standards must be similar in chemical composition. Standards can be prepared with known matrix composition of the samples or actual samples must be calibrated from analysis by other methods. The method used for quantitative analysis is generally the comparison of *homologous pairs* which are selected sets of lines of about the same wavelength and with about the same excitation sensitivity, so that both lines are equally affected by any variations in the analysis. One line in the homologous pair is usually from the spectrum of a known macroconstituent in the sample; the other is from the spectrum of the desired trace element.

A comparison of the sample requirements and absolute sensitivities between the flame, spark, and arc excitation sources is given in Table 9.4. The greater sensitivity of the electrical discharge methods is evident.

x-ray fluorescence. Extension of the excitation source to energies beyond optical emission leads to the method of x-ray fluorescence in which the term "fluorescence," as described in Section 5.3.1, denotes the emission of fluorescent x-rays produced in the sample by primary x-rays from the excitation source. This source is generally an x-ray tube, from which the x-rays with wavelengths between 0.3 and 5 Å are generally used. The components of a x-ray spectrometer are shown in Figure 9.1.

The resolution of the characteristic x-rays are made by diffracting crystals. Since the spacings between planes of natural crystals are of the same order of magnitude as the characteristic x-radiations, strong inference maxima are

**Table 9.3** Sensitivities with dc Arc Spectrochemical Analysis[a]

| Element | Approximate Limit of Sensitivity (%) | Element | Approximate Limit of Sensitivity (%) |
|---|---|---|---|
| Ag | 0.001 | Na | <0.001 |
| Al | 0.001 | Nb | 0.01 |
| Au | 0.3 | Nd | 0.1 |
| B (BO) | 0.5 | Ni | 0.01 |
| Ba | <0.001 | Os | 0.1 |
| Be | 0.01 | Pb | 0.001 |
| Bi | 0.001 | Pd | 0.01 |
| Br (BaBr) | 10 | Pt | 0.1 |
| Ca | <0.001 | Rb | 0.1 |
| Cd | 0.001 | Re | 0.1 |
| Ce | 0.05 | Rh | 0.1 |
| Cl (CaCl) | 1.0 | Ru | 0.01 |
| Co | 0.5 | Sb | 5.0 |
| Cr | 0.001 | Sc | 0.005 |
| Cs | 1.0 | Si | 1 |
| Cu | 0.001 | Sn | 0.1 |
| F (CaF) | 0.001 | Sr | <0.001 |
| Fe | 0.05 | Ta | 0.5 |
| Ga | 0.001 | Te | 8.0 |
| Ge | 0.1 | Th | 0.5 |
| Hg | 0.1 | Ti | 0.001 |
| In | 0.001 | Tl | 0.001 |
| Ir | 0.5 | U | 0.5 |
| K | 0.1 | V | 0.1 |
| La | 0.05 | W | 0.1 |
| Li | <0.001 | Y | 0.001 |
| Mg | 0.001 | Zn | 0.001 |
| Mn | 0.001 | Zr | 0.005 |
| Mo | 0.001 | | |

[a] From S. Siggia, *Survey of Analytical Chemistry* (McGraw-Hill, New York 1968), pp. 13–17.

obtained according to Bragg's law, when

$$n\lambda = 2d \sin \theta \qquad (2)$$

where $n$ = integer order of the diffraction,

$\lambda$ = wavelength of the incident radiation,

$d$ = distance between crystal planes,

$\theta$ = angle between the incident radiations and the reflecting plane.

**Table 9.4** Comparison of Flame and Electrical Excitation Sources[a]

| Excitation Source | Sample Requirement | Absolute Sensitivities ($\mu g$) | | | | | | |
|---|---|---|---|---|---|---|---|---|
| | | Co | Zn | Be | Mn | B | Cu | Mo |
| Flame (Lundegardh) | 2-ml solution | 2–20 | 6000 | ... | 0.6 | ... | 1 | ... |
| Flame (Beckman) | 2-ml solution | 2 | 400 | 50 | 0.02 | 2 | 0.02 | 6 |
| Spark (porous cup) | 1-ml solution | 2 | 25 | 0.02 | 0.6 | 0.5 | 0.6 | ... |
| Dc-arc (cathode excitation) | 5-mg solids | 0.01 | 0.5–1.5 | 0.01–0.05 | 0.05–0.3 | 0.05 | ... | 0.05–0.005 |
| Dc-arc (anode excitation) | 50-mg solids | 0.1–0.3 | 0.3–5 | 0.1 | 0.2 | 0.05 | 0.005–0.02 | 0.3 |

[a] Adapted from V. A. Fassel in J. H. Yoe and H. J. Koch, *Trace Analysis* (Wiley, New York, 1957).

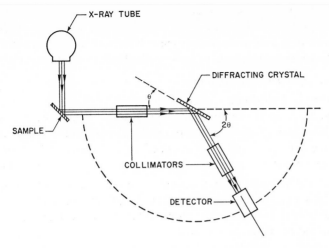

**Figure 9.1** Schematic diagram of an x-ray fluorescence spectrometer.

Crystals in common use include the alkali halides (e.g., LiF), quartz, topaz, and other such minerals. Most of the diffraction intensity is present in the first-order line ($n = 1$). The $d$ spacings for these crystals range from 1.36 Å for topaz to 5.32 Å for $NH_4H_2PO_4$. These crystals in an air system are not useful for analysis of elements with $Z < 20$ because of the excessive absorption of the low· energy characteristic x-rays. The use of helium reduces the lower limit measurable to $Z \simeq 12$ and vacuum-path instruments enable elements down to $Z \simeq 9$ to be measured. However, another limitation to low-$Z$ element analysis is the lack of crystals with $d$ spacings large enough to diffract the desired x-rays in the instrument. Detectors used in x-ray fluorescence analyzers are generally scintillation, flow proportional, or Geiger detectors.

The advantages of x-ray fluorescence as a trace-element method is its specificity for the chemical elements, the speed of analysis, and simple resolution of the spectra; these are beneficial for computer-controlled analysis. Many are nondestructive. The disadvantages, besides that of low-$Z$ sample difficulties, include those of sensitivity and need for comparitor standards, especially when samples of thickness greater than about 100 $\mu$ are analyzed. Matrix effects, such as absorption of primary and emitted x-rays, can be important, especially for macroconstituents with $Z = Z - 1$ of the desired element. The sensitivity of x-ray fluorescence is probably the most limiting problem for trace-element analysis. Without prior chemical concentration, sensitivities determined by signal-to-noise ratios of good detectors are of the order of $10^{-4}$ g/g. With sample concentration the sensitivities can be increased to about $10^{-6}$ g/g. Accuracies attainable at minor element concentrations ($10^{-3} - 10^{-5}$ g/g) are better than $\pm 5\%$ with adequate standards duplicating the matrix as well as the desired element effects. The method, like most emission spectrometry methods, is readily adaptable to automatic operation.

### Absorption Spectroscopy Methods

COLORIMETRY. Colorimetry (also known as photometry or spectrophotometry) methods are based on the light absorptive capacity of solutions containing low concentrations of colored ions or suspensions. These methods are among the oldest and still most useful of trace-element analysis methods. The two steps involved in a colorimetric chemical analysis are the following:

1. Chemical development of a suitably colored solution of the desired element, after separation, if necessary, from interfering elements.

2. Measurement, in a suitable monochromatic optical system, of the light transmitted through the solution.

Photometer or spectrophotometer systems are based on the Lambert

**Table 9.5**   Examples of Colorimetric Analysis[a]

| Element | Reagent | Approximate[b] Sensitivity |
|---------|---------|:--------:|
| Ag | p-Diethylaminobenzylidene-rhodanine | 2 |
| Al | Aurintricarboxylate | 1 |
| As | Molybdate-hydrazine sulfate | 2 |
| Au | p-Diethylaminobenzylidene-rhodanine | 2 |
| B | Substituted anthraquinonylamines | ?2 |
| Bi | Na-diethyldithiocarbamate and NaCN | ?2 |
| Cd | Dithizone | 1 |
| Co | Nitroso-R salt | 1 |
| Cr | Diphenylcarbazide | 1 |
|  | as $CrO_4^{2-}$ | 2 |
| Cu | Dithizone | 1 |
|  | Na-diethyldithiocarbamate | 3 |
| F | Aluminon-eriochrome cyanine | 1–2 |
| Fe | o-Phenanthroline | 1 |
|  | thiocyanate | 2 |
| Ga | Rhodamine B | 1 or 2 |
| Ge | Molybdate—$FeSO_4$ | 2 |
| Hg | Dithizone | 2 |
| K | Na-tetraphenylboron | mg quantities |
| Li | Thoron | 2 |
| Mn | as $MnO_4^-$ | 3 |
| Mo | Thiocyanate-stannous chloride | 3 |
| Nb | Hydrogen peroxide in $H_3PO_4$–$H_2SO_4$ | 3 |

## Table 9.5 (Continued)

| Element | Reagent | Approximate[b] Sensitivity |
|---------|---------|---------------------------|
| Ni | Dimethylglyoxime | 2 |
| Os | Thiourea | 4 |
| P | Molybdivanadophosphate | 2–1 |
| Pb | Dithizone | 1 |
| Pd | p-Nitrosodiphenylamine | 4 |
| | 3-hydroxy-1-(p-sulfenyl)-3-phenyltriazene | ?2 |
| Pt | Stannous chloride | 3 |
| Re | Thiocyanate-stannous chloride | 2 |
| S (as SO₄) | Barium chloranilate | ?2 |
| Sb | Rhodamine B | 1 |
| Si | Molybdate | 3 |
| Ta | Gallic acid | ? |
| Ti | Disodium-1,2-dihydroxy-benzene-3,5-disulfonate | 1 |
| | Hydrogen peroxide | 4 |
| V | Phosphotungstic acid | 3 |
| W | Thiocyanate-stannous chloride | 4 |
| Zn | Dithizone | 1 |

[a] From E. A. Vincent, Analysis by Gravimetric and Volumetric Methods, Flame Photometry, Colorimetry, and Related Techniques, Chapter III, in Smales and Wager, *Methods in Geochemistry* (Interscience, New York, 1960).

[b]

| Class | Minimum Detectable Amount, ($\mu$g/cm$^2$) |
|-------|--------------------------------------------|
| 1 | 0.001–0.005 |
| 2 | 0.005–0.01 |
| 3 | 0.01–0.05 |
| 4 | >0.05 |

(or Bouguer)-Beer law of optical absorption:

$$\frac{I}{I_0} = 10^{-kCd} \tag{3}$$

where $I$ = intensity of transmitted light,
$I_0$ = intensity of incident light,
$k$ = extinction (absorption) coefficient,
$C$ = concentration of the dilute colored solution,
$d$ = thickness of solution normal to the beam of light.

Thus for the fixed path length $d$ in spectrophotometers the concentration of the colored ion or suspension is directly proportional to the *optical density*, $\log_{10} I_0/I$. Spectrophotometers are generally calibrated to relate the concentration $C$ to the transmittance $I/I_0$ in percent. Measurements of transmittance are attainable with a precision of about 0.2% at values of 80 to 90%. For maximum reliability concentrations are generally adjusted to obtain transmittance values in the range between 20 to 60%.

A selection of some colorimetric reactions and their approximate sensitivity for trace-element analysis is given in Table 9.5. The sensitivities are grouped into four ranges of detection level. Variables that affect the accuracy of the transmittance and thus the analysis include the measurement wavelength, chemical system, spectral band width, and temperature.

FLUORIMETRY.  Fluorescence induced in dilute solutions by ultraviolet light is another useful phenomenon for trace-element analysis. The fluorescence process in crystalline materials was described in Section 5.3.1. The relationship between the intensity of fluorescence and the concentration of a fluorescent material in solution may be derived from the Lambert (Bouguer)-Beer law (3), in which the fraction of light *absorbed* is given by

$$1 - \frac{I}{I_0} = 1 - 10^{-kCd} \tag{4}$$

from which

$$\Delta I = I_0 - I = I_0(1 - 10^{-kCd}) \tag{5}$$

Since the intensity of the fluorescence $F$ is proportional to the amount of light absorbed,

$$F = KI_0(1 - 10^{-kCd}) \tag{6}$$

If $kCd$ is small ($<0.05$), the term $(1 - 10^{-kCd})$ can be expanded according to (7-1), and, neglecting terms of power greater than 1, the relationship between fluorescence and concentration is given by

$$F = (2.3KI_0kd)C \tag{7}$$

For values of $kCd > 0.05$ the relation between $F$ and $C$ is determined by instrument calibration.

Examples of fluorimetric analysis data are given in Table 9.6. For the pertinent elements the method has been shown to be sufficiently sensitive for trace-element analysis. A major limitation of the method is the constant hazard of sample contamination. Such common laboratory materials as filter paper, corks, rubber stoppers, and alcohol contain materials that can increase or quench the fluorescence of the desired elements.

Table 9.6   Examples of Fluorometric Analysis[a]

| Element | Reagent | Fluorescence Wavelength $(m\mu)$ | Sensitivity $(\mu g)$ |
|---------|---------|---------------------------------|-----------------------|
| Al | Pontachrome BBR | 635–700 | 0.2 |
| Al | Morin | 525–570 | 0.0005 |
| Al | Oxine | 450–610 | 0.1 |
| Be | Quinizarin | 570–640 | 1.0 |
| Be | Morin | 510–615 | 0.004 |
| Zr | Flavonol | 420–520 | 0.1 |
| F | Al-Alizarin Garnet R | 550–600 | 0.1 |
| U | NaF melt | 540–550 | 0.0001 |
| B | Benzoin | 440–630 | 0.2 |
| Tb | $TbCl_3$ (no addition) | 525–555 | 0.005% |
| Li | Oxine | 490–570 | 3.0 |

[a] Adapted from C. E. White, Fluorometry, Chapter 7 in Yoe and Koch, Eds. *Trace Analysis* (Wiley, New York, 1957).

ATOMIC ABSORPTION SPECTROSCOPY.   One of the newer methods for trace-element analysis of great promise is atomic absorption spectroscopy. The method is a flame emission method that uses the absorption process to increase the sensitivity of the flame-photometric determination. It was noted in the description of flame photometry that only a small fraction of the vaporized sample atoms are excited to produce the emission spectrum. Thus almost all of the atoms of the desired element atomized in the flame are unexcited atoms. In atomic absorption spectroscopy, resonance radiation of the desired element is directed through the flame. This radiation is efficiently absorbed by the atoms of the desired element in the flame-atomized sample. A monochromatic light source is made in the form of a hollow cathode tube which contains the desired element. The light emitted by these tubes can be obtained mono-chromatically, with half-widths, of the order of 0.01 A, which gives the absorption process a high specificity for the desired element and results in a quantitative method in which the amount of absorption is proportional to the amount of the desired element in the sample. The method is not only sensitive but it eliminates another problem of flame photometry in its independence of flame temperature.

Figure 9.2 shows a schematic diagram of an atomic absorption spectrometer. The unit is generally available as an accessory to a flame spectro-photometer. The total instrument is thus more expensive than the flame spectrophotometer but still less expensive than the arc-spark or the x-ray fluorescence spectrometer.

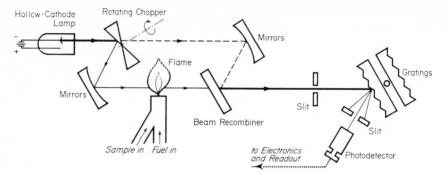

**Figure 9.2** Schematic diagram of an atomic absorption spectrometer (Courtesy of Perkin-Elmer Co.)

Table 9.7 lists some of the determinations that can be obtained by atomic absorption spectroscopy. This method is useful for a large number of elements that can be determined in concentrations as low as 0.1 to 0.01 $\mu$g/ml, with adequate accuracies, if interferences are removed or chemically reduced by complexing. The major limitation to this method is that it can determine only one element at a time. In fact a separate cathode tube, costing more than $100, is required for each element. As with other emission or absorption techniques, adequate calibration of the system with standards in suitable matrix form is required.

### 9.1.2   Criteria for Choice of Activation Analysis

The development of analytical chemistry for the determination of trace elements has a rather short history. It is only recently that the impetus for this development came from researches in high-purity materials, biology and medicine, forensic investigations, and transport processes in all types of environment. In this short period of time, however, many excellent trace-element analytical methods have been developed, most of which are based on the atomic characteristics of the elements. Some of the more pertinent are the spectrochemical methods described in Section 9.1.1. Radioactivation analysis, in common with only a few other methods, such as Mössbauer and nuclear magnetic resonance, is based on the nuclear characteristics of the atoms of the elements. With such a large variety of analytical methods for trace elements, it is indeed difficult to establish a set of criteria that clearly shows that any one of these methods is superior to the others in a general way.

Radioactivation analysis has been singled out from among the few analytical chemistry methods based on the properties of the atomic nucleus as the general method of elemental analysis. The literature already contains reports

**Table 9.7** Relative Detection Limits with Atomic-Absorption Spectroscopy[a]

| Metal | Relative Detection Limit[b] ($\mu$g/ml) | $\lambda$ (Å) | Slit (Å) | mA | Metal | Relative Detection Limit[b] ($\mu$g/ml) | $\lambda$ (Å) | Slit (Å) | mA |
|---|---|---|---|---|---|---|---|---|---|
| Ag | 0.01 | 3281 | 7 | 12 | Na | 0.005 | 5890 | 40 | 500 |
| Al[d] | 0.1 | 3093 | 0.7 | 40 | Nb[d] | 20 | 4059 | 2 | 40 |
| As | 0.5 | 1937 | 7 | 12 | Nd[d,f] | 2 | 4634 | 2 | 25 |
| Au | 0.1 | 2428 | 20 | 14 | Ni[h] | 0.01 | 2320 | 0.7 | 15 |
| B[d] | 15 | 2497 | 7 | 30 | Os | | | | |
| Ba[d,f] | 0.1 | 5536 | 40 | 20 | Pb[h] | 0.01 | 2170 | 2 | 10 |
| Be[d] | 0.003 | 2349 | 2 | 50 | Pd | 0.5 | 2476 | 2 | 20 |
| Bi | 0.02 | 2231 | 2 | 10 | Pr[d,f] | 10 | 4951 | 2 | 25 |
| Ca[d] | 0.003 | 4227 | 20 | 10 | Pt | 0.5 | 2659 | 2 | 25 |
| Cd | 0.01 | 2288 | 7 | 4 | Rb[c,f] | 0.005 | 7800 | 40 | 400 |
| Ce[d] | . . . | 5200 | | | Re[d] | 1.5 | 3460 | 2 | 30 |
| Co[c,h] | 0.007 | 2407 | 2 | 15 | Rh | 0.03 | 3435 | 2 | 20 |
| Cr | 0.005 | 3579 | 2 | 10 | Ru | 0.3 | 3499 | | |
| Cs[c,f] | 0.05 | 8521 | 40 | 400 | Sb | 0.2 | 2175 | 7 | 20 |
| Cu | 0.005 | 3247 | 7 | 10 | Sc[d] | 0.2 | 3912 | 7 | 40 |
| Dy[d,f] | 0.2 | 4212 | 2 | 30 | Se | 0.5 | 1961 | 2 | 20 |
| Er[d,f] | 0.2 | 4008 | 2 | 30 | Si[d,h] | 0.2 | 2516 | 2 | 30 |
| Eu[d,f] | 0.2 | 4594 | 2 | 20 | Sm[d,f] | 5 | 4297 | 2 | 25 |
| Fe | 0.01 | 2483 | 2 | 40 | Sn[g,h] | 0.1 | 2246 | 7 | 8 |
| Ga | 0.07 | 2874 | 20 | 4 | Sr | 0.01 | 4607 | 13 | 10 |
| Gd[d,f] | 4 | 3684 | 2 | 20 | Ta[d] | 6 | 2715 | 2 | 30 |
| Ge[d] | 2 | 2651 | 2 | 16 | Tb[d,f] | 2 | 4326 | 2 | 25 |
| Hf[d] | 15 | 3072 | 2 | 30 | Te | 0.3 | 2143 | 7 | 16 |
| Hg[e] | 0.2 | 2537 | 20 | 200 | Th[f] | | | | |
| Ho[d,f] | 0.3 | 4163 | 2 | 25 | Ti[d,h] | 0.1 | 3643 | 2 | 40 |
| In | 0.05 | 3040 | 7 | 6 | Tl | 0.2 | 2768 | 20 | 12 |
| Ir[d] | 4 | 2850 | 7 | 40 | Tm[d] | 0.1 | 4094 | | |
| K[c] | 0.005 | 7665 | 40 | 400 | U[d] | 30 | 3514 | 2 | 40 |
| La[d] | 80 | 3928 | 0.7 | 30 | V[d,h] | 0.1 | 3184 | 7 | 30 |
| Li | 0.005 | 6708 | 40 | 15 | W[d] | 3 | 4008 | 7 | 50 |
| Lu[d] | 50 | 3312 | | | Y[d,f] | 0.3 | 4077 | 2 | 25 |
| Mg | 0.0005 | 2852 | 20 | 6 | Yb[d] | 0.04 | 3988 | 2 | 20 |
| Mn | 0.005 | 2795 | 20 | 10 | Zn | 0.002 | 2138 | 20 | 10 |
| Mo | 0.1 | 3133 | 2 | 30 | Zr[d] | 5 | 3601 | 2 | 55 |

[a] From S. Siggia, *Survey of Analytical Chemistry* (McGraw-Hill, New York, 1968), p. 22.
[b] Perkin-Elmer model 303 atomic absorption spectrometer.
[c] Osram spectral lamp is used.
[d] A nitrous oxide–acetylene flame is required.
[e] General Electric OZ4 mercury germicidal lamp.
[f] In presence of high concentrations of another ionizable metal.
[g] Using an air-hydrogen flame.
[h] Using Perkin-Elmer high-brightness lamps.

on uses of activation analysis for determinations of more than 70 of the chemical elements. Although the development of activation analysis was accelerated by the nuclear scientists who had access to the initial nuclear irradiation facilities and the latest nuclear radiation measurement equipment, much recent development has been made by the metallurgist, the geoscientist, the biologist, the medical doctor, the criminologist, and others who have come to regard activation analysis as an analytical tool. This method is also one of the few analytical analysis methods that is taught as an academic subject in several universities. This progress has been made in spite of the several limitations described in Chapter 8. Not all trace element analyses are easily obtained by radioactivation analysis; many others which can be so obtained are more easily made by spectrochemical analysis. The decision to use activation analysis as *the* method for a specific requirement should thus be based on some criteria for choice of analytical method.

In general, the choice of a method for trace-element analysis should involve the consideration of some or all of the following criteria:

1. Range of concentration of the desired elements.
2. Required or desired accuracy.
3. Delay time tolerable.
4. Interferences present in the matrix.
5. Number or frequency of required analyses.
6. Availability or economic justification of pertinent equipment.
7. Skill and experience of the analyst.

Furthermore, the choice of a method for a particular analytical requirement may be worthy of continuous or frequent review because of two time-dependent properties of analytical chemistry:

1. Existing methods of analytical chemistry are under constant improvement and development; for example, an interference that dictates the elimination from consideration of a method otherwise satisfactory may subsequently be found to be easily removed by some new chemical reagent.

2. New methods of chemical analysis are frequently evolved; for example, although radioactivation analysis is a comparatively new method, such methods as nuclear resonance, atomic, and Mössbauer absorption spectroscopy have since been developed.

There are indeed many advantages of activation analysis that justify its choice as a trace-element analytical method, foremost among which is the sensitivity of the method. The comparison of its sensitivity with other methods is described in Section 9.1.3. Another major advantage is its selectivity. This advantage results from the great number of combinations of nuclear reactions and nuclear radiations available for each element to avoid troublesome interferences in most cases. Radiochemical separation can increase selectivity and sensitivity even further. Other advantages, which are sometimes the criteria for choice of activation analysis, are the ability to obtain the analysis with nondestruction of the sample, speed, economy, or convenience of the analysis, and the potential ability to regulate and automate the determination of elements on a continuous on-line, multisample basis.

### 9.1.3 Comparison of Activation Analysis with Other Methods

One of the earlier comparisons of activation analysis with other trace-element analytical methods was made by Meinke* in 1955. At that time the development of activation analysis was proceeding at an accelerating rate because of the increasing availability of higher flux thermal-neutron irradiation facilities. In his comparison Meinke selected the sensitivities reported by Leddicotte and Reynolds† and five typical trace-element analytical methods then in general use. The comparison was based primarily on *sensitivity*, although it was noted that such considerations as interference, speed, and multielement analysis were also important. His comparison of sensitivities is reproduced in Table 9.8. These values were obtained with the introduction of gross normalizing factors for each method to adjust the values for the analysis of solutions and soluble salts to the same units as the activation analysis units. They were therefore given as "order of magnitude" values only.

A more detailed comparison for specific elements was added from analysis of the data in Table 9.8. By bar-graphic presentation of method versus log sensitivity (in $\mu$g/ml) Meinke noted that the elements for which activation analysis was decidedly more sensitive included the rare earths, manganese, indium, rhenium, and iridium. Other elements whose sensitivity comparison indicated a potential for activation analysis included aluminum, vanadium, copper, gold, arsenic, and antimony. Among the elements for which activation analysis was not considered advisable, unless some special condition required it, were iron, calcium, lead, and bismuth.

A more recent comparison of trace-element analytical methods has been made by Smales.‡ In addition to sensitivity, he lists as the factors important for a comparison multielement analysis, specificity, accuracy, and reagent and surface contamination hazards. He noted the frequently overlooked problem that reagent blanks in activation (and other methods of) analysis

---

* W. W. Meinke, Trace-Element Sensitivity: Comparison of Activation Analysis with Other Methods, *Science* **121**, 177–184 (1955).

† G. W. Leddicotte and S. A. Reynolds, Neutron Activation Analysis: A Useful Analytical Method for Determination of Trace Elements, AECD-3489 (January 1953).

‡ A. A. Smales, The Place of Activation Analysis in a Research Establishment Dealing with Pure Materials, in *Modern Trends in Activation Analysis* (Texas A&M University, College Station, 1965), pp. 186–188.

Table 9.8  Comparison of Sensitivity by Neutron Activation
Analysis and Spectrochemical Methods[a]

| Element | Z | Neutron Activation Analysis | Spectrochemical Methods | | | |
|---------|---|-----------------------------|------------------------|-----|-----|-----|
| | | | Copper Spark | Dc Arc | Flame Photometry | Colorimetry |
| Li | 3 | | 0.002 | | 0.02 | |
| Be | 4 | | 0.002 | | 250 | 0.04 |
| B | 5 | | 0.1 | | 10 | |
| Na | 11 | 0.00035 | 0.1 | 20 | 0.0002 | |
| Mg | 12 | 0.03 | 0.01 | 0.1 | 1 | 0.06 |
| Al | 13 | 0.00005 | 0.1 | 0.2 | 20 | 0.002 |
| Si | 14 | 0.05 | 0.1 | 2 | | 0.1 |
| P | 15 | 0.001 | 20 | 50 | | 0.001 |
| S | 16 | 0.2 | | | | |
| Cl | 17 | 0.0015 | | | | 0.04 |
| K | 19 | 0.004 | 0.1 | | 0.01 | |
| Ca | 20 | 0.19 | 0.1 | | 0.03 | |
| Sc | 21 | 0.0001 | 0.005 | | | |
| Ti | 22 | | 0.1 | | 2 | 0.03 |
| V | 23 | 0.00005 | 0.05 | | 2 | 0.2 |
| Cr | 24 | 0.01 | 0.05 | 2 | 1 | 0.02 |
| Mn | 25 | 0.00003 | 0.02 | 0.2 | 0.1 | 0.001 |
| Fe | 26 | 0.45 | 0.5 | 0.2 | 2 | 0.05 |
| Co | 27 | 0.001 | 0.5 | | 10 | 0.025 |
| Ni | 28 | 0.0015 | 0.1 | 4 | 10 | 0.04 |
| Cu | 29 | 0.00035 | | 0.2 | 0.1 | 0.03 |
| Zn | 30 | 0.000 | 2 | 20 | 2000 | 0.016 |
| Ga | 31 | 0.00235 | 1 | | 1 | |
| Ge | 32 | 0.002 | | | | 0.08 |
| As | 33 | 0.0001 | 5 | 10 | | 0.1 |
| Se | 34 | 0.0025 | | | | |
| Br | 35 | 0.00015 | | | | |
| Rb | 37 | 0.0015 | 0.2 | | 0.1 | |
| Sr | 38 | 0.03 | 0.5 | | 0.1 | |
| Y | 39 | 0.0005 | 0.01 | | 50 | |
| Zr | 40 | 0.015 | 0.1 | | | 0.13 |
| Nb | 41 | 0.5 | 0.2 | | 20 | 50 |
| Mo | 42 | 0.005 | 0.05 | | 30 | 0.1 |
| Ru | 44 | 0.005 | | | 10 | 0.2 |
| Rh | 45 | | | | 1 | 0.2 |
| Pd | 46 | 0.00025 | 0.5 | | 1 | 0.1 |
| Ag | 47 | 0.0055 | | 0.1 | 0.5 | 0.1 |

Table 9.8 (*continued*)

| Element | Z | Neutron Activation Analysis | Spectrochemical Methods | | | |
|---|---|---|---|---|---|---|
| | | | Copper Spark | Dc Arc | Flame Photometry | Colorimetry |
| Cd | 48 | 0.0025 | 2 | 4 | 20 | 0.01 |
| In | 49 | 0.000005 | 1 | | 1 | 0.2 |
| Sn | 50 | 0.01 | | 0.2 | 10 | |
| Sb | 51 | 0.0002 | 5 | 4 | | 0.03 |
| Te | 52 | 0.005 | 0.5 | | 100 | 0.5 |
| I | 53 | 0.0001 | | | | |
| Cs | 55 | 0.0015 | 0.5 | | 1 | |
| Ba | 56 | 0.0025 | 0.1 | | 3 | |
| La | 57 | 0.0001 | 0.05 | | 5 | |
| Ce | 58 | 0.005 | 0.5 | | 20 | 0.25 |
| Pr | 59 | 0.0001 | 0.2 | | 100 | |
| Nd | 60 | 0.005 | 0.2 | | 50 | |
| Sm | 62 | 0.00003 | 0.2 | | 100 | |
| Eu | 63 | 0.0000015 | 0.02 | | | |
| Gd | 64 | 0.001 | 0.1 | | 10 | |
| Tb | 65 | 0.0002 | | | | |
| Dy | 66 | 0.0000015 | 0.5 | | 10 | |
| Ho | 67 | 0.00002 | 0.2 | | | |
| Er | 68 | 0.001 | 0.5 | | | |
| Tm | 69 | 0.0001 | 0.05 | | | |
| Yb | 70 | 0.0001 | 0.1 | | | |
| Lu | 71 | 0.000015 | 2 | | | |
| Hf | 72 | 0.001 | 0.5 | | | |
| Ta | 73 | 0.00035 | 1 | | | |
| W | 74 | 0.00015 | 0.5 | | | 0.4 |
| Re | 75 | 0.00003 | 2 | | | 0.05 |
| Os | 76 | 0.001 | | | | 1 |
| Ir | 77 | 0.000015 | 5 | | | 2 |
| Pt | 78 | 0.005 | 0.02 | | | 0.2 |
| Au | 79 | 0.00015 | 0.2 | | 200 | 0.1 |
| Hg | 80 | 0.0065 | 5 | 2 | 100 | 0.08 |
| Tl | 81 | 0.03 | | 0.2 | 1 | |
| Pb | 82 | 0.1 | 0.05 | 0.2 | 20 | 0.03 |
| Bi | 83 | ~0.02 | 0.2 | 0.2 | 300 | 1 |
| U | 92 | 0.0005 | 1 | | 10 | 0.7 |

[a] From W. W. Meinke, Trace-Element Sensitivity: Comparison of Activation Analysis with Other Methods, *Science* **121**, 177–184 (1955). Sensitivities in $\mu$g/ml based on gross normalizing factors introduced to adjust values to those for neutron activation analysis with a reactor flux of $10^{13}$ n/cm$^2$-sec.

**Table 9.9** Comparison of Selected Methods for Analysis of High-Purity Materials[a]

| | Several Elements Simultaneously | Sensitivity (ppm) | Specificity | Accuracy | Freedom from Contamination, Reagents Blanks, etc. | Possibility of Overcoming Surface contamination | Comments |
|---|---|---|---|---|---|---|---|
| Absorption spectroscopy | No | 0.1–1 | Reasonable | Good | Bad | Bad | |
| Fluorescence spectroscopy | No | 0.1 | Reasonable | Fairly good | Bad | Bad | |
| Emission spectroscopy   Direct | Yes | 1 | Good | Needs standards | Good | Good | |
|   After chemical concentration | Yes | 0.1 | Good | Reasonable | Bad | Bad | |
| Fluorescent x-ray spectroscopy | Yes | 10–100 | Good | Needs standards | Good | Possible by precleaning | |
| Mass spectrometry   Isotope dilution | Limited | Very high | Good | Reasonable | Bad | Bad | |
|   Vacuum spark | Yes | 0.01 | Good | Needs standards | Good | Good | |
|   Gas analysis | Yes | Recently improved | Fair ($N_2$-CO) | Good | Good | Good | |

| | | | | | | | For chemical form |
|---|---|---|---|---|---|---|---|
| Resonance spectroscopy — Nuclear magnetic | No | Poor | | | | | For chemical form |
| Electron spin | Possibly | Fair | | | | | For chemical form |
| Vacuum-fusion (extraction) | Oxygen Hydrogen Nitrogen | 1 | Reasonable | Reasonable | Good | Bad | |
| Gas chromatography | Yes | 1 | Good | Good | | | |
| Electrochemical methods | Fair | 0.1 | Reasonable | Good | Bad | Bad | |
| Radiotracers | No | high | Good | ? | Good | Bad | Not direct determination but useful for transfer, etc. |
| Radioactivation | In some cases | Very high | Good | Good | Good | Good | |

[a] From A. A. Smales, The Place of Activation Analysis in a Research Establishment Dealing with Pure Materials, in *Modern Trends in Activation Analysis* (Texas A&M University, College Station, 1965), pp. 186–188.

may well be the limiting factor rather than its absolute sensitivity. The comparison of his selected methods for trace-element analysis of high-purity materials is summarized in Table 9.9.

It was noted that for requirements of qualitative survey of all possible impurities, spark-source mass spectrometry is probably the first choice of method, followed closely by emission spectroscopy. For specified impurities, however, activation analysis was noted to excel in terms of general sensitivity, especially because of the alleviation of any post-irradiation contamination problems. For the specific requirement of determining the exact location of an impurity in a given sample he nominated the method of fluorescent x-ray spectroscopy with the electron microprobe.

Among the elements for which Smales describes activation analysis as important in establishing "analyzed samples" in a research establishment dealing with pure materials he includes aluminum, silicon, calcium, iron, and zinc. In the category of "indispensable activation methods of analysis" he notes the determination of zirconium with a sensitivity of 0.1 $\mu$g (even in the presence of its chemical homolog, hafnium) oxygen (by fast neutron or $^3$He ion activation), and carbon (by photoactivation).

In an interesting *military* conclusion, Smales gives the following quotation:

"Activation analysis is clearly one of the 'corps elites' in the vanguard of the army of methods available to the modern analytical chemist. Its shock-troop role is unexcelled, and of course in certain instances it has captured whole areas with its speed and direct methods. No doubt, too, that it is playing its part in the more general artillery role, and this aspect will clearly increase. But its officers must be aware of the development of competing arms of the whole force; otherwise they may be outgunned if not outgeneralled. Perhaps one great advantage of the activation-analysis corps is its possession of the nuclear weapon!"*

## 9.2 SURVEY OF USES OF ACTIVATION ANALYSIS FOR TRACE ELEMENTS

The beginning of the activation analysis literature is credited to Hevesy and Levi* who in 1936 described the use of 2.3 Ci of radium in a Ra–Be neutron

---

* A. A. Smales, The Place of Activation Analysis in a Research Establishment Dealing with Pure Materials, in *Modern Trends in Activation Analysis* (Texas A&M University, College Station, 1965), p. 188.
† G. V. Hevesy and H. Levi, The Action of Neutrons on the Rare Earth Elements, *Kgl. Danske Videnskab. Mat-fys. Medd.* **14**, (5), 3–34 (1936).

source to activate rare-earth elements. The first U.S. publication by Seaborg and Livingston* in 1938 described the determination of trace quantities of gallium in high-purity iron by charged-particle activation.

An illustration of the growth of activation analysis is given effectively by the "literature growth curve" used as the *motif* of the NBS bibliography on activation analysis. This "growth" curve, reproduced in Figure 9.3, shows a "doubling period" of three years since about 1950.

Activation analysis has been applied to the determination of elements in almost every aspect of the material world accessible to man. A rational division of these aspects has been made by Lenihan,† who related the applications of activation analysis to the study of the nature of the environment and of man's interaction with it. Lenihan chose as the six components of man's environment cosmic, geological, biological, internal biological, artificial, and social.

A somewhat different, though by no means better, division of man's environment might consist of the physical, biological, and industrial material

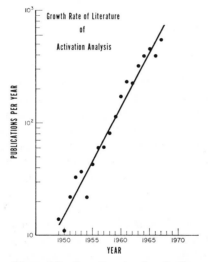

**Figure 9.3** Growth rate of activation analysis literature showing a "doubling period" of about three years. [From *Activation Analysis: A Bibliography*, National Bureau of Standards Technical Note 467 (1968).]

* G. T. Seaborg and J. J. Livingood, Artificial Radioactivity as a Test for Minute Traces of Elements, *J. Am. Chem. Soc.* **60**, 1784–1786 (1938).

† J. M. A. Lenihan, Activation Analysis in the Contemporary World, in *Modern Trends in Activation Analysis* (National Bureau of Standards, Gaithersburg, Md., 1969), pp. 1–32.

parts of the world. For convenience the physical realm is considered as a set of concentric spheres with decreasing radius of influence defining the cosmosphere, atmosphere, hydrosphere, and lithosphere.

The "spheres of interest" in man's biological environment, which may be termed the biosphere, are considered to include the life processes existing in the lower atmosphere, the upper layer of the oceans, and the upper crust of the earth.

In the context of a *biosphere* man's environment is considered to include biochemistry, biology, internal medicine, and environmental sciences.

The industrial applications of activation analysis have taken many forms, from the routine determination of trace elements in all types of industrial material to several special applications. Several of the latter applications are described in Section 9.3. Although a detailed examination of the uses of activation analysis for trace-element determinations in these many aspects of man's environment is beyond the scope of this chapter, a summary of just a few of them will emphasize how extensively the method is being adapted.

### 9.2.1   The Physical World

#### The Cosmosphere

The origin, composition, and behavior of the universe and how the chemical elements formed in it have been questions that have intrigued man since earliest recorded history. Until recently almost nothing was known about the chemical composition of the universe beyond man's reach into the earth's atmosphere. The development of emission spectrochemical analysis for earth samples enabled the first estimates of the elemental composition of stellar matter to be made. The measurement of the hydrogen lines in starlight not only confirmed its presence but aided significantly in the understanding of thermonuclear processes as a universe energy source. The development of other telemetric analytical methods may significantly increase our determination of the chemical nature of universe matter. The only other means of achieving such knowledge is to analyze samples of universe matter. Such samples can be obtained in two ways:

(a) Waiting until meteoritic samples fall upon earth; (b) sending analytical systems to universe bodies to measure the composition *in-situ* or gathering samples by landed spacecraft for analysis after return to earth.

METEORITES.   Meteoritic matter has fallen on earth in sizes ranging from tiny stones, weighing less than 0.07 g, to giant boulders, weighing more than 60 tons. The elemental composition of meteorites has been extensively studied not only to obtain information about the evolution of the solar systems but also for evidence of the chemical composition of the earth's interior which may never be directly accessible.

Several general observations have been made as a result of the extensive chemical analysis of meteorites:

1. No chemical element has been found in meteorites that does not also occur on earth.

2. No substances have been discovered that require the presence of living organisms.

3. Of the 92 elements in the periodic table (excluding the transuranium elements with $Z > 92$) only the elements technicium ($Z = 43$), promethium ($Z = 61$), astatine ($Z = 85$), and francium ($Z = 87$) have not yet been identified in meteorites.

The abundances of the elements in the cosmosphere have been deduced from chemical analysis of meteorites in combination with solar and stellar spectral data and theoretical considerations. The abundances of the chemical elements, relative to silicon, in "cosmic" matter are shown in Figure 9.4. The relative scale covering more than 12 orders of magnitude emphasizes the characteristic major abundances of hydrogen and helium compared with all other elements. Other general features are the decrease in abundance with increasing atomic number and the odd-even effect of nuclear stability noted so early in the development of nuclear physics.

In comparing the elemental analysis of meteoritic matter (fallen on earth) with the elemental analysis of the earth, it must be noted that two problems exist that make such comparisons difficult. The first is the physical history of the meteorite's fall through the earth's atmosphere. The intense heating may cause losses of the volatile components of the meteor. The second is the tendency of meteorites to fall into several classes of chemical composition, these classes based, in turn, on the tendency of the elements to concentrate in geochemical groups.

| | |
|---|---|
| Siderophile (iron phase) | association with metallic iron. |
| Chalcophile (troilite) | an affinity for sulfur. |
| Lithophile (silicate phase) | an affinity for oxygen. |
| Atmophile (volatiles) | occurs as gaseous components. |

The elements that constitute these classifications are listed in Table 9.10.

Meteorites are generally classified into three principal types, according to major element composition: *irons*, *stony irons*, and *stones*, the irons consisting primarily of metallic iron, the stones, of mineral silicates, and the stony irons intermediate in composition between the other two. Analysis of some 700 fallen meteorites shows the distribution of the three to be about 6% for iron

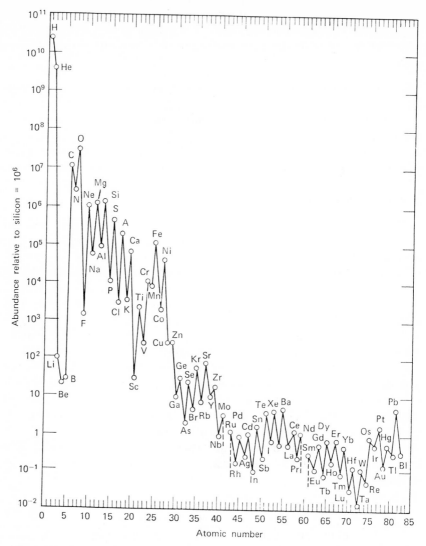

**Figure 9.4** "Cosmic" abundance diagram of the elements, relative to $10^6$ atoms of silicon. [From L. H. Ahrens, *Distribution of the Elements in Our Planet* (McGraw-Hill, New York, 1965), p. 14.]

meteorites, 2% for stony irons, and 92% for stones. The stony meteorites are generally subdivided further into two subtypes, *chondrites* and *achondrites*, the former being the more abundant and most closely approximating the chemical composition of the earth's crust.

The elements present in chondrites with composition greater than 1% are oxygen (34.8), iron (25.1), silicon (17.8), magnesium (14.4), sulfur

**Table 9.10**   Geochemical Classification of the Elements

| Siderophile | | | | | | | | |
|---|---|---|---|---|---|---|---|---|
| | Fe | Co | Ni | Ge | Mo | Ru | Rh | Pd |
| | Sn | Re | Os | Ir | Pt | Au | | |
| Chalcophile | | | | | | | | |
| | S | Cu | Zn | Ga | As | Se | Ag | Cd |
| | In | Sb | Te | Hg | Tl | Pb | Bi | |
| Lithophile | | | | | | | | |
| | Li | Be | B | C | O | F | Na | Mg |
| | Al | Si | P | Cl | K | Ca | Sc | Ti |
| | V | Cr | Mn | Br | Rb | Sr | Y | Zr |
| | Nb | I | Cs | Ba | RE | Hf | Ta | W |
| | Ra | Th | U | | | | | |
| Atmophile | | | | | | | | |
| | H | He | N | Ne | Ar | Kr | Xe | Rn |

(2.1), calcium (1.4), nickel (1.3), and aluminum (1.3). The majority of the elements are found in meteorites at concentrations of only a few grams per ton and require trace-element analytical methods for their accurate determination. The importance of determining these trace concentrations has led to a considerable development of their analysis by spectrochemical methods. Radioactivation analysis has become an important analytical method in meteorite chemical analysis. Early determinations were made for the elements gallium, palladium, rhenium, and gold. Measurements by activation analysis are now reported for more than 50 of the elements. A summary of these data is given in Table 9.11.

REMOTE ANALYSIS.   The rapid development of interest in space exploration has spurred the search for analytical techniques for *in situ* elemental analysis; for example, considerable attention has been given to methods for determining the chemical composition of the lunar surface by instrument packages delivered by manned or unmanned spacecraft. Hislop and Wainerdi* have listed possible analytical methods as (a) x-ray emission, (b) x-ray diffraction, (c) mass spectroscopy, (d) alpha-particle scattering, and (e) neutron methods. The neutron methods might consist of any one or combination of capture gamma-ray analysis, inelastic neutron scattering, thermal neutron die-away techniques, and activation analysis.

The alpha-particle scattering method has already proved successful for partial analysis of the lunar surface. A review of these experiments has been given by Turkevich, Patterson, and Franzgrote.† It is interesting that the

---

* J. S. Hislop and R. E. Wainerdi, Extraterrestrial Neutron Activation Analysis, *Anal. Chem.* **39**, 28A–39A (February 1967).

† A. L. Turkevich, J. H. Patterson, and E. J. Franzgrote, The Chemical Analysis of the Lunar Surface, *Am. Scientist* **56**, 312–343 (1968).

**Table 9.11** Elements Determined in Meteorites by Activation Analysis[a]

| Element | Chondrites ($\mu$g/g) | Achondrites ($\mu$g/g) | Irons ($\mu$g/g) | Stony Irons ($\mu$g/g) |
|---|---|---|---|---|
| Ar | 30–80 | <0.6–40 | | |
| K | 800 | <10–340 | 0.006–2.2 | |
| Ca | 11,500 | 8,000–77,000 | | |
| Sc | 8.6 | 17–45 | 0.0004–0.002 | |
| V | 67 | | <0.2 | |
| Cr | 2200 | 1,000–3,200 | 6–185 | <5 |
| Co | 150–900 | | | 1,100 |
| Ni | 5,000–19,000 | | | 19,000 |
| Cu | 75–100 | | 120–430 | 140–230 |
| Ga | | | 0.4–100 | |
| Ge | | | 0.2–360 | 50 |
| As | | | 4–16 | 30 |
| Se | 6–13 | 0.002–6 | | |
| Rb | 1–3 | | <0.0003 | |
| Rh | 0.2 | | | |
| Pd | | | 1–10 | |
| Ag | 0.04–0.13 | | | |
| In | <0.001 | | | |
| Sb | | | 0.01–0.8 | 0.2 |
| Te | 0.5–0.9 | <0.007 | | |
| I | 0.04–0.5 | | | |
| Cs | 0.01–0.1 | | | |
| Ba | 3.5 | 44 | <0.001 | |
| La | 0.32 | | | |
| Ce | 0.51 | | | |
| Pr | 0.12 | | | |
| Nd | 0.63 | | | |
| Sm | 0.22 | | | |
| Eu | 0.083 | 0.01–0.77 | | |
| Gd | 0.34 | | | |
| Tb | 0.051 | | | |
| Dy | 0.37 | | | |
| Ho | 0.075 | | | |
| Er | 0.21 | | | |
| Tm | 0.038 | | | |
| Yb | 0.19 | | | |
| Lu | 0.36 | | | |
| Ta | 0.021 | | 0.0024 | |
| W | 0.14 | | 1.7 | |

**Table 9.11** (*Continued*)

| Element | Chondrites ($\mu$g/g) | Achondrites ($\mu$g/g) | Irons ($\mu$g/g) |
|---------|-----------|-----------|-----------|
| Re | | | 0.22 |
| Os | | | 2.0 |
| Au | 0.1–0.29 | | 0.1–10 |
| Hg | 0.012–0.09 | 0.08 | |
| Tl | 0.0004 | 0.0007 | |
| Pb | 0.16 | 0.5 | |
| Bi | 0.003 | <0.001 | |
| Th | 0.04 | 0.006–0.6 | 0.00003 |
| U | 0.01 | 0.15 | <0.0001 |

[a] From H. J. M. Bowen and D. Gibbons, *Radioactivation Analysis* (Oxford, London, 1963), pp. 146–147.

principle for this new remote analytical method is based on one of the earliest methods used in nuclear physics. In 1909 it was noted that when α-particles impinged on a thin foil some were scattered through large angles, Indeed, this observation led to Rutherford's description of the current "practical model" of the atom. Lord Rutherford, in 1919, also noted that protons can be produced by nuclear reactions with α-particles. These two phenomena are

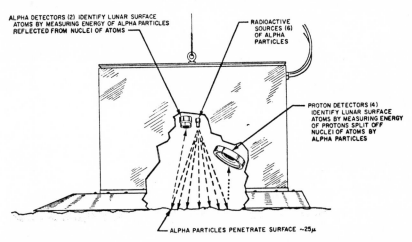

ALPHA DETECTORS (2) IDENTIFY LUNAR SURFACE ATOMS BY MEASURING ENERGY OF ALPHA PARTICLES REFLECTED FROM NUCLEI OF ATOMS

RADIOACTIVE SOURCES (6) OF ALPHA PARTICLES

PROTON DETECTORS (4) IDENTIFY LUNAR SURFACE ATOMS BY MEASURING ENERGY OF PROTONS SPLIT OFF NUCLEI OF ATOMS BY ALPHA PARTICLES

ALPHA PARTICLES PENETRATE SURFACE ~25μ

**Figure 9.5** The sensor head of the alpha scattering lunar chemical analysis instrument. The cutaway diagram shows the geometric relations between the six $^{242}$Cm alpha sources and the alpha and proton detectors. [From A. L. Turkevich, J. H. Patterson, and E. J. Franzgrote, The Chemical Analysis of The Lunar Surface, *Am. Scientist*, **56**, 312–343 (1968).]

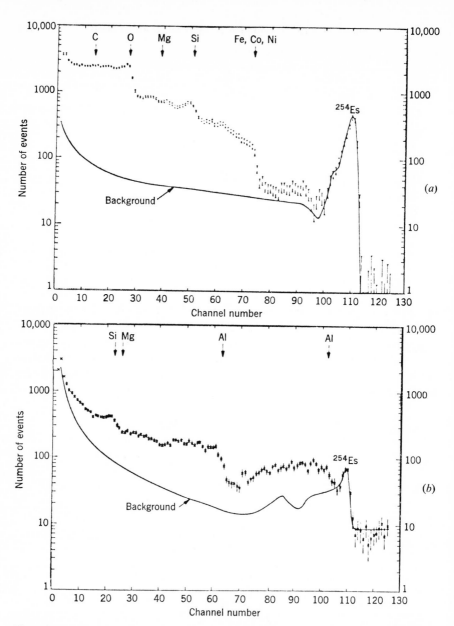

**Figure 9.6** The spectral data for the first alpha-particle analysis of the lunar surface. The data were obtained during 900 min of operation on the Surveyor V Mission. The number of events registered in the alpha (*a*) and proton (*b*) modes of the instrument are plotted, together with their associated (1σ) error bars, on a logarithmic scale as a function of channel number (energy). The background rates (converted to a comparable time) are shown by the solid curves. The peaks at approximately channel 112 in both modes are due to small amounts of $^{254}$Es placed close to each detector for internal energy calibration. Characteristic features in some elemental spectra are indicated by arrows. [From A. L. Turkevich, J. H. Patterson, and E. J. Franzgrote, The Chemical Analysis of the Lunar Surface, *Am. Scientist* **56**, 312–343 (1968).]

**Table 9.12** Composition of Lunar Surface Material at the
Surveyor Landing Sites[a]

| Element | Mass Range (A) | Surveyor Mission Number | | |
| | | V (atom %) | VI (atom %) | VII (atom %) |
| --- | --- | --- | --- | --- |
| Oxygen | 16 | 61.1 | 59.3 | 61.8 |
| Sodium | 23 | 0.5 | 0.6 | 0.5 |
| Magnesium | 24–26 | 2.8 | 3.7 | 3.6 |
| Aluminum | 27 | 6.4 | 6.5 | 9.2 |
| Silicon | 28–30 | 17.1 | 18.5 | 16.3 |
| "Calcium" | 39–44 | 5.5 | 5.2 | 6.9 |
| "Titanium" | 45–51 | 2.0 | 1.0 | 0.0 |
| "Iron" | 52–61 | 3.8 | 3.9 | 1.6 |

[a] Surveyors V and VI landed in equatorial mare regions (*Mare Tranquillitatis* and *Sinus Medii*). Surveyor VII landed in the lunar highlands near the crater Tycho. (Data courtesy of Ernest J. Franzgrote, Jet Propulsion Laboratory, California Institute of Technology, July 1970.)

combined in the lunar-surface analysis instrument. The sensor for this instrument, containing 163-d $^{242}$Cm as the alpha-particle source, is shown in Figure 9.5. The Surveyor Missions (V, VI, and VII), flown since September 1967, have used this method for lunar-surface analysis. The spectral data for the $\alpha$-scattering and proton responses obtained for the Surveyor V mission are shown in Figure 9.6. The results reported for the three Surveyor missions are summarized in Table 9.12. These data have been interpreted by Jackson and Wilshire* as compositionally similar to terrestrial basalts. Although this method is simple (and has proven successful and reliable), it is limited in its use by its relative insensitivity (minimum detection of about 0.5 atom %) and lack of resolution for hydrogen and elements with $Z > \sim25$.

An alternate method for lunar and planetary surface analysis by combining four neutron methods into a single package has been considered. For a single package the instrumentation may consist of a 14-MeV neutron generator and a gamma-ray spectrometer. A neutron generator is considered superior to an isotopic neutron source, since it can be turned off during flight to avoid the activation of the equipment on the spacecraft and since a background spectrum can be taken of the surface before activation. A description

* E. D. Jackson and H. G. Wilshire, Chemical Composition of the Lunar Surface at the Surveyor Landing Sites, *J. Geophys. Res.* **73**, 7621–7629 (1968).

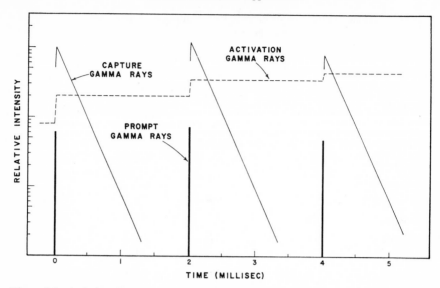

**Figure 9.7** A timing diagram for a combination neutron experiment for remote analysis [From R. L. Caldwell et al. Combination Neutron Experiment for Remote Analysis, *Science* **152**, 457–465 (1966).]

of the four neutron methods has been given by Caldwell et al.* The system is based on measuring time and energy distributions of the gamma rays resulting from fast neutron bursts. Figure 9.7 illustrates the timing diagram for gamma-ray intensities for successive neutron pulses of 20 $\mu$sec duration at a rate of 500/sec. Prompt gamma rays from inelastic scattering appear only during the bursts. Figure 9.7 shows that the intensity of the capture gamma rays from the compound nuclei builds up and decays exponentially after each burst with a characteristic decay constant $\lambda$. The value chosen for this case was $\lambda^{-1} \simeq 200$ $\mu$sec, a typical value for rock materials. The activation gamma rays are based on a mean life of 1 sec for the activation products. The time behavior of these gamma rays is characterized by $e^{-t}$, where $t$ is in seconds. Thus the build up during 20-$\mu$sec pulses is essentially linear and the decay between bursts is negligible. Figure 9.7 shows that although the intensities of prompt and capture gamma rays from successive bursts are independent, the intensity of activation gamma rays is cumulative. After the last burst the activation gamma rays will be proportional to $e^{-t}$.

Figure 9.8 shows the spectral features of the three types of gamma-radiation for the five most abundant elements expected on the lunar surface. The height

* R. L. Caldwell, W. R. Mills, L. S. Allen, P. R. Bell, and R. L. Heath, Combination Neutron Experiment for Remote Analysis, *Science* **152** 457–465 (1966).

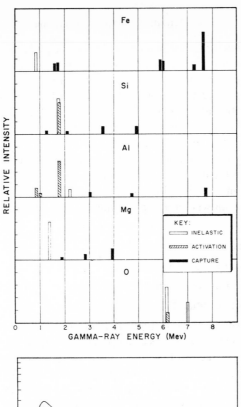

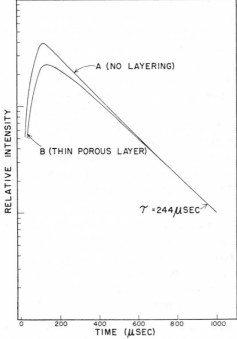

**Figure 9.8** Relative intensity characteristics of the four neutron-interaction methods to be combined for remote analysis. [From R. L. Caldwell et al., Combination Neutron Experiment for Remote Analysis **152**, 457–465 (1966).]

of the bars for each element is proportional to $\sigma\Gamma$, where $\sigma$ is the cross section for the appropriate process and $\Gamma$ is the number of gamma rays of a given energy produced per interaction. Thus the analysis of the energy spectra of prompt, capture, and activation gamma rays with time discrimination should afford a qualitative analysis of the elements. To distinguish between different rock types on the lunar surface quantitative results, which can be obtained from element ratios, are required.

The die-away technique, described in Section 7.3.3, complements the three spectral measurements. The die-away decay constant $\lambda$ depends on two processes: absorption of thermal neutrons and leakage of thermal neutrons out of the surface. Caldwell et al. approximates $\lambda$ by the expression

$$\lambda = \nu \Sigma + DB^2 \tag{8}$$

where $\nu =$ thermal neutron velocity,

   $\Sigma =$ macroscopic cross section,

   $D =$ diffusion coefficient,

   $B^2 =$ effective "buckling."

The product $DB^2$ is not sensitive to rock type and is considered constant. The velocity depends on surface temperature at time of measurement. The macroscopic cross section is a function of the composition of the rock and has been estimated for thermal and epithermal neutrons for acidic, basaltic, and aerolitic rock types. If both the thermal and epithermal decay constants can be measured, the rock type can be readily determined, since the ratios of these two cross sections vary significantly. The die-away experiment is especially sensitive to the hydrogen content (and therefore it can give an indication of the presence of water), since hydrogen has the maximum efficiency for moderating the fast neutrons to thermal energy.

A more detailed discussion of the possibility of fast neutron activation analysis for remote use was given by Hislop and Wainerdi. Their "turnaround" system is illustrated in Figure 9.9; the system can be rotated about a central axis to alternate the position of either the small neutron generator or the 3 × 3 in. NaI detector close to the "sample" surface.

THE APOLLO MOON SAMPLES.   The feat of man's landing on the moon has made possible the return of moon samples to the earth for laboratory determination of the chemical composition of lunar material. The samples of the moon's surface returned by the Apollo 11 and 12 space missions have received the most careful chemical analysis available by man in more than 200 laboratories throughout the world.* One of the first estimates of the chemical composition of lunar matter was reported by Morrison (1970) for

---

* A full description of the lunar sample analysis program is given in the summary of the Apollo 11 Lunar Science Conference in *Science* **167**, 447–784 (1970).

seven rock samples and one soil sample analyzed by spark-source mass spectrography and neutron activation analysis. Approximately 60 elements per sample were determined. Table 9.13 lists the results of the neutron activation analyses for 12 major elements and 29 trace elements in these samples compared with the analysis of USGS standard W-1. The results indicate an apparent uniformity of composition among the samples. In a comparison with solar, meteoritic, and terrestrial abundances Morrison noted a depletion of volatile elements and an enrichment of the rare earths, Ti, Zr, Y, and Hf. Although the over-all results indicate a general similarity of the lunar material to basaltic achondrites, the differences imply detailed geochemistry processes special to the history of the lunar material.

*The Atmosphere*

The atmosphere, extending with decreasing density to a height of several hundred miles from the earth's crust, may be considered a simple geochemical shell, its composition in the first 60 km consisting essentially of nitrogen, oxygen, and argon. These three elements constitute 99.94% (by weight) of the dry atmosphere. The remaining 0.06% consists of the minor gaseous constituents listed in Table 9.14. The role of these minor constituents, of the many gaseous and aerosol trace constituents, and of the variable amounts of water vapor in the atmosphere are of considerable importance in regulating climate and life on the earth. Carbon dioxide, whose concentration is considerably less than 1%, is of vital importance to plant life. Ozone, in increasing concentration with altitude, acts as a chemical filter for the absorption of dangerous levels of ultraviolet radiation from space. Water vapor acts as the part of the hydrologic cycle which effects the circulation of the atmosphere. Radon, emanating from the earth's surface, distributes its radioactive daughter products throughout the troposphere. Man-made aerosols can be considered as a pollutant of the atmosphere. The burning of fuels for heat and transportation add sufficient aerosols to the atmosphere to become objectionable at times. This injection of artificial aerosols may play a role in climatic change. Studies are also underway to examine the effects on weather modification by deliberate injection of aerosol particles that have high efficiency as ice nucleation particles.

Little use has been made of radioactivation analysis for the determination of the trace gaseous constituents of the atmosphere. As in most other analytical methods, the problems of sampling and containment or concentration of large air samples are formidable ones. Filtering of aerosols from large volumes of air is a relatively easy operation, and several studies of the composition of aerosol material have been carried out. The very small concentrations of aerosols in the atmosphere make activation analysis a useful method.

**Figure 9.9** A turnaround system for remote activation analysis. Part 1,a photo of the prototype. [From J. S. Hislop and R. E. Wainerdi, Extraterrestrial Neutron Activation Analysis, *Anal. Chem.* **39**, 28A–39A (February, 1967).]

The concentrations of natural aerosols vary with time throughout the atmosphere due to discontinuous imputs and removals and to the irregularities in atmospheric circulation. The total concentration of particulates in air may range from less than 100 to more than 2000 $\mu g/m^3$. Only a few publications report on the chemical composition of aerosol particulates. A summary is given in Table 9.15.

The data on Cl, Br, and I given in Table 9.15 have been used to study air-sea interactions. The high bromine values at Cambridge, Mass., were attributed* to local pollution of bromine from automobile exhaust. A correlation

* R. L. Lininger, R. A. Duce, J. W. Winchester, and W. R. Matson, Chlorine, Bromine, Iodine, and Lead in Aerosols from Cambridge, Massachusetts, *J. Geophys. Res.* **71**, 2457–2463 (1966).

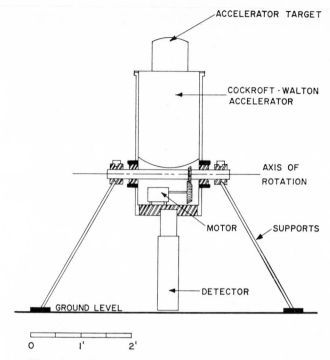

**Figure 9.9** (*Continued*) Part 2, a schematic of the apparatus.

of bromine to the lead on the filters was noted, the latter element being determined by an anodic-stripping voltammetry method. An extensive examination of the areal distribution of air pollutants was made in a 24-hr collection of particulates at 22 locations throughout the Chicago Metropolitan area.* In addition to the elements listed in Table 9.15, measurements were reported for aluminum and vanadium. Another extensive examination of particulates in air was reported for several cities in England.†

Dudey, Ross, and Noshkin‡ collected five large samples of marine aerosols

* S. S. Brar et al., Thermal Neutron Activation Analysis of Airborne Particulate Matter in Chicago Metropolitan Area, *Modern Trends in Activation Analysis* (National Bureau of Standards, Gaithersburg, 1968), pp. 560–568.

† J. R. Keane and E. M. R. Fisher, Analysis of Trace Elements in Air-Borne Particulates by Neutron Activation and Gamma-Ray Spectrometry, *Atmospheric Environment* **2**, 603–614 (1968).

‡ N. D. Dudey, L. E. Ross, and V. E. Noshkin, Application of Activation Analysis and Ge (Li) Detection Techniques for the Determination of Stable Elements in Marine Aerosols, *Modern Trends in Activation Analysis* (National Bureau of Standards, Gaithersburg, 1968), pp. 569–577.

| Element | Range in 7 Rock Samples | Soil |
|---------|-------------------------|------|
| **Major Elements (wt%)** | | |
| Al | 4.0–6.2 | 7.3 |
| Ti | 5.0–6.7 | 4.1 |
| Fe | 13.2–15.7 | 12.5 |
| Mg | 2.8–5.5 | 4.6 |
| Ca[b] | 7.9–8.6 | 7.5 |
| Na | 0.28–0.35 | 0.33 |
| K | 0.052–0.29 | 0.11 |
| Mn | 0.16–0.20 | 0.16 |
| Cr | 0.14–0.24 | 0.20 |
| Zr | 0.036–0.072 | 0.039 |
| Ni | 0.004–0.008 | 0.017 |
| V | 0.0040–0.0068 | 0.0078 |
| **Trace Elements (ppm)** | | |
| Sc | 64–97 | 60 |
| Co | 14–42 | 40 |
| Cu | 3.7–18 | 9.9 |
| Zn | 2.1–30 | 22 |
| Ga | 3.5–5.1 | 4.6 |
| As | 0.03–0.09 | 0.07 |
| Br | 0.06–0.3 | 0.2 |
| Rb | 1.2–5.7 | 4.4 |
| Sr | 130–180 | 200 |
| Mo | 0.4–0.7 | 0.7 |
| Sb | 0.005–0.01 | 0.005 |
| Cs | 0.06–0.3 | 0.2 |
| Ba | 96–300 | 220 |
| La | 11–35 | 22 |
| Ce | 34–96 | 50 |
| Nd | 43–120 | 46 |
| Sm | 14–28 | 18 |
| Eu | 1.6–3.0 | 1.9 |
| Gd | 17–31 | 20 |
| Tb | 3.5–6.8 | 3.8 |
| Ho | 7–10 | 6 |
| Tm | 1.2–2.8 | 1.2 |

**Table 9.13** (Continued)

| Element | Range in 7 Rock Samples | Soil |
|---------|------------------------|------|
| Yb | 15–28 | 12 |
| Lu | 1.5–2.6 | 1.4 |
| Hf | 11–18 | 9 |
| Ta | 1.0–2.2 | 1.3 |
| W | 0.13–0.42 | 0.25 |
| Th | 1.1–4.8 | 2.3 |
| U | 0.14–0.60 | 0.48 |

[a] Adapted from G. H. Morrison, Multielement Analysis of Lunar Soil and Rocks Returned by Apollo 11, *The Chemist* **47**, 142–146 (1970).

[b] New value by communication from G. H. Morrison, 1970.

aboard a ship in the open ocean. They determined 20 trace elements in these marine aerosols; their data are reproduced in Table 9.16.

An analysis has also been made of the elemental composition of aerosols in the stratosphere by activation analysis. Friend et al.* reported the results

**Table 9.14**  Average Composition of the Dry Atmosphere[a]

| Gas | Composition (weight %) |
|-----|------------------------|
| $N_2$ | 75.51 |
| $O_2$ | 23.15 |
| Ar | 1.28 |
| $CO_2$ | 0.046 |
| Ne | 0.00125 |
| Kr | 0.00029 |
| $CH_4$ | 0.000094 |
| $N_2O$ | 0.00008 |
| He | 0.000072 |
| $O_3$ | 0.00007 |
| Xe | 0.000036 |
| $H_2$ | 0.0000035 |

[a] From B. Mason, *Principles of Geochemistry* (Wiley, New York, 1966).

* J. P. Friend, Ed., The High Altitude Sampling Program, Vol. 5, DASA-1300 (1961).

**Table 9.15**   Some Determinations of Aerosol Elements by
Activation Analysis

| Location | Element Concentration | | | | |
|---|---|---|---|---|---|
| | Na ($\mu$g/m³) | Mn ($\mu$g/m³) | Cl ($\mu$g/m³) | Br (ng/m³) | I (ng/m³) |
| Washington, D.C.[a] | 0.51 | 0.04 | | 220 | |
| Hawaii[b] | | | 1–4 | 12–24 | 0.4–2 |
| Cambridge, Mass.[b] | | | 1–6 | 20–800 | 2–10 |
| Barrow, Alaska[b] | | | <0.02–4 | 1–30 | 0.3–10 |
| Chicago, Ill.[c] | 0.08–0.67 | 0.1–0.9 | 0.6–5.9 | 41–323 | |
| Chilton, England[d] | 0.85–1.09 | 0.016–0.033 | 2.1–2.6 | 29–43 | |

[a] C. M. Gordon and R. E. Larson, Activation Analysis of Aerosols, *J. Geophys. Res.* **69**, 2881–2885 (1964).
[b] J. W. Winchester and R. A. Duce, Coherence of Iodine and Bromine in the Atmosphere of Hawaii, Northern Alaska, and Massachusetts, *Tellus* **18**, 287–292 (1966).
[c] S. S. Brar et al., Thermal Neutron Activation Analysis of Airborne Particulate Matter in Chicago Metropolitan Area, *Modern Trends in Activation Analysis* (National Bureau of Standards, Gaithersburg, 1968), pp. 560–568.
[d] J. R. Keane and E. M. R. Fisher, Analysis of Trace Elements in Air-Borne Particulates, by Neutron Activation and Gamma-Ray Spectrometry, *Atmospheric Environment*, **2**, 603–614 (1968).

of activation analysis determinations of aerosol samples collected on U-2 aircraft nose sample filters flown from an airbase in North Dakota. With filtered volumes of air of the order of $2 \times 10^4$ m³ nine trace elements were identified and determined. Their data are summarized in Table 9.17.

Radioactivation analysis has been used to examine the composition of aerosols near active volcanos; for example, Gordon, Jones, and Hoover* examined a steam-ash cloud from the fissure-type Surtsey volcano, sampling on filter paper by aircraft at an altitude of 7500 ft. The concentrations downwind of the volcano at several altitudes, compared with the concentrations in the Surtsey cloud, for sodium, chlorine, and manganese, are listed in Table 9.18. A factor of 90 for scandium concentration in the steam-ash cloud compared with the downwind aerosols was also noted.

Activation analysis has also been used to measure the presence of materials injected into the atmosphere for cloud-seeding weather modification studies.

* C. M. Gordon, E. C. Jones, and J. I. Hoover, Atmospheric Aerosols Near the Surtsey Volcano, Report of NRL Progress, 1–3 (1964).

**Table 9.16** Trace Element Concentrations in Marine Aerosols[a]

| Element | Mean Concentration ($\mu$g/m$^3$) |
|---------|-----------------------------------|
| Ce | $3.0 \times 10^{-4}$ |
| La | $1.5 \times 10^{-4}$ |
| Sm | $3.4 \times 10^{-5}$ |
| Eu | $4.4 \times 10^{-6}$ |
| Tb | $8.9 \times 10^{-6}$ |
| Yb | $5.0 \times 10^{-6}$ |
| Hf | $9.0 \times 10^{-6}$ |
| Sc | $4.1 \times 10^{-5}$ |
| Cr | $4.2 \times 10^{-4}$ |
| Fe | $2.3 \times 10^{-1}$ |
| Co | $8.7 \times 10^{-5}$ |
| Cu | $1.4 \times 10^{-1}$ |
| Zn | $2.5 \times 10^{-3}$ |
| Ga | $1.4 \times 10^{-4}$ |
| Se | $3.6 \times 10^{-5}$ |
| In | $1.5 \times 10^{-4}$ |
| Sb | $8.0 \times 10^{-5}$ |
| Cd | $3.0 \times 10^{-4}$ |
| Au | $7.3 \times 10^{-7}$ |
| Hg | $1.4 \times 10^{-5}$ |

[a] From N. D. Dudey, L. E. Ross, and V. E. Noshkin, Application of Activation Analysis and Ge(Li) Detection Techniques for the Determination of Stable Elements in Marine Aerosols, *Modern Trends in Activation Analysis* (National Bureau of Standards, Gaithersburg, 1968), pp. 569–577.

**Table 9.17** Stratospheric Aerosols Determined by Activation Analysis[a]

| Element | Approximate Concentration[b] (ng/m$^3$) |
|---------|------------------------------------------|
| Cu | 0.2 |
| Zn | 2 |
| P | 30 |
| Cr | 0.6 |
| Fe | 12 |
| Ca | 120 |
| Co | 0.02 |
| Ni | 1.5 |
| Mn | 0.2 |

[a] From J. P. Friend, Ed., The High Altitude Sampling Program, Vol. 5, DASA-1300 (1961).
[b] Based on two northward and southward flights in the stratosphere above North Dakota.

**Table 9.18**    Aerosols Near Surtsey Volcano Determined by
Activation Analysis[a]

| Location | Altitude (ft) | Concentration ($\mu$g/m³) | | |
|---|---|---|---|---|
| | | Na | Cl | Mn |
| ~150 miles downwind | 10,000 | 0.10 | 0.36 | 0.0014 |
| of Surtsey | 8,000 | 0.13 | 0.11 | 0.0028 |
| | 6,000 | 0.17 | 0.61 | 0.0041 |
| | 5,000 | 0.33 | 0.64 | 0.0020 |
| | 4,000 | 0.65 | 0.64 | 0.0036 |
| | 2,000 | 1.3 | 2.8 | 0.015 |
| | 1,000 | 2.0 | 3.7 | 0.017 |
| | 500 | 2.0 | 2.7 | 0.020 |
| In Surtsey cloud | 7,500 | 59.6 | 86.0 | 1.2 |

[a] From C. M. Gordon, E. C. Jones, and J. I. Hoover, Atmospheric Aerosols Near the Surtsey Volcano, Report of NRL Progress, 1–3 (1964).

Silver iodide, because of its similarity to ice as condensation nuclei, is generally used as a cloud-seeding material. Warburton and Young[*] estimated that precipitation in the form of rain, hail, and snow would have AgI concentrations in the range $10^{-10}$ to $10^{-12}$ g/ml. They developed a neutron activation analysis procedure as a sensitive and positive nondestructive method of detection, using the 24-s $^{110}$Ag product radionuclide for gamma ray spectroscopic determination. Their method, which has a concentration detection limit of ~$2 \times 10^{-11}$ g/ml for 1-l. samples, showed silver present in rain and hail collected in areas in which cloud seeding was being conducted and no silver in snow collected in areas of no seeding activity.

### The Hydrosphere

Mason[†] describes the hydrosphere as the discontinuous shell of water—fresh, salt, and solid—at the surface of the earth. It constitutes the oceans with their connected seas and gulfs, the lakes, the waters of the rivers and streams, ground water, and snow and ice. The hydrosphere is estimated to

[*] J. A. Warburton and L. G. Young, Neutron Activation Measurements of Silver in Precipitation from Locations in Western North America, *J. Applied Meteor.* **7**, 444–448 (1968).
[†] B. Mason, *Principles of Geochemistry* (Wiley, New York, 1966), pp. 195–196.

contain about 273 l. of water/cm² of earth's surface:

|  | Liters | Kilograms |
|---|---|---|
| Seawater | 268.45 | 278.11 |
| Fresh water | 0.1 | 0.1 |
| Continental ice | 4.5 | 4.5 |
| Water vapor | 0.003 | 0.003 |

It is noted that seawater constitutes about 98% of the mass of the hydrosphere, and thus its chemical composition determines the average composition of the hydrosphere. The salinity of seawater, a function of the weight of dissolved solids and the density, is essentially constant at about 3.5% throughout the world. The elemental composition of seawater is given in Table 9.19. Values for a significant number of trace elements are missing; many of these elements are readily determined by activation analysis.

**Table 9.19**   Elements Present in Seawater[a]

| Element | Abundance (mg/l.) | Element | Abundance (mg/l.) | Element | Abundance (mg/l.) |
|---|---|---|---|---|---|
| Li | $1.7 \times 10^{-1}$ | Ni | $2 \times 10^{-3}$ | La | $2.9 \times 10^{-6}$ |
| Be | $6 \times 10^{-7}$ | Cu | $3 \times 10^{-3}$ | Ce | $1.3 \times 10^{-6}$ |
| B | 4.6 | Zn | $1 \times 10^{-2}$ | Pr | $6.4 \times 10^{-7}$ |
| C | 28 | Ga | $3 \times 10^{-5}$ | Nd | $2.3 \times 10^{-6}$ |
| N | $5 \times 10^{-1}$ | Ge | $7 \times 10^{-5}$ | Sm | $4.2 \times 10^{-7}$ |
| F | 1.3 | As | $3 \times 10^{-3}$ | Eu | $1.1 \times 10^{-7}$ |
| Na | 10,500 | Se | $4 \times 10^{-4}$ | Gd | $6.0 \times 10^{-7}$ |
| Mg | 1,350 | Br | 65 | Dy | $7.3 \times 10^{-7}$ |
| Al | $1 \times 10^{-2}$ | Rb | $1.2 \times 10^{-1}$ | Ho | $2.2 \times 10^{-7}$ |
| Si | 3.0 | Sr | 8.0 | Er | $6.1 \times 10^{-7}$ |
| P | $7 \times 10^{-2}$ | Y | $3 \times 10^{-4}$ | Tm | $1.3 \times 10^{-7}$ |
| S | 885 | Nb | $1 \times 10^{-5}$ | Yb | $5.2 \times 10^{-7}$ |
| Cl | 19,000 | Mo | $1 \times 10^{-2}$ | Lu | $1.2 \times 10^{-7}$ |
| K | 380 | Ag | $4 \times 10^{-5}$ | W | $1 \times 10^{-4}$ |
| Ca | 400 | Cd | $1.1 \times 10^{-4}$ | Au | $4 \times 10^{-6}$ |
| Sc | $4 \times 10^{-5}$ | In | $<10^{-2}$ | Hg | $3 \times 10^{-5}$ |
| Ti | $1 \times 10^{-3}$ | Sn | $8 \times 10^{-4}$ | Tl | $<10^{-5}$ |
| Cr | $5 \times 10^{-5}$ | Sb | $5 \times 10^{-4}$ | Pb | $3 \times 10^{-5}$ |
| Mn | $2 \times 10^{-3}$ | I | $6 \times 10^{-2}$ | Bi | $2 \times 10^{-5}$ |
| Fe | $1 \times 10^{-2}$ | Cs | $5 \times 10^{-4}$ | Th | $5 \times 10^{-5}$ |
| Co | $1 \times 10^{-4}$ | Ba | $3 \times 10^{-2}$ | U | $3 \times 10^{-3}$ |

[a] From B. Mason, *Principles of Geochemistry* (Wiley, New York, 1966) pp. 195–196.

An extensive investigation of the distribution of several trace elements in seawater by activation analysis has been reported by Schutz and Turekian,* who developed methods for the determination of 18 trace elements:

| Au | Se | Sb | Ag | Co | Ni | Hg | Cs | Rb |
|----|----|----|----|----|----|----|----|----|
| Cr | Zr | Hf | Ta | Zn | Fe | Sc | Ba | Sr |

Among these elements sufficient analyses of the first six were completed to examine their distribution in the world's oceans. These values and the results of some earlier activation analyses are listed in Table 9.20.

**Table 9.20**    Some Trace-Element Concentrations in Seawater Determined by Activation Analysis

| | Concentration ($\mu$g/Kg) | | |
|---|---|---|---|
| | Schutz[a] | | Fukai[b] |
| Element | Mean | Range | Value |
| Sc | <0.004 | | |
| Co | 0.27 | 0.035–4.1 | |
| Ni | 5.4 | 0.43–43 | |
| As | | | 3 |
| Se | 0.09 | 0.052–0.12 | |
| Rb | | | 120 |
| Sr | | | 8000 |
| Ag | 0.29 | 0.055–1.5 | |
| Sb | 0.33 | 0.18–1.1 | |
| Cs | | | 0.5 |
| Ba | | | 6.3 |
| Hf | <0.008 | | |
| Ta | <0.0025 | | |
| Au | 0.011 | 0.004–0.027 | 0.03 |
| U | | | 2 |

[a] D. F. Schutz and K. K. Turekian, *loc. cit.*, 1965.
[b] R. Fukai and W. W. Meinke, Trace Analysis of Marine Organism, *Limnology and Oceanography* **4,** 398–408 (1959).

* D. F. Schutz and K. K. Turekian, The Investigation of the Geographical and Vertical Distribution of Several Trace Elements in Sea Water Using Neutron Activation Analysis, *Geochim. Cosmochim. Acta* **29,** 259–313 (1965).

**Table 9.21**  Neutron Activation Analysis of Drinking Waters[a]

| Element | Concentration ($\mu$g/l.) | | | | | | |
|---|---|---|---|---|---|---|---|
| | Atlanta, Ga. | Charlottes-ville, Va. | Chicago, Ill. | Cincinnati, Ohio | Denver, Colo. | New York, N.Y. | Oak Ridge, Tenn. |
| Br | 170 | 50 | 320 | 120 | 180 | 70 | 180 |
| Cl | 12,000 | 12,100 | 6,600 | 5,100 | 20,800 | 6,700 | 3,100 |
| Cu | 3 | 7 | 4 | 10 | 17 | 5 | 3 |
| As | <0.02 | <0.02 | <0.02 | <0.02 | <0.02 | <0.02 | <0.02 |
| Mn | 2.8 | 40 | 0.4 | 2.8 | 9.8 | 4.9 | 1.4 |
| Zn | <10 | 38 | 28 | <10 | <10 | 47 | <10 |
| Ba | 5 | <5 | 10 | <5 | <5 | 6 | 21 |
| Sr | 140 | 80 | 290 | 610 | 12.3 | 100 | 380 |
| Na | 3,800 | 4,000 | 6,500 | 1,600 | 21,200 | 1,600 | 6,400 |
| Rb | <200 | <200 | 200 | <200 | <200 | <200 | <200 |
| K | 720 | 530 | 1,900 | 3,600 | 2,100 | 670 | 1,400 |
| Mg | 1,100 | 8,500 | 20,700 | 3,600 | 4,500 | 5,000 | 2,800 |
| I | 27 | 18 | 27 | 27 | 7 | 4 | 2 |
| Si | 6,600 | 1,700 | 1,850 | 4,250 | 4,200 | 3,800 | 1,700 |
| P | 1.7 | 14 | 3.7 | 29 | 11 | 3.1 | 5 |
| Sb | 57 | 44 | 30 | 16 | 23 | 40 | 30 |
| Ca | 11,600 | 12,300 | 34,600 | 63,300 | 23,500 | 8,500 | 36,000 |
| Th | <1 | <1 | <1 | <1 | <1 | <1 | 20 |

[a] From R. L. Blanchard, G. W. Leddicotte, and D. W. Moeller, Water Analysis by Neutron Activation, *J. Am. Water Works Assoc.* **51**, 967–980 (1959).

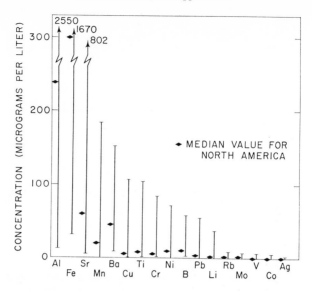

**Figure 9.10** Range of trace element concentrations found in the world's large rivers. [From W. H. Durum and J. Haffty, Implications of the Minor Element Content of Some Major Streams of the World, *Geochim. Cosmochim. Acta* **27**, 1–11 (1963).]

Surprisingly little data are available for the activation analysis determination of trace elements in surface and ground waters. An early review of this potential was made by Blanchard, Leddicotte, and Moeller* in 1959. Their data for 18 trace elements in drinking waters at seven locations in the United States are summarized in Table 9.21. A review of minor element content of some major streams in the world is reported by Durum and Haffty†. Their data, with median values for North American rivers, are shown in Figure 9.10. The measurements were made by spectrographic methods. Values of zero were reported for zinc and vanadium, since concentrations were below the level of detection in more than half the observations. These elements have excellent sensitivity properties for radioactivation analysis. Linstedt and Kruger‡ have examined the seasonal transport of vanadium throughout the Colorado River and its tributaries by activation analysis

* R. L. Blanchard, G. W. Leddicotte, and D. W. Moeller, Water Analysis by Neutron Activation, *J. Am. Water Works Assoc.* **51**, 967–980 (1959).

† W. H. Durum and J. Haffty, Implications of the Minor Element Content of Some Major Streams of the World, *Geochim. Cosmochim. Acta,* **27**, 1–11 (1963).

‡ K. D. Linstedt and P. Kruger, Vanadium Concentrations in Colorado River Basin Waters, *J. Am. Water Works Assoc.* **61**, 85–88 (1969).

**Table 9.22**  Variability of Vanadium Concentrations In the Colorado River and Its Tributaries[a]

| Sources | 1966 | | | | | | | | | 1967 | | | | | |
|---|---|---|---|---|---|---|---|---|---|---|---|---|---|---|---|
| | April | May | June | July | Aug | Sept | Oct | Nov | Dec | Jan | Feb | March | April | May | June |
| Tributaries | | | | | | | | | | | | | | | |
| Flaming Gorge | | | | | | | | | | | | | | | |
| Green River | [b]... | 1.0 | 1.8 | 0.6 | 0.4 | 1.3 | 0.7 | 0.9 | 0.9 | 1.2 | 0.7 | 0.9 | 0.7 | 0.6 | 0.6 |
| Cedar Hill | | | | | | | | | | | | | | | |
| Animas River | 0.4 | 0.3 | 0.2 | 0.2 | ... | ... | 0.2 | 0.4 | 0.3 | 0.4 | ... | ... | ... | ... | ... |
| Shiprock | | | | | | | | | | | | | | | |
| San Juan River | 2.1 | 0.6 | ... | ... | 11.3 | ... | 3.9 | ... | ... | 7.3 | 1.3 | ... | 9.0 | 7.3 | 25.1 |
| Loma | 3.6 | 4.3 | 3.9 | 5.9 | 4.8 | 6.2 | 3.5 | ... | ... | ... | ... | 8.9 | 4.8 | 1.9 | 3.3 |
| The Colorado River | | | | | | | | | | | | | | | |
| Page | 3.4 | 3.9 | 4.1 | ... | ... | ... | ... | ... | 3.5 | 4.6 | 2.8 | 3.8 | 3.9 | 2.7 | 2.6 |
| Hoover Dam | 2.8 | 3.2 | 3.2 | 2.9 | 3.1 | 2.3 | 2.4 | 2.5 | 3.7 | 3.3 | 3.9 | 3.1 | 2.8 | 3.0 | 2.1 |
| Parker Dam | 3.1 | 3.7 | 1.7 | 1.9 | ... | 2.5 | 2.7 | 2.7 | 3.2 | 3.8 | 2.0 | 2.6 | 2.7 | 3.6 | 3.7 |
| Yuma | 3.0 | 2.1 | ... | 2.2 | 1.9 | ... | 1.7 | 1.8 | 1.5 | 1.7 | 1.7 | ... | 2.1 | 1.9 | 3.0 |

[a] From K. D. Linstedt and P. Kruger, Vanadium Concentrations in Colorado River Basin Waters, *J. Am. Water Works Assoc.* **61**, 85–88 (1969).

[b] Samples lost or not collected.

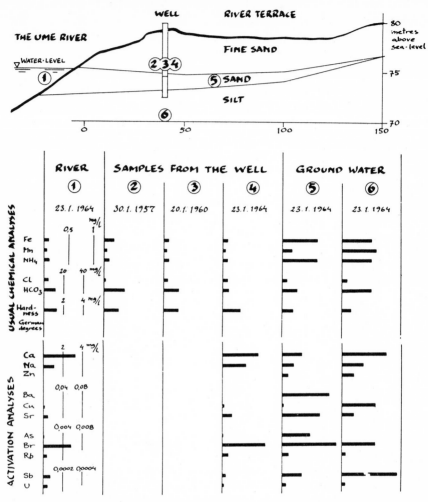

**Figure 9.11**  Activation analysis data used to solve a ground-water problem at Bran on the Ume River in Sweden. [From O. Landström and C. G. Wenner, Neutron Activation Analysis of Natural Water Applied to Hydrogeology, Aktiebolaget Atomenergi Report AE-204, Stockholm, Sweden (1965).]

measurement. An indication of the variations observed is noted in Table 9.22.

Neutron activation analysis has been applied to hydrogeologic studies by Landström and Wenner.* The analysis of trace elements in ground-water

---

* O. Landström and C. G. Wenner, Neutron-Activation Analysis of Natural Water Applied to Hydrogeology, Report No. AE-204, Aktiebolaget Atomenergi, Stockholm, Sweden, 1965.

samples has been used to examine such hydrogeologic problems as the quality and capacity of water supplies, the origin and existence of ground-water streams, and ground-water exchange with rivers. An example of their activation analysis data to determine the origin of water from a well is illustrated in Figure 9.11 (see Problem 9.3).

## The Lithosphere

The structure of the earth is generally considered as consisting of an iron core, a homogeneous silicate mantle, and a thin heterogeneous silicate crust. Because of the geologic and geochemical importance of the crust to man, its chemical composition is of considerable interest. The evolution of rock materials is described by the rock cycle in which the basic rock types are igneous, metamorphic, and sedimentary. Since the upper crust of the earth is 95% igneous, the average chemical composition of the crust is determined by such rocks. The major elements that constitute the continental crust have been estimated from 5159 chemical analyses evaluated by Clark and Washington.* The data, given in Table 9.23, represent an igneous rock intermediate between

**Table 9.23** Major Element Composition of the Upper Crust[a]

| Element in Oxide Form | Average in Continental Crust[b] (%) | Average in Whole Crust[c] (%) |
|---|---|---|
| $SiO_2$ | 60.18 | 55.2 |
| $Al_2O_3$ | 15.61 | 15.3 |
| $Fe_2O_3$ | 3.14 | 2.8 |
| FeO | 3.88 | 5.8 |
| MnO | ... | 0.2 |
| MgO | 3.56 | 5.2 |
| CaO | 5.17 | 8.8 |
| $Na_2O$ | 3.91 | 2.9 |
| $K_2O$ | 3.19 | 1.9 |
| $TiO_2$ | 1.06 | 1.6 |
| $P_2O_5$ | 0.30 | 0.3 |

[a] Neglecting water and minor constituents [From B. Mason, *Principles of Geochemistry* (Wiley, New York, 1966).]
[b] For the composition of continental areas.
[c] For the crust as a whole, including the submarine areas.

* F. W. Clarke and H. S. Washington, The Composition of the Earth's Crust, U.S. Geol. Survey Professional Paper 127, 1924.

granite and basalt, the two types that make up the bulk of all igneous rocks. A significant difference is noted between igneous rocks contained in the continental and the submarine regions.

The determination of minor and trace elements in rocks has increased in scope with the development of modern analytical methods. Analyses have been made for virtually all of the elements. To assist in obtaining "best" values for comparison of different analytical methods and for different laboratories two reference rock samples were prepared in 1949 for wide distribution as analytical comparison samples. By 1965 analyses, as summarized by Fleischer,* had been reported for 74 elements in 256 publications from 105 laboratories in 22 countries. Table 9.24 lists the average elemental composition of the earth's crust and the two standard samples, as of 1966. Table 9.25 summarizes some recent values reported for activation analysis determinations.

To fulfill a need for other reference-rock materials, and because the supply of standard rock G-1 was exhausted, the U.S. Geological Survey prepared 200-to-300-lb samples of six other silicate rocks. A description of these samples is reported by Flanagan.†

G-2     Westerly granite from Rhode Island (a replacement for G-1).

GSP-1  Granodiorite from Colorado.

AGV-1 Andesite from Oregon.

PCC-1 Periodotite from California.

DTS-1 Dunite from Washington.

BCR-1 Basalt from Washington-Oregon.

Samples are available from U.S. Geological Survey in Washington, D.C. to laboratories that wish to participate in the analysis. A summary of some of the results of analysis of these rocks by activation analysis is given in Table 9.26.

Activation analysis may play a role in several kinds of geochemical studies. Results have already been reported for trace-element determinations in studies of marine and lacustrine deposits in bore-hole samples, continental drift by geochemical surveying methods, vulcanology, and cosmic spherules found in deep oceanic sediments.

---

* M. Fleischer, Summary of New Data on Rock Samples G-1 and W-1, 1962–1965, *Geochim. Cosmochim. Acta* **29**, 1263–1283 (1965).

† F. J. Flanagan, U.S. Geological Survey Silicate Rock Standards, *Geochim. Cosmochim. Acta* **31**, 289–308 (1967).

**Table 9.24** Average Concentrations of Elements in Crystal Rocks[a]

| Atomic Number | Element | Crustal Average | Granite (G-1) | Diabase (W-1) |
|---|---|---|---|---|
| 1 | H | 1,400 | 400 | 600 |
| 3 | Li | 20 | 24 | 12 |
| 4 | Be | 2.8 | 3 | 0.8 |
| 5 | B | 10 | 2 | 17 |
| 6 | C | 200 | 200 | 100 |
| 7 | N | 20 | 8 | 14 |
| 8 | O | 466,000 | 485,000 | 449,000 |
| 9 | F | 625 | 700 | 250 |
| 11 | Na | 28,300 | 24,600 | 15,400 |
| 12 | Mg | 20,900 | 2,400 | 39,900 |
| 13 | Al | 81,300 | 74,300 | 78,600 |
| 14 | Si | 277,200 | 339,600 | 246,100 |
| 15 | P | 1,050 | 390 | 650 |
| 16 | S | 260 | 175 | 135 |
| 17 | Cl | 130 | 50 | |
| 19 | K | 25,900 | 45,100 | 5,300 |
| 20 | Ca | 36,300 | 9,900 | 78,300 |
| 21 | Sc | 22 | 3 | 34 |
| 22 | Ti | 4,400 | 1,500 | 6,400 |
| 23 | V | 135 | 16 | 240 |
| 24 | Cr | 100 | 22 | 120 |
| 25 | Mn | 950 | 230 | 1,320 |
| 26 | Fe | 50,000 | 13,700 | 77,600 |
| 27 | Co | 25 | 2.4 | 50 |
| 28 | Ni | 75 | 2 | 78 |
| 29 | Cu | 55 | 13 | 110 |
| 30 | Zn | 70 | 45 | 82 |
| 31 | Ga | 15 | 18 | 16 |
| 32 | Ge | 1.5 | 1.0 | 1.6 |
| 33 | As | 1.8 | 0.8 | 2.2 |
| 34 | Se | 0.05 | | |
| 35 | Br | 2.5 | 0.5 | 0.5 |
| 37 | Rb | 90 | 220 | 22 |
| 38 | Sr | 375 | 250 | 180 |
| 39 | Y | 33 | 13 | 25 |
| 40 | Zr | 165 | 210 | 100 |
| 41 | Nb | 20 | 20 | 10 |
| 42 | Mo | 1.5 | 7 | 0.05 |
| 44 | Ru | 0.01 | | |

(Continued overleaf)

**Table 9.24** (*continued*)

| Atomic Number | Element | Crustal Average | Granite (G-1) | Diabase (W-1) |
|---|---|---|---|---|
| 45 | Rh | 0.005 | | |
| 46 | Pd | 0.01 | 0.01 | 0.02 |
| 47 | Ag | 0.07 | 0.04 | 0.06 |
| 48 | Cd | 0.2 | 0.06 | 0.3 |
| 49 | In | 0.1 | 0.03 | 0.08 |
| 50 | Sn | 2 | 4 | 3 |
| 51 | Sb | 0.2 | 0.4 | 1.1 |
| 52 | Te | 0.01 | | |
| 53 | I | 0.5 | | |
| 55 | Cs | 3 | 1.5 | 1.1 |
| 56 | Ba | 425 | 1,220 | 180 |
| 57 | La | 30 | 120 | 30 |
| 58 | Ce | 60 | 230 | 30 |
| 59 | Pr | 8.2 | 20 | 2 |
| 60 | Nd | 28 | 55 | 15 |
| 62 | Sm | 6.0 | 11 | 5 |
| 63 | Eu | 1.2 | 1.0 | 1.1 |
| 64 | Gd | 5.4 | 5 | 4 |
| 65 | Tb | 0.9 | 1.1 | 0.6 |
| 66 | Dy | 3.0 | 2 | 4 |
| 67 | Ho | 1.2 | 0.5 | 1.3 |
| 68 | Er | 2.8 | 2 | 3 |
| 69 | Tm | 0.5 | 0.2 | 0.3 |
| 70 | Yb | 3.4 | 1 | 3 |
| 71 | Lu | 0.5 | 0.1 | 0.3 |
| 72 | Hf | 3 | 5.2 | 1.5 |
| 73 | Ta | 2 | 1.6 | 0.7 |
| 74 | W | 1.5 | 0.4 | 0.45 |
| 75 | Re | 0.001 | 0.0006 | 0.0004 |
| 76 | Os | 0.005 | 0.0001 | 0.0004 |
| 77 | Ir | 0.001 | 0.006 | |
| 78 | Pt | 0.01 | 0.008 | 0.009 |
| 79 | Au | 0.004 | 0.002 | 0.005 |
| 80 | Hg | 0.08 | 0.2 | 0.2 |
| 81 | Tl | 0.5 | 1.3 | 0.13 |
| 82 | Pb | 13 | 49 | 8 |
| 83 | Bi | 0.2 | 0.1 | 0.2 |
| 90 | Th | 7.2 | 52 | 2.4 |
| 92 | U | 1.8 | 3.7 | 0.52 |

[a] Values in ppm. [From B. Mason, *Principles of Geochemistry* (Wiley, New York, 1966.]

**Table 9.25**  Elements Determined in Standard Rock Samples by
Activation Analysis[a]

| Element | Activation Analysis[b] | | Preferred Values[b] | | Method |
|---------|------|------|------|------|--------|
|         | G-1  | W-1  | G-1  | W-1  |        |
| Ag      | 0.04    | 0.06     |      |      | Activation analysis |
| As      | 0.67    | 2.24     | 0.8  | 2.2  | Various |
| Au      | 0.005   | 0.007    |      |      | Activation analysis |
| Ba[c]   | ...     | ...      | 1220 | 180  | Various |
| Br      | 0.48    | 0.50     |      |      | Activation analysis |
| Ce[c]   | 134     | 24.4     | 600  | 70   | Various |
|         | 330     | 36.6     |      |      |         |
|         | 150     | 15.1     |      |      |         |
| Co      | 2.3     | 52       |      |      | Activation analysis |
| Cr      | 20.7    | 116.3    | 22   | 120  | Various |
| Cs      | 2.35    | ...      | 1.5  | 1.1  | Various |
|         | 1.6     | ...      |      |      |         |
| Cu      | 13      | 110      | 13   | 110  | Various |
| Dy      | 1.33    | 3.3      |      |      |         |
|         | 2.51    | 7.7      |      |      |         |
|         | 2.52    | 4.38     |      |      |         |
| Er      | 1.4     | 2.57     |      |      |         |
| Eu      | 0.8     | ...      |      |      | Activation analysis |
|         | 0.95    | 1.12     |      |      |         |
|         | 1.42    | 2.2      |      |      |         |
|         | 1.04    | 1.09     |      |      |         |
|         | 1.36    | 1.29     |      |      |         |
| F[c]    | 1063    | 489      | 730  | 200  | Chemical |
| Ga      | 19.6    | 18.3     | 18   | 16   | Various |
|         | 20.8    | 16.5     |      |      |         |
| Gd      | 4.88    | 4.2      |      |      |         |
| Hf      | 7.6     | ...      |      |      |         |
| Hg      | 0.34    | 0.17     |      |      |         |
| Ho      | 0.50    | 1.35     |      |      |         |
|         | 0.41    | 0.86     |      |      |         |
| In      | 0.03    | 0.08     |      |      | Activation analysis |
| Ir      | 0.0063  | ...      |      |      | Activation analysis |
|         | $\leq$0.00007 | $\leq$0.00005 |  |  |     |
| La[c]   | 102     | 11.7     | 120  | 30   | Various |
|         | 142     | 26       |      |      |         |
|         | 129     | 20       |      |      |         |
|         | 92.4    | 9.3      |      |      |         |

(*Continued overleaf*)

Table 9.25 (*continued*)

| Element | Activation Analysis | | Preferred Values | | |
| | G-1 | W-1 | G-1 | W-1 | Method |
|---|---|---|---|---|---|
| Lu | 0.17 | 0.33 | | | |
| | 0.12 | 0.35 | | | |
| Mo | 7.0 | 1.3 | 7 | 0.5 | Various |
| Nd | 54.6 | 15.1 | 80 | 50 | Spectroscopy |
| | 55.2 | 20.2 | | | |
| O | 48.52% | 44.61% | | | |
| | 48.63% | 44.58% | | | |
| Os | 0.0002 | 0.00046 | | | |
| | 0.00005 | 0.000026 | | | |
| Pd | <0.01 | 0.02 | 0.001 | 0.02 | Activation analysis |
| | 0.0016 | 0.016 | | | |
| Pr | 20.9 | 3.68 | | | |
| | 12.9 | 3.51 | | | |
| Re | 0.0007 | 0.0007 | | | Activation analysis |
| | 0.00056 | 0.00042 | | | |
| Sb | 0.4 | 1.1 | | | Activation analysis |
| | 0.25 | 0.96 | | | |
| Sc | 2.4 | 36.3 | 3 | 34 | Activation analysis |
| | 3.0 | 33.3 | | | |
| Sm | 8.6 | 3.79 | 11 | 5 | Activation analysis |
| | 13.2 | 6.6 | | | |
| | 8.25 | 3.46 | | | |
| Sn | 3.4 | 3.4 | 4 | 3 | Various |
| | 3.5 | ... | | | |
| Sr[c] | 250 | 175 | 250 | 180 | Various |
| Ta | 2 | 0.7 | 1.6 | 0.7 | Activation analysis |
| | 0.9 | | | | |
| | 1.3 | 0.69 | | | |
| Tb | 0.05 | 0.75 | 0.6 | 0.8 | Activation analysis |
| | 0.64 | 0.81 | | | |
| Th | 42 | 2.2 | 52 | 2.4 | Various |
| Ti[c] | 1300 | 5400 | 1500 | 6400 | Chemical |
| Tm | 0.20 | 0.355 | 0.2 | 0.3 | |
| | 0.16 | 0.33 | | | |
| U | 3.1 | 0.55 | 3.7 | 0.52 | Various |
| | 4.7 | 0.65 | | | |
| V[c] | ... | ... | 16 | 240 | Various |
| W | 0.47 | 0.46 | 0.4 | 0.45 | Activation analysis |
| | 0.50 | 0.58 | | | |
| Y | 12.5 | 23.8 | 13 | 25 | Various |
| | 13.0 | 27.1 | | | |
| | 13.3 | 28.0 | | | |

Table **9.25** (*continued*)

| Element | Activation Analysis | | Preferred Values | | |
| | G-1 | W-1 | G-1 | W-1 | Method |
| --- | --- | --- | --- | --- | --- |
| Yb | 0.625 | 2.10 | 1 | 3 | Spectroscopy |
| | 0.94 | 2.23 | | | |
| Zn[c] | 44 | 85 | 45 | 82 | Various |
| | 45.7 | 82.8 | | | |

[a] From M. Fleischer, Summary of New Data on Rock Samples G-1 and W-1, 1962–1965, *Geochim. Cosmochim. Acta* **29,** 1263–1283 (1965).    [b] Values in ppm.
[c] Agreements considered poor, further measurements are suggested.

**Table 9.26**   Some Activation Analysis Determination in the New U.S. Geological Survey Standard Rock Samples[a]

| Element[b] | Rock Type | | | |
| | G-2 | GSP-1 | AGV-1 | BCR-1 |
| --- | --- | --- | --- | --- |
| Na(%) | 2.95 | 2.1 | 3.15 | 2.40 |
| K(%) | 4.0 | 4.5 | 2.2 | 1.2 |
| Rb | 129 | 255 | 61 | <105 |
| Cs | 1.4 | 0.8 | 1.3 | 1.5 |
| Ba | 1800 | 1110 | 1180 | 650 |
| La | 81 | 171 | 33 | 23 |
| Ce | 144 | 390 | 57 | 46 |
| Sm | 8.7 | 23.2 | 5.4 | 5.9 |
| Eu | 1.37 | 2.0 | 1.55 | 1.95 |
| Tb | 0.52 | 1.3 | 0.77 | 1.0 |
| Tm | <0.4 | 0.5 | ~0.4 | ~0.4 |
| Yb | 0.8 | 2.0 | 1.6 | 3.2 |
| Lu | 0.18 | 0.17 | 0.37 | 0.60 |
| Th | 25.9 | 125 | 7.0 | 6.7 |
| Zr | 250 | 450 | 175 | 210 |
| Hf | 7.8 | 15 | 5.4 | 4.7 |
| Ta | 1.0 | 1.4 | 1.0 | 0.9 |
| Mn | 212 | 264 | 640 | 1300 |
| Co | 4.3 | 7.0 | 14.7 | 36.3 |
| Fe | 1.72 | 2.8 | 4.3 | 9.0 |
| Sc | 3.5 | 6.0 | 11.7 | 32.5 |
| Cr | 4.6 | <18 | 8.6 | 19 |
| Sb | <0.6 | 4.1 | 5.2 | 1.1 |

[a] From G. E. Gordon et al., Instrumental Activation Analysis of Standard Rocks with High-Resolution Gamma-ray Detectors, *Geochim. Cosmochim. Acta* **32,** 369–396 (1968).
[b] Concentration in ppm unless (%) indicated.

### 9.2.2   The Biosphere

The biosphere is generally considered in two ways: one encompasses those portions of the atmosphere, hydrosphere, and lithosphere that are capable of supporting life; the other encompasses all living matter—plants, animals, and microorganisms. Mason* notes that the latter consideration includes insects and spores collected at high altitudes, living organisms dredged from ocean bottoms at great depths, and bacteria found in oil-well brines in upper crustal layers. Although these organic materials are of special importance in specific investigations, they represent but a tiny fraction of the living mass supported in the narrow zone at the surface of the hydrosphere and the lithosphere. The biosphere considered as the sum total of living matter is insignificant in mass compared with the hydrosphere; the mass ratio is about 1:69,100. It is, however, a sphere of great chemical activity and, of course, of considerable interest to those concerned with the biological realm. In this context the biological realm of man includes the following:

1. Biochemistry.
2. Biology.
3. Internal medicine.
4. The environmental sciences.

Mason gives an early (1937) estimate of the major and some minor elements that constitute the organisms of the biosphere. These elements are shown in Figure 9.12 as invariable and variable constituents. It is noted that water is the major component of all organic matter (about 50% in wood, about 66% in vertebrates, and about 99% in marine invertebrates). The metabolic role of the invariable elements may be classified as energy elements (C, O, H, N), macronutrients (P, Ca, Mg, K, S, Na, Cl), and micronutrients (Fe, Cu, Mn, Zn for both animals and plants; B, Mo, Si for plants only; and Co, I for animals only). Some variable elements, such as V, F, and Br, are of biological interest. Many trace elements are biologically toxic above fairly low levels. The variability of elemental composition of different types of organism is noted from the data in Table 9.27 for man, vegetation, and marine invertebrates.

### Biochemistry

Among the materials in biochemistry whose elemental compositions are of interest are biogenic deposits. Such deposits may be bioliths in the form of sedimentary rocks (such as limestone) or carbon compounds (such as coal or petroleum). Extensive applications of activation analysis for the analysis of

---

* B. Mason, *Principles of Geochemistry* (Wiley, New York, 1966), Chapter 9.

| | | | | | | | | | | | | | | | | | He |
|---|---|---|---|---|---|---|---|---|---|---|---|---|---|---|---|---|---|
| H ≡ | | | | | | | | | | | | | | | | | He |
| Li — | Be — | | | | | | | | | | | B — | C ≡ | N ≡ | O ≡ | F — | Ne |
| Na = | Mg = | | | | | | | | | | | Al — | Si — | P ≡ | S = | Cl = | A |
| K = | Ca = | Sc — | Ti = | V = | Cr — | Mn — | Fe — | Co — | Ni — | Cu — | Zn — | Ga — | Ge — | As — | Se — | Br = | Kr |
| Rb — | Sr — | Y — | Zr | Nb | Mo — | | Ru | Rh | Pd | Ag — | Cd — | In | Sn — | Sb | Te | I — | Xe |
| Cs — | Ba — | La-Lu — | Hf | Ta | W | Re | Os | Ir | Pt | Au | Hg — | Tl | Pb — | Bi | | | |
| Ra — | | | Th | U | | | | | | | | | | | | | |

Invariable { Primary (1–60%) ≡ ; Secondary (0.05–1%) = ; Microconstituents (<0.05%) — }

Variable { Secondary = ; Microconstituents — }

**Figure 9.12** The periodic table of the elements classified according to their distribution as percentage body weights of organisms. [From B. Mason, *Principles of Geochemistry* (Wiley, New York), 1966, Chapter 9.]

**Table 9.27** Elemental Composition of Different Organisms[a]

| Element | Organism | | |
|---|---|---|---|
| | Man (w/o) | Vegetation (alfalfa) (w/o) | Marine Invertebrate (%) |
| O | 62.81 | 77.90 | 79.90 |
| C | 19.37 | 11.34 | 6.10 |
| H | 9.31 | 8.72 | 10.21 |
| N | 5.14 | 8.25 | 1.52 |
| Ca | 1.38 | 0.58 | 0.04 |
| S | 0.64 | 0.10 | 0.14 |
| P | 0.63 | 0.71 | 0.13 |
| Na | 0.26 | ... | 0.54 |
| K | 0.22 | 0.17 | 0.29 |
| Cl | 0.18 | 0.07 | 1.05 |
| Mg | 0.04 | 0.082 | 0.03 |
| Fe | 0.005 | 0.0027 | 0.007 |
| Si | 0.004 | 0.0093 | 0.007 |
| Zn | $2.5 \times 10^{-3}$ | $3.5 \times 10^{-4}$ | ... |
| Rb | $9 \times 10^{-4}$ | $4.6 \times 10^{-4}$ | ... |
| Cu | $4 \times 10^{-4}$ | $2.5 \times 10^{-4}$ | $6 \times 10^{-4}$ |
| Sr | $4 \times 10^{-4}$ | Various | ... |
| Br | $2 \times 10^{-4}$ | $5 \times 10^{-5}$ | $9 \times 10^{-4}$ |
| Sn | $2 \times 10^{-4}$ | Various | ... |
| Mn | $1 \times 10^{-4}$ | $3.6 \times 10^{-4}$ | ... |
| I | $1 \times 10^{-4}$ | $2.5 \times 10^{-6}$ | $2 \times 10^{-4}$ |
| Al | $5 \times 10^{-5}$ | $2.5 \times 10^{-3}$ | ... |
| Pb | $5 \times 10^{-5}$ | Various | ... |
| Ba | $3 \times 10^{-5}$ | Various | ... |
| Mo | $2 \times 10^{-5}$ | $1 \times 10^{-4}$ | $2.4 \times 10^{-5}$ |
| B | $2 \times 10^{-5}$ | $7 \times 10^{-4}$ | $1.5 \times 10^{-3}$ |
| As | $5 \times 10^{-6}$ | Various | ... |
| Co | $4 \times 10^{-6}$ | $2 \times 10^{-6}$ | $6 \times 10^{-6}$ |
| Li | $3 \times 10^{-6}$ | ... | ... |
| V | $2.6 \times 10^{-6}$ | $1.6 \times 10^{-5}$ | $4 \times 10^{-5}$ |
| Ni | $2.5 \times 10^{-6}$ | $5 \times 10^{-5}$ | ... |
| F | ... | $1.5 \times 10^{-4}$ | ... |
| Ti | ... | $9 \times 10^{-5}$ | $2 \times 10^{-5}$ |

[a] From B. Mason, *Principles of Geochemistry* (Wiley, New York, 1966), p. 231.

trace elements in such biogenic deposits are included in the literature of geochemistry.

### Biology

The extent of the use of radioactivation analysis in the fields of biology and medicine is readily observed from the contents of the proceedings* of a symposium on this subject in 1967 and the hundreds of references† to published literature that describes activation analysis methods for the study of

**Table 9.28**   Determination of Essential Elements in Plants by Activation Analysis[a]

| Element | Determination Using $\gamma$-Spectrometry | Determination using Radiochemistry | Remarks |
|---------|-------------------------------------------|------------------------------------|---------|
| B | Measure $\alpha$ from (n,$\alpha$) reaction | (p,n) reaction suitable | Colorimetry adequate |
| C | possible using ($\gamma$,n) reaction | Not reported | Unsuitable |
| Ca | Possible using $^{49}$Ca | Difficult using $^{49}$Ca/$^{49}$Sc | Other methods adequate |
| Cl | Possible using $^{38}$Cl | Possible | Other methods adequate |
| Co | Difficult using $^{60}$Co | Possible | Competitive |
| Cu | Unsatisfactory using $^{64}$Cu | Very sensitive | Competitive |
| Fe | Possible using $^{59}$Fe | Possible | Other methods adequate |
| H | Unsuitable | Unsuitable | Unsuitable |
| K | Unsatisfactory | Possible | Other methods adequate |
| Mg | Unsatisfactory | Difficult using $^{27}$Mg | Other methods preferable |
| Mn | Easy | Very sensitive | Competitive |
| Mo | Not reported | Possible | Competitive |
| N | Possible using $^{16}$N | Not reported | $\gamma$-spectrometry is fast |
| Na | Easy | Very sensitive | Other methods adequate |
| O | Possible; only ($\gamma$,n) results reported | Not reported | Other methods adequate |
| P | Unsuitable | Very sensitive | Competitive |
| S | Unsuitable | Not reported | Other methods unsatisfactory |
| Se | Possible, but not reported | Sensitive | Competitive |
| Si | Unsuitable | Possible | Competitive |
| Zn | Possible using $^{65}$Zn | Sensitive | Competitive |

[a] From H. J. M. Bowen, Activation Analysis in Botany and Agriculture, in *Nuclear Activation Techniques in the Life Sciences* (International Atomic Energy Agency, Vienna, 1967), pp 287–299.

* *Neutron Activation Techniques in the Life Sciences* (International Atomic Energy Agency, Vienna, 1967).
† *Activation Analysis: A Bibliography*, National Bureau of Standards Technical Note 467 Parts I and II (1968).

microorganisms, organic components, soil science, botany, agriculture, plants, animals, and external and internal medicine. This tremendous literature, in fact, makes review of the applications for this survey extremely difficult. A summary of some of the advances in activation analysis in botany and agriculture is given by Bowen.* His review of the potential of both instrumental and radiochemical methods of activation analysis, in comparison to other methods, is given in Table 9.28. Data on trace-element concentrations for some 12 elements in marine algae determined by activation analysis are given in Table 9.29.

**Table 9.29**   Determination of Trace Elements in Marine Algae
by Activation Analysis[a]

| Element | Type of Algae | Dry algae (ppm) | Seawater (ppm) |
|---------|---------------|-----------------|----------------|
| As | Green | 1.2–5.4 | 0.003 |
| Au | Green | 0.0035–0.02 | 0.000004 |
| Ba | Brown | 6–19 | 0.03 |
| Ba | Green | 0.4–2.5 | 0.03 |
| Ba | Red | 0.6–6 | 0.03 |
| Co | Brown | 0.2–3 | 0.0001 |
| Co | Red | 0.5 | 0.0001 |
| Cs | Brown | 0.04–0.1[b] | 0.0005 |
| Cs | Red | 0.03–0.05[b] | 0.0005 |
| Mo | Red | 1 | 0.01 |
| Rb | Brown | 5.6–9.6 | 0.12 |
| Rb | Red | 2.4–4.3[b] | 0.12 |
| Rb | Green | 0.011–0.016 | ? |
| Sr | Brown | 430–1240 | 8.0 |
| Sr | Green | 38–87 | 8.0 |
| Sr | Red | 19–130 | 8.0 |
| U | ? | 0.065[c] | 0.003 |
| V | Green | 1.3–3.1 | 0.002 |
| V | Red | 16 | 0.002 |
| W | Green | 0.03–0.04 | 0.0001 |

[a] From H. J. M. Bowen, Activation Analysis in Botany and Agriculture, in *Nuclear Activation Techniques in the Life Sciences* (International Atomic Energy Agency, Vienna, 1967), pp. 287–299.
[b] Recalculated from fresh weight data assuming 75% water content.
[c] Assuming ash weight is 5% of dry weight.

* H. J. M. Bowen, Activation Analysis in Botany and Agriculture, in *Nuclear Activation Techniques in the Life Sciences, op. cit.*, pp. 287–297.

Table 9.30    Survey of Elements Determined in Animal Samples
by Radioactivation Analysis[a]

| Material | Elements Found |
|---|---|
| Milk (dried) | As |
| Mammalian tissues (mouse, rat, rabbit, etc.) | As Au B Co Cu Ga Hg I K Mn Mo Na P Se Sn Zn |
| Marine animals | Ba Dy K Na Sr |
| Fresh-water animals (trout) | I V |
| Lamb tissues (wool, liver) | Al Au Dy Mn Mo Se V W Zn |
| Calf heart (including subcellular fractions) | Ag As Ba Br Ca Cd Ce Co Cr Cs Cu Fe Hg La Mo P Rb Sb Sc Se Sm W Zn |
| Ox bone | Cr |
| Goat tissues | Se |
| Cow milk | Se |
| Snake venom | Cu Zn |
| Marine organisms | As Au Br Ca Cu Dy I Mg Mn Re U V W Zn |
| Snail tissues | Al Br Cl Cu Mg Mn Na Sr |
| Fresh-water mollusk | Ag Cd Co Cr Fe Hg Mn Sc Sr Th Zn |
| Beetle wings | P |

[a] From N. Spronk, Nuclear Activation in the Animal Sciences, in *Nuclear Activation Techniques in the Life Sciences* (International Atomic Energy Agency, Vienna, 1967), pp 335–352.

In a companion survey paper Spronk* reviews the applications of activation analysis in the animal sciences. Since 1946 all types of animal material have been analyzed. These samples are either related in some way to human life (e.g., meat, blood, excreta) or to experimental animals (e.g., mouse, rat, rabbit, or chicken). A list of some of the animal materials examined and the trace elements determined is given in Table 9.30. Activation analysis has been used extensively in zoological research in such areas as (a) dynamics and interaction of major and minor constituents, (b) mineral metabolic cycle studies, (c) radioecological studies, and (d) toxicological studies.

The use of activation analysis in the medical sciences has increased considerably in recent years, especially as the role of trace elements in human biochemistry is discovered and explored. The determination of one or more minor or trace elements simultaneously in single samples allows the trace-element patterns to be used in comparisons under normal and pathological conditions. A review paper by Taylor† notes that such activation analysis

* N. Spronk, Nuclear Activation in the Animal Sciences, *op. cit.*, pp. 335–352.
† D. M. Taylor, "Activation Analysis in Human Biochemistry," in *Nuclear Activation Techniques in the Life Sciences* (International Atomic Energy Agency, Vienna, (1967), pp. 391–401).

studies may indicate disturbances of the metabolism of certain elements in a specific disease or may uncover inter-relationships between two or more trace elements.

Activation analysis has been used to determine trace-element concentrations in such human materials as blood, cerebrospinal fluid, urine, tissues, and teeth. Medical studies involving activation analysis have included the disturbances of copper and manganese metabolism in some forms of cirrhosis of the liver, the role of iodine in thyroid physiopathology, the metabolism of bromine and the volume of extracellular liquid, diurnal variations in plasma trace elements, the electrolytic balance of the inner ear, and the distribution of elements in dental enamel. Although the details of these studies are beyond this survey, it is of interest to note some of the variability that is found in trace-element concentrations in medical samples. The data from a study of trace elements in human enamel are given in Table 9.31. Activation analysis made it possible to establish concentrations in a single sound tooth. In many cases the standard deviation exceeds the mean. It was noted that nonessential trace elements (Sb, Hg, As) showed a skew distribution, whereas the essential elements (Cu, Mn, Zn) showed a normal distribution.

### Internal Medicine

A new technique in activation analysis was introduced into the medical literature in 1964 by Anderson et al.,* who reported on the use of *in vivo*

**Table 9.31**   Trace Elements in Human Enamel[a]

| Element | No. of Samples | Maximum | Minimum | Median | Mean | Standard Deviation | Remarks |
|---------|----------------|---------|---------|--------|------|--------------------|---------|
| Antimony | 61 | 0.66 | 0.005 | 0.39 | 0.078 | 0.115 | ... |
| Arsenic | 75 | 0.63 | 0.003 | 0.050 | 0.070 | 0.085 | ... |
| Cadmium | 6 | ... | ... | ... | ... | ... | <0.03 |
| Copper | 103 | 39.7 | 1.59 | 7.88 | 10.11 | 7.84 | ... |
| Manganese | 62 | 2.01 | 0.30 | 0.73 | 0.83 | 0.37 | ... |
| Mercury | 59 | 18.1 | 0.14 | 2.13 | 3.22 | 3.62 | ... |
| Molybdenum | 11 | 0.12 | 0.026 | 0.054 | 0.054 | 0.027 | ... |
| Vanadium | 23 | ... | ... | ... | ... | ... | <0.01 |
| Zinc | 130 | 992 | 58 | 339 | 365 | 182 | ... |

[a] From G. S. Nixon, H. Smith, and H. D. Livingston, Trace Elements in Human Tooth Enamel, in *Nuclear Activation Techniques in the Life Sciences* (International Atomic Energy Agency, Vienna, 1967), pp 455–462.

* J. Anderson, S. Orborn, R. Tomlinson, R. Newton, J. Rundo, L. Salmon, and J. Smith, Neutron-Activation Analysis in Man In Vivo, *Lancet*, 1201. (December 1964).

activation analysis to estimate the total body sodium by whole-body exposure to 14 MeV neutrons. *In vivo* activation analysis has become a useful method for measuring total body content of several vital trace elements as well as the total amount of specific elements in specific body organs. These measurements of elemental mass cannot be easily made in living man by any other method. Several research programs are underway to develop this method,

**Table 9.32** Elements of Internal Biological Significance[a]

| Bulk Elements | | Essential Elements | |
|---|---|---|---|
| Element | Concentration (g/70 kg-man) | Element | Concentration (mg/70 kg-man) |
| O | 45,500 | Zn | 2300 |
| C | 12,600 | Cu | 150 |
| H | 7,000 | I | 30 |
| N | 2,100 | Mn | 20 |
| Ca | 1,050 | Mo | 5 |
| P | 700 | Co | 3 |
| S | 175 | Se | ? |
| K | 140 | | |
| Na | 105 | | |
| Cl | 105 | | |
| Mg | 35 | | |
| Fe | 4 | | |

*In-Vivo* Activation Analysis

| Possible Reactions | Product Half-Life |
|---|---|
| $^{16}O(n,p)^{16}N$ | 7 sec |
| $^{12}C(\gamma,n)^{11}C$ | 20 min |
| $^{1}H(n,\gamma)^{2}H$ | prompt $\gamma$ |
| $^{14}N(n,2n)^{13}N$ | 10 min |
| $^{48}Ca(n,\gamma)^{49}Ca$ | 9 min |
| $^{31}P(n,\alpha)^{28}Al$ | 2.3 min |
| $^{23}Na(n,\gamma)^{24}Na$ | 15 hr |
| $^{37}Cl(n,\gamma)^{38}Cl$ | 37 min |

[a] From J. M. A. Lenihan, Activation Analysis in the Contemporary World, in Modern Trends in Activation Analysis (National Bureau of Standards, Gaithersburg, Md., 1969), pp. 1–32.

among which are the measurement of total iodine content of thyroid glands and total body content of sodium, chlorine, calcium, and other elements.

Lenihan* has reviewed some of these applications in medical research. His data on bulk and essential elements in man and some possible reactions for *in vivo* activation analysis are summarized in Table 9.32. The analysis involves an irradiation of the patient with d,t generator- or cyclotron-produced neutrons and radiation measurement in a whole-body counter.

Two general problems associated with *in vivo* activation analysis are calibration of the analysis because of the nonhomogeneity of irradiation flux and the radiological hazard to the patient. The flux inhomogeneity problem is treated in several ways: (a) the use of a phantom (a plastic mannikin) filled with appropriate solution, (b) isotope dilution methods with tracer doses of the product radionuclide, and (c) stable tracer isotopes. A combination of the second and third methods is used for iodine determination in the thyroid. The tracer used is the long-lived radioactive iodine isotope $1.7 \times 10^7$-y $^{129}$I, which emits low energy $\gamma$-rays. This isotope is also activatable with thermal neutrons to 12.6-h $^{130}$I. Thus after administration of a dose of $^{129}$I the tracer amount in the thyroid can be measured by external counting. The $^{129}$I then serves as an internal flux monitor for the *in vivo* activation. The thyroid iodine is estimated from the measured activities of the $^{128}$I and $^{130}$I products. A review of activation analysis methods for determination of $^{129}$I levels in the biosphere is available.†

The radiological hazard due to *in vivo* activation analysis has been examined; for example, a thyroid irradiated for 10 min with a thermal neutron flux of about $10^7$ n/cm²-sec receives a radiation dose of about 4 rads. This dose is less than that received from the standard thyroid uptake test with 12 mCi $^{131}$I as the source. To keep the total radiation dose to a patient no greater than is acceptable in diagnostic radiology many researchers limit the maximum whole-body dose to 0.1 rad.

### Environmental Sciences

The use of activation analysis in exploring the chemical aspects of man's environment has developed in several fields. One is in the general problem of environmental pollution. Some of the efforts in the study of air and water pollution have been described in the respective geospheres.

Another aspect of the social environment is that of public health. The epidemiological applications have already been reviewed. Trace-element

---

* J. M. A. Lenihan, Activation Analysis in the Contemporary World, in *Modern Trends in Activation Analysis* (National Bureau of Standards, Gaithersburg, Md, 1969), pp. 1–32.
† B. Keisch, R. C. Koch, and A. S. Levine, Determination of Biospheric Levels of $^{129}$I by Neutron-Activation Analysis, in *Modern Trends in Activation Analysis* (Texas A&M University, College Station, 1965), pp. 284–290.

determinations can be made for large numbers of samples from population groups large enough for accurate statistical analysis. Such studies are of especial interest in cancer, cirrhosis of the liver, heart failures, and dental caries.

Another concern in public health is the ultimate fate of toxic materials introduced into the natural and industrial environments. The increased use of nondegradable pesticides, insecticides, herbicides, defoliating agents, and other such materials into the biosphere has caused considerable concern about their role in killing many species of wildlife and their eventual hazard to man. A review of some of the applications of activation analysis of foods for pesticide residues has been given by Guinn.* Included are studies of bromine residues from nematocides, bromine residues from methyl bromide fumigations, DDT residues in foods, and mercury-contaminated wheat.

A review of some of the applications in the field of industrial hygiene has been given by Lenihan and Smith.† Included are examples of occupational hazards associated with the handling of toxic materials, such as compounds of arsenic and mercury.

### 9.2.3 Industrial Applications

The adaptation of activation analysis practices in the industrial realm requires, in addition to the technical advantages of high sensitivity and nondestructiveness, the economic advantages of low cost and speed. The added advantage of automatic analysis is, of course, of great interest, but its potentially greater advantage as an industrial tool is its use for in-plant process control.

Manufacturers of material products are potential users of activation analysis methods. The techniques are available for process research, testing, process control, and product-quality improvement. Obviously in so large a range of industrial requirements the methods of activation analysis may be applied from the routine analysis of trace elements in a variety of industrial products to highly specialized techniques. The general principles of trace-element analysis have already been reviewed for a great many kinds of sample. As two indications of some of the special techniques that have been developed for industrial applications, the use of activation analysis will be examined for the determination of oxygen content in metallurgy and the use of neutron sondes for oil, gas, and water well logging. The development of on-stream

* V. P. Guinn, Neutron Activation Analysis of Foodstuffs for Pesticide Residues, *World Review of Pest Control* **3,** 138–147 (1964).

† J. M. A. Lenihan and H. Smith, Activation Analysis and Public Health, in *Nuclear Activation Techniques in the Life Sciences* (International Atomic Energy Agency, Vienna, 1967), pp. 601–614.

or cyclic activation analysis methods for process control is described further in Section 9.3.4.

### Oxygen in Metals

The determination of the oxygen content in metals, such as iron and steel, aluminum, beryllium, and titanium, is an important but difficult process. The oxygen content is considered an important variable affecting the deterioration of the physical properties of metals, such as corrosion resistance, fatigue life, diffusion ability, and surface properties. The demand for rapid and reliable oxygen determinations has been especially strong because of increased production and new rapid methods of steelmaking; for example, a basic oxygen furnace can produce more than 300 tons of steel in less than an hour. To ensure quality of the required composition, analyses must be made during the "heat" for several key elements, among them oxygen, and reported to the furnace operators in a matter of minutes.

Activation analysis techniques have proved successful in providing such rapid, accurate analyses. Methods have been developed by using fast neutrons and charged particles; for example, Veal and Cook* described a rapid method for oxygen by fast-neutron activation. They used the reaction $^{16}O(n,p)^{16}N$ and measured the gamma rays of $E > 5$ MeV from the 7.4-s $N^{16}$. A counting delay of 0.2 sec after activation eliminated interference from the 0.022-sec product from the $^{12}C(n,p)^{12}B$ reaction, but interferences from fluorine and boron are limiting. Fluorine is especially troublesome, since it results in the same product radionuclide by the $^{19}F(n,\alpha)^{16}N$ reaction. A fluorine content of 10% that of oxygen increases the $^{16}N$ production by 100% for 14-MeV neutrons. Oxygen content in samples free of fluorine (and excessive boron) is determined by comparison to the oxygen content in standard benzoic acid.

Wood and Pasztor† described a fast-neutron activation analysis oxygen analyzer that compared favorably with the vacuum-fusion techniques used for the routine determination of oxygen in steels. Significant advantages are the use of larger samples and faster analysis. The first advantage is especially important for molten steels in which the oxygen is less uniformly distributed compared with finished steels. A block diagram of the analyzer is shown in Figure 9.13. Mylar standards with known oxygen contents are used for calibration. The system yields approximately 2200 count/mg $O_2$ at $10^{11}$ n/sec output. The oxygen content (w/o) of a sample is given by the comparitor

* D. J. Veal and C. F. Cook, A Rapid Method for the Direct Determination of Elemental Oxygen by Activation with Fast Neutrons, *Anal. Chem.* **34**, 178–184 (1962).
† D. E. Wood and L. C. Pasztor, A Comparison of Neutron-Activation Analysis and Vacuum-Fusion Analysis of the Oxygen Content of Steel, in *Modern Trends in Activation Analysis* (Texas A&M University, College Station, 1965).

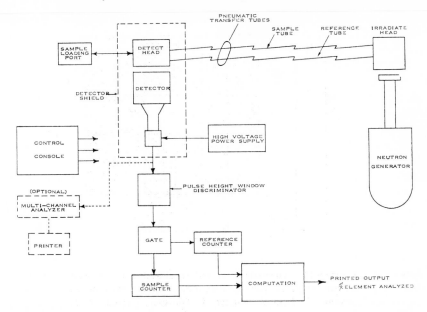

**Figure 9.13**  Block diagram of a fast-neutron activation analysis oxygen analyzer. [From D. E. Wood and L. C. Pasztor, A Comparison of Neutron Activation Analysis and Vacuum-Fusion Analysis of the Oxygen Content of Steel, in *Modern Trends in Activation Analysis* (Texas A&M University, College Station, 1965).]

analysis as

$$O_2(w/o) = \frac{1}{CM}\left(\frac{S - b}{R} - B\right) \tag{9}$$

where $C$ = calibration factor for geometry and counter sensitivities,

   $M$ = weight of sample in grams,

   $S$ = count of the sample,

   $b$ = background count,

   $R$ = count for Lucite reference,

   $B$ = ratio of count for the empty bottle to reference counts.

For a typical analysis, with $C = 0.333$, $M = 10$ g, $S = 3050$, $b = 50$, $R = 60,000$, and $B = 0.0167$, oxygen contents of 0.01 w/o are readily determined in a total analysis time of less than 2 min. The reproducibility of the method for routine analysis of low-oxygen steels (30–300 ppm) at the 95% confidence limit is $\pm 10\%$ for an average of two measurements.

Automatic equipment for the determination of the oxygen content in bulky solid materials has been developed for in-plant industrial application.

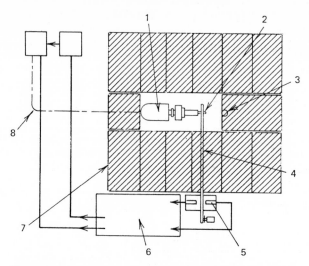

**Figure 9.14** Schematic of an industrial system for in-plant oxygen analysis. 1. Neutron generator. 2. Sample in irradiated position. 3. Neutron counter. 4. Transport mechanism. 5. Gamma counter. 6. Control desk. 7. Shield tanks. 8. High-voltage cable. [From Gray and Metcalf, Industrial Applications of Neutron Activation, in *Modern Trends in Activation Analysis* (Texas A&M University, College Station (1965), pp. 86–90.]

Gray and Metcalf* have described a system that can determine oxygen in steel samples weighing more than 100 g, using a conveyor belt as the transport mechanism between the neutron generator and the gamma-ray spectrometer. A diagram of an industrial-plant system is shown in Figure 9.14. The latest model of this industrial equipment offers a sensitivity for 100-g steel samples better than 900 counts/mg $O_2$ with a total analysis time of less than 1 min.

Oxygen can be determined in metal surfaces by the use of the less penetrating tritons produced in charged-particle accelerators. Barrandon and Albert† have discussed the calibration of oxide thickness, using the equivalent thickness concept for 2-MeV tritons produced in a 3-MeV Van de Graaff accelerator. They assume that the thickness of the oxide film on the surface of a standard $\Delta x_s$, and on the surface of a sample $\Delta x_e$, is sufficiently thin so that the triton energy is considered constant through the layer. The cross

* A. L. Gray and A. Metcalf, Industrial Applications of Neutron Activation, in *Modern Trends in Activation Analysis* (Texas A&M University, College Station, 1965), pp. 86–90.
† J. N. Barrandon and P. Albert, Determination of Oxygen Present at the Surface of Metal by Irradiation with 2-Mev Tritons, in *Modern Trends in Activation Analysis* (National Bureau of Standards, 1968), pp. 298–305.

section through the layer is thus constant and given by

$$\sigma(E) = \sigma(E_0) + \left(\frac{d\sigma}{dE}\right)_{E_0} \Delta E = \sigma(E_0)\left[1 + \frac{d \log \sigma(E_0)}{dE} \Delta E\right] \quad (10)$$

where

$$\frac{d \log \sigma(E_0)}{dE} \Delta E \ll 1 \quad (11)$$

For short irradiation times, compared with the half-life of the product $^{18}F$, the initial activities of standard and sample will be

$$A_s = Q_s n_s \int_0^{\Delta x_s} \sigma_x \, dx \quad (12)$$

$$A_e = Q_e n_e \int_0^{\Delta x_e} \sigma_x \, dx \quad (13)$$

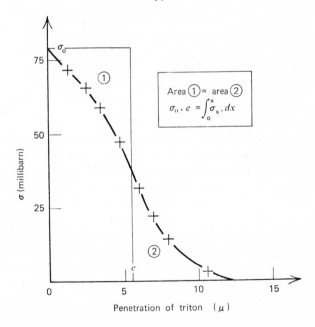

**Figure 9.15** The "equivalent thickness" determination of 2-MeV tritons in $ZrO_2$. [From J. N. Barrandon and P. Albert, Determination of Oxygen Present at the Surface of Metals by Irradiation with 2 MeV Tritons, in *Modern Trends in Activation Analysis* (National Bureau of Standards, 1968) pp. 298–305.]

where          $Q$ = integrated irradiation beam flux,

                 $n$ = concentration of oxygen atoms,

     $\sigma_0 = \sigma(E_0)$ = cross section of the reaction for incident particle energy $E_0$,

     $\sigma_x = \sigma(E_x)$ = cross section of the reaction for particle energy $E_x$, where the range of the particles is $x$.

The equivalent thickness $e$ is defined from Figure 9.15 as the thickness at which the activity would be uniform through the range $x$ as

$$e = \frac{\int_0^{\Delta x_s} \sigma_x \, dx}{\sigma_0} \tag{14}$$

**Table 9.33**    Surface Oxygen in Pure Metals[a]

| Zirconium | |
| --- | --- |
| Treatment | Oxygen Content ($\mu$g/cm²) |
| Initial (rolled) | 1.46 |
| Annealing 1 hr | 1.6–1.74 |
| Chemically polished | 1.22 |
|   23N HF (18%) | |
|   11N HNO$_3$ (82%) | |
|     Dried with alcohol | 1.90 |
|     Dried with acetone | 2.15–2.30 |
| Mechanically polished | |
|   dried with acetone | 5.8–6 |

| Aluminum | | |
| --- | --- | --- |
| Treatment | Oxygen Content ($\mu$g/cm²) | Oxide Thickness Al$_2$O$_3$ (Å) |
| Double electrolysis (99.99%) | | |
|   industrially rolled | 1.65 | 85 |
| Etched | 1.1 | 58 |
| Simple electrolysis (99.7%) | | |
|   rolled and etched (lab.) | 0.9 | 47 |
| Double electrolysis (99.999%) | | |
|   rolled and etched (lab.) | 0.48 | 25 |

[a] From Barrandon and Albert, *loc. cit.*, 1968.

which gives the sample oxide film thickness as

$$\Delta x_e = e \cdot \frac{A_e n_s Q_s}{A_s n_e Q_e} \tag{15}$$

Some typical results that relate the surface oxygen content as a function of the metal treatment for zirconium and aluminum are given in Table 9.33.

### Well-Logging Techniques

Neutron activation analysis has been added to the list of geophysical measurement methods employed to probe and study the complex geologic structure of the earth's crust. Well logging is the name given to the investigations in which sensor probes are lowered into observation or production holes to measure some property of the various geologic formations encountered on the way down or up and the evaluation of the response in terms of the formations. Sensors that measure electric, magnetic, density, and other physical properties are in common use. Well logging was also an early application of applied nuclear technology.* The development of the small neutron generator and scintillation detector allowed the concomitant development of neutron sondes which can be lowered into boreholes with a diameter of less than 6 in. The development of a pulsed neutron-source technique for well logging was described by Tittman and Nelligan.† A laboratory mockup of their logging sonde is shown in Figure 9.16. The use of the neutron dieaway method has been applied to well logging; for example, by the response to chlorine as described by Youmans et al.‡

The response of a gamma-ray spectrum to a neutron activation pulse in well logging requires interpretation in terms of changes in the geochemical composition around the borehole. Field experience has been acquired in such evaluations as oil content, lithology, and porosity. A logging system has been described by Hoyer and Rumble§ which consists of a pulsed high-energy neutron generator, a gated Na(Tl) crystal-photomultiplier gamma-ray detector, and surface equipment to record the spectra. By pulsed operation of the generator and simultaneous gating of the detector, only prompt gamma

---

* B. Pontecorvo, Neutron Well Logging—A New Geological Method Based on Nuclear Physics, *Oil Gas J.* **40**, 32 (1941).
† J. Tittman and W. B. Nelligan, Laboratory Studies of a Pulsed Neutron-Source Technique in Well Logging, Tech. Note 2061, *J. Petr. Tech.* 63–66 (July 1960).
‡ A. H. Youmans, E. C. Hopkinson, R. A. Bergan, and H. E. Oshry, Neutron Lifetime, A Nuclear Log, *J. Petr. Tech.* 319 (March 1964).
§ W. A. Hoyer and R. C. Rumble, Field Experience in Measuring Oil Content, Lithology and Porosity with a High-Energy Neutron-Induced Spectral Logging System, *J. Petr. Tech.* 801–807 (July 1965).

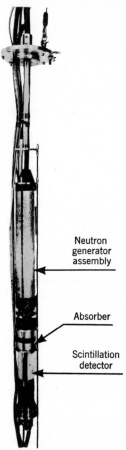

Neutron
generator
assembly

Absorber

Scintillation
detector

**Figure 9.16** A labora-
tory mockup of a pulsed
neutron sonde used for
well logging studies.
[From Tittman and
Nelligan, Laboratory
Studies of a Pulsed
Neutron-Source Tech-
nique in Well Logging,
Tech. Note 2061, *J.
Petr. Tech.*, 63–66
(July 1960).]

rays from the high-energy neutron reactions are detected. Repeated cycles of
5 $\mu$sec counts and 995 $\mu$sec deadtime result in the accumulation of a gamma-
ray spectrum which can be analyzed to determine the elements present in the
formation.

The oil content can be inferred from the measured C/O ratio. An example
of a continuous logging for C/O ratios in a limestone formation is shown in
Figure 9.17*a*.

The formation lithology evaluation can include measurements of Si for
sandstone, Ca for limestone, Mg for dolomite, S for anhydrite, and Al for
shale. A gamma-ray spectrum obtained in a dolomite formation at a depth
of 5605 ft is shown in Figure 9.17*b*.

The response of the pulsed fast-neutron logging sonde to porosity has been
compared both with the response from a commercial neutron log and with

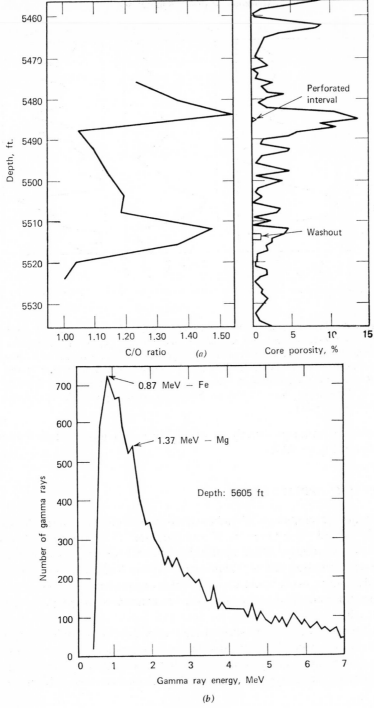

**Figure 9.17**

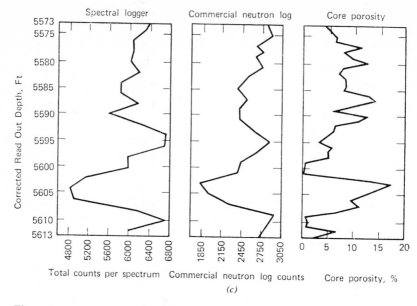

**Figure 9.17** Data obtained with a high-energy neutron-induced spectral logging system: (*a*) C/O ratios and core porosites in limestone; (*b*) gamma-ray spectrum in dolomite at a depth of 5605 ft; (*c*) porosity response compared with that of a commercial neutron log and core porosity data in limestone-dolomite. [From Hoyer and Rumble, Field Experience in Measuring Oil Content, Lithology, and Porosity with a High-Energy Neutron-Induced Spectral Logging System, *J. Petr. Tech.* 801–807 (July 1965).]

core porosities measured directly on small individual samples. These logs are illustrated in Figure 9.17*c*.

## 9.3  SPECIAL APPLICATIONS

The major developments of radioactivation analysis have been directed primarily toward the attainment of increased sensitivity for trace-element analysis. Concurrently, other applications have evolved in which trace-element determination was not the primary purpose. Radioactivation analysis also lends itself to many applications in which determination of one or more major or important minor elements is used to examine some property of the matrix itself.

The diversity with which such special applications of radioactivation analysis can be invented is illustrated in the four examples described in this section. Comparative analysis deals with the small differences in chemical composition of similar types of material. Surface analysis takes advantage of the small depths of penetration by projectile radiations into materials. Isotopic analysis makes use of the different nuclear properties of the stable isotopes of an element. On-line analysis involves the use of activation analysis of a major or minor element in a process stream material to assist in process or quality control.

### 9.3.1  Comparative Analysis

Comparative activation analysis has been referred to as a "nuclear finger-print" method, a way of using the radiation pattern of an activated sample to identify its chemical composition. In such analysis the determination of absolute concentrations of the constituents is not required—only the relative concentrations among key elements and their differences compared among different samples of a given kind of material.

An obvious application of comparative analysis is in the field of forensic science. The need for newer or better scientific methods for legal evidence is obvious to law enforcement officers. Yet Maletskos* has noted the random, laborious, and lengthy procedures needed to introduce a new scientific method in court. A new method must fulfill two conditions:

1. It must be sufficiently mature to be scientifically sound.
2. It must be presented and challenged in court.

Although activation analysis, it is generally agreed, has met the first condition, a series of steps or "precedents" is required, in challenges and appeals through lower courts and finally by appeal to the U.S. Supreme Court, before the method can be approved in the legal sense. It has been noted, for example, that the internationally accepted method of fingerprint identification had its initial study in 1859, its avaialbility for use in 1877, but its legal recognition in the United States not until 1911.

The first use of neutron activation analysis in a court trial occurred on March 16, 1964,† when the U.S. Internal Revenue Service presented $\gamma$-ray spectra of neutron-activated soil samples scraped from the bottom of a tractor-trailer seized in New York, which contained 2418 gallons of illicit liquor. These data matched identically the data obtained from soil samples taken around a still in Georgia, where it was believed the liquor had been loaded. These data helped to establish the connection to the satisfaction of judge and jury and a conviction resulted.

* C. J. Maletskos, On the Introduction of New Scientific Methods in Court, in Proceedings First International Conference on Forensic Activation Analysis, GA-8171 (General Atomic Corporation, La Jolla, Calif., 1966).

† U.S. Atomic Energy Commission Press Release G-112, May 13, 1964.

**Table 9.34** Neutron-Activation Analysis Results
on Gunhand from Single Firings of
a 0.38-Caliber Revolver[a]
(same gun, same kind of ammunition,
different people)

|  | Ba ($\mu$g) | Sb ($\mu$g) |
|---|---|---|
|  | 1.03 | 0.22 |
|  | 0.58 | 0.17 |
|  | 0.99 | 0.27 |
|  | 0.37 | 0.19 |
|  | 0.40 | 0.12 |
|  | 0.32 | 0.26 |
|  | 0.27 | 0.18 |
|  | 0.38 | 0.22 |
|  | 0.82 | 0.28 |
|  | 0.54 | 0.26 |
| Mean: | 0.57 ± 0.27 | 0.22 ± 0.05 |

[a] From D. E. Bryan, V. P. Guinn, and D. M. Settle, New Developments in the Application of Neutron Activation Analysis to Problems in Scientific Crime Detection, in *Modern Trends in Activation Analysis* (Texas A&M, College Station, 1965), pp. 140–145.

In addition to the use in combatting "moonshining," a number of other forensic applications have been developed, some of which include identification of drugs and narcotics (e.g., opiates, marijuana, and barbiturates), gunshot residues (e.g., Ba and Sb from the primer), hair (e.g., for identification or suspicion of poisoning of persons), and such evidential materials as paint, glass, paper, grease, tire rubber, and plastics. An example of a detailed analyses of trace elements in various gunshot residues is given in Table 9.34. Other data with similar tables of analyses for 19 commercial plastics, 9 tire rubbers, and 13 chassis-lubricant samples are reported by Ruch et al.[*] A survey and extensive bibliography of activation analysis in forensic science has been prepared by Jervis.[†]

[*] R. Ruch, J. Buchanan, V. Guinn, S. Bellanca, and R. Pinker, Neutron Activation in Scientific Crime Prevention, *J. Forensic Sci.* **9**, 119–133 (1963).
[†] R. E. Jervis, Activation Analysis in Forensic Science, in *Nuclear Activation Techniques in the Life Sciences* (International Atomic Energy Agency, Vienna, 1967), pp. 645–659.

Several other uses of "activation fingerprinting" besides law enforcement have been developed for comparative activation analysis, some of which have industrial significance; for example, manufacturers can characterize the materials in some patented products as a means of tracing or detecting illegal rebranding. Specific products can be "tagged" by the addition of trace quantities of unusual elements with good nuclear properties for activation analysis. Such tagged products are useful for inventory control or for protection against unjustified claims of damage.

Another problem for which nuclear fingerprint techniques may be useful is the identification of oil slicks in offshore waters. The cataloging of nuclear analysis of characteristic trace elements in oils produced throughout the world could, with the additional knowledge of marine traffic, help pinpoint a source of an oil slick to an offending ship or a leak from an offshore oil well.

Comparative activation analysis is also useful in the archeological sciences. Studies have been reported on such investigations as the age and place of origin of oil paintings, variations in silver and copper content of Roman coins, excavated pottery and other artifacts, classification of ancient Greek marbles, Damascus steels, ancient glasses, coins, and ceramics.* Many other possibilities exist.

### 9.3.2 Surface Analysis

Nuclear methods have proved to be of great utility in the chemical analysis of surfaces. Two such applications have already been described—the α-scattering method for the lunar surface analysis in Section 9.2.1 and the triton activation method for the determination of oxygen in metal surfaces in Section 9.2.3. The general methods of analysis of solid surfaces by nuclear scattering and nuclear reactions have been described by Rubin, Passell, and Bailey.† The determination of chemical composition in only the surface-layer is achieved by the use of charged-particles, whose penetration in solids is generally in the range of a few atomic layers to several microns. The several types of interaction radiations are available for analytical purposes, namely, scattered radiation, prompt radiation, and activation-product radiation. Although these methods are best suited for the light elements from which adequate radiation yields can be obtained with low energy ($<2$ MeV) incident particles, elements as heavy as lead ($Z = 82$) have been determined. These nuclear methods also offer two advantages in sensitivity: (a) surface

---

* Compiled from *Activation Analysis: A Bibliography*, National Bureau of Standards, Technical Note 467, 2 parts, 1968.
† S. Rubin, T. O. Passell, and L. E. Bailey, Chemical Analysis of Surfaces by Nuclear Methods, *Anal. Chem.* **29**, 736–743 (1957).

concentrations as small as $10^{-8}$ g/cm$^2$ are detectable and (b) film thickness and variations of concentration with depth can be measured with a resolution of about $10^{-2}$ $\mu$.

The elastic scattering method is based on the classical Rutherford formula for the elastic scattering cross section which, for elements of $Z > 30$, is accurate to within a few percent:

$$\frac{d\sigma}{d\Omega} = \left(\frac{zZe^2}{4E_0}\right)^2 \sin^{-4}\frac{\theta}{2} = 1.30 \times 10^{-27} \frac{(zZ)^2}{E_0^2 \sin^4 \theta/2} \tag{16}$$

where $d\sigma/d\Omega$ = scattering cross section in cm$^2$/sr,

     $z$ = charge on the irradiating particle,

     $Z$ = charge on the target nucleus,

     $E_0$ = energy of the irradiating particle,

     $\theta$ = angle of scattering from the forward direction.

For lighter elements experimentally determined scattering cross sections are used. When possible, analytical results are obtained by comparison with measurements on reference samples of known surface concentration.

For an irradiation of $I_0$ incident particles per second, $n$ scattering nuclei/cm$^3$, and $d$ cm, the effective target thickness, the number of particles scattered through an angle $\theta$ and observed by a counter subtending the solid angle $\Omega$ in steradians is given by

$$I = I_0 nd \frac{d\sigma}{d\Omega} \tag{17}$$

Figure 9.18 shows a schematic drawing of a magnetic spectrometer used to measure scattered protons from a target irradiated in a 2-MeV Van de Graaff generator. This apparatus has been used for several interesting studies; for example, a scattered proton momentum profile was made for a steel sample exposed to the detonation of an explosive. Figure 9.19 shows the data that result from an irradiation of 1.4-MeV protons, measured at a scattering angle of 150°. The data show the "front edge" of the plateau due to scattering from the iron in the steel. This front edge, noted by the vertical arrow labeled Fe, represents the maximum energy of protons scattered elastically from iron nuclei. The detonation products of elements of $Z > 26$ are evident to the right of the iron cutoff. The narrow peaks are indicative of thin layers on the surface with concentrations of the order of $\mu$g/cm$^2$.

Another interesting application reported by Rubin et al. was the study of elements contained in aerosols, such as smog. Their proton-scattering analysis of aerosols deposited by electrostatic precipitation on clean aluminum foil is shown in Figure 9.20. The figure shows the resolution for the determined elements—carbon, oxygen, silicon, and sulfur. The peaks correspond to

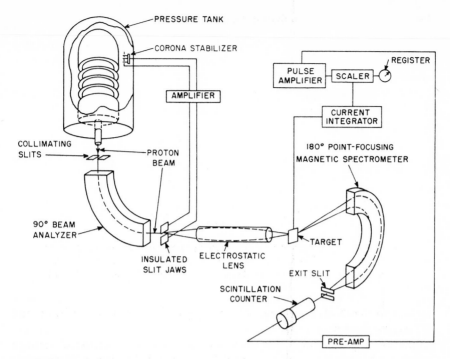

**Figure 9.18** Apparatus used for chemical analysis of surfaces by charged-particle scattering measurements. [From S. Rubin, T. O. Passell, and L. E. Bailey, Chemical Analysis of Surfaces by Nuclear Methods, *Anal. Chem.* **29**, 736–743 (1957).]

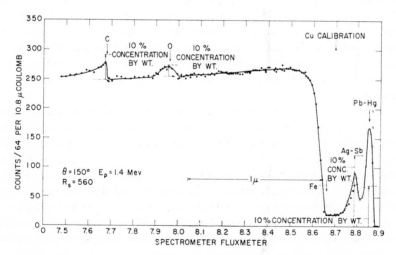

**Figure 9.19** Momentum profile of scattered protons from a steel sample exposed to detonation of an explosive. The 10% concentration levels are calculated from the scattering cross section of each element relative to iron. [From S. Rubin, T. O. Passell, and L. E. Bailey, Chemical Analysis of Surfaces by Nuclear Methods, *Anal. Chem.* **29**, 736–743 (1957).]

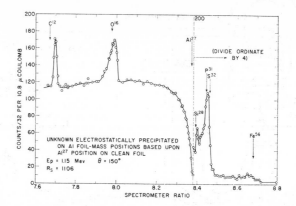

**Figure 9.20**  Momentum profile of scattered protons from an aerosol deposited on aluminum foil. [From S. Rubin, T. O. Passell, and L. E. Bailey, Chemical Analysis of Surfaces by Nuclear Methods, *Anal. Chem.*, **29**, 736–743 (1957).]

surface concentrations of 5 $\mu$g/cm$^2$ for oxygen, 0.13 $\mu$g/cm$^2$ for silicon, and 0.3 $\mu$g/cm$^2$ for sulfur.

The use of prompt and activation gamma radiation is also useful for such surface concentration studies, especially for the light elements, since the sensitivity decreases rapidly with increases in atomic number. Some of the applications reported by Rubin et al. include the determination of fluorine in opal glass by the prompt high-energy gamma rays resulting from the $^{19}$F(p,$\alpha\gamma$) $^{16}$O reaction, sodium in glass, using the gamma-rays from 15-h $^{24}$Na, and carbon and oxygen deposited in the detonation of an explosive charge by the characteristic proton spectra resulting from the (d,p) reactions of the two elements. The pulse-height spectrum of protons from a 1.4 MeV deuteron irradiation of a steel sample exposed to an explosive charge is shown in Figure 9.21.

The use of scattered alpha particles for the analysis of surfaces has been reported by Peisach and Poole.* They used semiconductor detectors in lieu of a magnetic spectrometer for the measurements and thus obtained directly the energy spectrum of the scattered particles instead of the momentum spectrum obtained by a magnetic spectrometer. They also note that another advantage of $\alpha$-particle scattering is the greater sensitivity for surface layer analysis. The relative energy loss per collision for an alpha particle with atom $M$ is $4[(M + 1)/(M + 4)]^2$ times that of a proton under similar measuring conditions. This method was used to determine gold film thickness on aluminum,

---

* M. Peisach and D. O. Poole, Analysis of Surfaces by Scattering of Accelerated Alpha Particles, *Anal. Chem.* **38**, 1345–1350 (1966).

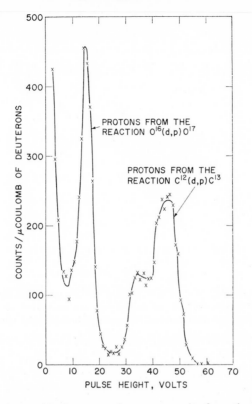

500 ┤

400 ┤

COUNTS/μCOULOMB OF DEUTERONS

300 ┤

PROTONS FROM THE
REACTION $O^{16}(d,p)O^{17}$

PROTONS FROM THE
REACTION $C^{12}(d,p)C^{13}$

200 ┤

100 ┤

0 ┴

0   10   20   30   40   50   60   70

PULSE HEIGHT, VOLTS

**Figure 9.21** Spectrum of proton energies from the
1.4-MeV deuteron irradiation of a steel sample exposed to an explosive charge. [From S. Rubin, T. O.
Passell, and L. E. Bailey, Chemical Analysis of
Surfaces by Nuclear Methods, *Anal. Chem.* **29**, 736–
743 (1957).]

stainless steel, tin, and mica surfaces. The calibration of the equipment to
relate alpha counts to gold film thickness is shown in Figure 9.22. In comparing
the results of these measurements to those obtained by neutron activation
analysis, they noted that in the range 1 to 10 $\mu g/cm^2$ the methods differed by
less than ±0.2 $\mu g/cm^2$. Other applications included the measurement of light
element coatings, thick surface coatings, and geologic samples.

Surface contamination measurements are useful in high-purity metals
production; for example, Blake et al.* described the determination of oxygen,

* K. R. Blake, T. C. Martin, I. L. Morgan, and C. D. Houston, The Measurement of
Surface Contamination of High-Purity Beryllium Samples, in *Modern Trends in Activation
Analysis* (Texas A&M University, College Station, 1965).

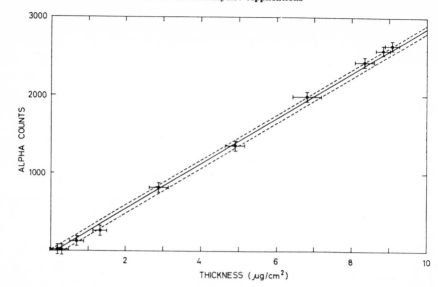

**Figure 9.22** Measurement of the thickness of gold films on aluminum by alpha-particle scattering; the thickness errors of the data points are associated with the errors from neutron activation analysis comparisons. (From Peisach and Poole, 1966.)

nitrogen, and carbon in high-purity beryllium. The surface oxygen contamination was calculated from total-oxygen-content measurements made by fast-neutron activation analysis. The carbon and nitrogen surface contamination was measured by using 1.8 MeV deuterons from a Van de Graaff accelerator. The calculation of the actual weight of the element in the surface layer requires a knowledge of the range of 1.8-MeV deuterons in the sample and standard materials. Since the range of low-energy charged particles is difficult to measure, Blake et al. derived a relation for the relative amount of the carbon and nitrogen in the beryllium sample surfaces compared with standard samples as

$$\frac{W_x}{W_B} = N_x \frac{R_s}{R_B} \frac{\rho_s}{\rho_B} \frac{C_x}{C_s} \tag{18}$$

where  $W_x$ = weight of unknown material,
  $W_B$ = weight of beryllium,
  $R_s$ = range in standard,
  $R_B$ = range in beryllium,
  $\rho_s$ = density of standard,
  $\rho_B$ = density of beryllium,
  $C_x$ = count rate from unknown material,
  $C_s$ = count rate from standard,
  $N_x$ = fraction of unknown material in standard.

Using the relationship of (4–26) for the relative range of the same particle in different materials, (18) can be simplified by substitution of

$$\frac{\rho_s R_s}{\rho_B R_B} = \frac{\sqrt{A_s}}{\sqrt{A_B}} \tag{19}$$

where $A_s$ = atomic weight of standard,
$\quad\ A_B$ = atomic weight of beryllium,
to

$$\frac{W_x}{W_B} = N_x \frac{\sqrt{A_s}}{\sqrt{A_B}} \frac{C_x}{C_s} \tag{20}$$

Further examples of analysis of surfaces by charged particles with semiconductor detectors were reported by Anders[*] who evaluated by computer program, the energies of elastically scattered particles for a fixed incident particle energy and a fixed scattering angle of 135°. Anders also calculated the energies of emitted particles undergoing nuclear interactions of the type shown in Section 1.5.3, in which the $Q$-value of the reaction is given by

$$Q = (M_a + M_A - M_b - M_B)c^2 - \epsilon \tag{21}$$

where $b$ and $B$ refer to the ejected particle and residual nucleus, respectively, and $\epsilon$ is the value of the energy difference between the level in which the residual nucleus is left after the reaction and the ground state of the nuclide. The energy of the ejected particle is given by

$$E = E_T B \left[ \cos\theta \pm \sqrt{\frac{D}{B} - \sin^2\theta} \right]^2 \tag{22}$$

where $E_T$ = total energy available,
and

$$B = \frac{M_a M_b (E_a/E_T)}{(M_a + M_A)(M_b + M_B)} \tag{23}$$

$$D = \frac{M_A M_B}{(M_a + M_A)(M_b + M_B)} \left( 1 + \frac{M_a Q}{M_A E_T} \right) \tag{24}$$

Table 9.35 lists values for energies of the ejected particles from (d,p) and (d,α) reactions for some elements irradiated with 2 MeV deuterons with measurements at an angle of 135° with respect to the incident beam. Applications of the method were made for comparative studies of thin polyvinyl chloride (PVC)—polyvinyl acetate (PVA) copolymer (VYNS) films, films

---

[*] O. U. Anders, Use of Charged Particles from a 2-megavolt Van de Graff Accelerator for Elemental Surface Analysis, *Anal. Chem.* **38**, 1442–1452 (1966).

**Table 9.35** Energies of Ejected Particles from
(d,p) and (d,$\alpha$) Reactions[a]

| Target Nucleus | Proton Energy (MeV) | Alpha-Particle Energy (MeV) |
|---|---|---|
| $^6$Li | 4.859 | 9.551 |
| $^7$Li | 0.870 | 6.875 |
| $^9$Be | 5.042 | 4.389 |
| $^{10}$Be | 9.247 | 11.309 |
| $^{11}$B | 2.256 | 5.602 |
| $^{12}$C | 3.702 | 0.054 |
| $^{14}$N | 9.158 | 10.304 |
| $^{16}$O | 3.193 | 3.190 |
| $^{18}$O | 3.094 | 4.193 |
| $^{19}$F | 5.560 | 8.770 |
| $^{20}$Ne | 5.742 | 3.251 |
| $^{23}$Na | 6.024 | 6.733 |
| $^{24}$Mg | 6.395 | 2.823 |
| $^{27}$Al | 6.846 | 6.848 |
| $^{28}$Si | 7.579 | 2.536 |
| $^{31}$P | 7.124 | 8.314 |
| $^{32}$S | 8.021 | 5.516 |
| $^{35}$Cl | 7.808 | 8.612 |
| $^{37}$Cl | 5.459 | 8.264 |
| $^{40}$Ca | 7.667 | 5.604 |
| $^{45}$Sc | 8.099 | 8.731 |
| $^{55}$Mn | 6.727 | 9.188 |

[a] For 2-MeV deuterons with measurement at an angle of 135° with respect to the incident beam. [From Anders, *loc. cit.*, (1966).]

coated with thin lead deposits, chromate pickled magnesium, and plastics, metals, and rocks. The resolution attainable by this method is illustrated in Figure 9.23, which shows the spectrum of protons and scattered deuterons for nine identified peaks from reactions of 2 MeV deuterons with a lead coated thin film of VYNS measured at an angle of 135°. Three of the nine reactions result in a residual nucleus in an excited state.

### 9.3.3 Isotopic Analysis

The determination of isotopic abundance has become useful in many different kinds of scientific and industrial study. The nuclear-energy industry is

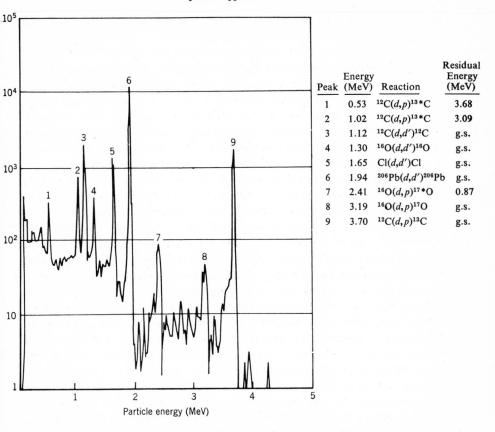

**Figure 9.23** Spectrum of charged-particle energies emitted at an angle of 135° from a lead coated thin film of VYNS when irradiated with 2-MeV deuterons. [From Anders, (1966).]

intimately involved with uranium enrichments, isotopic composition of fissile and fertile elements, production of transuranium nuclides, and isotopic enrichment or composition of allied materials, such as hydrogen, lithium, and boron. Isotopic analysis is also useful for tracing purposes in physical and biological systems. Separated or enriched isotopes are becoming readily available for such studies.

Until recently mass spectrometry has been the primary means for measuring isotopic composition of natural and enriched elements. Radioactivation analysis has now become another important method for isotopic analysis. The method takes advantage of the difference in nuclear properties of the stable isotopes of non-monoisotopic elements. Differences can be sought in activation cross section as a function of irradiation particle type and energy and/or in radionuclide produced by element or half-life or radiation type and energy.

The development of Ge(Li) detectors for high resolution of $\gamma$-ray spectra permits the simultaneous determinations of several $\gamma$-ray total energy peaks emitted by different radionuclides in a single sample. Instrumental radio-activation analysis is generally required for isotopic analysis, since radio-chemical separation methods are of little use for such analysis except to reduce the general backgrounds from other radioelements in the sample. Activation reactions that result in chemical transmutations still allow increase sensitivity with radiochemical separations of the product radionuclides.

With instrumental activation analysis, isotopic ratios can be determined for two or more isotopes of an element if their induced radioactivities are readily identified. The analysis of a single $\gamma$-ray spectrum for isotopic composition avoids many of the usual sources of error in both the laboratory processing (e.g., sample weighing, dilutions, and chemical yield determinations) and activation (e.g., flux inhomogeneities, and differences in flux suppression in samples and standards). Generally only the errors inherent in the radiation measurement and spectra analysis are important.

Table 9.36 lists some of the elements for which isotopic analyses have been reported and some of the purposes for the measurements. The purposes can be grouped in several types; for example, radiogenic abundances, search for natural long-lived radionuclides, enrichments in natural systems, and enrichments in laboratory or industrial processes.

Merz and Herr* described several searches for radiogenic-origin of stable isotopes believed to result from radioactive decay of long-lived radioisotopes found in nature. The isotopic composition of the elements, Hf, Os, W, and Ru, was examined for evidence of the following natural radioactive-decay processes:

$$^{176}\text{Lu} \xrightarrow{\beta^-} {}^{176}\text{Hf}$$

$$^{187}\text{Re} \xrightarrow{\beta^-} {}^{187}\text{Os}$$

$$^{180}\text{Ta} \xrightarrow{\beta^-} {}^{180}\text{W} \ (?)$$

$$^{98}\text{Tc} \xrightarrow{\beta^-} {}^{98}\text{Ru} \ (?)$$

$$^{99}\text{Tc} \xrightarrow{\beta^-} {}^{99}\text{Ru} \ (?)$$

The radiogenic origin of $^{176}\text{Hf}$ was shown from the large amount of $^{176}\text{Hf}$ determined by neutron activation of $\text{HfO}_2$ in a gadolinite mineral containing lutecium compared with a "normal" $\text{HfO}_2$ in a zircon mineral. The excess

---

* E. Merz and W. Herr, Microdetermination of Isotopic Abundance by Neutron Activation, in *Proc. 2nd UN Intern. Conf. Peaceful Uses Atomic Energy*, **28**, 984, 491–495 (1958).

**Table 9.36** Some Uses of Isotopic Activation Analysis

| Element | Isotopes | Purpose |
|---------|----------|---------|
| H | $^2$H | Abundance in natural waters |
| He | $^3$He | Isotopic abundance |
| Li | $^6$Li | Isotopic abundance |
| O | $^{18}$O | Photosynthesis studies |
| Si | $^{32}$Si | Search for $^{32}$Si in natural silicon |
| Ar | $^{40}$Ar | Radiogenic abundance |
| K | $^{39}$K/$^{40}$K | Isotopic abundance |
| Ca | $^{48}$Ca | Abundance in nature |
| Zn | $^{68}$Zn/$^{64}$Zn | Ratios in rocks and minerals |
| Br | $^{79}$Br/$^{81}$Br | Enrichment studies; Composition in meteorites |
| Ru | $^{98}$Ru/$^{99}$Ru | Search for natural Tc |
| I | $^{129}$I | Occurrence in nature |
| Ba | $^{135}$Ba/$^{131}$Ba | Composition in meteorites |
| RE's | Many | Fractionation studies |
| Hf | $^{176}$Hf | Radiogenic abundance |
| W | $^{180}$W | Search for unstable $^{180}$Ta |
| Os | $^{187}$Os | Radiogenic abundance |
| Hg | $^{202}$Hg/$^{196}$Hg | Composition in meteorites |
| Pb | $^{204}$Pb/$^{208}$Pb | Th-Pb age determinations |
| Th | $^{232}$Th/$^{230}$Th | Isotopic composition |
| U | $^{235}$U/$^{238}$U | Isotopic composition; Enrichment of $^{235}$U |
| Pu | $^{241}$Pu | Accountability in residue solutions |

$^{176}$Hf was determined from its ratio with $^{180}$Hf which was present equally in both samples.

The natural radioactivity of $^{187}$Re was reported in 1948. Its half-life of $4.3 \times 10^{10}$ yr allows a geological-age dating method for minerals containing rhenium by the measurement of the radiogenic $^{187}$Os present. In such minerals the $^{187}$Os abundance is nearly 100%, whereas in "normal" osmium the isotopic abundance of $^{187}$Os is 1.64%.

The search for radioactive $^{180}$Ta ($T_{1/2} > 1 \times 10^{12}$ yr) was made by activation of tungsten extracted from old tantalites. The "normal" isotopic abundance of $^{180}$W is 0.14%. In relation to a constant $^{187}$W specific activity the possible enrichment of $^{180}$W was concluded to be less than 2%.

The longest-lived isotopes of technicium are $1.5 \times 10^6$-y $^{98}$Tc and $2.12 \times 10^5$-y $^{99}$Tc. Thus technicium is not found in nature. A search for their stable decay products, $^{98}$Ru and $^{99}$Ru, was carried out in ruthenium extracted from 24 kg of Precambrian minerals and two iron meteorites. The results of the neutron activation analysis normalized to ruthenium from platinum ores set upper limits for the possible radiogenic content of $^{98}$Ru at 2.0 × $10^{-12}$ g/g mineral and of $^{99}$Ru at 9 × $10^{-12}$ g/g mineral.

In reviewing the use of neutron activation in geochemistry, Herr* also notes the geochronology dating methods that use $^{40}$K/$^{40}$Ar and $^{87}$Rb/$^{87}$Sr ratios with activation analysis measurements of the $^{40}$Ar and $^{87}$Sr, respectively. These ratio dating methods have also proved to be useful for determining the results of cosmic radiation on meteorites. Examples are the determination of $^3$He and $^{45}$Sc produced by cosmic radiation in iron meteorites. It was observed that the scandium content is nearly proportional to the helium content, which indicates that the helium content is not lost by possible intermediate heating.

The very heavy elements, for example, uranium, can be analyzed isotopically by a variety of nuclear reactions and reaction-product radionuclides. A procedure for determining the isotopes $^{234}$U, $^{235}$U, and $^{238}$U in natural uranium was reported by Seyfang and Smales.† Thermal neutron irradiation of uranium results in the following major reactions:

$$^{234}U(n,\gamma)^{235}U \ (T_{1/2} = 8.9 \times 10^8 \ \text{yr})$$

$$^{235}U(n,f) \ \text{fission products}$$

$$^{238}U(n,\gamma)^{239}U \xrightarrow{\beta^-} {}^{239}Np \xrightarrow{\beta^-} {}^{239}Pu$$

The determination of the 0.0061% abundance of $^{234}$U is difficult in short irradiations because of the long half-life of $^{235}$U. Any one or more of the many fission products are suitable for measurement of $^{235}$U. The decay chain of $^{239}$U allows for radiochemical measurement of $^{239}$Np or $^{239}$Pu at a suitable time after the irradiation. With refinement of the radiochemcial method Seyfang‡ expressed the precision of the determination of $^{235}$U in "depleted" uranium by radioactivation with a coefficient of variation ( = $\pm\sigma \cdot 100/\sigma^2$ %) of about ±0.5%. Comparison of the precision of the radioactivation method to mass spectrometric determinations is given in Table 9.37.

The use of high-resolution $\gamma$-ray spectrometry for simultaneous measurement of several $\gamma$-ray peaks of different activation and fission products allows rapid, nondestructive analysis of $^{235}$U and $^{238}$U in uranium samples.

---

* W. Herr, "Neutron Activation Applied to Geochemistry," in *Radioactivation Analysis* (Butterworth, London, 1960), pp. 35–52.

† A. P. Seyfang and A. A. Smales, The Determination of Uranium-235 in Mixtures of Naturally Occurring Uranium Isotopes by Radioactivation, *Analyst* **78**, 394–405 (1953).

‡ A. P. Seyfang, An Improvement in the Determination of Uranium-235 by Radioactivation, *Analyst* **80**, 74–76 (1955).

**Table 9.37**  Comparison of Radioactivation and
Mass Spectrometry for the Determination of $^{235}$U
in Depleted Uranium[a]

| Radioactivation ($\% \pm 3\sigma$) | Mass Spectrometry ($\% \pm 3\sigma$) |
|---|---|
| 0.391 ± 0.006 | 0.388 ± 0.007 |
| 0.388 ± 0.006 | 0.392 ± 0.007 |
| 0.385 ± 0.006 | 0.390 ± 0.007 |
| 0.391 ± 0.006 | 0.390 ± 0.007 |
| 0.391 ± 0.006 | 0.391 ± 0.007 |
| 0.374 ± 0.006 | 0.373 ± 0.007 |
| 0.389 ± 0.008 | 0.395 ± 0.007 |
| 0.387 ± 0.008 | 0.388 ± 0.007 |
| 0.398 ± 0.008 | 0.396 ± 0.007 |

[a] From A. P. Seyfang, An Improvement in the Determination of Uranium-235 by Radioactivation, *Analyst* **80,** 74–76.

**Table 9.38**  Gamma-Ray Photopeaks Useful for
$^{238}$U/$^{235}$U Ratio Measurements[a]

| Analysis | Nuclide | Photopeak Energy (MeV) |
|---|---|---|
| $^{238}$U | $^{239}$Np | 0.105 |
|  |  | 0.210 |
|  |  | 0.278 |
| $^{235}$U | $^{99}$Mo–$^{99}$Tc | 0.140 |
|  | $^{143}$Ce | 0.290 |
|  | $^{133}$I | 0.530 |
|  | $^{91}$Sr–$^{91m}$Y | 0.556 |
|  | $^{97}$Zr–$^{97}$Nb | 0.658; 0.743 |
|  | $^{132}$Te–$^{132}$I | 0.773 |

[a] From M. Mantel, J. Gilat, and S. Amiel, Isotopic Analysis of Uranium by Neutron Activation and High Resolution Gamma-Ray Spectrometry, in *Modern Trends in Activation Analysis* (National Bureau of Standards, 1968) pp. 79–88.

Mantel, Gilat, and Amiel* have described the resolution of multiple $\gamma$-ray peaks from $^{239}$Np and six fission products for the analysis of 20–25-mg samples of uranium. The radionuclides and their photopeak energies are listed in Table 9.38. This rapid method yields a precision of about $\pm 0.6\%$ for a single determination of $^{235}$U in natural uranium.

The isotopic analysis of lithium isotopes is of general interest. Several ways to determine lithium isotopes have been developed. The major non-nuclear methods are mass spectrometry and spectrometric measurement based on the isotope shift of a characteristic resonance line. Three methods have been reported for determining lithium isotopes by neutron activation.

Kaplan and Wilzbach† described a method that employs the nuclear reaction

$$^{6}\text{Li}(n,\alpha)^{3}\text{H} \tag{25}$$

and measures the $^{3}$H produced. A precision within $\pm 1\%$ was reported.

Coleman‡ described a method that uses the secondary nuclear reactions

$$^{16}\text{O}(t,n)^{18}\text{F} \tag{26}$$

with the production of tritons by the nuclear reaction in (25). This method has the advantages of not requiring long irradiations at high neutron flux to produce sufficient $^{3}$H for accurate measurement. The positrons from the 112-m $^{18}$F product are readily measured without chemical separations. An error of 1 to 2% was reported for 5-mg natural lithium samples.

Amiel and Welwart§ described still another method that uses the secondary nuclear reaction

$$^{18}\text{O}(t,\alpha)^{17}\text{N} \tag{27}$$

also following the production of tritons by the nuclear reaction in (25). This reaction results in the production of a radionuclide that decays by neutron emission:

$$4.14\text{-s }^{17}\text{N} \xrightarrow{\beta^{-}} {}^{17*}\text{O} \xrightarrow{n} {}^{16}\text{O}$$

---

* M. Mantel, J. Gilat, and S. Amiel, Isotopic Analysis of Uranium by Neutron Activation and High Resolution Gamma-Ray Spectrometry, in *Modern Trends in Activation Analysis* (National Bureau of Standards, 1968), pp. 79–88.

† L. Kaplan and K. E. Wilzbach, Lithium Isotope Determination by Neutron Activation, *Analyst* **85**, 285–288 (1960).

‡ R. F. Coleman, Isotopic Determination of Lithium by Neutron Activation, *Analyst* **85**, 285–288 (1960).

§ Amiel and Y. Welwart, Lithium and Lithium-6 Analysis by Counting Delayed Neutrons, *Anal. Chem.* **35**, 566–570 (1963).

Errors reported for single measurements with natural water solutions $[f(^{18}O) = 0.002]$ are about $10\%$. The sensitivity can be improved with the use of $^{18}O$-enriched water.

Table 9.36 lists several studies related to relative isotopic abundances of elements in meteorites and environmental samples, among which are noted the measurements of Reed[*] for mercury, lead, uranium, and barium, Corliss[†] for $^{48}Ca$ in natural materials, and Filby[‡] for zinc in rocks and minerals. Filby considers the possibility that under the right conditions isotopic fractionation of elements may occur in nature. His results for zinc in the standard rock samples G-1 and W-1 show no significant difference in the $^{68}Zn/^{64}Zn$ ratio relative to standard zinc.

The use of charged-particle activation analysis for calcium isotopic analysis developed as a tracer technique for biomedical research has been reported by Peisach and Pretorius.[§] Isotopic analysis for $^{48}Ca$ and $^{43}Ca$ are made with the nuclear reactions

$$^{48}Ca(p,n)^{48}Sc$$

$$^{43}Ca(p,n)^{43}Sc$$

Measurements may be made by gamma-ray spectrometry of the scandium radionuclides or by time-of-flight spectroscopy of the prompt neutrons. The measurement of calcium isotope ratios by either method negates the need to determine the total calcium content separately.

A final example of isotopic activation analysis is given for its use in determining enrichments by physical processes. The determination of enriched $^{235}U$ needed for nuclear fuels has already been discussed. Cameron et al.[‖] have compared the results of isotopic abundance analysis by activation and mass spectrometric methods for Br isotopes following enrichments by electrolytic fractionation of molten $PbBr_2$. Their results, which show the degree

[*] G. W. Reed, Discussion, in *Radioactivation Analysis Symposium* (Butterworth, London, 1960), pp. 48–51.

[†] J. T. Corliss, Determination of Calcium-48 in Natural Calcium by Neutron Activation Analysis, *Anal. Chem.* **38**, 810–813 (1966).

[‡] R. H. Filby, Determination of the Zinc-68–Zinc-64 Ratio in Rocks and Minerals by Neutron Activation Analysis, *Anal. Chem.* **36**, 1597–1600 (1964).

[§] M. Peisach and R. Pretorius, The Determination of Stable Calcium Isotopes by Charged-Particle Irradiation, in *Modern Trends in Activation Analysis* (National Bureau of Standards, 1969), pp. 306–315.

[‖] A. E. Cameron, W. Herr, W. Herzog, and A. Lunden, Isotopen—Anreicherung beim Brom durch elektrolytische Uberfuhrung in geschmolzenem Bleibromid, *Z. Naturforsch* **11a**, 203 (1956).

**Table 9.39**   $^{79}Br$ Enrichment by Electrolytic Fractionation[a]

| Sample | Content of $^{79}Br$ (%) | | |
|---|---|---|---|
| | From $^{80m}Br$ by Neutron Activation | From $^{82}Br$ by Neutron Activation | By Mass Spectrometry |
| 1 | 55.5 ± 1.0 | 55.6 ± 1.0 | 56.500 ± 0.008 |
| 2 | 51.4 ± 1.0 | 51.6 ± 0.9 | 51.699 ± 0.007 |
| 3 | 52.7 ± 1.0 | 53.4 ± 0.9 | 53.621 ± 0.009 |
| 4 | 51.2 ± 1.5 | 51.0 ± 1.0 | 50.868 ± 0.004 |
| Normal | | | 50.414 ± 0.002 |

[a] From E. Merz and W. Herr, Microdetermination of Isotopic Abundance by Neutron Activation, in *Proceedings, 2nd International Conference on Peaceful Uses of Atomic Energy*, **28**, 984, 491–495.

of $^{87}Br$ enrichment attained and the comparison of sensitivities for the two methods of analysis, are given in Table 9.39.

### 9.3.4   On-Line Activation Analysis

The development of small charged-particle accelerators and fast-neutron generators and the improvements in gamma-ray measurement have led to the increased use of activation analysis methods for the monitoring of plant streams. Small compact neutron generators are especially useful because of their relatively low cost, adequate neutron fluxes, and the greater penetrating power of fast neutrons. With high-efficiency gamma-radiation detectors the on-line activation systems can analyze larger samples directly in the process stream, nondestructively, and with greater reliability in sample representation and irradiation homogeneity.

On-line systems generally depend on the detection of either prompt radiation or radiation from short-lived radionuclides so that the analysis can be completed quickly and automated. These conditions avoid the production of undesirable levels of long-lived radioactivity in the process material. Anders* has tabulated the elements that produce a radionuclide with a half-life of less than 1 min and decays with prominent gamma-radiations. These data for the

* O. U. Anders, Activation Analysis for Plant Stream Monitoring, *Nucleonics* **20**, 78–83 (February 1962).

**Table 9.40** Elements Suitable for Plant Stream Activation[a]

| Element | Activation Product | $T_{\frac{1}{2}}$ (sec) | Gamma Energy (MeV) | |
|---------|-------------------|------|------|------|
| O | $^{16}$N | 7.4 | 6.1 | |
| F | $^{16}$N | 7.4 | 6.1 | |
| | $^{20}$F | 10.7 | 1.6 | |
| Na | $^{23}$Ne | 38 | 0.44 | |
| Sc | $^{46m}$Sc | 19.5 | 0.14 | |
| Ge | $^{75m}$Ge | 49 | 0.14 | 0.15 |
| | $^{77m}$Ge | 54 | 0.16 | 0.21 |
| Se | $^{77m}$Se | 17.5 | 0.16 | |
| Br | $^?$Se not $^{80m}$Br | 4.5 | 0.21 | |
| Rb | $^{86m}$Rb | 62 | 0.56 | |
| | ? | <60 | 0.24 | |
| Y | $^{89m}$Y | 14 | 0.91 | |
| Rh | $^{104}$Rh | 42.1 | 0.55 | |
| Ag | $^{110}$Ag | 24.2 | 0.66 | |
| In | $^{114m}$In | 2.5 | 0.15 | 0.19 |
| Er | $^{167m}$Er | 2.5 | 0.06 | 0.21 |
| Yb | $^?$Yb | 6 | 0.12 | 0.25 |
| Hf | $^{179m}$Hf | 19 | 0.16 | 0.22 |
| W | $^{183m}$W | 5.5 | 0.06 | 0.105 |
| Ir | $^{191m}$Ir | 4.9 | 0.04 | 0.13 |
| Au | $^{197m}$Au | 7.2 | 0.07 | 0.28 |

[a] From O. U. Anders, Activation Analysis for Plant Stream Monitoring, *Nucleonics* **20**, No. 2, 78–82 (1962).

elements that can be determined by on-stream activation analysis are listed in Table 9.40.

There are several limitations that require evaluation for specific applications. On-line applications of activation analysis are generally suited only for major, perhaps minor, but not trace elements in the process stream material. Fast neutron reactions are generally more suited to the detection of light elements, and in many cases give rise to (n,2n) reactions that form positron-emitting radionuclides. Such radiation results in the characteristic annihilation radiation of 0.51 MeV which can also be produced by the annihilation of $e^-$ — $e^+$ pairs in the pair-production process of high-energy gamma-radiation emitted from other radionuclides in the sample. Fast neutrons can also produce (n,p) reactions from which radionuclides follow by secondary proton reactions.

A description of a neutron-activation stream analyzer was given by Anders. Figure 9.24 shows the principles of operation. The two vessels for irradiation

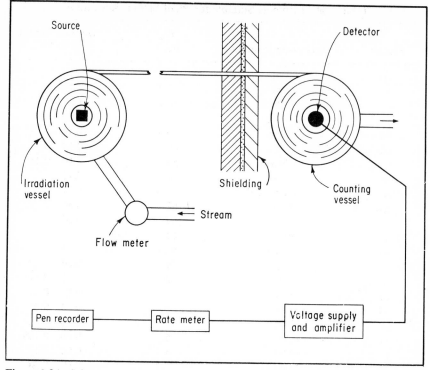

**Figure 9.24** Schematic diagram of an activation analysis on-stream analyzer. [From O. U. Anders (1962).]

and counting are equipped with baffles or stream guides to prevent channeling. They furnish a fluid reservoir to provide a reasonable sample size but store the stream for not more than about three half-lives of the induced activity. The fluid should take only a fraction of a half-life to transfer between irradiation and counting vessels. The measurement output can be converted to an analog signal by a count-rate meter to drive a pen recorder or be used in digital form for computer processing.

Figure 9.25 shows the addition of a second counter and counting vessel to the stream analyzer that detects any changes in pumping speed. This system provides a neutron-activation flow-rate meter if the design parameters and the half-life of the induced product are known. The flow rate is determined by counting the activity in each of the counting vessels, with the two counting rates adjusted for the relative counting efficiencies of the two detectors.

The flow rate $F$ is given by

$$F = \frac{V}{\Delta t} = \frac{V \cdot \lambda}{\log kA/B} \tag{28}$$

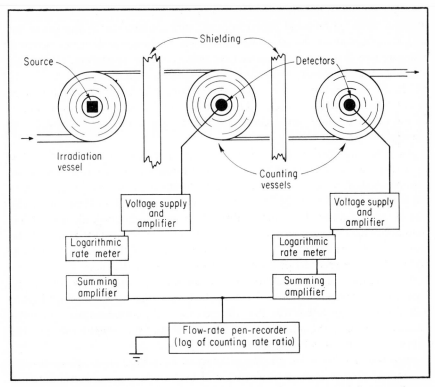

**Figure 9.25** Schematic diagram of an activation analysis on-stream analyzer used as a flow-rate meter. [From O. U. Anders (1962).]

where $V$ = volume of the stream represented by half the sum of holdup volumes of the two counting vessels plus volume of connection between them,

$\Delta t$ = transfer time between the two vessels,

$\lambda$ = half-life of the counted radionuclide,

$k$ = efficiency factor,

$A$ = counting rate of first counter,

$B$ = counting rate of second counter.

On-line activation analysis has been applied in several ways. Martin, Morgan, and Hall* described a closed-loop system for the analysis of coals. Measurements are made of the carbon, oxygen, silicon, and aluminum

* T. C. Martin, I. L. Morgan, and J. D. Hall, Nuclear Analysis System for Coal, in *Modern Trends in Activation Analysis* (Texas A & M University, College Station, 1965), pp. 71–75.

content of coal on a conveyor belt. This system utilizes a fast neutron generator and two gamma-ray spectrometers, one monitors near the generator and the other about 40 sec downstream. The carbon and oxygen content is measured by the prompt gamma rays from inelastic scattering, the silicon and aluminum content, by the gamma rays from their respective activation products. Feasibility studies have shown that the silicon and aluminum content can be empirically related to the total ash content, the carbon content, to the calorific value of the coal, and the oxygen content, to the moisture content of wet coals.

Another application for on-stream activation analysis was reported by Ljunggren and Christell* for the continuous monitoring of boron (as sodium perborate) content of a process stream of detergent. Their system consisted of a 0.5 Ci $^{210}$Po-Be isotopic neutron source and measured the boron content either by neutron absorption or $^{7}$Li gamma radiation. Their data indicated that sodium perborate contents in the range of interest could be readily measured by either measurement method and that for 12% content (relevant for detergents) precisions of 6 to 17% were attainable.

An activation analysis method which can sort copper ores on a go-no go basis in time periods of 0.1 to 1 sec has been designed by Ramdohr.† Figure 9.26 illustrates an apparatus that determines, by plastic-scintillator coincidence counting of the annihilation radiation from $^{64}$Cu, whether or not the copper ore content is between 0.2 and 0.3%, and sorts the ore accordingly.

Improved methods for on-stream analysis have been developed with the introduction of sample recirculation and cyclic activation. In recirculation activation analysis the sample is recycled between a continuously emitting neutron source and a shielded detector. This method gives significant gains in sensitivity compared with the standard "open-loop" method of on-stream activation analysis.

Ashe, Berry, and Rhodes‡ described a closed-loop activation analysis system; a schematic is shown in Figure 9.27. With the detector measuring the radioactivity of the fluid continuously throughout the analysis period, they show that the detector count $C_n$ after a total of $n$ cycles is

$$C_n = S_0 n G \tag{29}$$

---

* K. Ljunggren and R. Christell, Continuous Determination of Boron in a Process Stream Using a Low Level Neutron Source, *Atompraxis* **10**, 259–263 (1964).

† H. F. Ramdohr, Anwendung der Aktivierungsanalyse zur Kupfererzsortierung, *Kerntechnik* **5**, 204–206 (1963).

‡ J. B. Ashe, P. F. Berry, and J. R. Rhodes, On-Stream Activation Analysis Using Sample Recirculation, in *Modern Trends in Activation Analysis* (National Bureau of Standards, 1969), pp. 913–917.

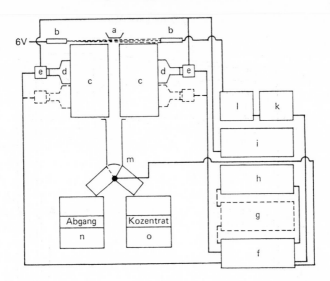

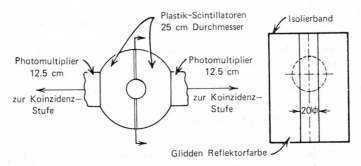

**Figure 9.26** *Top:* schematic diagram of an ore sorter: (*a*) feeder, (*b*) light path, (*c*) plastic scintillator, (*d*) photomultiplier, (*e*) preamplifier, (*f*) coincidence circuit, (*g*) total activity counter, (*h*) coincidence counter, (*i*) high-voltage supply, (*k*) timer, (*l*) photorelay, (*m*) sorter valve, (*n*) wastes tank, (*o*) concentrates tank. *Bottom:* top view and section of the plastic scintillation detector. [From Ramdohr (1963).]

where $S_0$ = the activation factor for a single cycle,

$G$ = the gain factor over the signal obtained in open-loop conditions (for the same total analysis time $t$).

These two factors are determined by

$$S_0 = k \frac{NV}{\lambda} (1 - e^{-\lambda t_a}) e^{-\lambda t_d} (1 - e^{-\lambda t_c}) \qquad (30)$$

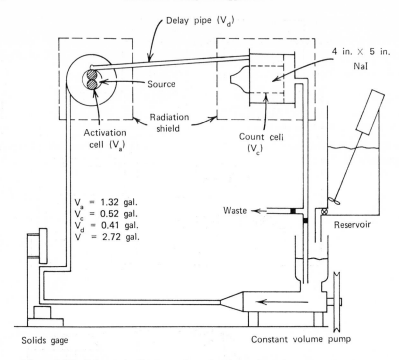

**Figure 9.27** Schematic diagram of a recirculating loop for on-stream activation analysis. [From J. B. Ashe, P. F. Berry, and J. R. Rhodes (1968).]

where $k$ = a constant which is the product of the effective neutron flux in the activation cell, the cross section for production of the radiation to be detected, and the detector efficiency including geometrical factors,

$N$ = the number of nuclei per cm³ available for the reaction,

$V$ = the volume of the loop in cm³,

$\lambda$ = the decay constant

$t_a$ = effective time in the activation cell,

$t_d$ = effective delay time,

$t_c$ = effective time in the counting cell,

and

$$G = \frac{1}{1 - e^{-\lambda T}}\left[1 - \frac{e^{-\lambda T}(1 - e^{-\lambda t})}{n(1 - e^{-\lambda T})}\right] \tag{31}$$

where $T$ equals the time for one cycle.

The actual gain factor achievable is related directly to $n$ by (31) but is also affected by second-order effects due to partial activation of the contents of the activation cell in each cycle and any incomplete passage of material through the counting cell at the end of each cycle. It is noted that for half-lives short compared with the recirculation time $T$ the gain factor tends to unity. For $T_{1/2} = T$ the gain factor tends toward 2 and for $T_{1/2} \gg T$, $G$ increases with $n$, with the limit

$$\lim_{\lambda T \to 0} G = \frac{n+1}{2} \tag{32}$$

Analysis of the flow rate in relation to the observed counting rate shows that the ratio $I_t$ of the total count in the $n$th cycle to the circulation time $T$ can be expressed as

$$I_t = \frac{S_0}{T} \frac{1 - e^{-n\lambda T}}{1 - e^{-\lambda T}} \tag{33}$$

Thus after an initial linear increase with time the average count rate reaches a steady value for a specific flow rate $Q$ given by

$$I^{\infty} = \frac{kN(1 - e_a^{-\lambda}t_a)e^{-\lambda}t_d(1 - e^{-\lambda}t_c)}{(1 - e^{-\lambda}t_a)} Q \tag{34}$$

For $T_{1/2} \gg T$, the steady count rate is independent of flow rate and is proportional to

$$\frac{t_a t_c}{T} = \frac{V_a V_c}{T} \tag{35}$$

The concept of cyclic activation analysis, as described by Givens, Mills, and Caldwell,* involves the use of a pulsed radiation source and cyclic counting of the induced radionuclides between successive bursts of radiation. Figure 9.28 shows a timing diagram for cyclic activation analysis. The cycle

* W. W. Givens, W. R. Mills, and R. L. Caldwell, Cyclic Activation Analysis, in *Modern Trends in Activation Analysis* (National Bureau of Standards, 1968), pp. 139–147.

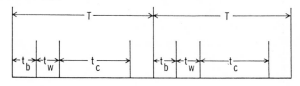

$t_b$ – duration of neutron burst

$t_w$ – waiting period

$t_c$ – counting period

$T$ – time between neutron bursts; $T \geq t_b + t_w + t_c$

**Figure 9.28** Timing diagram for cyclic activation analysis. [From Givins, Mills, and Caldwell (1968).]

consists of an irradiation burst, delay period, and count period of duration $t_b$, $t_w$, and $t_c$, respectively. The cycle period is $T \geq t_b + t_w + t_c$.

The detector response for the $n$th counting period, by analogy to (33), is given by

$$I_n = kN(1 - e^{-\lambda t_b})e^{-\lambda t_w}(1 - e^{-\lambda t_c})\left(\frac{1 - e^{-n\lambda T}}{1 - e^{-\lambda T}}\right) \tag{36}$$

and the cumulative detector response for $n$ successive cycles, as

$$I_T = \sum_{n'=1}^{n} I_{n'}$$

$$= kN(1 - e^{-\lambda t_b})e^{-\lambda t_w}(1 - e^{-\lambda t_c})\left[\frac{n}{1 - e^{-\lambda T}} - \frac{e^{-\lambda T}(1 - e^{-n\lambda T})}{(1 - e^{-\lambda T})^2}\right] \tag{37}$$

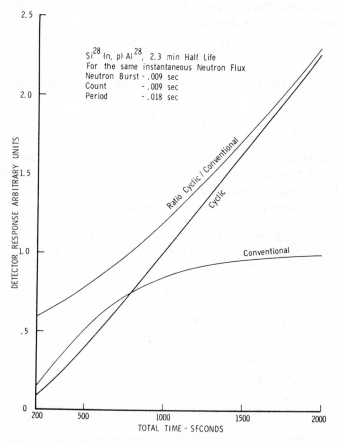

**Figure 9.29** Comparison of conventional and cyclic activation analysis for silicon. [From Givins, Mills, and Caldwell (1968).]

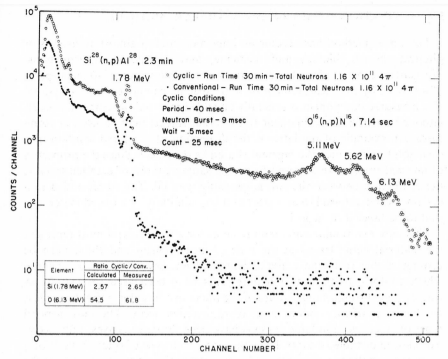

**Figure 9.30** Comparison of the gamma-ray spectra from conventional and cyclic activation analysis of silicon and oxygen in a pure limestone (SiO₂). [From Givins, Mills, and Caldwell (1968).]

The detector response may be maximized by suitable selection of the parameters in (37); for example, for a given $T_{1/2}$ and $T$ a maximum is achieved with $t_w = 0$ and $t_b = t_c = T/2$. For a given $T_{1/2}$ there is also an optimum $T$ equal to $2.4 \times T_{1/2}$.

The total response decreases rapidly for $T > 2.4 \times T_{1/2}$ and is relatively constant for $T < 2.4 \times T_{1/2}$. Thus cyclic activation analysis is best suited to induced radioactivities with short half-lives up to several minutes. Figure 9.29 shows a calculated detector response as a function of total analysis time for cyclic and open-loop activation analysis for silicon, based on the 2.3-m ²⁸Al product. Beyond 800 sec total time the detector response from cyclic activation increases almost linearly with time, whereas the conventional method reaches a maximum but smaller value. An actual measurement comparing spectra obtained for cyclic and convention activation of silicon and oxygen in a pure sandstone (SiO₂) is shown in Figure 9.30. The method is considered suitable for such applications as process control and *in-situ* analysis.

## 9.4  ACTIVATION ANALYSIS AS A TRACER METHOD

The tracer method has become an important tool in almost all aspects of physical science, biology and medicine, and engineering investigations. The general concept of the tracer method is the observation of some property of a system by the measurement of some specific component of the system. The measured component is generally an added material (an external tracer), although an original component (an internal tracer) may be used in many cases. An example of the latter is the dispersion of a colored solution in a clear solution in which the pigment is used as the tracer. Since the properties of internal tracers are required to be the same as those of external tracers, the distinction between the two is generally ignored. The tracer added to a system is often referred to as a label or a tag. Similarly, the traced material is said to be labeled or tagged.

There are two fundamental requirements for a tracer: (a) it must represent the material being traced at the time of measurement and (b) it must be distinguishable from the material being traced. Several types of material which conform to these two requirements have been used successfully as tracers. The distinguishing feature of the tracer can be color, refractive index, conductivity, density, odor, taste, or radioactive decay. The most general types of tracer used in large systems have been fluorescent dyes, radioactive isotopes, and stable isotopes The choice of which type of tracer is best for a given investigation is generally made after evaluation of the properties of the tracers and the conditions of the investigation; for example, fluorescent dyes are generally chosen for studies of large bodies of water streams, rivers, or estuaries. These dyes are quite suitable in clear waters but become less so as the turbidity or presence of suspended solids increases.

An extensive literature that describes the uses of fluorescent dyes is available. These dyes are organic pigments whose molecules can be excited to fluorescence at wavelengths of 480–560 m$\mu$. Fluorometers for measuring the fluorescence have been developed with sensitivities of about one part per billion in fresh and estuarine waters without prior chemical treatment. The most useful of the fluorescent dyes are Rhodamine B and Rhodamine WT.

Radioactive tracers are being used in increasing numbers of studies because of the large number of advantages they provide. Primarily they satisfy the two fundamental requirements in that they can be made to represent almost any given material and can be easily measured with radiation equipment in any concentration range. The major limiting factor of radioactive tracers is generally the real or imaginary health hazard involved in their use in large-scale systems. An alternate choice, one in which the major advantage of great sensitivity is retained but without the problem of radiation hazard, is the substitution of stable isotopes as tracers with activation analysis measurement on samples taken from the system. These two methods are examined further.

### 9.4.1 The Radiotracer Method

The two fundamental assumptions of isotopic radiotracing were described by Wahl and Bonner*:

1. Before a radioactive atom decays, its chemical, physical, and biological behavior is based solely on the chemical identity ($Z$) of the tracer element and is *essentially* identical to those of its stable isotopes of the same element.

2. When a radioactive atom does decay, its radiation can be quantitatively measured.

The first assumption concerns the similarity of isotopes of an element. The word essentially is italicized to emphasize that there are slight differences in behavior of isotopes, known as the isotope effect, due to the mass differences between the isotopes. These differences are measurable for such properties as diffusion coefficients, chemical reaction rate constants, and equilibrium constants. The effects are generally greatest for the light elements (e.g., hydrogen). It is noted that a great manufacturing process exists solely for the separation of the 0.7 % $^{235}U$ from $^{238}U$ by the slight differences in the diffusional properties of compounds of these two isotopes. For practical tracing purposes, however, radiotracers are considered completely identical to their stable isotopes. The second assumption considers not only the validity of the radioactive decay equation [see (1–31)] but also the validity of the linear response of radiation detection equipment with source strength. Thus for the decay constant $\lambda$ and over-all counting efficiency $\epsilon$, constant for a given detection system and geometry, the detector response is proportional to the amount of tracer in the counted sample.

It should be noted that many biological or engineering tracer studies do not involve isotopic tracing. Nonisotopic tracing, in which the tracer may not remain with the bulk material being traced, is often useful. The detection of leak points, phase boundaries, or other separation tracing requires nonhomogeneity of tracer and bulk material.

#### General Considerations

The general principles and techniques of radiotracing have been described by Gardner and Ely.† They list several advantages of radiotracers:

* A. C. Wahl and N. A. Bonner, *Radioactivity Applied to Chemistry* (Wiley, New York, 1951).
† R. P. Gardner and R. L. Ely, *Radioisotope Measurement Applications in Engineering* (Reinhold, New York, 1967).

1. Radioisotopes are available for almost every element in the periodic table—in most cases more than one radioisotope exists for each element.

2. With proper synthesis the radiotracer can be incorporated into complicated natural products such as coal or crude oil.

3. Radioisotopes are measurable with great sensitivity; in some cases as little as $10^{-16}$ g is detectable.

4. Radiation measurement yields great selectivity; interference from colored materials, suspended solids, etc., is generally negligible.

5. Detection of radiation is unambiguous, as described for the second fundamental assumption, although such properties as refractive index and color may not be linear.

6. Radioisotopes are easy to measure, easier than stable isotopes as measured by mass spectroscopy or activation analysis; radioisotopes can be detected *in-situ* (e.g., in the human body or inside an opaque pipe).

7. Maximum measurement precision can be calculated from the statistics of radioactive decay.

8. Radiotracers can be added in high specific activity, thus causing negligible interferences in the bulk material properties, such as weight, volume, density, and color.

In addition to the problem of the isotope effect, Gardner and Ely list several other potential problems of radiotracers:

1. Radiotracer purity, pure radiotracers minimize measurement requirements for chemical separations and/or radiation spectroscopy. Three types of radioactive impurity are:

genetic (e.g., $^{140}Ba \rightarrow {}^{140}La$),

isotopic (e.g., $^{80}Br + {}^{82}Br$),

impurities (e.g., from irradiation of impurities in the tracer material).

2. Radioisotope exchange in which the radioactive atoms exchange with stable isotopes in another compound, for example, in the use of tritiated water for ground-water studies, hydrated minerals (e.g., $MnCl_2 \cdot 4H_2O$) can become involved in the exchange reaction:

$$HTO + MnCl_2 \cdot 4H_2O \underset{k_2}{\overset{k_1}{\rightleftharpoons}} H_2O + MnCl_2 \cdot 3H_2O \cdot HTO \qquad (38)$$

If the reaction rate constant $k_1$ is greater than $k_2$, the tritiated water may not be representative of the unbound water when hydrated compounds are presented in the subsurface media.

3. Radiation effect, especially in biological systems. The radiation added in the form of the radiotracer may influence the biological processes being traced.

4. Radiocolloid effect, which is possible in trace levels of radioisotopes. Carrier-free or high specific activity radioisotopes may exhibit properties of colloids (e.g., adsorb to glass surfaces or suspended solids). When this effect is detrimental, the addition of carrier may avoid the problem.

### Design Parameters for Radiotracer Studies

General design parameters which evaluate the limits of radioactivity suitable for a given tracer study have been described by Fluharty.* These limits are based on the constraints on radiation levels due on one hand to the hazardous nature of radiation and on the other to the maximum sensitivity with which a diluted radioactive material can be measured. In practice, the amount of radioactivity used in a tracer study is a compromise between the two limiting levels. If $A(\max)$ is the greatest amount of radiotracer tolerable for an experiment and $A(\min)$ is the minimum amount detectable, a range factor can be defined as

$$R = \frac{A(\max)}{A(\min)} \tag{39}$$

where $R$ is the total activity range available within the study. Figure 9.31

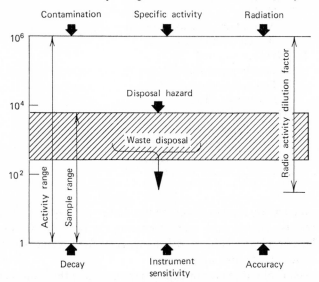

**Figure 9.31** Schematic limits on the use of radiotracers. The vertical distance is equal to the log ratio of the activity to the minimum detectable activity. (Courtesy U.S. Atomic Energy Commission.)

* R. Fluharty, Tracer Experiments, Chapter 10, in J. R. Bradford, Ed., *Radioisotopes in Industry* (Reinhold, New York, 1953).

**Table 9.41**   Constraints on Radiotracer Levels

---

Factors Affecting Maximum Level

   1.  Radiation considerations
        Health hazards
        Radiation effects on experiment materials
   2.  Contamination hazards
        Special facilities for handling radioactive materials
        Interference between experiments
        Increase in background levels
        Possibility of ruining laboratory
   3.  Specific activity
        Quantity of stable element required
        Quantity required cannot be detected
   4.  Activity availability
   5.  Tracer cost

Factors Affecting Minimum Level

   1.  Yield factor
        Half life
        Absorption prior to detection
        Overall counting efficiency
   2.  Accuracy required
        Counting statistics
        Technique accuracy
        Instrument reliability
   3.  Time
        Sample assay time
        Sample preparation time
        Number of samples required

---

shows the radioactivity level in tracer studies determined by several upper and lower constraints. The vertical scale represents the ratio of this level to the minimum detection level. Table 9.41 lists many of the specific aspects of the upper and lower level constraints.

A major technical parameter in experiment design is the dilution factor

$$D = \frac{A_i}{A_f} \tag{40}$$

where $i$ and $f$ refer to the initial and final conditions of the tracer study. The tracer has an initial concentration (or specific activity) in a given volume

$v_i$ of $C_i = A_i/v_i$. After mixing with volume $V$ of the matrix, the final concentration is

$$C_f = \frac{A_i}{v_i + V} = \left(\frac{v_i}{v_i + V}\right) C_i \qquad (41)$$

This relationship is the basis of the special radiotracer technique of isotope dilution analysis, described in the next topic.

In the design of a specific radiotracer experiment other factors are to be considered:

1. Radiotracer half-life: must be compatible with the time period needed for the study, from time of insertion into the system until time of last radiation measurement. Longer half-life is associated with a greater radiation safety requirement.

2. Type and energy of radiations: must be compatible with the instruments available, the need to make measurements in the laboratory or *in-situ*, the need for chemical treatment, the physical state of the samples, and the possible use of multiple tracers with different half-life or radiation type or energy.

3. Compatibility with the system: includes such considerations as the chemical, physical, or biological form of tracer for example, oxidation state, complex ions or molecules, solubility, particle size, phase, and enzyme action.

4. Availability and cost: for both tracer and analytical requirements.

## Some Types of Application

Radiotracers have been used in almost every branch of science and technology. The references in Section 9.5.4 describe a great many of these uses; several examples of analytical chemistry techniques follow.

ISOTOPE DILUTION ANALYSIS. This method involves the dilution of an isotopic radiotracer with the element to be determined. The basis is given by (9.41) in which the specific activity of the tracer and final extract are compared. The method is especially suitable to biological processes in which the compound can be extracted in pure form but not quantitatively. The tracer is incorporated into a compound with specific activity $S_i$, which is then added to the unknown and mixed to obtain a uniform concentration. A portion of the compound to be analyzed is extracted by suitable means and the specific activity $S_f$ is measured. If the tracer is present in $W_t$ g of the known compound, the weight of the compound in the original sample $W_u$ is given by

$$W_u = W_t \left(\frac{S_i}{S_f} - 1\right) \qquad (42)$$

RADIOMETRIC ANALYSIS.  This method involves a stoichiometric change in phase by using a radioactive reagent to determine the nonactive element in a sample or, conversely, a nonactive reagent to determine a radiotagged element in the sample. Radiometric titration, illustrated by the determination of either silver or chloride by AgCl precipitation titration, is the most widely used form; for example, for a routine determination of chloride a $AgNO_3$ reagent containing the radionuclide 7.5-d $^{111}Ag$ can be used. With increment additions of substoichiometric amounts of $AgNO_3$ and counting a fixed aliquot of the supernate following centrifugation of the AgCl precipitate, the solution would show background levels until the stoichiometric amount of $AgNO_3$ was exceeded. Thereafter the solution would increase in activity with further increments. The intersection of the two straight lines determines the end point of the titration. Figure 9.32$a$ shows a typical analysis by this method. If $3.5 \times 10^5$-y $^{36}Cl$ were added to the solution before the addition of increments of nonactive $AgNO_3$, the titration curve shown in Figure 9.32$b$ would be obtained. For a fixed set of analytical procedures in which the slopes of the respective lines in Figure 9.32 can be determined, routine analysis can be run simply and rapidly with

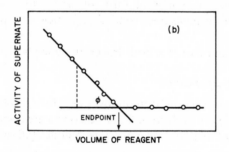

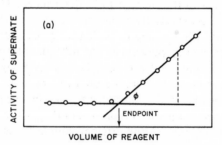

**Figure 9.32**  Radiometric titration curves: ($a$) for radiotracer added to reagent, ($b$) for radiotracer added to sample.

the addition of a single increment of reagent and a single radiation measurement after the centrifugation (see Problem 9.5).

AUTORADIOGRAPHY. The interactions of radiation with photographic emulsions that result in fogging or blackening of the film have been developed as a tracer method for locating the position, transport, or dispersion of a material within a complicated matrix. This method has been of value in metallurgy, for example, in determining the microsegregation of impurities in metals and in biology in determining the distribution of phosphorus in a leaf.

ACTIVATION. Neutron or charged-particle activations have been used to introduce radiotracers into materials. This technique is especially useful in cases in which the introduction of extraneous radioactive materials is either undesirable or impossible. One of the earliest successes of radioisotope utilization was the determination of piston-ring wear by activating the rings in a nuclear reactor and measuring the buildup of the radioactivity in the oil of a running engine.

Besides these applications in chemical analysis, studies have been made in science, biology, and engineering that involve applications in the areas of fluid flow and dispersion, uptake and mixing, wear and corrosion, and process characteristics. Some specific examples, described in detail in the references in Section 9.5.4, include interface detection in pipelines, leak detection, flow patterns, flow rates, pollution dispersion, sedimentation rates, mixing times, residence times, pharmaceutical uptake rates, and many others.

## 9.4.2 Stable Tracers with Activation Analysis

The substitution of stable isotopes for radioactive isotopes in tracing, used in conjunction with activation analysis measurement, removes the need to introduce radioactive materials into the traced system while retaining the selectivity and sensitivity of radiation measurement. Such tracers have been called *activable tracers*.

### General Considerations

Activable tracers must meet the two fundamental requirements of tracers. It is relatively simple to prepare activable tracers to represent the material being traced. With activation analysis it is also relatively simple to distinguish the tracer. The activable tracing method is generally carried out in three steps:

1. The addition of the stable tracer, in sufficient weight (and in appropriate chemical, physical, and/or biological form), to provide the dilution factor required by the system,

2. Sampling taken at predetermined intervals and/or locations,
3. Measurement of the tracer by activation analysis.

It is possible in many experiments to combine steps (2) and (3) by *in-situ* or on-line activation analysis at some convenient location in the system.

The design parameters for activable tracer studies are generally similar to those for radiotracers. The total weight range available for a given study may be defined from (40) by replacing the radioactivities with corresponding tracer weights. The range is generally greater, since $W(max)$ is not limited by the radiation hazard, although it may be limited by possible toxicity, high cost, or massiveness of the required tracer; $W(min)$ is limited essentially by the activation analysis sensitivity. In general practice the amount of stable tracer required for a study is determined by $W(min)$ and the anticipated dilution factor.

The two major considerations in the selection of a suitable activable tracer are (a) its background concentration in the system, (b) its nuclear properties. The background concentration should be as close as possible to the maximum sensitivity of the activation analysis. In many types of experiment the value of $W(min)$ is determined as a function of the average background concentration, (e.g., twice the background). The nuclear properties include the natural abundance, available enrichments, cross sections, and radiation properties of the activation product. Among the latter are the half-life and type and energy of the radiations. The properties of the system materials must also be considered. The system matrix should not be excessively activable such that the tracer cannot be easily measured, with or without chemical separations.

### Applications

Activation analysis of stable isotope tracers has extensive applications and is becoming increasingly important as methods for less expensive routine activation analysis are developed. An indication of the scope of the applications possible may be gleaned from the few illustrative examples cited here from biological, environmental, and industrial uses.

BIOLOGY AND MEDICINE STUDIES.    Activable tracer measurement has been employed extensively in biological and medical research. An obvious benefit is the removal of the radiation environment in the biological system, but the associated risk is the potential toxicity due to the amounts of stable tracer needed for adequate sensitivity. A method for *in-vivo* tracer application was described by Lowman and Krivet,* who compared the *in-vivo* plasma

---

* J. T. Lowman and W. Krivet, New In-Vivo Tracer Method with the Use of Nonradioactive Isotopes and Activation Analysis, *J. Lab. Clin. Med.* **61**, 1042–1048 (1963).

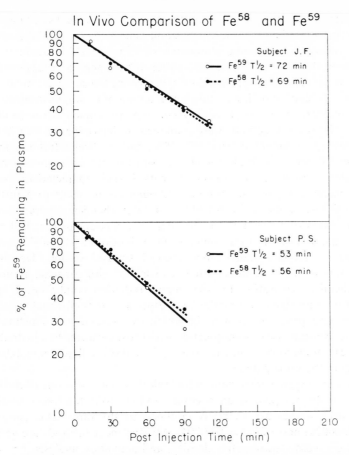

**Figure 9.33** *In vivo* plasma clearance rates for simultaneously injected stable and radioactive isotopes of iron in two human subjects. [From Lowman and Krivit (1963).]

clearance rates of the stable isotope $^{58}$Fe and the radioisotope $^{59}$Fe injected simultaneously into human subjects. Pre- and postinjection blood samples were drawn periodically and the radiotracer and neutron-activated stable tracer retentions were determined. The plasma clearance curves for the two tracers for two normal adults are shown in Figure 9.33. The data show that the two isotopes gave clearance curves that agree within the limits of variation for counting statistics alone and with differences in clearance rates less than those expected for duplicate radiotracer measurements. They demonstrated the identical *in-vivo* utilization of the stable and radioactive isotopes.

Chromium has been successfully used as a tracer for red cells in blood by Johnson, Tothill, and Donaldson.* The stable isotope $^{50}$Cr is administered as $Na_2CrO_4$. Although the natural abundance of $^{50}$Cr of 4.3% is adequate for measurement, the use of available 97%-enriched $^{50}$Cr increases the activation sensitivity by a factor of more than 20. To determine the survival time of the red cells, small blood samples are taken at intervals after administration and the $^{50}$Cr concentration is determined by x-ray spectroscopic measurement of activated $^{51}$Cr. Several methods were examined to determine the $^{51}$Cr in the presence of large amounts of $^{24}$Na, $^{38}$Cr, $^{42}$K, and the Bremsstrahlung from high-energy $^{32}$P beta-rays. Radiochemical separations were also investigated, since purified $^{51}$Cr can also be measured by its 5.0-keV characteristic x-ray. The radiochemical separation method, although costly, leads to a further gain in sensitivity of about a factor of 3 to 4. With chemical separation and irradiation fluxes of $10^{12}$ n/cm$^2$ sec for 20 hr, the amount of enriched $^{50}$Cr needed per survival test is 1 mg, which is equivalent to 85 $\mu$Ci of $^{51}$Cr nominally used in standard radioisotope red-cell survival studies. However, 1 mg $^{50}$Cr was sufficient to measure red-cell survival with adequate accuracy without chemical separation for the same irradiation conditions. It was noted that the activable tracer method was especially useful for such patients as children and pregnant women, in which exposure to radionuclides is undesirable. Another use is in conjunction with the radiotracer $^{51}$Cr in double-label studies; for example to measure the survival of two different populations of red cells in the same patient.

The element oxygen is important in biological systems. It is one of the few elements that do not possess an adequate radioisotope for tracer studies. Oxygen tracer studies are generally confined to the rare stable isotopes $^{17}$O ($f = 0.00039$) and $^{18}$O ($f = 0.00204$). Measurement methods are generally based on mass spectrometry, density, and activation analysis. Nuclear magnetic resonance techniques are also applicable to $^{17}$O measurement.

A study of the chemical incorporation of oxygen into algae by photosynthesis, reported by Fogelström-Fineman et al.,† illustrates the use of activation analysis for the tracing of oxygen in organic systems. Their method utilizes paper chromatographic separations and proton activation of the $^{18}$O tracer to produce 111-m $^{18}$F. The tracer was 20% enriched $^{18}$O in water. The sensitivity for $^{18}$O by the (p,n) reaction with an irradiation of 1 to 10 $\mu$A of 4-MeV protons for a few minutes was of the order of 0.1 to 1 $\mu$g. The

---

* P. F. Johnson, P. Tothill, and G. W. K. Donaldson, Stable Chromium as a Tracer for Red Cells, with Assay by Neutron Activation Analysis, *Intern. J. Appl. Radiation Isotopes* **20,** 103–108 (1969).

† I. Fogelström-Fineman, O. Holm-Hansen, B. M. Tolbert, and M. Calvin, A Tracer Study with O¹⁸ in Photosynthesis by Activation Analysis, *Intern. J. Appl. Radiation Isotopes,* **2,** 280–286 (1957).

experimental objective was the quantitative determination of oxygen incorporated into different compounds in algae grown in $^{18}$O-enriched water for short periods of time. Separation of the compounds was achieved by paper chromatography, in which the compounds are found as spots on the paper at different locations. Since chromatographic papers contain oxygen or other interfering elements, the compounds were transferred from the paper chromatogram to a metal sheet by an ingenious thermal transfer process. The apparatus and the result of a test of the transfer reliability, using a $^{14}$C-labeled compound and autoradiographic location and intensity measurements, are shown in Figure 9.34. The results of tests which evaluated the sensitivity of the activation analysis for $^{18}$O determination are shown in Figure 9.35. The algae *Chlorella Pyrenoidosa* was grown in $^{18}$O-rich water, and samples of varying amounts of algae and $^{18}$O content were analyzed together with three samples of untagged algae. The results show the spot containing 2 $\mu$g of algae and 0.1 $\mu$g $^{18}$O can still be detected against the background.

ENVIRONMENTAL STUDIES.    Several applications of activable tracers have been reported for studies of atmospheric and hydrospheric particle transport, air pollution, and waste-water effluents in the environment. Activable tracers are especially desirable in light of the usually high dilution factors experienced in environmental transport and mixing processes and the hazardous levels of radioactive materials that would be required to achieve these dilution factors.

Gatz, Dingle, and Winchester* describe the use of indium as an atmospheric tracer with detection by neutron activation. Indium dispersed as fine particles produced from pyrotechnic flares from aircraft at the base of rain clouds is measured in rainwater collected on the ground. The sensitivity of the method is determined by the natural background of indium in rainwater. In rain samples collected serially from two untagged rains in Oklahoma the indium concentrations were $6 \pm 3$ ng/l with reagent blanks of about 2 ng/l. Rain samples from clouds tagged with 200 g indium contained 7 to 24 ng/l. The chemical treatment in the field consisted of coprecipitating the indium and a lanthanum internal standard with $Fe(OH)_3$ onto a membrane filter. After irradiation the indium and lanthanum activities were determined by comparator analysis. The weight of indium in the rainfall samples is given by

$$W(In_r) = W(In_s) \frac{(In/La)_r}{(In/La)_s} \tag{43}$$

where the subscripts $r$ and $s$ refer to the rain and standard samples, respectively.

* D. F. Gatz, A. N. Dingle, and J. W. Winchester, Detection of Indium as an Atmospheric Tracer by Neutron Activation, *J. Appl. Meteor.* **8,** 229–235 (1969).

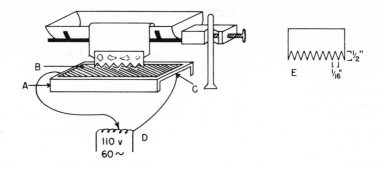

Direction of chromatogram development ⟶

a: Autoradiogram of a paper chromatogram of C¹⁴-labeled sugar phosphate developed in butanol–propionic acid. Exposure time 12 hr.

b: Autoradiogram of the metal sheet to which the paper chromatogram in "a" had been eluted. Exposure time 12 hr.

**Figure 9.34** An apparatus for transferring a developed one-dimensional chromatogram to a metal sheet: (a) metal supporting plate; (b) tantalum sheet; (c) heating tape under sheet; (d) variable voltage control; (e) brass sheet used to serrate edges of chromatogram, and an example of a transfer of paper chromatogram pattern to a metal sheet. [From I. Fogelström-Fineman et al. (1957).]

Nakasa and Ohno* were able to trace the effluent gas from a steam-power station stack over a distance of more than 10 km. Their objective was the tracing of the sources of air pollution from both industrial and domestic wastes in industrial areas. The tracer chosen for $SO_2$ releases in stack gas was cobalt sulfate. This tracer was chosen because cobalt

(a) has a relatively large activation cross section,

---

* H. Nakasa and H. Ohno, Application of Neutron Activation Analysis to Stack-Gas Tracing, in *Radioisotope Tracers in Industry and Geophysics* (International Atomic Energy Agency, Vienna, 1967), pp. 239–250.

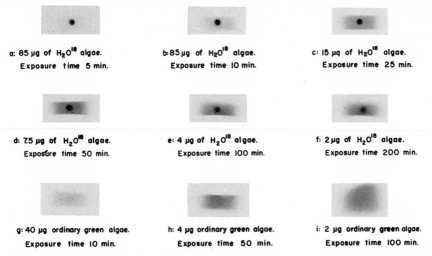

a: 85 µg of $H_2O^{18}$ algae.
Exposure time 5 min.

b: 85 µg of $H_2O^{18}$ algae.
Exposure time 10 min.

c: 15 µg of $H_2O^{18}$ algae.
Exposure time 25 min.

d: 7.5 µg of $H_2O^{18}$ algae.
Exposure time 50 min.

e: 4 µg of $H_2O^{18}$ algae.
Exposure time 100 min.

f: 2 µg of $H_2O^{18}$ algae.
Exposure time 200 min.

g: 40 µg ordinary green algae.
Exposure time 10 min.

h: 4 µg ordinary green algae.
Exposure time 50 min.

i: 2 µg ordinary green algae.
Exposure time 100 min.

**Figure 9.35** Autoradiograms with exposures started 2 hr after proton irradiation of various amounts of [18]O-enriched and natural green algae. [From I. Fogelström-Fineman, O. Holm-Hansen, B. Tolbert, and M. Calvin (1957).]

(b) yields a product which emits energetic easily detected gamma-rays; that is, detected quantitatively from the 2.50-MeV sum peak without chemical separation,

(c) has insignificant natural concentration in the atmosphere,

(d) is nontoxic,

(e) is readily adaptable to field working conditions.

Cobalt sulfate can be dispersed by spraying its solution into a stack-gas stream with an atomizer and high-pressure air. The particles become part of the stack-gas aerosols and can be sampled downstream by air-filter samplers. Measurement of three natural aerosol samples in Tokyo showed background concentrations of 1 to 6 ng/m³. Figure 9.36 shows a ground-level distribution of tracer $CoSO_4$ in a test and a comparison of $SO_2$ concentrations determined both by chemical analysis and by $CoSO_4$ tracing. On the basis that the $CoSO_4$ will behave in the same way as the stack-gas during its diffusion in the atmosphere, the ground-level concentration $C$ is related to the emission rate $Q$ by

$$C_s = C_t \cdot \frac{Q_s}{Q_t} \qquad (44)$$

where the subscripts $s$ and $t$ refer to $SO_2$ and $CoSO_4$ tracer, respectively. The $SO_2$ concentration from a particular source can be calculated if $C_t$ is

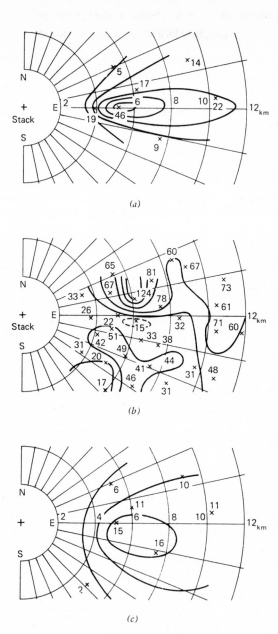

**Figure 9.36** (*a*) Test result for ground-level distribution of cobalt sulfate tracer (in ng/m³); (*b*) ground-level distribution of SO₂ measured by chemical analysis (in ppb); (*c*) same distribution as (*b*) derived from cobalt sulfate tracing (in ppb). [From H. Nakasa and H. Ohno (1967).]

measured and $Q_s$ and $Q_t$ are known. If the total ground-level $SO_2$ concentration $C'_s$ is measured at the sampling station, the background concentration of $SO_2$ from other sources $C_B$ is given by

$$C_B = C'_s - C_s = \frac{C'_s - C_t Q_s}{Q_t} \qquad (45)$$

Thus the relative intensity $P$ of the particular source to the surrounding air pollution is

$$P = \frac{C_s}{C'} = \frac{C_t Q}{C'_s Q_t} \qquad (46)$$

Activable tracers have also been used in natural water environments. Cappadona* described the use of stable tracers for measuring the movement of solids in such studies as erosion of coasts, littoral drifts, and deposition of sand in shallow waters. For the latter application the movement of sand was traced at a beach at Palermo, Sicily. Silver was chosen as the tracer. To obtain an average concentration of 30 $\mu g/kg$ of sand, in a coastal sand volume of $9 \times 10^4$ m³ ($2.5 \times 10^8$ kg), a source of 7.5 kg of Ag was required. The silver was added to 400 kg of sand and scattered in the sea along the littoral. The transport is determined by activation analysis of many 1-kg samples of sand taken along the coast. The system can be calibrated with the use of 7.6-d $^{111}$Ag as an internal radiotracer. A standard sand sample is prepared with $^{111}$Ag of known specific activity. A small aliquot $C_1^x$ is standardized by the $^{111}$Ag counting rate, so that the amount $P_1$ of Ag carrier in the aliquot is known. This aliquot is added to the 1-kg sand sample from which the silver is chemically removed. The extracted sample $C_2^x$ is standardized again by counting the $^{111}$Ag. These counts determine the chemical yield $R$ of the sample and therefore the amount of silver $P_2$ contained in the aliquot $C_2^x$. After neutron irradiation, together with a standard containing the same amount of silver $P_2$, the amount of silver tracer in the sand sample is given

$$x = P_x R - P_1 \qquad (47)$$

where $P_x$ is the total weight of $P_2$ and Ag extracted from the sand sample.
Channell and Kruger† examined the potential of the rare earth elements as

* C. Cappadona, Measurements of Movements of Solid Substances in Water by Means of Stable Tracers and Activation Analysis, in *Modern Trends in Activation Analysis* (National Bureau of Standards, 1969), pp. 72–75.
† J. K. Channell and P. Kruger, Post-Sampling Activation Analysis of Stable Nuclides for Estuary Water Tracing, in *Modern Trends in Activation Analysis* (National Bureau of Standards, 1968), pp. 600–606.

estuary water tracers. The selection of the rare earths was based on the properties of

(a) low natural background in estuarine waters,
(b) favorable nuclear properties,
(c) persistence in the estuarine environment,
(d) reasonable cost.

The activation analysis of preirradiation concentrated samples with group separation and gamma-ray spectroscopy following neutron activation at low fluxes of $10^{11}$ n/cm²-sec was sufficient to determine lanthanum and cerium at natural background concentrations of 180 and 12 ng/l, respectively, in Central San Francisco Bay. A comparison has also been made* of tracing costs for stable lanthanum, radiotracers $^{140}$La and $^3$H, and fluorescent dye Rhodamine WT. The cost of the tracing was divided into two parts, the analytical costs and the tracer costs. These costs are separated, since the ratio of amount of tracer added to the number of samples analyzed may vary considerably. Three scales of tracer studies were examined:

1. A small-scale study, such as local outfall dispersion with a dilution factor to trace $10^6$ acre feet (1 MAF) of water at the minimum detection amount (MDA) with injection over one tidal cycle ($\sim$12 hr) and persistence for two tidal cycles after injection.

2. An intermediate-size study, such as a sewage effluent outfall line with continuous injection over two complete tidal cycles to trace 10 MAF at MDA for 10 tidal cycles ($\sim$5 days).

3. A large-scale study, such as the tidal exchange rate through the Golden Gate of San Francisco Bay, which may require the tracing of 100 MAF at MDA for about 20 tidal cycles.

This large-scale study has meaning in that the estimated residence time of water in the San Francisco Bay is of the order of 15 to 20 tidal cycles. Figure 9.37 shows the time-concentration curves observed in the Corps of Engineers San Francisco Bay Model.† The average concentration $\bar{C}$ flowing out of the bay during the observation period is

$$\bar{C} = \frac{W}{V_{tp}T} \tag{48}$$

where $W$ = mass of tracer added to the estuary,
$\quad V_{tp}$ = mean tidal prism volume,
$\quad T$ = number of tidal cycles.

---

* J. K. Channell and P. Kruger, Activable Rare Earth Elements as Estuarine Water Tracers, *Proc. 5th Intern. Conf. Water Pollution Research*, San Francisco, July 1970.

† R. L. O'Connell and C. M. Walter, Hydraulic Model Tests of Estuarial Waste Dispersion, *J. Sani. Engr. Div. (ASCE)* **89**, No. SA1, 51–65 (1963).

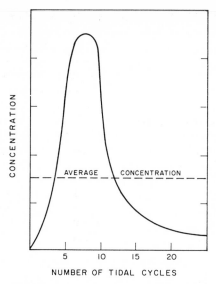

**Figure 9.37** A typical concentration-tidal cycle curve.

If $\bar{C}$ is taken as 2 × MDA for 40 tidal cycles for a tidal prism volume of 1.14 MAF, the amount of tracer required would correspond to a total traced volume of 91 MAF at MDA.

The cost comparisons for the three scales of tracing are summarized in Table 9.42 It is noted that $^{140}$La is the cheapest tracer for small-scale and intermediate-scale studies but that $^3$H is the cheapest tracer for large-scale studies, although $^{140}$La is not much more expensive. Because of the low analytical costs for measuring fluorescent dyes, the tracing costs for Rhodamine WT rise linearly with the amount of water tagged and become excessive in large studies. Thus to avoid the excessive costs of the dyes and the external radiation hazards of injecting kilocurie quantities of radionuclides into estuarine waters, stable lanthanum appears to be the best tracer for large-scale studies. An interesting possibility might be a combined-tracer study with $^{140}$La in permissible amounts together with sufficient stable lanthanum to cover the needed duration of the study. The $^{140}$La can be used to monitor *in-situ* the initial dispersion of the stable tracer.

INDUSTRIAL STUDIES. The special application of industrial on-stream activation analysis, reviewed in Section 9.3.4, may be considered as an example of internal stable isotope tracing. Other developments are being made in which activable nuclides are added as external tracers to the systems to be studied. Two such studies are examined.

Table 9.42    Tracing Costs for Model Field Studies[a]

|  | Scale of Study | | |
| --- | --- | --- | --- |
|  | Small | Intermediate | Large |
| **Conditions** | | | |
| Volume tagged (MAF × MDA) | 1 | 10 | 100 |
| Duration (days) | 1.5 | 5 | 10 |
| Number of samples | 100 | 100 | 200 |
| **Stable Lanthanum** | | | |
| Weight needed (lb) | 314 | 3,500 | 41,200 |
| Tracer Cost ($) | 360 | 4,020 | 43,100 |
| Analysis Cost ($) | 7,000 | 7,000 | 13,000 |
| Total Cost | 7,360 | 11,020 | 56,100 |
| **Rhodamine WT** | | | |
| Weight needed (lb) | 945 | 10,550 | 124,000 |
| Tracer Cost ($) | 1,965 | 22,000 | 258,000 |
| Analysis Cost ($) | 100 | 100 | 200 |
| Total Cost | 2,065 | 22,100 | 258,200 |
| **Tritium** | | | |
| Activity needed (Ci) | 280 | 2,830 | 28,300 |
| Tracer Cost ($) | 560 | 4,750 | 34,000 |
| Analysis Cost ($) | 3,000 | 4,500 | 6,000 |
| Total Cost | 3,560 | 9,250 | 40,000 |
| **$^{140}$La** | | | |
| Activity needed (Ci) | 71 | 2,900 | 270,000 |
| Tracer Cost ($) | 620 | 4,000 | 53,000 |
| Analysis Cost ($) | 100 | 100 | 200 |
| Total Cost | 720 | 4,100 | 53,200 |

[a] J. K. Channell and P. Kruger, Activable Rare Earth Elements as Estuarine Water Tracers, *Proc. 5th Intern. Conf. Water Pollution Research*, San Francisco, Calif., July, 1970.

One involves a new method of investigating wear with the use of activable tracers, described by Radvan, Revenska-Kostzyuk, and Vezranovski.*

* M. Radvan, B. Revenska-Kostzyuk, and E. Vezranovski, Use of a New Method Involving Labelling with Non-Radioactive Elements and Activation Analysis to Investigate Wear, in *Radioisotope Tracers in Industry and Geophysics* (International Atomic Energy Agency, Vienna, 1967), pp. 71–79 (in Russian).

The method has been used to investigate the wear of bearings in agricultural machines under operating conditions and the wear of fireproof materials in modern steel production. The tracers used were rare-earth or lanthanum oxides, which have favorable nuclear properties and which can be incorporated into cast-iron and polyamide bearings in sufficiently low amounts to cause no structural changes. Separation methods were developed to extract the rare-earth tracers from the oil or grease of agricultural machinery and from the large quantities of iron used in the investigations concerning the effect of vacuum extraction and the use of induction mixers on the passage of fire-resistant particles into the steel. The separation from the iron is achieved by adding calcium as a carrier for the rare earths and precipitating the oxalates while the iron forms a soluble complex ion. The sensitivity that has been achieved by this method is of the order of 1 $\mu$g of labeled material. The results of these investigations are noted in Figure 9.38. The top curves show a comparison of the wear in bearings made by three measurement methods using a radioactive bushing and a nonradioactive bushing with activation of the wear material preceding or following chemical separation. The bottom curve shows the dependence of the amount of inclusions appearing in the steel as a function of induction-mixer operating time. The amount of lanthanum can be determined directly by comparison of the gamma-ray spectrum with a standard lanthanum sample, correcting only for $^{24}$Na, in the energy range of 1.51 to 1.72 MeV. This method showed that the lanthanum measured in the inclusions was less than $4 \times 10^{-9}$ g/g of steel.

Another use of activable tracers in industrial plants was reported by Chatters and Peterson* who developed a method for the identification and tracing of wood fibers in pulp and paper mills. The tracing method involves treating a small sample of wood fibers with a metallic salt, for example, rare-earth sulfate, dispersing the tracer pulp into the mill system at any desired point, and collecting fiber samples downstream in the process lines and effluent streams. Measurement of the tracer pulp is made by neutron activation analysis. The rare-earth element lanthanum was once again found to be a suitable activable tracer. It has been used most successfully in fiber tracing because of its favorable nuclear properties, adherence to wood fibers, low natural abundance in wood fibers and mill processing lines, and economic feasibility.

One of the major problems of an external tracer added to tag a small amount of a process material is the need to ensure that the tracer will remain with the tagged process material throughout the operations being investigated. In the case of the wood fibers an extensive investigation of several

* R. M. Chatters and R. L. Peterson, Use of Tracers in In-Plant Evaluation of Processes and in Pollution Control of Effluent from Pulp and Paper Mills into Streams, in *Radioisotope Tracers in Industry and Geophysics* (International Atomic Energy Agency, Vienna, 1967), pp. 251–258.

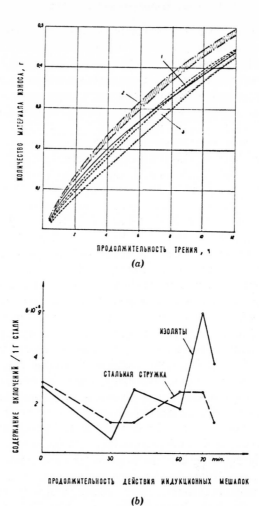

**Figure 9.38** (*a*) Relationship of the amount of wear material (ordinate, in grams) to the duration of sliding friction of the bearing (abscissa, in hours) for experiments with curve 1, radioactive bushings; curve 2, nonradioactive bushings with chemical treatment of the wear material preceded by activation; curve 3, nonradioactive bushings with chemical treatment following activation. (*b*) Relationship of the amount of inclusions (ordinate, in μg/g steel) to the duration of induction mixer operation (abcissa, in minutes); the dashed line is for steel cuttings and the solid line for electrolytic insulators. [From Radvan et al. (1967).]

tagging materials and methods of attachment to wood-pulp fibers was made. The tracer metal component may become bonded with the cellulose or be precipitated within the fiber lumen. Lanthanum, chosen as the best tracer, precipitates from solution into rosin size and adheres firmly to wood fibers. Stabilized concentrations of about 2.5 mg La/g pulp are achievable. The tagged fiber was used in several in-plant experiments for the following purposes:

1. To ascertain the detectability of lanthanum-tracer pulp when mixed with a greater mass of untreated pulp.

2. To evaluate the effects of chemical processing additives on the lanthanum-treated pulp.

3. To evaluate the effects of varying physical processes and various environments on the lanthanum-treated pulp.

4. To follow the flow of treated wood fibers qualitatively and quantitatively throughout the pulp and paper mill processing lines and into the finished paper product and effluent streams.

These in-plant tests indicated that the various chemical and physical conditions to which the tagged fibers were exposed in the paper mill had no deleterious effects on the added lanthanum-treated pulp. The treated pulp was found to be quantitatively detectable at all collection points in the plant, even when initial mixing ratios of 3 lb of treated fibers to 40,000 lb of untreated fibers were used. All of these experiments were conducted without interruption of, or interference with, the paper mill's normal operating procedures.

## 9.5  BIBLIOGRAPHY

### 9.5.1  Trace Element Analysis

GENERAL

J. P. Cali, *Trace Analysis in Semiconductor Materials* (Pergamon, New York, 1964).

W. W. Meinke, Trace-Element Sensitivity: Comparison of Activation Analysis with Other Methods, *Science* **121**, 177–184 (1955).

W. W. Meinke and B. F. Scribner, Eds., Trace Characterization, Chemical and Physical, National Bureau of Standards Monograph 100, 1967. (Available from Superintendent of Documents, U.S. Government Printing Office, Washington, D.C.)

G. H. Morrison, *Trace Analysis, Physical Methods* (Wiley-Interscience, New York, 1965.)

W. F. Pickering, *Fundamental Principles of Chemical Analysis* (Elsevier, Amsterdam, 1966).

S. Siggia, *Survey of Analytical Chemistry* (McGraw-Hill, New York, 1968).

A. A. Smales, The Place of Activation Analysis in a Research Establishment Dealing with Pure Materials, in *Modern Trends in Activation Analysis* (Texas A&M University, College Station, 1965), pp. 186–188.

A. A. Smales and L. R. Wager, *Methods in Geochemistry* (Interscience, New York, 1960).

H. Yoe and H. J. Koch, *Trace Element Analysis* (Wiley, New York, 1957).

SPECTROCHEMICAL METHODS

L. H. Ahrens and S. R. Taylor, *Spectrochemical Analysis*, 2nd ed. (Addison-Wesley, Reading, 1961).

L. S. Birks, *X-ray Spectrochemical Analysis* (Interscience, New York, 1959).

W. R. Brode, *Chemical Spectroscopy*, 2nd ed. (Wiley, New York, 1943).

F. Burriel-Marti and J. Ramirez-Munoz, *Flame Photometry* (Elsevier, Amsterdam, 1957).

J. A. Dean, *Flame Photometry* (McGraw-Hill, New York, 1960).

W. T. Elwell and J. A. F. Gidley, *Atomic Absorption Spectrometry* (Macmillan, New York, 1962).

D.·M. Hercules, *Fluorescence and Phosphorescence Analysis: Principles and Applications* (Wiley, New York, 1966).

N. H. Nachtrieb, *Principles and Practice of Spectrochemical Analysis* (McGraw-Hill, New York, 1950).

J. A. Radley and J. Grant, *Fluorescence Analysis in Ultraviolet Light*, 4th ed. (Van Nostrand Co., New York, 1954).

E. B. Sandell, *Colorimetric Determination of Traces of Metals*, 2nd ed. (Interscience, New York, 1950).

F. D. Snell, C. T. Snell, and C. A. Snell, *Colorimetric Methods of Analysis*, Vol. III (Van Nostrand Co, Princeton, 1959).

F. D. Snell, *Colorimetry*, 3rd ed. (Van Nostrand, Princeton, 1948–1959), 4 vols.

## 9.5.2  Summaries of Applications

COSMOSPHERE

R. L. Caldwell, W. R. Mills, L. S. Allen, P. R. Bell, and R. L. Heath, Combination Neutron Experiment for Remote Analysis, *Science* **152**, 457–465 (1966).

F. Heide, *Meteorites* (translated by E. Anders and E. du Fresne) (University of Chicago Press, 1964).

J. S. Hislop and R. E. Wainerdi, Extraterrestrial Neutron Activation Analysis, *Anal. Chem.* **39**, 28A–39A (February 1967).

E. D. Jackson and H. G. Wilshire, Chemical Composition of the Lunar Surface at the Surveyor Landing Sites, *J. Geophys. Res.* **73**, 7621–7629 (1968).

B. Mason, *Meteorites* (Wiley, New York, 1962).

A. L. Turkevich, J. H. Patterson, and E. J. Franzgote, The Chemical Analysis of the Lunar Surface, *Amer. Scientist* **56**, 312–343 (1968).

ATMOSPHERE

S. S. Brae, et al., Thermal Neutron Activation Analysis of Airborne Particulate Matter in Chicago Metropolitan Area, in *Modern Trends in Activation Analysis* (National Bureau of Standards, Gaithersburg, 1968), pp. 560–568.

N. D. Dudey, L. E. Ross, and V. E. Noshkin, Application of Activation Analysis and Ge (Li) Detection Techniques for the Determination of Stable Elements in Marine Aerosols, in *Modern Trends in Activation Analysis* (National Bureau of Standards, Gaithersburg, 1968), pp. 569–577.

J. P. Friend, Ed., The High Altitude Sampling Program, Vol. 5, DASA-1300, 1961.

C. M. Gordon, E. C. Jones, and J. I. Hoover, Atmospheric Aerosols Near the Surtsey Volcano, Report of NRL Porgress, 1–3, 1964.

C. M. Gordon and R. E. Larson, Activation Analysis of Aerosols, *J. Geophys. Res.* **69**, 2881–2885 (1964).

J. R. Keane and E. M. R. Fisher, Analysis of Trace Elements in Air-Borne Particulates by Neutron Activation and Gamma-Ray Spectrometry, *Atmospheric Environment* **2**, 603–614 (1968).

R. L. Lininger, R. A. Duce, J. W. Winchester, and W. R. Matson, Chlorine Bromine, Iodine, and Lead in Aerosols from Cambridge, Massachusetts, *J. Geophys. Res.* **71**, 2457–2463 (1966).

J. A. Warburton and L. G. Young, Neutron Activation Procedures for Silver Analysis in Precipitation, *J. Appl. Meteor.* **7**, 433–443 (1968).

J. A. Warburton and L. G. Young, Neutron Activation Measurements of Silver in Precipitation from Locations in Western North America, *J. Applied Meteor.* **7**, 444–448 (1968).

HYDROSPHERE

R. L. Blanchard, G. W. Leddicotte, and D. W. Moeller, Water Analysis by Neutron Activation, *J. Am. Water Works Assoc.* **51**, 967–980 (1959).

W. H. Durum and J. Haffty, Implications of the Minor Element Content of Some Major Streams of the World, *Geochim. Cosmochim. Acta* **27**, 1–11 (1963).

R. Fukai and W. W. Meinke, Trace Analysis of Marine Organisms: A Comparison of Activation Analysis and Conventional Methods, *Limnology and Oceanography* **4**, 398–408 (1959).

O. Landström and C. G. Wenner, Neutron-Activation Analysis of Natural Water Applied to Hydrogeology, Report AE-204, Aktiebolaget Atomeneigi, Stockholm, Sweden, 1965.

K. D. Linstedt and P. Kruger, Vanadium Concentrations in Colorado River Basin Waters, *J. Am. Water Works Assoc.* **61**, 85–88 (1969).

B. Mason, *Principles of Geochemistry*, 3rd ed. (Wiley, New York, 1966), Chapter 7.

D. F. Schutz and K. K. Turekian, The Investigation of the Geographical and Vertical Distribution of Several Trace Elements in Sea Water Using Neutron Activation Analysis, *Geochim. Cosmochim. Acta* **29**, 259–313 (1965).

LITHOSPHERE

L. H. Ahrens, *Distribution of the Elements in Our Planet* (McGraw-Hill Paperbacks, New York, 1965).

H. S. M. Bowen and D. Gibbons, *Radioactivation Analysis* (Oxford, London, 1963), Chapter 10.

F. W. Clarke and H. S. Washington, The Composition of the Earth's Crust, U.S. Geol. Survey Professional Paper 127, 1924.

F. J. Flanagan, U.S. Geological Survey Silicate Rock Standards, *Geochim. Cosmochim. Acta* **31**, 289–308 (1967) and U.S. Geological Survey Standards—II. First Compilation of Data for the new U.S.G.S. Rocks, *Geochim. Cosmochim. Acta* **33**, 81–120 (1969).

M. Fleischer, Summary of New Data on Rock Samples G-1 and W-1, 1962–1965, *Geochim. Cosmochim. Acta* **29**, 1263–1283 (1965) and for additional data 1965–1967, *Geochim. Cosmochim. Acta* **33**, 65–79 (1969).

E. M. Lobanov, Activation Analysis, *Proc. First All-Union Coordinating Conference*, Tashkent, USSR, October 1962, AEC-tr-6639, 1966. (Available from Clearinghouse, Springfield, Va. 22151, $3.00).

J. W. Winchester, Radioactivation Analysis in Inorganic Geochemistry, Chapter 1, in F. A. Cotton, Ed., *Progress in Inorganic Chemistry*, Vol. 2 (Interscience, New York, 1960).

BIOSPHERE

Activation Analysis on Man—A New Tool for Medical Research, *Scientific Research* 53–56 (July 1967).

J. Anderson, S. Osborn, R. Tomlinson, R. Newton, J. Rundo, L. Salmon, and J. Smith, Neutron-Activation Analysis in Man In Vivo, *Lancet* 1201 (December 1964).

H. J. M. Bowen, Activation Analysis in Botany and Agriculture, in *Nuclear Activation Techniques in the Life Sciences* (International Atomic Energy Agency, Vienna, 1967), pp. 287–297.

V. P. Guinn, Neutron Activation Analysis of Foodstuffs for Pesticide Residues, *World Review of Pest Control* **3**, 138–147 (1964).

V. P. Guinn, Ed., Proceedings of the First International Conference on Forensic Activation Analysis, Report No. GA-8171, 1966, available from Gulf General Atomic Corp. La Jolla, Calif.

B. Keisch, R. C. Koch, and A. S. Levine, Determination of Biospheric Levels of $I^{129}$ by Neutron-Activation Analysis, in *Modern Trends in Activation Analysis* (Texas A&M University, College Station, 1965), pp. 284–290.

J. M. A. Lenihan, Activation Analysis in the Contemporary World, in *Modern Trends in Activation Analysis* (National Bureau of Standards, Gaithersburg, Md., 1969), pp. 1–32.

J. M. A. Lenihan and H. Smith, Activation Analysis and Public Health, in *Nuclear Activation Techniques in the Life Sciences* (International Atomic Energy Agency, Vienna, 1967), pp. 601–614.

G. S. Nixon, H. Smith, and H. D. Livingston, Trace Elements in Human Tooth Enamel, in *Nuclear Activation Techniques in the Life Sciences* (International Atomic Energy Agency, Vienna, 1967), pp. 455–462.

Nuclear Activation Techniques in the Life Sciences (International Atomic Energy Agency, Vienna, 1967).

N. Spronk, Nuclear Activation in the Animal Sciences, in *Nuclear Activation Techniques in the Life Sciences* (International Atomic Energy Agency, Vienna, 1967), pp. 335–352.

D. M. Taylor, Activation Analysis in Human Biochemistry, in *Nuclear Activation Techniques in the Life Science* (International Atomic Energy Agency, Vienna, 1967), pp. 391–401.

H. W. Wagner, Ed., *Principles of Nuclear Medicine* (Saunders, Philadelphia, 1968) Chapter 20.

### 9.5.3  Special Applications

COMPARATIVE ANALYSIS

R. E. Jervis, Activation Analysis in Forensic Science, in *Nuclear Activation Techniques in the Life Sciences* (International Atomic Energy Agency, Vienna, 1967), pp. 645–659.

C. J. Maletskos, On the Introduction of New Scientific Methods in Court, in *Proc., First International Conference on Forensic Activation Analysis*, GA-8171 (General Atomic Corp., La Jolla, California, 1966).

R. Ruch, J. Buchanan, V. Guinn, S. Bellanca, and R. Pinker, Neutron Activation in Scientific Crime Prevention, *J. Forensic Sci.* **9**, 119-133 (1963).

U.S. Atomic Energy Commission Press Release G-112, May 13, 1964.

SURFACE ANALYSIS

O. U. Anders, Use of Charged Particles from a 2-Megavolt Van de Graaf Accelerator for Elemental Surface Analysis, *Anal. Chem.* **38**, 1442–1452 (1966).

J. N. Barrondon and P. Albert, Determination of Oxygen Present at the Surface of Metals by Irradiation with 2 Mev Tritons, in *Preprints, Modern Trends in Activation Analysis* (National Bureau of Standards, Gaithersburg, Md., 1968), pp. 298–305.

K. R. Blake, T. C. Martin, I. L. Morgan, and C. D. Houston, The Measurement of Surface Contamination of High-Purity Beryllium Samples, in *Proc., Modern Trends in Activation Analysis* (Texas A&M University, College Station, 1965), pp. 76–81.

M. Peisach and D. O. Poole, Analysis of Surfaces by Scattering of Accelerated Alpha Particles, *Anal. Chem.* **38**, 1345–1350 (1966).

S. Rubin, T. O. Passell. and L. E. Bailey, Chemical Analysis of Surfaces by Nuclear Methods, *Anal. Chem.* **29**, 736–743 (1957).

ISOTOPIC ANALYSIS

S. Amiel and Y. Welwart, Lithium and Lithium-6 Analysis by Counting Delayed Neutrons, *Anal. Chem.* **35**, 566–570 (1963).

A. E. Cameron, W. Herr, W. Herzog, and A. Lunden, Isotopen-Anreicherung beim Brom durch elektrolytische Uberfuhrung in geschmolzenem Bleibromid, *Z. Naturforsch.* **11a**, 203 (1956).

R. F. Coleman, Isotopic Determination of Lithium by Neutron Activation, *Analyst* **85**, 285–288 (1960).

J. T. Corliss, Determination of Calcium-48 in Natural Calcium by Neutron Activation Analysis, *Anal. Chem.* **38**, 810–813 (1966).

R. H. Filby, Determination of the Zinc-68–Zinc-64 Ratio in Rocks and Minerals by Neutron Activation Analysis, *Anal. Chem.* **36**, 1597–1600 (1964).

W. Herr, Neutron Activation Applied to Geochemistry, in *Radioactivation Analysis* (Butterworth, London, 1960), pp. 35–52.

L. Kaplan and K. E. Wilzbach, Lithium Isotope Determination by Neutron Activation, *Analyst* **85**, 285–288 (1960).

M. Mantel, J. Gilat, and S. Amiel, Isotopic Analysis of Uranium by Neutron Activation and High Resolution Gamma-Ray Spectrometry, in *Modern Trends in Activation Analysis* (National Bureau of Standards, 1968), pp. 79–88.

E. Merz and W. Herr, Microdetermination of Isotopic Abundance by Neutron Activation, in *Proc. 2nd UN Intern. Conf. Peaceful Uses of Aomic Energy*, **28**, 984, 491–495 (1958).

M. Peisach and R. Pretorius, The Determination of Stable Calcium Isotopes by Charged-Particle Irradiation, in *Modern Trends in Activation Analysis* (National Bureau of Standards, 1969), pp. 306–315.

G. W. Reed, Discussion, in *Radioactivation Analysis Symposium* (Butterworth, London, 1960), pp. 48–51.

A. P. Seyfang and A. A. Smales, The Determination of Uranium-235 in Mixtures of Naturally Occurring Uranium Isotopes by Radioactivation, *Analyst* **78**, 394–405 (1953).

A. P. Seyfang, An Improvement in the Determination of Uranium-235 by Radioactivation, *Analyst* **80**, 74–76 (1955).

ON-LINE ACTIVATION ANALYSIS

O. U. Anders, Activation Analysis for Plant Stream Monitoring, *Nucleonics* **20**, (2), 78–83 (1962).

J. B. Ashe, P. F. Berry, and J. R. Rhodes, On-Stream Activation Analysis using Sample Recirculation, in *Modern Trends in Activation Analysis* (National Bureau of Standards, 1968), pp. 120–125.

W. W. Givens, W. R. Mills, and R. L. Caldwell, Cyclic Activation Analysis, *op. cit.*, pp. 139–147.

K. Ljunggren and R. Christell, Continuous Determination of Boron in a Process Stream Using a Low Level Neutron Source, *Atompraxis* **10**, 259–263 (1964).

T. C. Martin, I. L. Morgan, and J. D. Hall, Nuclear Analysis System for Coal, in *Modern Trends in Activation Analysis* (Texas A&M University, College Station, 1965), pp. 71–75.

H. F. Ramdohr, Anwendung der Aktivierungsanalyse zur Kupfererzsortierung, *Kerntechnik* **5**, 204–206 (1963).

## 9.5.4   Stable-Isotope Tracing

RADIOISOTOPE TRACING METHODS

J. R. Bradford, Ed., *Radioisotopes in Industry* (Reinhold, New York, 1953).

J. F. Duncan and G. B. Cook, *Isotopes in Chemistry* (Clarendon, Oxford, 1968).

R. L. Ely, Jr., Radioactive Tracer Study of Sewage Field in Santa Monica Bay, *IRE Trans. Professional Group Nucl. Sci.*, **NS-4**, November 1, 49–50 (March 1957).

G. E. Francis, W. Mulligan, and A. Wormall, *Isotopic Tracers* (University of London, 1954).

R. P. Gardner and R. L. Ely, *Radioisotope Measurement Applications in Engineering* (Reinhold, New York, 1967).

G. Hevesy, *Radioactive Indicators* (Interscience, New York, 1948).

M. D. Kamen, *Radioactive Tracers in Biology* (Academic, New York, 1947).

J. Kohl, R. D. Zentner, and H. R. Lukens, *Radioisotope Applications Engineering* (Van Nostrand, Princeton, N.J., 1961).

R. T. Overman and H. M. Clark, *Radioisotope Techniques* (McGraw-Hill, New York, 1960).

G. K. Schweitzer and I. B. Whitney, *Radioactive Tracer Techniques* (Van Nostrand New York, 1949).

C. W. Sheppard, *Basic Principles of the Tracer Method* (Wiley, New York, 1962).

A. C. Wahl and N. A. Bonner, *Radioactivity Applied to Chemistry* (Wiley, New York, 1951).

ACTIVABLE TRACER STUDIES

C. Cappadona, Measurements of Movements of Solid Substances in Water by Means of Stable Tracers and Activation Analysis, in *Modern Trends in Activation Analysis* (National Bureau of Standards, 1969), pp. 72–75.

J. K. Channell and P. Kruger, Post-Sampling Activation Analysis of Stable Nuclides for Estuary Water Tracing, in *Modern Trends in Activation Analysis* (National Bureau of Standards, 1968), pp. 600–606.

J. K. Channell and P. Kruger, Activable Rare Earth Elements as Estuarine Water Tracers, *5th Intern. Conf. Water Pollution Research* (Pergamon, Amsterdam, 1971).

R. M. Chatters and R. L. Peterson, Use of Tracers in In-Plant Evaluation of Processes and in Pollution Control of Effluent from Pulp and Paper Mills into Rivers, in *Radioisotope Tracers in Industry and Geophysics* (International Atomic Energy Agency, Vienna, 1967), pp. 251–258.

I. F. Fineman, O. H. Hansen, B. M. Tolbert, and M. Calvin, Tracer Study with $^{18}O$ in Photosynthesis by Activation Analysis, *Intern. J. Appl. Radiation Isotopes* **2**, 280–286 (1957).

D. F. Gatz, A. N. Dingle, and J. W. Winchester, Detection of Indium as an Atmospheric Tracer by Neutron Activation, *J. Appl. Meteor.* **8**, 229–235 (1969).

P. F. Johnson, P. Tothill, and G. W. K. Donaldson, Stable Chromium as a Tracer for Red Cells, with Assay by Neutron Activation Analysis, *Intern. J. Appl. Radiation Isotopes* **20**, 103–108 (1969).

J. T. Lowman and W. Krivet, New In-Vivo Tracer Method with the Use of Non-radioactive Isotopes and Activation Analysis, *J. Lab. Clin. Med.* **61**, 1042–1048 (1963).

H. Nakasa and H. Ohno, Application of Neutron Activation Analysis to Stack-Gas Tracing, in *Radioisotope Tracers in Industry and Geophysics* (International Atomic Energy Agency, Vienna, 1967), pp. 239–250.

M. Radvan, B. Revenska-Kostzyuk, and E. Vezranovski, Use of a New Method Involving Labelling with Non-Radioactive Elements and Activation Analysis to Investigate Wear, in *Radioisotope Tracers in Industry and Geophysics* (International Atomic Energy Agency, Vienna, 1967), pp. 71–79 (in Russian).

## 9.6  PROBLEMS

1.  A sample of oil, free of fluorine and boron, is analyzed for oxygen with a 14-MeV neutron irradiation and measurement of the 6.17-MeV gamma ray of $^{16}N$ using a benzoic acid standard as a comparator. The procedure consists of a 72-sec irradiation of the sample, followed by a count for 72 sec. The background is determined by an additional 72-sec count. A standard benzoic acid sample is treated identically. The empty rabbits

are also irradiated to correct for their oxygen content. A typical set of data for neutron source strength and counting data is

| | Counts | | | Source Strength |
| --- | --- | --- | --- | --- |
| | First 72 sec | Second 72 sec | Net for Empty Rabbit | (n/sec) |
| Oil sample 80.0 mg | 188 | 14 | 43 | $1.65 \times 10^9$ |
| Benzoic acid 87.1 mg ($= 22.82$ mg $O_2$) | 4292 | 8 | 49 | $1.48 \times 10^9$ |

Determine the weight percent of oxygen in the oil sample.

2. A dose of 1 $\mu$Ci $1.7 \times 10^7$-y $^{129}I$ is irradiated in a thermal neutron flux of $10^7$ n/cm²-sec for 20 min just before ingestion by a patient requiring a thyroid iodine measurement. An external scintillation counter registers 940 cpm. The patient's thyroid is then irradiated with a flux of $10^7$ thermal neutrons for 20 min, and from decay curve analysis the initial activities of $^{128}I$ and 12.6-h $^{130}I$ were 680 and 720 cpm, respectively. Assuming complete retention of the added $^{129}I$ and equal counting efficiencies for the two activation radionuclides, estimate the amount of iodine in this patient's thyroid.

3. The well illustrated in Figure 9.11 is about 50 m from the Ume River. The origin of the water in the well was expected to be from one of three possible sources: infiltration from the river, a ground-water stream from the land side, or a mixture of both sources. From the data given in the figure determine the source of the well water.

4. To determine the volume of a pond 1.0 kg of $La_2O_3$ was dissolved and thoroughly mixed in the pond. A liter of pond water was taken before and after the $La_2O_3$ addition. Both were evaporated to dryness, encapsulated, and irradiated together with a 0.05 mg sample of pure $La_2O_3$ in a thermal reactor of neutron flux $10^{12}$ n/cm²-sec for 2 days. The three samples were then dissolved and La was quantitatively separated from each sample and counted in a gamma-ray spectrometer. The spectrometer efficiency for the 1.6-MeV gamma-ray which follows each of the 100% beta decays is 20%. The net counting rates corrected to the end of the irradiation were

No. 1, the preaddition sample        $A^0 = 440$ cpm,
No. 2, the postaddition sample       $A^0 = 1780$ cpm,
No. 3, the 0.05 mg pure $La_2O_3$ sample   $A^0 = 1970$ cpm.

Determine the volume (in liters) of the pond.

5. Derive equations for the endpoint, $E$ (in ml), of a radiometric titration for the single addition of reagent, $V$, and single radiation measurement in terms of the measured supernate activity, $A_s$, the background activity, $A_b$, and the angle $\phi$ for the two cases shown in Figure 9.32.

6. In a radioactive tracer study in Santa Monica Bay in California (see Ely, 1957), 20 Ci of 85-d $^{46}$Sc was injected over a period of 1 hr at a flow rate of $5 \times 10^6$ gal/hr into a sewage disposal plant with an outfall 1 mile out in the bay, 50 ft below surface. The activity measured at the surface, $A_f{}^0$, was 2500 cpm with a watertight gamma-ray scintillation detector having an over-all efficiency, $\epsilon = 27$ cpm/dpm-ml and a background of 19 cpm. Determine the actual dilution factor for this experiment and the maximum dilution factor for a minimum detection level of twice the background.

7. Assume that the experiment in Problem 6 could be duplicated with $^{45}$Sc as an activable tracer added to the effluent over the same time period and flow rate, a 10-gal surface-water sample taken over the same time period, $Sc_2O_3$ precipitated to remove most activatable major elements, the oxides irradiated in a thermal neutron flux of $10^{12}$ n/cm$^2$-sec for 1 hr ($\sigma = 24$ b), and $^{46}$Sc radiochemically separated with a chemical yield of 80%, and with $A^0 = 2500$ cpm in a detector of $\epsilon = 40\%$. Determine the amount of stable scandium (in kg) required to give the same actual dilution factor of 25 found in Problem 6.

# Index

513